IET ELECTROMAGNETIC WAVES SERIES 43

Series Editors: Professor P.J.B. Clarricoats
Professor J.R. Wait
Professor E.V. Jull

Open Electromagnetic Waveguides

Other volumes in this series:

Volume 1 **Geometrical theory of diffraction for electromagnetic waves, 3rd edition** G.L. James
Volume 10 **Aperture antennas and diffraction theory** E.V. Jull
Volume 11 **Adaptive array principles** J.E. Hudson
Volume 12 **Microstrip antenna theory and design** J.R. James, P.S. Hall and C. Wood
Volume 15 **The handbook of antenna design, volume 1** A.W. Rudge, K. Milne, A.D. Oliver and P. Knight (Editors)
Volume 16 **The handbook of antenna design, volume 2** A.W. Rudge, K. Milne, A.D. Oliver and P. Knight (Editors)
Volume 18 **Corrugated horns for microwave antennas** P.J.B. Clarricoats and A.D. Oliver
Volume 19 **Microwave antenna theory and design** S. Silver (Editor)
Volume 21 **Waveguide handbook** N. Marcuvitz
Volume 23 **Ferrites at microwave frequencies** A.J. Baden Fuller
Volume 24 **Propagation of short radio waves** D.E. Kerr (Editor)
Volume 25 **Principles of microwave circuits** C.G. Montgomery, R.H. Dicke and E.M. Purcell (Editors)
Volume 26 **Spherical near-field antenna measurements** J.E. Hansen (Editor)
Volume 28 **Handbook of microstrip antennas, 2 volumes** J.R. James and P.S. Hall (Editors)
Volume 31 **Ionospheric radio** K. Davies
Volume 32 **Electromagnetic waveguides: theory and applications** S.F. Mahmoud
Volume 33 **Radio direction finding and superresolution, 2nd edition** P.J.D. Gething
Volume 34 **Electrodynamic theory of superconductors** S.A. Zhou
Volume 35 **VHF and UHF antennas** R.A. Burberry
Volume 36 **Propagation, scattering and diffraction of electromagnetic waves** A.S. Ilyinski, G. Ya.Slepyan and A. Ya.Slepyan
Volume 37 **Geometrical theory of diffraction** V.A. Borovikov and B.Ye. Kinber
Volume 38 **Analysis of metallic antenna and scatterers** B.D. Popovic and B.M. Kolundzija
Volume 39 **Microwave horns and feeds** A.D. Olver, P.J.B. Clarricoats, A.A. Kishk and L. Shafai
Volume 41 **Approximate boundary conditions in electromagnetics** T.B.A. Senior and J.L. Volakis
Volume 42 **Spectral theory and excitation of open structures** V.P. Shestopalov and Y. Shestopalov
Volume 43 **Open electromagnetic waveguides** T. Rozzi and M. Mongiardo
Volume 44 **Theory of nonuniform waveguides: the cross-section method** B.Z. Katsenelenbaum, L. Mercader Del Rio, M. Pereyaslavets, M. Sorella Ayza and M.K.A. Thumm
Volume 45 **Parabolic equation methods for electromagnetic wave propagation** M. Levy
Volume 46 **Advanced electromagnetic analysis of passive and active planar structures** T. Rozzi and M. Farinai
Volume 47 **Electromagnetic mixing formulae and applications** A. Sihvola
Volume 48 **Theory and design of microwave filters** I.C. Hunter
Volume 49 **Handbook of ridge waveguides and passive components** J. Helszajn
Volume 50 **Channels, propagation and antennas for mobile communications** R. Vaughan and J. Bach-Anderson
Volume 51 **Asymptotic and hybrid methods in electromagnetics** F. Molinet, I. Andronov and D. Bouche
Volume 52 **Thermal microwave radiation: applications for remote sensing** C. Matzler (Editor)
Volume 53 **Principles of planar near-field antenna measurements** S. Gregson, J. McCormick and C. Parini
Volume 502 **Propagation of radiowaves, 2nd edition** L.W. Barclay (Editor)

Open Electromagnetic Waveguides

T. Rozzi and M. Mongiardo

The Institution of Engineering and Technology

Published by The Institution of Engineering and Technology, London, United Kingdom

First published 1997
Reprinted 2008

The Institution of Engineering and Technology
Michael Faraday House
Six Hills Way, Stevenage
Herts, SG1 2AY, United Kingdom

www.theiet.org

British Library Cataloguing in Publication Data
A CIP catalogue record for this book is available from the British Library

ISBN (10 digit) 0 85296 896 5
ISBN (13 digit) 978-0-85296-896-3

Printed in the UK by Short Run Press Ltd, Exeter
Reprinted in the UK by Lightning Source UK Ltd, Milton Keynes

Contents

Preface

Subject matter

In the environment of many modern microwave and millimetric components and subsystems in open transmission line the difference between guided propagation and radiation has become blurred: a discontinuity in open guide radiates, radiation propagates and causes long range interaction just as a propagating mode does; typically, both aspects are present in microstrip antenna feeds. An open dielectric waveguide of integrated optics ceases to be a perfect transmission media where an abrupt discontinuity is present, propagation in a dielectric bend takes place together with leakage of radiation.

Yet, the two aspects are still viewed as distinct physical problems by most researchers and students and handled by very different methods. In his 1972 book "Continuous transitions in open waveguides" V. V. Schevchenko first developed the concept of a unified method based on the complete spectrum of discrete and continuous modes in treating propagation and radiation waves on impedances surfaces and dielectric slabs.

The mathematics behind the idea of completeness and normalisation of the one–dimensional continuum dates,. of course, further back, as brilliantly exposed in the book by Friedman "Methods of mathematical physics".

One of the present authors developed systematically the use of the continuum in dealing with abrupt discontinuities and bends in open dielectric waveguides in the years of intense integrated optics research between 1975 and 1985: this material is contained in this book. Real life open trasmission media, however, are not dielectric slabs. Microstrip, slot–line, CPW inset guide as well as rib–guide in optics and high millimetrics, present a two-dimensional cross-section that, unlike for the optical cylindrical fiber, results in a non-separable problem. Apart from free–space expansions in plane and cylindrical waves the mathematics of the 2D-continuum that fits the guide boundary conditions can not be found in textbook: it took the authors several years from the mid-eighties to the mid-nineties to develop its formalism and its use: these are still in the making, as will be briefly discussed below.

Nevertheless it is now possible to deal with a via-hole in open microstrip, an air-bridge in CPW, or a system of dipoles in proximity of a slotline conceptually in exactly the same way as with a probe, a strip discontinuity, or a system of dipoles respectively, in closed waveguide and to obtain efficient numerical characterisations. The latter significant problems are treated in same detail in the book. Finally, a few words about two areas of development that could not possibly be discussed in the book in the appropriate manner.

Besides the usual rectangular, cylindrical and spherical system of coordinates there are other eight systems in which the wave equation separates. Some problems that appear nonseparable in the ordinary rectangular coordinate system, for instance that of a thin strip conductor or slot on an infinite conducting plane may become separable in another, in this case, the cylindrical elliptical system. In these situations, using the appropriate system greatly simplifies the complexity of the sprectum, reducing it to a pure discrete one in one dimension, with great numerical advantage.

Secondly, an important aspect of the continuum of open transmission media, e.g. microstrip, is the very direct link between near fields, currents and a fields that results now established by this approach. This feature, is felt, holds great potential for the direct synthesis of radiation patterns of planar antennas.

Aim of the book

This book is aimed at providing an extensive introduction to the practical use of spectral methods in solving real engineering problems in open electromagnetic waveguide environment. It is meant equally for the benefit of researchers in microwave and millimetric wave components as for its use as a textbook in specialistic courses.

Typically, both authors teach some of the more advanced material in the final (fifth) year M. Eng. courses in microwaves, optics and antennas at the Universities of Ancona and Perugia; whereas the introductory material is suitable for teaching at the level of intermediate electromagnetics.

Acknowledgments

Many people contributed to this book through discussion and support. First of all we wish to thank Prof. Jim Wait, editor of this IEE series, not only for making it possible, but also for his constant encouragement. His ideas and suggestions made the manuscript richer both in content and in form. The speedy production of this book is due to the efficient management of Ms Fiona MacDonald, Production Editor at the IEE.

The topic of this book is closely related to that of leaky waves, a subject which has been exploited during the years by several researchers and, in particular, by Prof. A. Oliner. The interest aroused by his work on this subject has had an undoubt impact in developing the theory illustrated in the rest of the book. Numerous conversations and exchange of ideas are gratefully acknowledged.

A particular thank goes to Prof. L. B. Felsen which has provided considerable encouragement in terminating this work. It is almost impossible to acknowledge all the time spent in discussions and the many suggestions provided on several subjects; in particular, the topic of alternative Green's function representations is of relevance since it allows field representations both in terms of leaky waves and/or modal spectra.

A final thank also goes to Prof. T. Itoh for so many discussions during conferences, workshops etc. that have also provided noticeable motivation toward the development of this work. His numerous contributions to the use of open waveguides in practical engineering components have been a constant reminder during the writing of this book.

How to read this book

In the concept of this book we faced a dilemma: should the structure of the book be from the general to the specific or vice versa? The obvious advantage of going from the general to the specific are that metodologies can be first developed in abstract and then applied to specific problems. Such a format, while being of benefit to seasoned researchers who wish to relate their own experience to a broader setting, is less user friendly for those who want to learn new techniques by working through engineering problems from the beginning to the end. Our choice has been to privilege the latter aspect, hence devising a pattern that, by starting from very simple problems, finally leads to the solution of very complex ones.

The book could well be divided into two main parts: the first four Chapters are of introductory type and are, in our opinion, useful to the relatively unfamiliar reader; they can be skipped from the more experienced researcher already aquainted with the topic. Chapters 5 to 8 are, instead, quite suitable also for the specialist since several techniques are introduced in these Chapters. In particular, Chapter 7 constitutes the main original contribution of the book since it contains the methodology on how to compute the complete spectrum of open non–separable waveguides. A few applications are illustrated in Chapter 8.

Although the titles of the various chapters should convey relatively clearly the subject addressed in each case, it is useful to provide a somewhat more detailed explanation of the content of each chapter.

Chapter 1 Provides definition of the type of waveguides of interest in this book and a short overview of their practical applications. This chapter can be used as a quick reference for the characteristics of some electromagnetic open waveguides.

Chapter 2 This chapter gives both a review of classical electromagnetic theory and a reference to some concepts frequently used throughout the following chapters. We believe that the section on spectral representation may be relevant for a simple introduction to continuous spectra. Moreover, the section "excitation by a source" tries to stimulate the interest of the reader toward the alternative Green's function representations.

Chapter 3 Slab waveguides, which are possibly the most simple and yet practically useful open waveguides, are considered in this chapter. It is shown how bound modes can occur and how to derive their complete spectra. The bound modes of slab waveguides are also introduced from a transverse resonance approach; whereas the continuous spectrum is derived from the transverse network concept.

A more complete understanding of the use of the modal spectra is obtained in Section 3.5 by studying the case of a field excited by a delta source. Finally, in Section 3.6, we consider the asymmetric three–layer waveguide which is a basic structure for practical integrated optics circuits.

Chapter 4 Bound modes of open waveguides play an important role in several applications. It is thus relevant to be able to calculate the propagation characteristics of such modes, even in an approximate manner. Two methods are introduced to this aim: the effective dielectric constant, useful for waveguides operating at "high" frequencies, and the conformal mapping technique, useful for waveguides operating at "low" frequencies.

Chapter 5 In this Chapter we introduce the techniques for dealing with discontinuities in open waveguides; essentially the classical modal formalism is extended to waveguides with infinite cross-section. Also considered are methods for dealing with interacting discontinuities and continuous transitions, such as bends. Since in this Chapter we study discontinuities in slab waveguides, the mathematical details can be kept to a minimum; nonetheless the examples chosen are of practical importance and numerical and experimental data are provided.

Chapter 6 This Chapter is mainly devoted to the transverse resonance technique applied to the problem of finding the bound modes of open waveguides of non–separable cross–section. The technique makes extensive use of the concepts presented in the previous Chapter. Transverse resonance approach was selected as the preferred way of calculating the bound modes since it provides considerable phenomenological insight. The technique is introduced by considering several examples relevant to practical applications. As in the rest of the book we start with simple examples and increase the difficulty as we progress.

Chapter 7 This Chapter contains the most novel and difficult material of the book. In fact, the computation of the complete spectrum of open waveguides with non-separable cross–section has only appeared in a few recent publications. Here we only deal with the theoretical formulation of the problem; the numerical solution being considered in the next Chapter.

Chapter 8 Some examples of applications are illustrated in this Chapter. These are only selected with the aim of familiarizing the reader with the methodology. In no way the treatment can be considered as exhaustive or complete; it is just an illustration of how the techniques introduced in the previous Chapters can be used in order to develop an efficient and phenomenological well consistent approach to the solution of electromagnetic problems in the presence of real life open waveguides.

Chapter 1

Introduction to open electromagnetic waveguides

An open waveguide may be defined as any physical device, with longitudinal axial symmetry and unbounded cross–section, capable of guiding electromagnetic waves. In contrast to closed waveguides, the electromagnetic field in the cross–section is not confined to a limited region of space, thus extending up to infinity. Examples of closed waveguides are well–known from basic electromagnetic courses [21]; they include coaxial cables, metallic waveguides (with rectangular or circular cross–section), etc. Their spectrum is also well–known, being made by an infinite, discrete set of modes, orthogonal and complete over the waveguide cross–section. On the other hand, any waveguide not entirely enclosed in a metal shielding, may be taken as an example of an open waveguide. It is also worth noting that, by cutting a slot parallel to the longitudinal axis in the metal enclosure of a closed waveguide, again an open waveguide is obtained. An open waveguide possesses a spectrum which is no longer discrete, but is constituted by a few bound modes, when present, plus a continuum of modes. In the next Chapter, in Section 2.8, we introduce the concept of the continuous spectrum of open waveguides, which is of fundamental importance throughout this book, by making reference to the well–known case of a closed 2D–waveguide of bounded cross–section with a well–defined modal spectrum, i.e. to a parallel plate waveguide. By removing to infinity one of the plates we obtain the case of a half–space, bounded on one side by the metal plate and unbounded on the other side.

According to the above definition of an open waveguide, free space too may be considered as a kind of open waveguide. In this case, it is useful to view the commonly used free space field expansions in spherical waves, cylindrical waves, plane waves etc., as modal expansions in waveguide. This type of interpretation, developed in the next Chapter, provides a certain insight into the modal fields of waveguides with unbounded cross–section.

Open electromagnetic waveguides may be either man–made, such as for instance microstrip lines or optical fibres, or naturally occur in the geophysical environment (see e.g. [33, 28]). There are many problems in communication, navigation, and applied geophysics where the system performance is dependent on the profile and the electrical properties of the earth's surface, which can be usefully modelled by using the concept of open waveguides [30, 31, 32]. Waveguides of this type of are quite relevant to a variety of fields and are used in several practical applications, although this book is mainly concerned with waveguides of the former type occurring in communications and remote sensing. A few of these applications are revised in the next Sections, which also provide some basic information about the types of commonly used open waveguides in optics, integrated optics, microwave and millimetre wave frequency ranges.

In view of their broad applicability and of their considerable theoretical interest, it is not surprising that open waveguides have received significant attention in the past (see references [7, 8, 23, 24, 14, 27, 26, 25, 16], to name just a few). Actually, when the open waveguide possesses a separable cross–section, it is now well understood how to derive its complete spectrum [23]. By "separable cross–section", here and in the following, we mean that waveguide boundaries are such as to fit one of the coordinate systems where the transverse Helmholtz equation may be solved by separation of variables [19]. As an example of separable waveguides we consider, in Chapter 3, the slab waveguide, for which we derive surface waves and continuum, also showing completeness and orthogonality properties of the modal spectrum. Further examples of separable waveguides are provided by the optical fibres (in this case the reader is referred to the excellent book [24]) or also a metallic strip, or a slot in a metallic plate, without dielectric, (such as that reported in [3]) etc.

The vast majority of open waveguides currently used in integrated optics, in the microwave frequency region and in millimetrics, do not allow solution of the Helmholtz equation in their cross–section by means of separation of variables. The evaluation of the complete spectrum of these open, nonseparable, waveguides and its application is a relatively new matter and is dealt with throughout this book. By contrast, it is noted that, for closed, nonseparable waveguides, the evaluation of the complete, discrete, spectrum follows a quite well known procedure. Nonetheless, it is the very availability of the complete spectrum of the open waveguide which allows one to use much of the same methodology, and to translate many useful concepts, employed in the analysis of closed waveguides such as described in the fundamental textbooks [7, 17, 22, 15].

Table 1.1 Radio frequency band designations and use.

Frequency	*Wavelength*	*Band designation*	*Typical service*
30-300 Hz 300-3000 Hz	10-1 Mm 1000-100 km	ELF extremely low frequency	
3-30 kHz	100-10 km	VLF very low frequency	Navigation, sonar
30-300 kHz	10-1 km	LF low frequency	Radio beacons, navigational aids
300-3000 kHz	1 km-100 m	MF medium frequency	AM broadcasting, maritime radio, Coast Guard communication, direction finding
3-30 MHz	100-10 m	HF high frequency	Telephone, telegraph, facsimile; shortwave international broadcasting; amateur radio; citizen's band; ship to coast and ship to aircraft communication
30-300 MHz	10-1 m	VHF very high frequency	Television, FM broadcast, air traffic control, police, taxicab mobile radio, navigational aids
300-3000 MHz	1 m-10 cm	UHF ultra high frequency	Television, radioprobes, satellite communication, surveillance radar, navigational aids
3-30 GHz	10-1 cm	SHF super high frequency	Airborne radar, satellite communication, common carrier land mobile communication, microwave links
30-300 GHz	1 cm- 1 mm	EHF extremely high frequency	Experimental, Radar, vehicular anti-collision radar

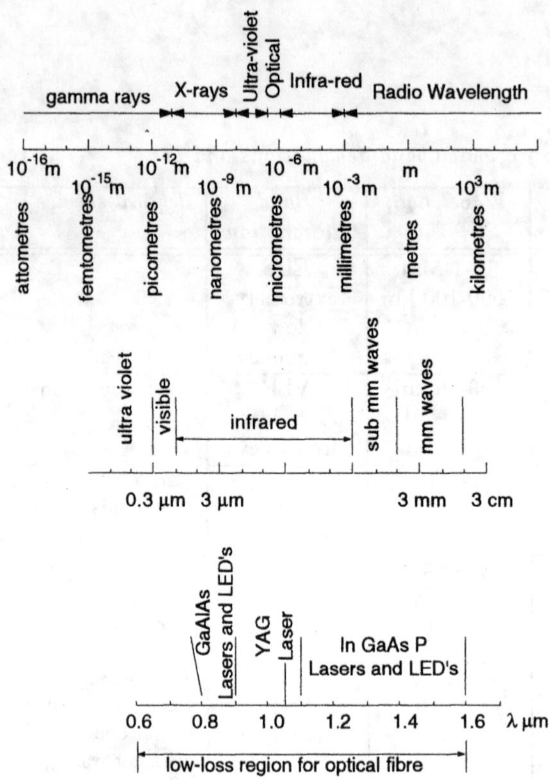

Figure 1.1 Electromagnetic wave spectrum.

1.1 Examples of open waveguides

The electromagnetic frequency spectrum is represented in Figure 1.1 in terms of the wavelength. In this figure is also shown the portion of spectrum where the transmission characteristics of optical fibres are best utilised [4]. Open waveguides used in optical communication systems and in integrated optics are generally operated in this frequency range. Another part of the spectrum intensively used is that corresponding to radio wavelengths, as detailed in Table 1.1 [6]. Open waveguides operating at microwave and millimetre waves are of particular interest for their wide range of applications. The frequency band designations employed in the microwave range are reported in Table 1.2.

As pointed out in [13], the wavenumber concept is perhaps more fundamental in electromagnetic wave theory than either of the more popular concepts of wavelength and frequency. The corresponding values of wavenumber k and frequency f are obtained from the wavelength λ by the following relationships

$$f = \frac{c}{\lambda} \qquad \lambda = \frac{2\pi}{k} \tag{1.1}$$

Table 1.2 Microwave frequency band designations.

Frequency	Wavelength	IEEE radar band designation	
		old	new
1-2 GHz	30-15 cm	L	D
2-3 GHz	15-10 cm	S	E
3-4 GHz	10-7.5 cm	S	F
4-6 GHz	7.5-5 cm	C	G
6-8 GHz	5-3.75 cm	C	H
8-10 GHz	3.75-3 cm	X	I
10-12.4 GHz	3-2.42 cm	X	J
12.4-18 GHz	2.42-1.67 cm	Ku	J
18-20 GHz	1.67-1.5 cm	K	J
20-26.5 GHz	1.5-1.13 cm	K	K
26.5-40 GHz	1.13 cm-7.5 mm	Ka	K
40-300 GHz	7.5-1.0 mm	mm	

where c is the free–space velocity of light,

$$c = \frac{1}{\sqrt{\mu_0 \varepsilon_0}} = 2.997925 \times 10^8 \text{m/s} \approx 3 \times 10^8 \text{m/s}$$

$$\begin{aligned} \mu_0 &= 4\pi \times 10^{-7} henry/metre \quad \left(\frac{V \cdot s}{A \cdot m}\right) \\ \varepsilon_0 &= \frac{1}{36\pi} \times 10^{-9} farad/metre \quad \left(\frac{A \cdot s}{V \cdot m}\right) \end{aligned} \tag{1.2}$$

while μ_0 is the magnetic permeability and ε_0 the dielectric permittivity of free–space.

Open waveguides may operate up to optical frequencies. For communication purposes, since it is generally desirable to employ as large a bandwidth as possible, it is advantageous to use the higher frequencies now available. Up to the millimetric range this is an added bonus, as circuit dimensions become smaller as frequency increases, thus allowing a saving of space and sometimes a reduction of costs when higher frequencies are used.

Unfortunately, as frequency increases so does the average atmospheric attenuation, as shown in Figure 1.2 [6]. This fact, together with problems related to electromagnetic interference, prevents the use of free–space as a transmitting media for very large amounts of data. Atmospheric attenuation is also important for many applications; for example, by choosing frequencies for which the atmosphere is opaque, one prevents detection of satellite–to–satellite communications by ground–based receivers. Similarly, terrestrial systems desiring to prevent signal overshoot in range may operate at a frequency of high atmospheric absorption [36].

Optical fibres, which are briefly introduced in Section 1.2, are nowadays widely used because of their low-loss, high bandwidth, transmission characteristics which make them ideally suited for carrying voice, data, and video signals between fixed points [4]. Existing optical fibres have now an attenuation of less than 0.2 dB/km for source wavelength near 1.55 μm [14]. Moreover, since optical fibres are made of a dielectric, the link between source and detector is nonmetallic; this avoids ground

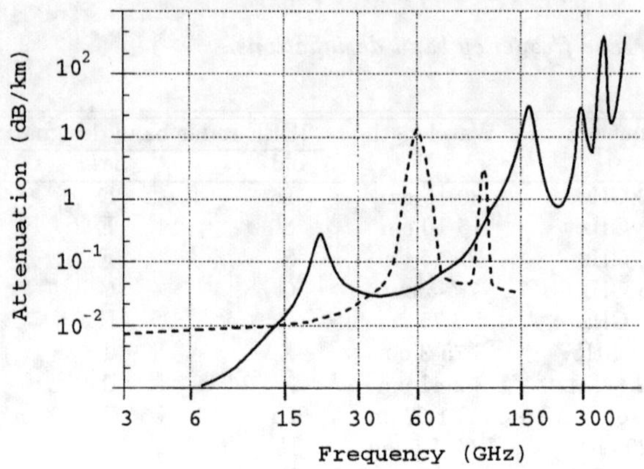

Figure 1.2 Attenuation by oxygen (dotted line) and water vapour (continuous line) at sea level. The water content is 7.5 g/m^3 and the temperature is $T = 20°C$.

loop pick–up problems resulting from electromagnetic interference.

Fibres, however, are not applicable to satellite communications, or to connect mobile systems, where wireless communication systems are required. The latter are commonly realised both in the microwave and millimetre–wave frequency ranges, upon consideration of the atmospheric attenuation. In this case, the preferred transmission lines are open waveguides such as microstrip lines, coplanar waveguides, etc. which are introduced in Section 1.4.

Nowadays, wireless information networks, optically–fed wireless communication systems and integration of optics and electronics for wireless communications are therefore receiving considerable attention and have led to the use of optically integrated circuits. The latter also make use of open waveguides, such as the rib waveguide, and are briefly introduced in Section 1.3.

1.2 Optical fibres

The fibres used for optical communication systems are an example of open waveguides; they are made of transparent dielectrics used to guide either visible or infrared light over long distances. Optical fibres are generally made of an inner cylinder of high–silica–content glass, SiO_2 doped with either GeO_2 or P_2O_5 to increase the refractive index, which is called the core [14, 11]. This is surrounded by a cylindrical shell of silica or P_2O_5 doped silica, of slightly lower refractive index, generally named the cladding (see Figure 1.3). For short–distance data communication and data links, where attenuation is not of primary concern, higher loss plastic–clad or totally plastic fibres may be used. The optical fibre is a lightguide where waves are guided by means of the total internal reflection that occurs within a medium of stratified index, i.e. within the core and the cladding.

Actually, the phenomenon of total internal reflection at the interface between two dielectric media has been known for quite a long time. It was in 1870 that Tyndall demonstrated the possibility of guiding light within a water jet [35]. Nonetheless, until the mid–1960s only simple dielectric rods were considered. The dielectric rods

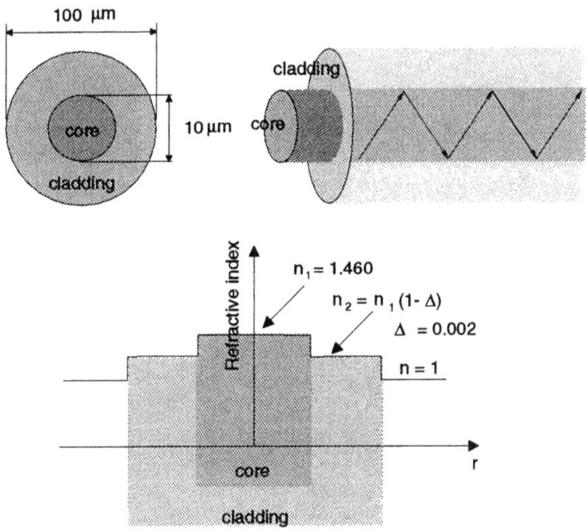

Figure 1.3 Step–index single–mode fibre.

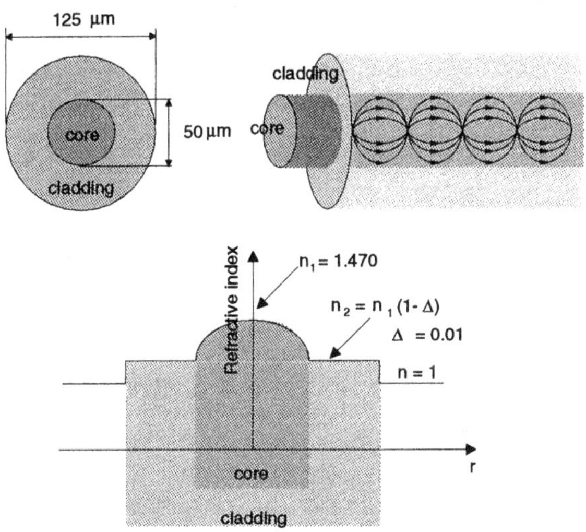

Figure 1.4 Graded–index multi–mode fibre.

had to be made impractically small to accommodate a single electromagnetic mode; moreover, perturbations at the surface of the rod cause high losses. It is just the presence of the cladding, first proposed in 1954 [29], which overcomes the above difficulties, making practical the idea of a communication system based on light propagation [12].

The fibres are often classified in terms of the refractive index profile of the core, as shown in Figures 1.3–1.5, and in terms of the number of modes propagating in the guide, i.e. single–mode and multi–mode fibres. According to Figures 1.3 and 1.5,

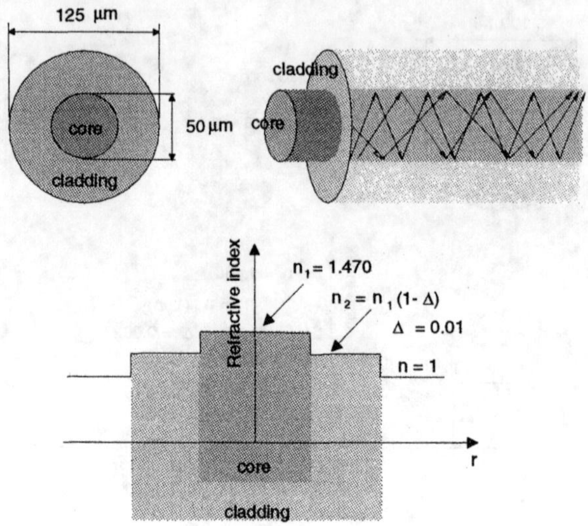

Figure 1.5 Step–index multi–mode fibre.

when the core has a uniform refractive index, n_1, we have the so–called "step–index fibre"; when the core has a non uniform refractive index, gradually decreasing from the centre toward the core–cladding interface as in Figure 1.4, the fibre is referred to as a "graded–index fibre".

Some advantages, constraints and applications of the various types of fibres are enumerated in Table 1.3 [4]. One of the advantages of multimode fibres is that incident waves impinging from different directions, thus possessing different **k** vectors, are critically reflected and captured by the fibre [10]. In this way, power emitted from a light–emitting diode (LED), or laser, can be coupled more efficiently into the multimode fibre than into a single–mode fibre, which accepts only a narrower range of **k** angles. A disadvantage of a multimode fibre is that modes possess different phase velocities. Two modes with different phase velocities interfere at the fibre–output end in a manner which varies with time, due to mechanical vibrations, temperature fluctuations, etc. However, when many modes are present, the total effects of these interferences statistically average out, thus allowing the use of these fibres with communication rates of several megabits per second over distances of a few kilometres [10]. The distance over which these fibres can operate is principally limited by dispersion problems. For long–distance propagation and high–speed communications, single–mode fibres have to be used. For the latter fibres, propagation distance is limited solely by the group velocity dispersion of the propagating mode, that can be compensated by material dispersion with opposite wavelength dependence. For this reason, when high–bit–rate optical communications are needed single mode fibres become necessary.

The low–loss and wide–band capabilities of optical fibres are best appreciated by comparing the signal attenuation versus frequency for three different transmission media as shown in Figure 1.6, [14]. In the latter figure, by denoting by P_i the power of a beam launched into one end of the optical fibre and by P_f the power received after a length of L km, the attenuation is given by

Table 1.3 Characteristics of the various types of fibres.

	Step–index single–mode	Graded–index multi–mode	Step index multi–mode
Source	Laser	Laser or LED	Laser or LED
Bandwidth	very very large > 3 GHz - km	very large 200 MHz to 3 GHz - km	large < 200 MHz - km
Splicing	very difficult	difficult	difficult
Application (example)	submarine cable system	telephone trunk between central offices	data links
Cost	less expensive	most expensive	least expensive

$$Attenuation = \frac{10\log_{10}(P_i/P_f)}{L}\ dB\ km^{-1} \tag{1.3}$$

It is apparent from this figure that, for signal frequencies above a few MegaHertz, optical fibres show lower losses than wire pairs or coaxial cables. Moreover, the attenuation does not depend on frequency. The low attenuation allows an increase in the distance between amplifiers in a communications system, thus sensibly reducing the cost of the system. Quite a realistic model of an optical fibre is one with a constant positive index extending to the outer radius of the fibre. The field solution for such a structure makes use of Bessel functions and is accomplished in a variety of texts to which the reader is referred (see e.g. [24, 4]). A simpler way to understand much of the physics of wave propagation in a fibre, may be gleaned from the simplified slab model of an optical guide, as treated in Chapter 3. The latter clearly shows mode dispersion as well as the dependence of the number of guided modes upon frequency.

1.3 Open waveguides for integrated optics

As we have seen, fibres may be advantageously employed to achieve signal transmission using lightwaves. However, when any processing of the signal is required, we have two possible alternatives at hand. The first is to convert the light signal into an electrical one at lower frequencies and then process it by standard means. The second alternative is to process the optical signal directly by means of some integrated optics (IO) circuitry [25, 11]. Actually, laser beams can also be processed by using discrete optical components such as prisms, lenses, mirrors, electro–optic modulators, etc. However, this type of equipment is unsuitable for most practical applications, since it requires vibration–proof mounting and a considerable amount of space, while also being quite expensive. To overcome the above problems, the basic concept of IO has been proposed by Anderson in 1965 [2] and developed thereafter, as illustrated in the historical overview of the first years of IO provided in

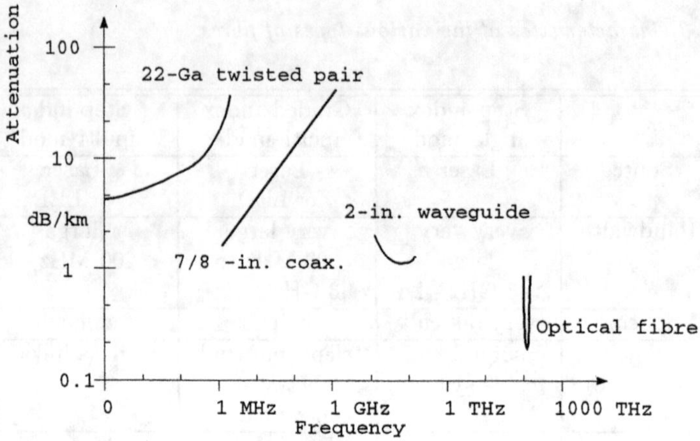

Figure 1.6 Attenuation comparison for different types of waveguides.

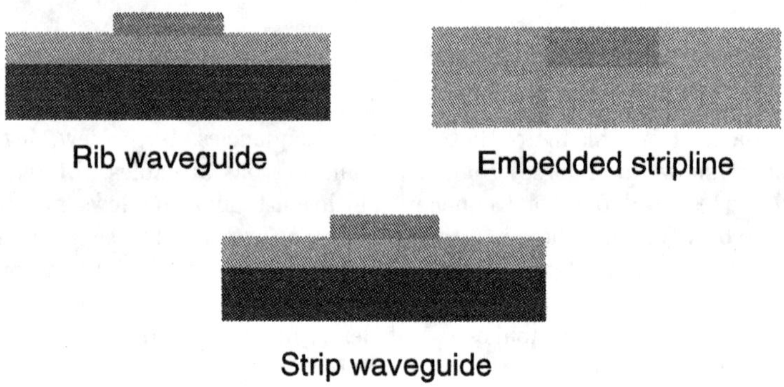

Figure 1.7 Cross–section of some common waveguides used in integrated optics.

[25, 18]. Thus, integrated optical circuits have the same advantages over discrete component optical systems as electrical integrated circuits have over hand–wired discrete component circuits. These advantages are: smaller size, reduced weight and costs, lower power requirement, improved reliability, mechanical robustness, etc.

The IO approach to signal transmission and processing offers significant advantages in both performance and costs when compared to conventional electrical methods. For example, optical integrated circuits have inherently the same large characteristic bandwidth as optical fibres, since in both cases the carrier medium is a lightwave. Moreover, by using IO one avoids the frequency limiting effects related to capacitive and inductive parasitics commonly found in electrical circuits operating at lower frequencies.

In IO the signal is generally carried by a planar waveguide in either slab or rib/strip form as shown in Figure 1.7. Note that in some cases, the embedded stripline is also referred to as a channel waveguide. In any of the above waveguides

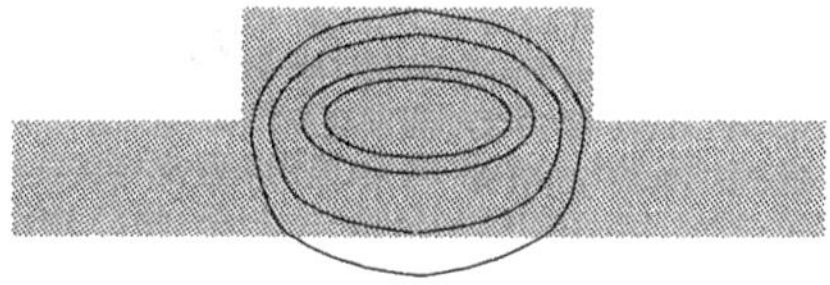

Figure 1.8 Approximate field contour profile for a rib waveguide.

Table 1.4 Substrates used for optical integrated circuits.

Passive	Active
quartz	gallium arsenide
lithium niobate	gallium aluminium arsenide
lithium tantalate	gallium arsenide phosphide
tantalum pentoxide	gallium indium arsenide
niobium pentoxide	II-V and II-VI direct bandgap semiconductors
silicon	

the field is mostly confined in the central part of the guide, i.e. in the rib or in the channel, as shown in Figure 1.8. The bound modes propagation characteristics of IO waveguides are fairly often derived by using the effective dielectric constant (EDC), which is introduced in Chapter 4. By using EDC, these types of guides are generally reduced to asymmetric planar slab waveguides which, in turn, are studied in Chapter 3. A detailed investigation of the bound modes of the image and rib waveguides is given in Chapter 6; while in Chapter 7 we derive the complete spectrum of the rib waveguide.

The guides used in IO are realised by a wide variety of techniques, as illustrated in [11]. Briefly, these techniques make use of sputtering of one type of glass onto another, in–diffusion of a layer of titanium deposited on a substrate of lithium niobate, and liquid phase epitaxy. This last technique can be used with semiconducting materials such as GaAs and GaAlAs. The choice of substrate material in IO is largely dependent on the type of approach used to realise the circuit. Essentially two types of approach are possible, namely the hybrid and the monolithic. In hybrid circuits, different devices are realised on different substrates, which are afterwards bonded together to optimise performance. In monolithic optical integrated circuits a single substrate is used for all devices, thus also for light sources. In these cases, optically active materials, such as the semiconductors listed in the right part of Table 1.4, [11], are needed.

Losses in IO waveguides are usually much larger than in optical fibres, being of the order of 0.1 dB mm^{-1}. One of the causes of losses is the large scattering from the waveguide/air interface which is often relatively rough. Another important source of losses occurs when bends are present. In order to prevent these losses the bends must have fairly large radii of curvature (of the order of millimetres). A quite exhaustive analysis of bends in rib waveguides can be found in [20], while this problem is also treated in Chapter 5 by the method of local modes.

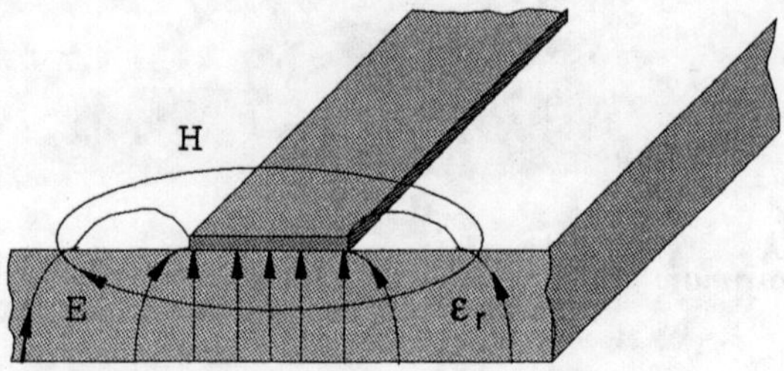

Figure 1.9 Quasi-static field of a microstrip line.

Table 1.5 Properties of dielectric substrates used for microwave and millimetre wave open waveguides (from [9]).

Material	ε_r	Loss tangent (@ 10 GHz)	Thermal conductivity W/cm/°C	Maximum Average Power (kilowatts) (@ 10 GHz)	Dielectric strength kV/cm
Sapphire	11.7	10^{-4}	0.4	5.10	4×10^3
Alumina	9.7	2×10^{-4}	0.3	5.17	4×10^3
Quartz (fused)	3.8	10^{-4}	0.01	0.523	10×10^3
Polystyrene	2.53	4.7×10^{-4}	0.0015	0.124	280
Beryllium oxide (BeO)	6.6	10^{-4}	2.5	75.7	-
GaAs ($\rho = 10^7\ \Omega$-cm)	12.3	16×10^{-4}	0.3	1.47	350
Si ($\rho = 10^3\ \Omega$-cm)	11.7	50×10^{-4}	0.9	2.23	300
Air	1	$\cong 0$	0.00024		30

1.4 Microwave and millimetric open waveguides

Open waveguides are widely used in microwave integrated circuits (MIC) and in their monolithic version (MMIC), [9]. The waveguides more commonly employed are planar transmission structures such as microstrip lines, slot lines, coplanar waveguides (CPW), coplanar strips, etc. These waveguides are realised by placing metallic strips, or cutting slots into metallic planes, which are generally backed by dielectric layers used both for mechanical support and in order to obtain better confinement of the electromagnetic fields. Types of dielectric commonly used in MIC/MMIC are listed, together with their main characteristics, in Table 1.5, [9].

The "planar" configuration is important in practical applications since, fairly often, the characteristics of the transmission line can be controlled by suitably changing the line dimensions on a single plane. As an example, microstrip impe-

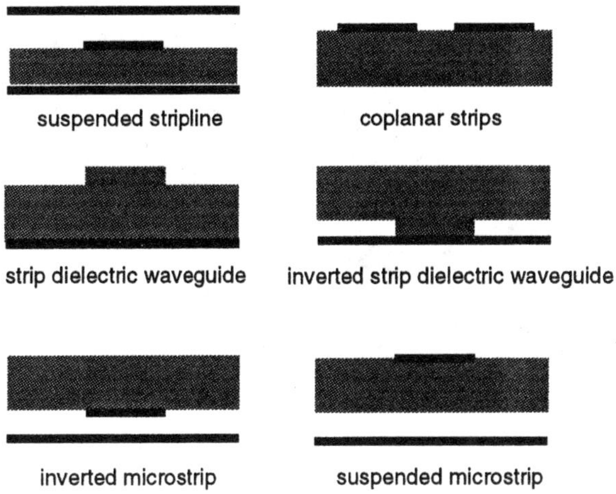

Figure 1.10 Some open waveguides used in the microwave and millimetre wave frequency range. Thick lines represent metallisation.

dance can be adjusted by varying the width of the strip; CPW impedance may be changed by varying the ratio between the width of the central strip and the width of the slots. In these cases, circuit fabrication can be carried out by using photolithographic techniques or by photoetching thin films.

The microstrip line is certainly the most used of the above waveguides; its mode of propagation is the quasi–TEM mode, shown in Figure 1.9, which, in fact, requires the presence of all six field components. The fact that a microstrip line cannot support a pure TE or TM field is shown in Chapter 4, where also the quasi–static method to find its propagation characteristic is illustrated. Several variations of microstrip configuration have been suggested for use in MIC and also to replace coaxial cables, e.g. in space applications where weight reduction is crucial [1]. Some examples of these transmission lines are listed in Figure 1.10.

Recently, slotlines and coplanar waveguides have received considerable attention as, among other advantages, these guides support a fundamental mode down to d.c. In particular, the slotline, proposed by Cohn in 1968 [5], has a fundamental mode which is non–TEM and almost transverse electric; moreover, being a two conductor structure, it has no low–frequency cut–off. The approximate field distribution for a slotline is shown in Figure 1.11; in Chapter 6, we find the bound modes of the slotline by the method of transverse resonance diffraction. The use of slotlines together with microstrip lines and coplanar waveguides allows the realisation of quite a large variety of components, such as those described in [9]. Slotlines are connected with microstrip circuits by etching the slotline in the ground plane of the microstrip, whereas their connection to CPW is realised on the same metallic surface.

Coplanar waveguides (CPW) have been proposed by Wen in 1969 [34]. Being a three conductor line, CPW has two distinct fundamental modes extending down to d.c.; these are commonly referred to as the even and odd CPW modes. The quasi–static field configuration of the even CPW mode is shown in Figure 1.12, while

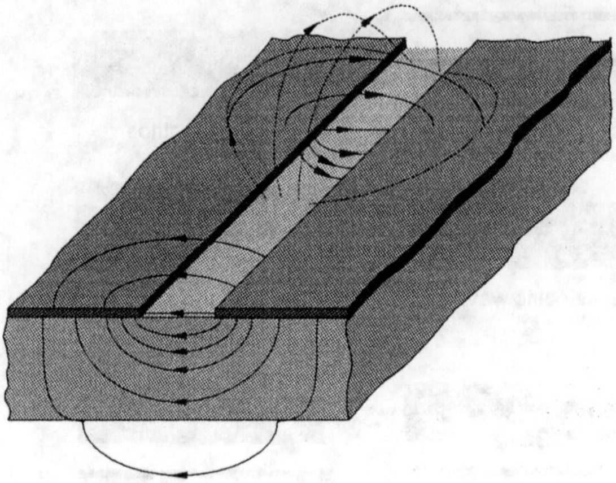

Figure 1.11 Slot line field configuration. The electric field is almost transverse, whereas the magnetic field has a longitudinal component.

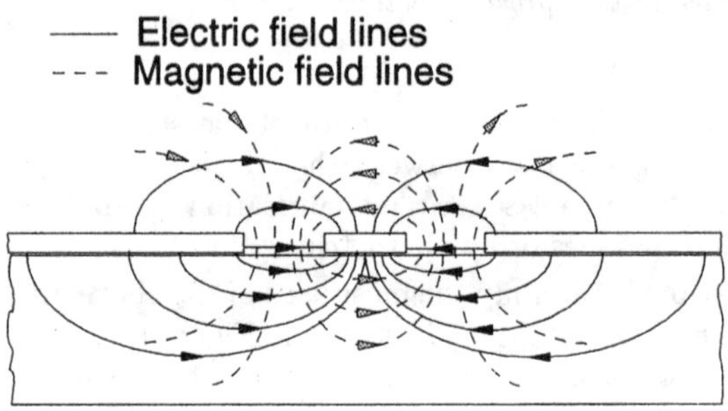

Figure 1.12 CPW even fundamental mode; electric and magnetic field distribution.

Figure 1.13 reports the odd mode field. The odd mode field, generally unwanted, is commonly rejected by placing, in suitable positions, air–bridges connecting the two lateral metallic planes, as shown in Figure 1.14. The necessity of using air–bridges is one of the most severe handicaps of CPW.

Naturally, all the above waveguides are used together with either lumped or active elements, mounted in series or in shunt configuration. As noted in Table 1.6 [9], some transmission lines allow easy mounting in one or another type of configuration. As an example, microstrip line allows easy mounting of series elements, whereas shunt elements, such as via holes, are difficult to place, requiring drilling of the dielectric layer(s). From this point of view, CPW allows both series and shunt mounting in an easy way. Also interesting is the microstrip loaded inset dielectric waveguide which offers features similar to those of CPW but avoiding the need for air–bridges.

Also noted in Table 1.6 is the presence of a row referring to radiation losses. In fact, although the above transmission lines are generally employed in their fundamental mode, when discontinuities occur a whole (continuous) spectrum of these open waveguides is excited. In these cases, part of the modes excited may give rise

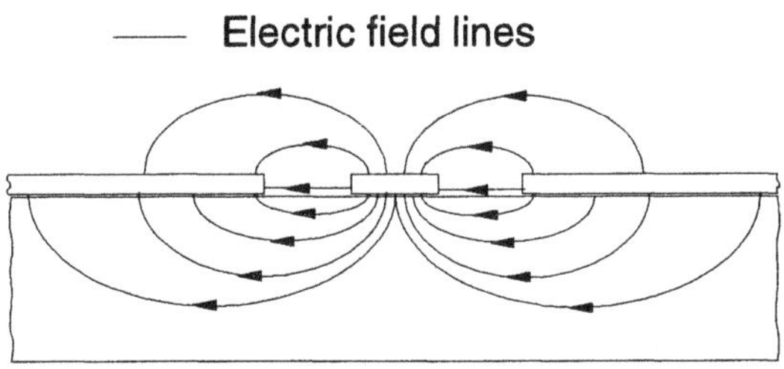

Figure 1.13 Electric field lines for the fundamental CPW odd (unwanted) mode.

Figure 1.14 CPW air–bridges used to suppress the odd mode.

to conspicuous radiative effects desirable for antenna applications or otherwise for MIC/compatibility applications.

In view of the practical importance of the bound modes of these lines, in Chapter 4 we introduce some approximate methods for finding the characteristic impedance and propagation constants relative to these lines. A more thorough full–wave treatment of the bound modes of the above lines is described in Chapter 6. The remaining part of the spectrum of these open waveguides, i.e. the continuum, which plays a fundamental role in discontinuity and radiation problems, is investigated in Chapters 7 and 8.

Table 1.6 Summary of the characteristics of some common waveguides used in microwave circuits. The letter "m" indicates that the limit is caused by higher order modes, while the letter "d" indicates that the limit is due to small dimensions (from [9]).

Characteristic	Microstrip	Slotline	Coplanar waveguide	Coplanar strips
Effective dielectric constant (ε_r= 10 and substrate thickness = 0.025 inch)	~ 6.5	~ 4.5	~ 5	~ 5
Power handling capabilities	High	Low	Medium	Medium
Radiation loss	Medium	High	Low	Medium
Unloaded Q	High	Low	Medium	Low (lower impedances) High (higher impedances)
Dispersion	Small	Large	Medium	Medium
Mounting of shunt components	Difficult	Easy	Easy	Easy
Mounting of series component	Easy	Difficult	Easy	Easy
Technological difficulties	Ceramic holes (for vias)	Double side etching	Air-bridges	-
Loss (dB/cm) ε_r = 10, h= 25 mil, Z_0= 50 f=10 GHz	0.04	0.15 (ε_r = 16 Z_0 = 75)	0.08 (h/W = 2)	0.83 (h/W = 2)
Lower limit for Z_0 (ohm) (ε_r= 10, h= 25 mil, freq= 10 GHz)	20 (m)	55 (d)	25 (m, d)	45 (m, d)
Upper limit for Z_0 (ohm)	110 (d)	300 (m)	155 (m, d)	280 (m, d)

Bibliography

[1] F. Alessandri, M. Mongiardo, R. Sorrentino, and J. Uher. Analysis of branch line coupler in suspended stripline with finite metallization thickness. *IEEE MTT-S International Symposium*, pages 1089–1092, 1993.

[2] D. B. Anderson. *Optical and Electro-optical Information Processing.* MIT Press, Cambridge, Mass., 1965.

[3] J. J. Bowman, T. B. A. Senior, and P. L. E. Uslenghi. *Electromagnetic and Acoustic Scattering by Simple Shapes.* Hemisphere publishing corp., New York, 1987.

[4] H. Cherin. *An Introduction to Optical Fibers.* McGraw-Hill, Singapore, 1983.

[5] S. B. Cohn. Slot line on a dielectric substrate. *IEEE Trans. on Microwave Theory Tech.*, 17(10):768–778, 1969.

[6] R. E. Collin. *Antennas and Radiowave Propagation.* McGraw-Hill, Singapore, 1985.

[7] R. E. Collin. *Field Theory of Guided Waves.* IEEE Press, New York, 1991.

[8] L. Felsen and N. Marcuvitz. *Radiation and Scattering of Waves.* Prentice Hall, Englewood Cliffs, NJ, 1973.

[9] K. C. Gupta, R. Garg, and I. J. Bahl. *Microstrip Lines and Slotlines.* Artech House, Washington, 1979.

[10] H. A. Haus. *Waves and Fields in Optoelectronics.* Prentice-Hall, Englewood Cliffs, New Jersey, 1984.

[11] R. G. Hunsperger. *Integrated Optics: Theory and Technology.* Springer-Verlag, Berlin, 1991.

[12] K. C. Kao and G. Hockam. Dielectric-fibre surface waveguides for optical frequencies. *Proc. IEEE*, 113:1151, 1966.

[13] J. A. Kong. *Electromagnetic Wave Theory.* John Wiley & Sons, Singapore, 1986.

[14] D. L. Lee. *Electromagnetic Principles of Integrated Optics.* John Wiley & Sons, Singapore, 1986.

[15] L. Lewin. *Theory of Waveguides.* John Wiley & Sons, New York, 1975.

[16] D. Marcuse. *Theory of Dielectric Optical Waveguides.* Academic Press, New York, 1974.

[17] N. Marcuvitz. *Waveguide Handbook.* McGraw-Hill, New York, 1951.

[18] S. E. Miller and A. G. Chynoweth. *Optical Fiber Communications.* Academic Press, New York, 1979.

[19] P. M. Morse and H. Feshback. *Methods of Theoretical Physics.* McGraw-Hill Book Company Inc, New York, 1950.

[20] E. C. M. Pennings. *Bends in Optical Ridge Waveguides: Modeling and Experiments.* PhD Thesis, Technische Universiteit Delft, 1990.

[21] S. Ramo, J. R. Whinnery, and T. Van Duzer. *Fields and Waves in Communication Electronics.* John Wiley & Sons Inc., New York, 1965.

[22] J. Schwinger and D. S. Saxon. *Discontinuities in Waveguides.* Gordon and Breach, New York, 1968.

[23] V. V. Shevchenko. *Continuous Transition in Open Waveguides.* Golem Press, Boulder, Colorado, 1971.

[24] E. W. Snyder and J. D. Love. *Optical Waveguide Theory.* Chapmann and Hall, 1983.

[25] (Ed.) T. Tamir. *Integrated Optics: Topics in Applied Physics*, volume 7. Springer- Verlag, Berlin, 1975.

[26] T. Tamir and A. A. Oliner. Guided complex waves - part 2. *Proc. IEE*, 110:325–333, February 1963.

[27] T. Tamir and A. A. Oliner. Guided complex waves, part 1: fields at the interface. *Proc. IEE*, 110:310–324, February 1963.

[28] F. T. Ulaby, R. K. Moore, and A. K. Fung. *Microwave Remote Sensing: Active and Passive.* Addison Wesley, Reading, MA, 1981.

[29] C. S. van Heel. A new method of transporting optical images without aberrations. *Nature*, 39:173, 1954.

[30] J. R. Wait. Diffraction and scattering of the electromagnetic groundwave by terrain features. *Radio Science*, 3(10):995–1003, October 1968.

[31] J. R. Wait. Propagation of electromagnetic waves over a smooth multisection curved earth — an exact theory. *Journal of Mathematical Physics*, 11(9):2851–2860, September 1970.

[32] J. R. Wait. Recent analytical investigations of electromagnetic ground wave propagation over inhomogeneous earth models. *IEEE Proceedings*, 62(8):1061–1071, August 1974.

[33] J. R. Wait. *Geo Electromagnetism.* Academic Press, New York, 1982.

[34] C. P. Wen. Coplanar waveguide: a surface strip transmission line suitable for nonreciprocal gyromagnetic device applications. *IEEE Trans. on Microwave Theory Tech.*, 17(12):1087–1090, 1969.

[35] J. Wilson and J. F. B. Hawkes. *Optoelectronics.* Prentice-Hall, Cambridge, 1989.

[36] J. C. Wiltse. Introduction and overview of millimeter waves. In J. C. Wiltse, editor, *Infrared and Millimeter Waves*, volume 4. Academic Press, Orlando, Florida, 1981.

Chapter 2

Electromagnetic theory

This Chapter provides both an overview of classical concepts in electromagnetic theory and a reference with respect to some concepts frequently used throughout the following Chapters.

2.1 Maxwell's equations

Large–scale electromagnetic phenomena are governed by equations linking electromagnetic field quantities introduced by James Clerk Maxwell in 1873. These equations were supported by a vast wealth of experimental evidence and by previous observations such as Coulomb's law of force between charges, Ampere's law on the interaction of current elements, the observations of Faraday on variable fields, etc. Moreover, Maxwell's equations were experimentally verified by Heinrich Hertz in 1888 and in 1905 Albert Einstein's special theory of relativity further asserted their rigorousness. A pleasant historical survey on the origin of these equations is provided in [9].

We assume the reader to be already familiar with Maxwell's equations as introduced by preliminary courses and we shall credit him with a general knowledge of the experimental facts and their theoretical interpretation.

2.1.1 Maxwell's equations in time–dependent form

It is customary to write Maxwell's equations in either differential form or in integral form. We shall first introduce the differential form; later we discuss briefly the constitutive relationship and then we introduce the integral form of Maxwell's equations.

Differential expressions

In three–dimensional vector notation, Maxwell's equations are

$$\nabla \times \mathbf{E}(\mathbf{r},t) = -\frac{\partial \mathbf{B}(\mathbf{r},t)}{\partial t} \qquad \text{Faraday's induction law} \tag{2.1}$$

$$\nabla \times \mathbf{H}(\mathbf{r},t) = \frac{\partial \mathbf{D}(\mathbf{r},t)}{\partial t} + \mathbf{J}(\mathbf{r},t) \qquad \text{Ampere's circuit law} \tag{2.2}$$

$$\nabla \cdot \mathbf{D}(\mathbf{r},t) = \rho(\mathbf{r},t) \qquad \text{Gauss's law} \tag{2.3}$$
$$\nabla \cdot \mathbf{B}(\mathbf{r},t) = 0 \qquad \text{Gauss's law (magnetic)} \tag{2.4}$$

where the various vectors are defined as:

$\mathbf{E}(\mathbf{r},t)$	electric field strength
$\mathbf{D}(\mathbf{r},t)$	electric displacement
$\mathbf{B}(\mathbf{r},t)$	magnetic flux density
$\mathbf{H}(\mathbf{r},t)$	magnetic field strength
$\mathbf{J}(\mathbf{r},t)$	electric current density
$\rho(\mathbf{r},t)$	electric charge density

Equations (2.1)–(2.4) are not independent since, for example, we may derive (2.4) by taking the divergence of (2.1). Another fundamental relationship may be derived by introducing (2.3) into the divergence of (2.2)

$$\nabla \cdot \mathbf{J}(\mathbf{r},t) = -\frac{\partial \rho(\mathbf{r},t)}{\partial t} \tag{2.5}$$

which provides the conservation law for electric charge and current densities.

Actually, the set of the three equations (2.1), (2.2) and (2.5) may be considered as the independent equations describing macroscopic electromagnetic fields, since the two Gauss equations (2.3) and (2.4) may be derived from this set.

In a three–dimensional space, the two curl equations provide six scalar equations which, in addition with the scalar conservation law for electric charge, provide a set of seven scalar equations. It is customary to consider as known currents and charges, also satisfying the conservation law (2.5). Therefore we are left with six scalar equations, corresponding to (2.1), (2.2) and four vector fields, i.e. with twelve scalar unknowns. It is thus necessary to supplement Maxwell's equations with the constitutive relations. On the other hand, it is apparent that the characteristics of the medium do not appear in equations (2.1) and (2.2), while it is rather obvious that the type of field generated by given sources should depend on the medium characteristics.

Constitutive relations

In general, constitutive relations may be written as

$$\begin{aligned} \mathbf{D} &= \mathcal{F}_d(\mathbf{E},\mathbf{H}) \\ \mathbf{B} &= \mathcal{F}_b(\mathbf{E},\mathbf{H}) \end{aligned} \tag{2.6}$$

where $\mathcal{F}_d$ and $\mathcal{F}_d$ are suitable functionals dependent on the medium considered, which may be classified as:

- nonlinear, when functionals depend on the electromagnetic field;
- inhomogeneous, when functionals depend on space coordinates;
- nonstationary, if functionals depend on time;
- spatial–dispersive, when functionals depend on spatial derivatives;

- time–dispersive, when functionals depend on time derivatives.

In the following, we will deal with linear, stationary and non–dispersive media. However, inhomogeneous media are considered since most of the open waveguides used in current practice require different dielectrics.

Another classification of different media is provided by the vector form of the constitutive relations. The simplest possibility arises when considering *isotropic media*, where the constitutive relations are given by:

$$\begin{aligned} \mathbf{D} &= \varepsilon \mathbf{E} \\ \mathbf{B} &= \mu \mathbf{H} \end{aligned} \tag{2.7}$$

with ε denoted as permittivity and μ permeability. In this case $\mathbf{E}$ is parallel to $\mathbf{D}$ and $\mathbf{B}$ is parallel to $\mathbf{H}$.

Anisotropic media are characterised by constitutive relations of the type

$$\begin{aligned} \mathbf{D} &= \underline{\underline{\varepsilon}}\,\mathbf{E} \\ \mathbf{B} &= \underline{\underline{\mu}}\,\mathbf{H} \end{aligned} \tag{2.8}$$

where $\underline{\underline{\mu}}$ is the permeability tensor and $\underline{\underline{\varepsilon}}$ is the permittivity tensor. The medium is called electrically anisotropic if it is described by the permittivity tensor $\underline{\underline{\varepsilon}}$, and magnetically anisotropic when it is described by the permeability tensor $\underline{\underline{\mu}}$. A medium can be both electrically and magnetically anisotropic. An interesting particular case is that of crystals, which may be described, by suitably choosing a particular coordinate system, the so called principal system, by using a tensor of the type:

$$\underline{\underline{\varepsilon}} = \begin{bmatrix} \varepsilon_x & 0 & 0 \\ 0 & \varepsilon_y & 0 \\ 0 & 0 & \varepsilon_z \end{bmatrix} \tag{2.9}$$

Cubic crystals, where $\varepsilon_x = \varepsilon_y = \varepsilon_z$, are isotropic; tetragonal, hexagonal and rhombohedral crystals have two parameters equal and the medium is called uniaxial. The principal axis that exhibits this anisotropy is also referred to as the optic axis. When all the three parameters are different, as in orthorhombic crystals, the medium is referred to as biaxial.

Finally, some media when placed in an electric field, or when placed in a magnetic field, become both polarised and magnetised. Such media are called *bianisotropic*, being characterised by constitutive relationships of the type

$$\begin{aligned} \mathbf{D} &= \underline{\underline{\varepsilon}}\,\mathbf{E} + \underline{\underline{\xi}}\,\mathbf{H} \\ \mathbf{B} &= \underline{\underline{\xi}}\,\mathbf{E} + \underline{\underline{\mu}}\,\mathbf{H} \end{aligned} \tag{2.10}$$

Integral form of the field equations

The properties of an electromagnetic field have been specified in differential form by using equations (2.1), (2.2), (2.3) and (2.4); such properties may also be expressed by an equivalent system of integral relations by using the two fundamental theorems of vector analysis, i.e. the theorem of divergence and Stokes' theorem [33].

Divergence theorem or Gauss theorem. Let $\mathbf{A}(\mathbf{r})$ be any vector function of position, continuous together with its first derivative throughout a volume V bounded by a surface S. The divergence theorem states that

$$\oint_S \mathbf{A}(\mathbf{r}) \cdot \mathbf{n}\, dS = \int_V \nabla \cdot \mathbf{A}(\mathbf{r})\, dV \tag{2.11}$$

As a matter of fact the Gauss theorem may also be used to define the divergence.

Curl theorem or Stokes theorem. Let $\mathbf{A}(\mathbf{r})$ be any vector function of position, continuous together with its first derivative throughout an arbitrary surface S bounded by a contour C, assumed to be resolvable into a finite number of regular arcs. The Stokes theorem states that

$$\oint_C \mathbf{A}(\mathbf{r}) \cdot d\ell = \int_S [\nabla \times \mathbf{A}(\mathbf{r})] \cdot \mathbf{n}\, dS \tag{2.12}$$

where $d\ell$ is an element of length along C and $\mathbf{n}$ is a unit vector normal to the positive side of the element area dS. This relationship may also be considered as an equation defining the curl.

By applying the curl theorem to (2.1) and (2.2) and by applying the divergence theorem to (2.3) and (2.4) we get:

$$\oint_C \mathbf{E}(\mathbf{r}) \cdot d\ell = -\int_S \frac{\partial \mathbf{B}(\mathbf{r},t)}{\partial t} \cdot \mathbf{n}\, dS \tag{2.13}$$

$$\oint_C \mathbf{H}(\mathbf{r}) \cdot d\ell = \int_S \frac{\partial \mathbf{D}(\mathbf{r},t)}{\partial t} \cdot \mathbf{n}\, dS + \int_S \mathbf{J}(\mathbf{r},t) \cdot \mathbf{n}\, dS \tag{2.14}$$

$$\int_V \nabla \cdot \mathbf{D}(\mathbf{r}) dv = \oint_S \mathbf{D}(\mathbf{r},t) \cdot \mathbf{n}\, dS = \int_V \rho\, dv \tag{2.15}$$

$$\int_V \nabla \cdot \mathbf{B}(\mathbf{r}) dv = \oint_S \mathbf{B}(\mathbf{r},t) \cdot \mathbf{n}\, dS = 0 \tag{2.16}$$

By defining the current I as

$$I = \int_S \mathbf{J}(\mathbf{r},t) \cdot \mathbf{n}\, dS$$

the charge Q as

$$Q = \int_V \rho\, dv$$

and the flux of the magnetic induction as

$$\Phi = \int_S \mathbf{B}(\mathbf{r},t) \cdot \mathbf{n}\, dS$$

we may write the previous equations as

$$\oint_C \mathbf{E}(\mathbf{r}) \cdot d\ell = -\frac{\partial \Phi}{\partial t} \tag{2.17}$$

$$\oint_C \mathbf{H}(\mathbf{r}) \cdot d\ell = \int_S \frac{\partial \mathbf{D}(\mathbf{r},t)}{\partial t} \cdot \mathbf{n}\, dS + I \tag{2.18}$$

$$\int_V \nabla \cdot \mathbf{D}(\mathbf{r})\, dv = Q \tag{2.19}$$

2.1.2 Complex quantities

Electromagnetic fields may be considered at a particular frequency and are known as time–harmonic or monochromatic waves or continuous waves. Time–harmonic fields are particularly important since they allow us to [19]

- eliminate the time dependence in Maxwell's equations;
- obtain the time–domain fields via Fourier transforms;
- cover the whole electromagnetic wave spectrum.

By adopting the time dependence $e^{j\omega t}$ to denote a time–harmonic field with angular frequency ω we let

$$\mathbf{E}(\mathbf{r},t) = \mathrm{Re}\left\{\mathbf{E}(\mathbf{r})e^{j\omega t}\right\} \tag{2.20}$$

where Re denotes the mathematical operator which selects the real part of a complex quantity.

In (2.20) we have used the same symbol to denote both the real quantity in the time domain, $\mathbf{E}(\mathbf{r},t)$, and the complex quantity, $\mathbf{E}(\mathbf{r})$, in the frequency domain. In the rest of the book we will generally refer to complex quantities unless otherwise explicitly stated.

Complex power

Let us consider two time–dependent quantities, $\mathcal{A}, \mathcal{B}$, which may be scalars or components of a vector. By using the phasor description (2.20) we may write

$$\begin{aligned} \mathcal{A} &= |A|\cos(\omega t + \alpha) = \mathrm{Re}\left(Ae^{j\omega t}\right) \\ \mathcal{B} &= |B|\cos(\omega t + \beta) = \mathrm{Re}\left(Be^{j\omega t}\right) \end{aligned} \tag{2.21}$$

where $A = |A|\,e^{j\alpha}$, $B = |B|\,e^{j\beta}$. The product of two such quantities is

$$\mathcal{A}\mathcal{B} = \frac{1}{2}\,|A|\,|B|\,[\cos(\alpha-\beta) + \cos(2\omega t + \alpha + \beta)] \tag{2.22}$$

By denoting the time average with a bar over the quantity, the time average of the above expression is

$$\overline{\mathcal{A}\mathcal{B}} = \frac{1}{2}\,|A|\,|B|\cos(\alpha-\beta) \tag{2.23}$$

On the other hand it is apparent that

$$AB^* = |A|\,|B|\,[\cos(\alpha-\beta) + j\sin(\alpha-\beta)] \tag{2.24}$$

and hence

$$\overline{\mathcal{A}\mathcal{B}} = \frac{1}{2}\mathrm{Re}\left\{AB^*\right\} \tag{2.25}$$

This identity defines complex power. As an example, we may identify $\mathcal{A}$ with a voltage $v(t) = \mathrm{Re}\left[Ve^{j\omega t}\right]$ and $\mathcal{B}$ with a current $i(t) = \mathrm{Re}\left[Ie^{j\omega t}\right]$. The power is given by $v(t)\,i(t)$, and the time–averaged power is given by

$$\overline{v(t)\,i(t)} = \frac{1}{2}\mathrm{Re}\left\{VI^*\right\}$$

We will make use of the above considerations in the next Section, after introducing the Poynting theorem in time and frequency domains.

2.1.3 Maxwell's equations in the frequency domain

By applying (2.20) to the field quantities appearing in (2.1), (2.2), (2.3) and (2.4) we obtain Maxwell's equations in the frequency domain. As an example, let us consider (2.1) for which we have:

$$\mathrm{Re}\left\{\left[\nabla \times \mathbf{E}(\mathbf{r}) + j\omega \mathbf{B}(\mathbf{r})\right] e^{j\omega t}\right\} = 0 \tag{2.26}$$

Since this equation is valid for *all time t*, we may make use of the above lemma and state that the quantity inside the square bracket must be equal to zero. By applying the same reasoning also to the other equations (2.2), (2.3) and (2.4) we get

$$\begin{aligned}
\nabla \times \mathbf{E}(\mathbf{r}) &= -j\omega \mathbf{B}(\mathbf{r}) && (2.27)\\
\nabla \times \mathbf{H}(\mathbf{r}) &= j\omega \mathbf{D}(\mathbf{r}) + \mathbf{J}(\mathbf{r}) && (2.28)\\
\nabla \cdot \mathbf{D}(\mathbf{r}) &= \rho(\mathbf{r}) && (2.29)\\
\nabla \cdot \mathbf{B}(\mathbf{r}) &= 0 && (2.30)
\end{aligned}$$

In the following, we make use of equivalence theorems which introduce magnetic current density, $\mathbf{M}(\mathbf{r})$, and magnetic charge distributions, $\rho_m(\mathbf{r})$. These quantities, although not physically present, help in the solution of several boundary value problems. When considering also magnetic currents and charges, the frequency–domain Maxwell's equations become

$$\begin{aligned}
\nabla \times \mathbf{E}(\mathbf{r}) &= -j\omega \mathbf{B}(\mathbf{r}) - \mathbf{M}(\mathbf{r}) && (2.31)\\
\nabla \times \mathbf{H}(\mathbf{r}) &= j\omega \mathbf{D}(\mathbf{r}) + \mathbf{J}(\mathbf{r}) && (2.32)\\
\nabla \cdot \mathbf{D}(\mathbf{r}) &= \rho(\mathbf{r}) && (2.33)\\
\nabla \cdot \mathbf{B}(\mathbf{r}) &= -\rho_m(\mathbf{r}) && (2.34)
\end{aligned}$$

2.1.4 Poynting theorem

A relation between the rate of change of the energy stored in the field and the energy flow was first derived by Poynting in 1884, and again in the same year by Heaviside. We may start by considering time–domain quantities and noting that $\mathbf{E} \cdot \mathbf{J}$ has the dimension of power per unit volume. This suggests multiplication of (2.2) by $\mathbf{E}$; in a like manner we multiply (2.1) by $\mathbf{H}$ and subtract the previous equation from the latter. Using the identity

$$\nabla \cdot (\mathbf{E} \times \mathbf{H}) = \mathbf{H} \cdot (\nabla \times \mathbf{E}) + \mathbf{E} \cdot (\nabla \times \mathbf{H}) \tag{2.35}$$

and integrating over a volume V bounded by a surface S yields

$$\int_S (\mathbf{E} \times \mathbf{H}) \cdot \mathbf{n}\, dS = -\int_V \mathbf{E} \cdot \mathbf{J}\, dv - \int_V \left(\mathbf{E} \cdot \frac{\partial \mathbf{D}}{\partial t} + \mathbf{H} \cdot \frac{\partial \mathbf{B}}{\partial t}\right) dv \tag{2.36}$$

By defining the time domain Poynting vector $\mathbf{P} = \mathbf{E} \times \mathbf{H}$ and the electric, W_e, and magnetic, W_m, power density as

$$\begin{aligned}
W_e &= \mathbf{E} \cdot \frac{\partial \mathbf{D}}{\partial t}\\
W_m &= \mathbf{H} \cdot \frac{\partial \mathbf{B}}{\partial t} && (2.37)
\end{aligned}$$

we may write the customary interpretation of (2.36) as

$$\int_S \mathbf{P} \cdot \mathbf{n}\, dS = -\int_V \mathbf{E} \cdot \mathbf{J}\, dv - \int_V (W_e + W_m)\, dv \tag{2.38}$$

which states that the intensity of energy flow inside the surface S corresponds to the rate of decrease of electric and magnetic energy and to the power dissipated in Joule heat, given by the first term on the right hand side.

Poynting theorem in frequency domain

The frequency domain version of the Poynting theorem may be obtained by starting from the following equations considered for complex quantities

$$\nabla \times \mathbf{E} = -j\omega \mathbf{B} \tag{2.39}$$

$$\nabla \times \mathbf{H}^* = -j\omega \mathbf{D}^* + \mathbf{J}^* \tag{2.40}$$

Upon multiplication of the first of the above equations by $\mathbf{H}^*$ and of the second equation by $\mathbf{E}$, after proceeding as in the previous case we obtain

$$\int_S (\mathbf{E} \times \mathbf{H}^*) \cdot \mathbf{n}\, dS = -\int_V \sigma\, |\mathbf{E}|^2\, dv + \int_V (j\omega \mathbf{E} \cdot \mathbf{D} - j\omega \mathbf{H}^* \cdot \mathbf{B})\, dv \tag{2.41}$$

where we have considered $\mathbf{J} = \sigma \mathbf{E}$ with σ denoting the conductivity. By introducing the frequency domain Poynting vector

$$\mathbf{P} = \frac{1}{2} \mathbf{E} \times \mathbf{H}^* \tag{2.42}$$

by taking the real and imaginary parts, while supposing that the medium is defined by the constants ε, μ, σ, we have:

$$\int_S \mathrm{Re}(\mathbf{P}) \cdot \mathbf{n}\, dS = -\frac{1}{2} \int_V \sigma\, |\mathbf{E}|^2\, dv \tag{2.43}$$

$$\int_S \mathrm{Im}(\mathbf{P}) \cdot \mathbf{n}\, dS = -\frac{\omega}{2} \int_V \left(\varepsilon\, |\mathbf{E}|^2 - \mu\, |\mathbf{H}|^2\right) dv \tag{2.44}$$

The first of the above equations expresses the energy dissipated while the second is proportional to the difference between magnetic and electric energies.

2.2 Boundary, edge and radiation condition

In order to obtain a unique solution of the Maxwell field equations, one should impose the appropriate boundary, radiation and edge conditions. In this context two facts should be particularly emphasised:

- the radiation condition applies to the total field, but not to each mode individually, since the latter need only be finite at infinity;
- the edge condition is a special kind of boundary condition that, in the presence of surface singularities, must be applied in order to ensure a unique solution of the field problem (see e.g. [10, p.89]) .

2.2.1 Boundary conditions

Let us consider a regular surface S of a medium discontinuity, as shown in Figure 2.1, where the subscripts 1 and 2 distinguish quantities in regions 1 and 2 respectively. By considering (2.13) and (2.14) as a consequence of a limit process it is easy to obtain the following conditions:

$$\mathbf{n} \times (\mathbf{H}_2 - \mathbf{H}_1) = \mathbf{J} \tag{2.45}$$

$$\mathbf{n} \times (\mathbf{E}_2 - \mathbf{E}_1) = -\mathbf{M} \tag{2.46}$$

where $\mathbf{J}$ and $\mathbf{M}$ are respectively the electric and magnetic surface current density distributions at the interface.

Similarly, by considering (2.15) and (2.16) and a small volume at the interface, after a limit process one obtains,

$$\mathbf{n} \cdot (\mathbf{B}_2 - \mathbf{B}_1) = -\rho_m \tag{2.47}$$

$$\mathbf{n} \cdot (\mathbf{D}_2 - \mathbf{D}_1) = \rho \tag{2.48}$$

where ρ and ρ_m are respectively the electric and magnetic surface charge density distributions on the interface.

If neither medium is perfectly conducting the tangential component of the fields $\mathbf{E}$ and $\mathbf{H}$ are continuous while their normal components undergo a jump corresponding to the discontinuity in the permittivity and permeability.

When medium 1 is a perfect electric conductor, the field inside this medium vanishes everywhere and induced electric charges and currents exist on the surface. In this case we have:

$$\mathbf{n} \times \mathbf{H}_2 = \mathbf{J} \tag{2.49}$$

$$\mathbf{n} \times \mathbf{E}_2 = 0 \tag{2.50}$$

$$\mathbf{n} \cdot \mathbf{B}_1 = 0 \tag{2.51}$$

$$\mathbf{n} \cdot \mathbf{D}_1 = \rho \tag{2.52}$$

which states the vanishing at the metal surface of the tangential components of $\mathbf{E}$ and of the normal component of $\mathbf{H}$.

In several cases, it is a convenient artifice to consider fields as generated from equivalent magnetic currents. Accordingly, the field generated by a magnetic current distribution in the immediate vicinity of a perfectly conducting surface is given by

$$\mathbf{n} \times \mathbf{E}_2 = -\mathbf{M} \tag{2.53}$$

2.2.2 Radiation condition

For an unbounded region it is necessary to specify the field behaviour on a surface at infinity. By assuming that all sources are contained in a finite region, only *outgoing waves* must be present at large distances from the sources. In other words, the field behaviour at large distances from the sources must meet the physical requirement that energy travel away from the source region. This requirement is the Sommerfeld "*radiation condition*" and constitutes a boundary condition on the surface at infinity. It assumes different expressions when dealing with 2D– or 3D–regions.

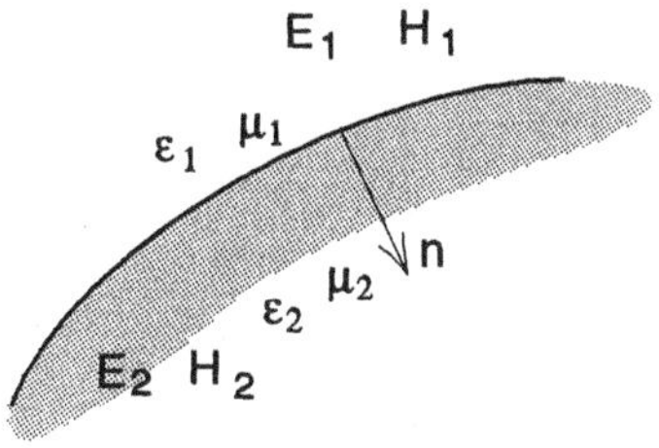

Figure 2.1 Interface between two media.

3D region. Let us denote with A any field component transverse to the radial distance r and with k the free–space wavenumber. The transverse field of a spherically diverging wave in a homogeneous isotropic medium decays as $1/r$ at large distances r from the source region; locally the spherical wave behaves like a plane wave travelling in the outward r direction. As such, each field component transverse to r must behave like $\exp(-jkr)/r$; this requirement may be phrased mathematically as

$$\lim_{r\to\infty} r\left(\frac{\partial A}{\partial r} + jkA\right) = 0 \tag{2.54}$$

Observe, for future use, that the above boundary condition is not self–adjoint in the Hermitian sense. The adjoint boundary condition would be

$$\lim_{r\to\infty} r\left(\frac{\partial A}{\partial r} - jkA\right) = 0 \tag{2.55}$$

corresponding to waves impinging from infinity. It is noted that, when using an eigenfunction representation, we need to represent our field by means of eigenfunctions of the given problem and eigenfunctions of the adjoint problem. This consideration may be applied to the example considered in the first Chapter, where we have obtained the eigenfunctions of a metallic plane. In that case, the eigenfunction may be considered as obtained from the superposition of an outgoing wave, with dependence $\exp(-jkx)$, and an incoming wave with amplitude $\exp(jkx)$. Thus, it is noted again that only the total field should satisfy the Sommerfeld radiation condition, while the eigenfunctions, obtained as a solution of the direct and the adjoint problems, may be finite at infinity.

2D region. Let us denote with ρ the radial variable in the transverse plane. The transverse field component A in a cylindrically diverging wave in a homogeneous isotropic medium decays as $1/\sqrt{\rho}$ at large distances ρ from the source region; locally it behaves like a plane wave travelling in the outward ρ direction. As such, each field component transverse to ρ must behave like $\exp(-jk\rho)/\sqrt{\rho}$; this requirement may be phrased mathematically as

$$\lim_{\rho\to\infty} \sqrt{\rho}\left(\frac{\partial A}{\partial \rho} + jkA\right) = 0 \tag{2.56}$$

The above equations apply to non–dissipative media. When the media are slightly lossy one may use the simpler requirement that all fields excited by sources in a finite region should vanish at infinity.

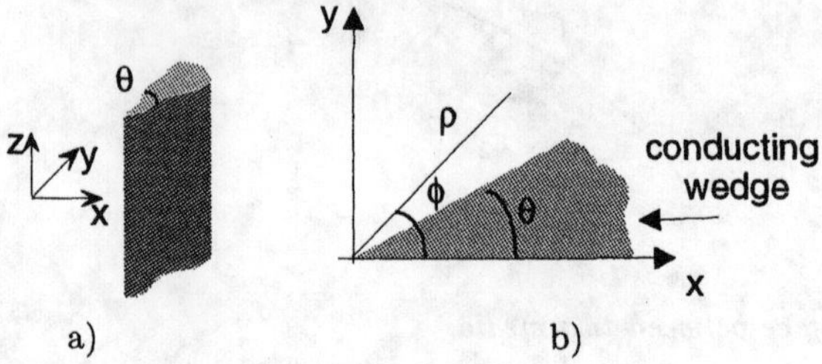

Figure 2.2 Perfectly conducting wedge infinite in the z–direction. a) 3D–view; b) geometry in the transverse plane also showing the cylindrical coordinate system.

2.2.3 Edge condition in two dimensions

It is well recognised that, in many cases, boundary and radiation conditions alone are not sufficient to determine the solution uniquely (see [2, p.385]), since it is possible to construct several different fields which satisfy these conditions [3]. As an example, let us consider a metallic wedge, infinite in the z–direction, such as that depicted in Figure 2.2. Due to the cylindrical geometry it is convenient to decompose the field into TE and TM waves with respect to z; moreover, it is also expedient to assume no variation along the z-coordinate. Let E_0 be any field which satisfies boundary and radiation conditions. Then the field

$$E_z = E_0 + C J_\nu(k\rho) \sin\left[\nu(\phi - \theta)\right] \tag{2.57}$$

satisfies Helmholtz's equation for any value of C, complies with radiation conditions and has the same boundary behaviour as E_0 since $E_z = 0$ when $\phi = 0, \theta$. However, an infinite set of solutions can be generated by giving different values to C [17, pp.531-532]. Therefore, it is necessary to apply an additional constraint in order to achieve a unique solution; such a constraint is represented by the edge condition [3, 24, 25]. We start by noting that *the electromagnetic energy density must be integrable over any finite domain* even if this domain contains singularities of the electromagnetic field. Differently stated, *the electromagnetic energy in any finite domain must be finite.* The sum of the electric and magnetic energies in a small volume V surrounding the edge is [6, p.24]

$$\frac{1}{2}\int_V \left(\varepsilon \mathbf{E}\cdot\mathbf{E}^* + \mu \mathbf{H}\cdot\mathbf{H}^*\right) \rho \, d\phi \, d\rho \, dz \tag{2.58}$$

In the vicinity of the edge the fields can be expressed as a power series in ρ; this series will have a dominant term ρ^α where α may be negative. Therefore, as ρ approaches zero the dominant term of the field components of (E, H) appearing in (2.58) grows as $\rho^{2\alpha}$. In order to obtain a finite value of the integral (2.58) $\alpha > -1$. The actual degree of singularity that a field experiences near the edge is dependent on the wedge configuration (i.e. on the angle θ in Figure 2.2). It is also noted that the field singularity does not depend on frequency, since in the proximity of the edge spatial derivatives of the fields are much larger than time derivatives, so that the latter can be neglected in Maxwell's equations.

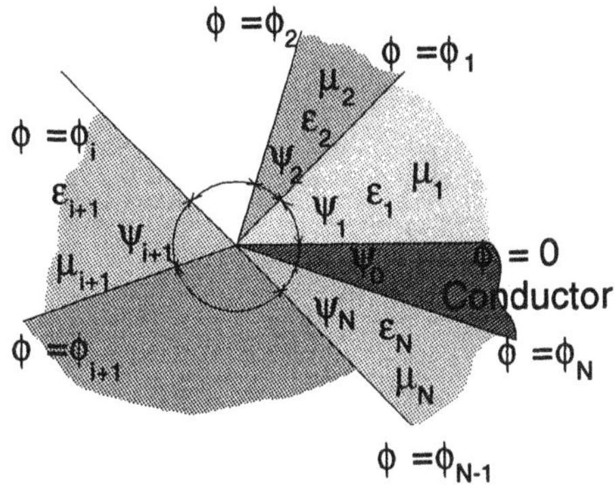

Figure 2.3 A perfectly conducting wedge and N dielectric wedges; the z-axis is along the common wedge and normal to the figure plane.

The exact knowledge of the type of field singularity near the edge is of considerable importance for numerical applications . One way to compute this quantity is the approach followed by [25]. Another interesting one, followed by [15], establishes a simple transmission line equivalent circuit of the wedge problem; by using a transverse resonance approach it is then possible to recover the value of field singularity, as outlined in the following. With reference to Figure 2.3 let us consider a TE field independent of the z–direction. By setting equal to zero the z–derivatives in Maxwell's equation in cylindrical coordinates, see page 76, we obtain the three–component field

$$
\begin{aligned}
E_\rho &= -j\omega\mu\frac{1}{\rho}\frac{\partial\psi}{\partial\phi} \\
E_\phi &= j\omega\mu\frac{\partial\psi}{\partial\rho} \\
H_z &= k^2\psi
\end{aligned}
\tag{2.59}
$$

where ψ is a suitable Hertzian magnetic potential which satisfies Helmholtz' equation in cylindrical coordinates. The field in (2.59) is TE with respect to the z–direction, briefly TE^z, but it is also TM with respect to the ϕ–direction (thus TM$^\phi$). As observed in Section 2.5.2, when several dielectrics are present the field must be described by two potentials unless the potentials are taken in the direction normal to the stratification. In this case the stratification is normal to the ϕ–direction. Hence, we describe the field in our structure as just TM$^\phi$ or TE$^\phi$. Since for a two–dimensional case TE^z coincides with TM$^\phi$, we may refer to this field by either of these names.

The ρ–dependence of the ψ potential can be written in general as a combination of Bessel functions $J_\nu(k\rho)$ and of Hankel functions $H_\nu^{(2)}(k\rho)$. The latter represent outward propagating cylindrical wave which, since no source is present at $\rho = 0$, should not be present either. Therefore, the ρ–dependence is completely described by $J_\nu(k\rho)$. In particular, near the edge we may use the small argument approximation of the Bessel function for the ρ–dependence times a trigonometric dependence on the angle ϕ with two as yet undetermined coefficients A_i, B_i; the suffix i denotes

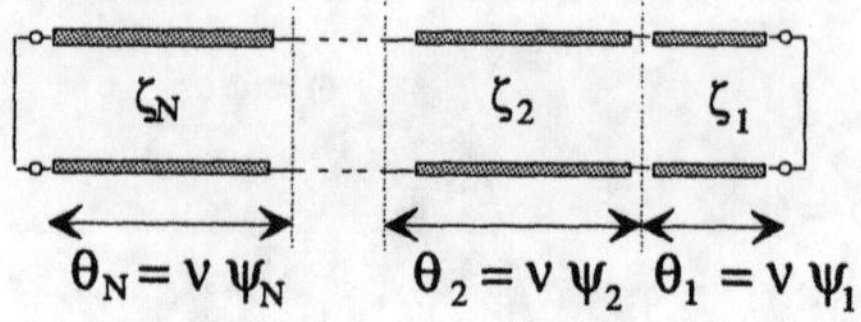

Figure 2.4 Transmission line equivalent for the wedge problem of the previous figure.

any of the possible solutions. We assume a potential of the form

$$\psi = \frac{1}{k_i^2}\rho^{\nu}\left[A_i \sin(\nu\phi) - B_i \cos(\nu\phi)\right] \tag{2.60}$$

which gives the following expressions for the fields

$$\begin{aligned} E_\rho &= -\frac{j\nu}{\omega\varepsilon_i}\rho^{\nu-1}\left[A_i \cos(\nu\phi) + B_i \sin(\nu\phi)\right] \\ H_z &= \rho^{\nu}\left[A_i \sin(\nu\phi) - B_i \cos(\nu\phi)\right] \end{aligned} \tag{2.61}$$

Hence, we identify α of the previous discussion with $\nu - 1$ of (2.61). Note that the latter equation represents a pair of fields transverse to the ϕ–direction; their ratio provides the TM$^\phi$ characteristic impedance and we may describe the ϕ–dependence by means of an equivalent transmission line. In particular by defining the following equivalent voltages and currents

$$\begin{aligned} V_i &= \frac{1}{\varepsilon_i}\left[A_i \cos(\nu\phi_i) + B_i \sin(\nu\phi_i)\right] \\ I_i &= \left[A_i \sin(\nu\phi_i) - B_i \cos(\nu\phi_i)\right] \end{aligned} \tag{2.62}$$

and by making use of the continuity of H_z and E_ρ one derives the following *ABCD* representation

$$\begin{pmatrix} V_{i+1} \\ I_{i+1} \end{pmatrix} = \begin{pmatrix} \cos(\nu\phi_i) & \zeta_i \sin(\nu\phi_i) \\ \zeta_i^{-1} \sin(\nu\phi_i) & \cos(\nu\phi_i) \end{pmatrix} \begin{pmatrix} V_i \\ I_i \end{pmatrix} \tag{2.63}$$

where $\zeta_i = 1/\varepsilon$ is the normalised TM$^\phi$ characteristic impedance. As a consequence, the structure of Figure 2.3 is represented by the cascaded transmission lines of Figure 2.4. The resonant frequencies of this transmission line give the sought values of ν. However, there is an infinity of solutions; as we have seen before, the edge condition demands that $\alpha > -1$, and since $\alpha = \nu - 1$, ν must be greater than zero. It is noted that when $1 > \nu > 0$ we have a singularity while for $\nu > 1$ the field remains finite near the edge. By introducing the load impedance $Z_i = V_i/I_i$ and $T_i = \tan(\nu\psi_i)$ and by using the standard formula for the impedance transformation along a transmission line

$$Z_i = \zeta_i \frac{Z_{i-1} + j\zeta_i T_i}{\zeta_i + jZ_{i-1}T_i} \qquad i = 1, 2, ..., N \tag{2.64}$$

we can evaluate the resonance condition by computing the total impedance of the transmission line circuit of Figure 2.4. It is now useful to apply (2.64) to derive the

Table 2.1 Behaviour of TE^z field near a metallic wedge.

Geometry	ν	E_ρ, E_ϕ	H_z
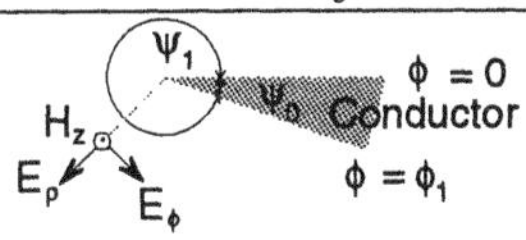	from $\nu = \frac{\pi}{\psi_1}$	$\rho^{\nu-1}$	ρ^{ν}
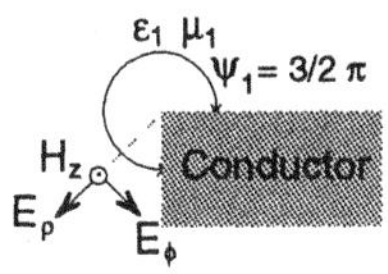	$\nu = \frac{2}{3}$	$\rho^{-1/3}$	$\rho^{2/3}$
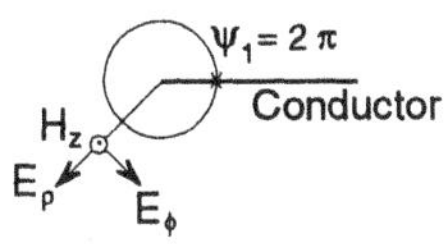	$\nu = \frac{1}{2}$	$\rho^{-1/2}$	$\rho^{1/2}$

field singularity in some common cases. As a first example, let us consider just one metallic wedge as in the first row of Table 2.1. In this case $N = 1$ and application of (2.64) with $Z_0 = Z_1 = 0$ yields $T_1 = 0$, that is

$$\nu = \frac{\pi}{\psi_1} \tag{2.65}$$

which is the well known result obtained by [23]. Particular cases, relevant for practical applications, are the 90°–metal edge and the infinitely thin metal conductor; these cases are listed in Table 2.1. When two dielectric wedges are present, the structure becomes that represented by the figure in the first row of Table 2.2. Here we have $N = 2$ and $Z_0 = Z_2 = 0$. Application of (2.64) yields the following transcendental equation

$$\frac{T_1}{\varepsilon_1} + \frac{T_2}{\varepsilon_2} = 0 \tag{2.66}$$

It is noted that for this case (TM^ϕ) only the dielectric constants play a role, whereas the difference of magnetic permeabilities does not introduce any singularity of the field. This is quite obvious since the only singular components of the field are those transverse to z and these are electric field components. Particular cases relevant to practical applications are a conducting half plane on a dielectric medium, and a 90°–dielectric corner backed by a conducting half–plane; these cases are listed in Table 2.2. The case of a 90°–dielectric wedge, the last one of Table 2.2, deserves particular attention. In fact, no closure condition is now present but rather a periodic transmission line occurs; this makes the evaluation of the resonance condition more difficult. Nevertheless, for the 90° corner, by suitably using symmetry we can substitute the structure of Figure 2.5a with that of Figure 2.5b. By applying the

Table 2.2 Behaviour of a TE^z field in the presence of dielectric wedges.

Geometry	ν	E_ρ, E_ϕ	H_z
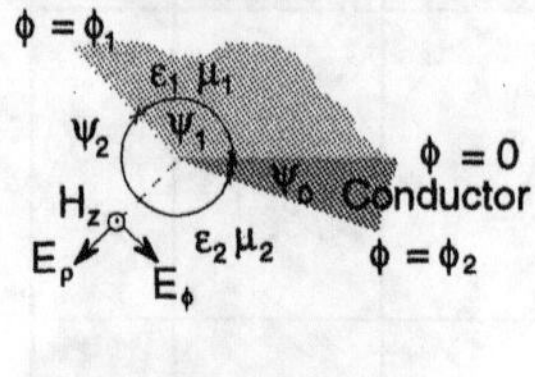	from (2.66)	$\rho^{\nu-1}$	ρ^{ν}
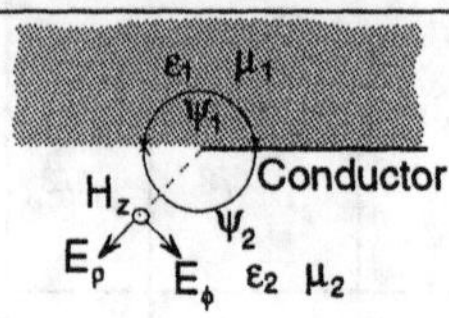	$\nu = \frac{1}{2}$	$\rho^{-1/2}$	$\rho^{1/2}$
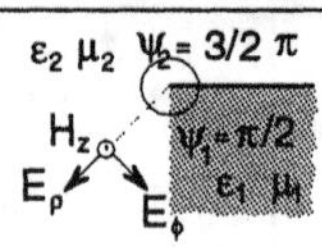	$\nu = \frac{1}{\pi}\cos^{-1}\left[\frac{\varepsilon_2-\varepsilon_1}{2(\varepsilon_2+\varepsilon_1)}\right]$	$\rho^{\nu-1}$	ρ^{ν}
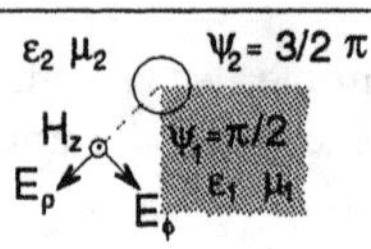	$\nu = \frac{2}{\pi}\cos^{-1}\left[\frac{\varepsilon_2-\varepsilon_1}{2(\varepsilon_2+\varepsilon_1)}\right]$	$\rho^{\nu-1}$	ρ^{ν}

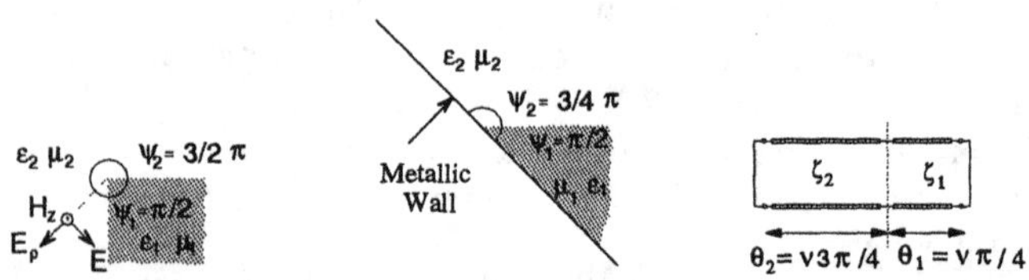

a) b) c)

Figure 2.5 Analysis of a 90°–dielectric corner by symmetry.

transmission line equivalence we get the circuit of Figure 2.5c which yields the field singularity, whereas by applying (2.66) to the configuration of Figure 2.5 we get the value of ν listed in Table 2.2.

A similar analysis can be carried out for TM^z (i.e. TE^ϕ) fields. By setting equal to zero the z–derivatives in (2.246) on page 76 we obtain the three–component field

$$H_\rho = j\omega\varepsilon\frac{1}{\rho}\frac{\partial\psi}{\partial\phi}$$

Table 2.3 Behaviour of a TM^z field near a metallic wedge.

Geometry	ν	H_ρ, H_ϕ	E_z
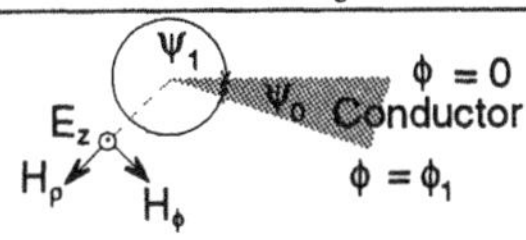	from $\nu = \frac{\pi}{\psi_1}$	$\rho^{\nu-1}$	ρ^{ν}
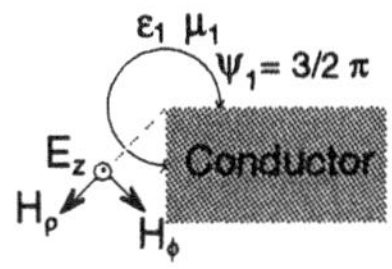	$\nu = \frac{2}{3}$	$\rho^{-1/3}$	$\rho^{2/3}$
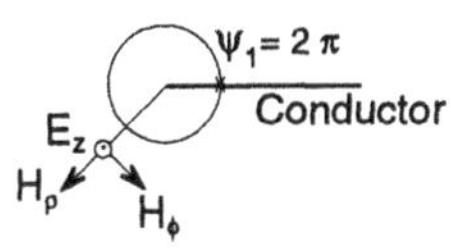	$\nu = \frac{1}{2}$	$\rho^{-1/2}$	$\rho^{1/2}$

$$
\begin{aligned}
H_\phi &= -j\omega\varepsilon \frac{\partial \psi}{\partial \rho} \\
E_z &= k^2 \psi
\end{aligned}
\tag{2.67}
$$

where ψ is a suitable Hertzian electric potential function. It is possible to apply the same reasoning used in equations (2.59) to (2.64), the only difference being the use in (2.64) of the normalised TE$^\phi$ characteristic impedance now given by $\zeta_i = \mu_i$. For reference purposes results similar to those listed in Tables 2.1 and 2.2 are also summarised for the TM^z case in Tables 2.3 and 2.4. When a metallic edge is present, the magnetic field undergoes the same type of singularity as the electric field does. As a consequence, currents flowing parallel to the edge on the perfect electric conductor also features a singular behaviour near the edge. Note that when a dielectric wedge is present, the magnetic field has a singular behaviour only if the magnetic permeabilities of the two media differ. We will make extensive use of field singularities when dealing with open waveguide discontinuities in Chapter 5 and following Chapters.

Table 2.4 Behaviour of a TM^z field in the presence of dielectric wedges.

Geometry	ν	H_ρ, H_ϕ	E_z
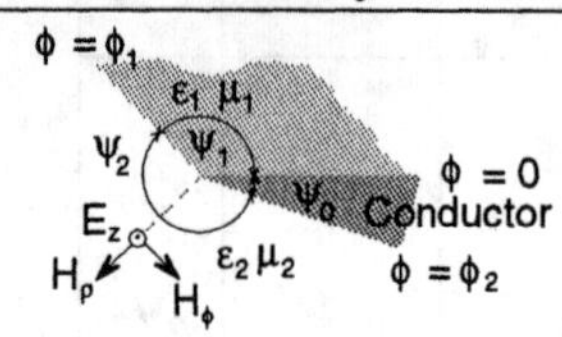	from $\mu_1 T_1 + \mu_2 T_2 = 0$	$\rho^{\nu-1}$	ρ^{ν}
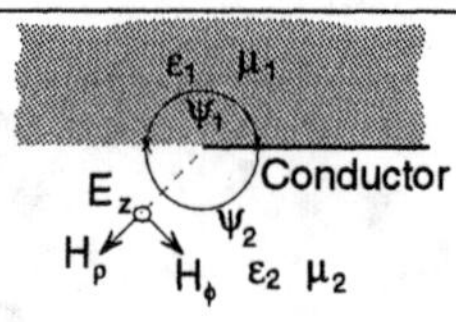	$\nu = \frac{1}{2}$	$\rho^{-1/2}$	$\rho^{1/2}$
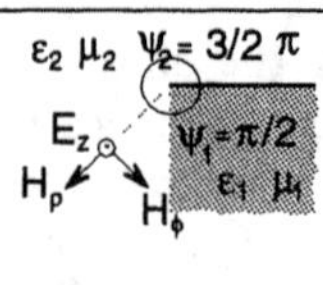	$\nu = \frac{1}{\pi}\cos^{-1}\left[\frac{\mu_2-\mu_1}{2(\mu_2+\mu_1)}\right]$	$\rho^{\nu-1}$	ρ^{ν}
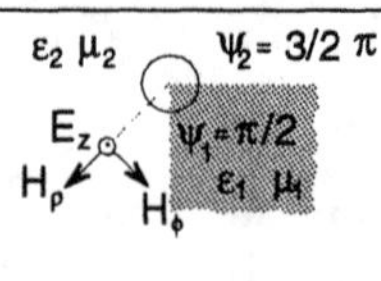	$\nu = \frac{2}{\pi}\cos^{-1}\left[\frac{\mu_2-\mu_1}{2(\mu_2+\mu_1)}\right]$	$\rho^{\nu-1}$	ρ^{ν}

2.3 Some theorems

2.3.1 Uniqueness theorem

The uniqueness theorem is the prerequisite for the application of most of the subsequent theorems, such as the equivalence principle, Huygens' principle, the image theorem, Babinet's principle, the induction theorem, etc. It indicates how a problem should be properly formulated in order to provide one and only one solution. In fact, uniqueness of the solution is a consequence of the proper imposition of the boundary conditions, since overdetermination, i.e. too many boundary conditions, may lead to no solution for a given problem, while a lack of boundary conditions may lead to multiple solutions.

For time–harmonic electromagnetic fields the uniqueness theorem states that when the sources and the tangential components of the electric *or* magnetic field are specified over the *whole* boundary surface of a given region, then the solution within this region is unique. This is actually true only if the medium is slightly lossy; otherwise it is possible to have a multiplicity of solutions as, for example, for a closed resonator.

The proof of the uniqueness theorem follows from considering two different solutions $\mathbf{E}_1$, $\mathbf{H}_1$ and $\mathbf{E}_2$, $\mathbf{H}_2$ in the volume V bounded by the surface S excited by the same system of sources. Let us define the difference fields $\delta\mathbf{E}$ and $\delta\mathbf{H}$ as

$$\begin{aligned} \delta\mathbf{E} &= \mathbf{E}_1 - \mathbf{E}_2 \qquad (a) \\ \delta\mathbf{H} &= \mathbf{H}_1 - \mathbf{H}_2 \qquad (b) \end{aligned} \tag{2.68}$$

By linearity and since the sources are the same, the difference fields satisfy the source–free Maxwell's equations

$$\begin{aligned} \nabla \times \delta\mathbf{E} &= -j\omega\mu\delta\mathbf{H} \qquad (a) \\ \nabla \times \delta\mathbf{H} &= j\omega\varepsilon\delta\mathbf{E} \qquad (b) \end{aligned} \tag{2.69}$$

where it has been assumed that the permittivity ε and the permeability μ are of the following form

$$\begin{aligned} \varepsilon &= \varepsilon_r - j\varepsilon_i \qquad (a) \\ \mu &= \mu_r - j\mu_i \qquad (b) \end{aligned} \tag{2.70}$$

and a small, positive, imaginary part is present. As noted in [19] the proof also holds when the imaginary parts are both negative. By scalar multiplying of (2.69a) by $\delta\mathbf{H}^*$ and the complex conjugate of (2.69b) by $\delta\mathbf{E}$ we obtain

$$\nabla \cdot (\delta\mathbf{E} \times \delta\mathbf{H}^*) = j\omega\varepsilon^* |\delta\mathbf{E}|^2 - j\omega\mu |\delta\mathbf{H}|^2 \tag{2.71}$$

The complex conjugate of (2.71) also holds, giving

$$\nabla \cdot (\delta\mathbf{E}^* \times \delta\mathbf{H}) = -j\omega\varepsilon |\delta\mathbf{E}|^2 + j\omega\mu^* |\delta\mathbf{H}|^2 \tag{2.72}$$

By adding (2.71) and (2.72), integrating over the volume V and applying the divergence theorem, we recover

$$\oint_S (\delta\mathbf{E} \times \delta\mathbf{H}^* + \delta\mathbf{E}^* \times \delta\mathbf{H}) \cdot d\mathbf{S} = -2\omega \int_V \left(\varepsilon_i |\delta\mathbf{E}|^2 + \mu_i |\delta\mathbf{H}|^2\right) dV \tag{2.73}$$

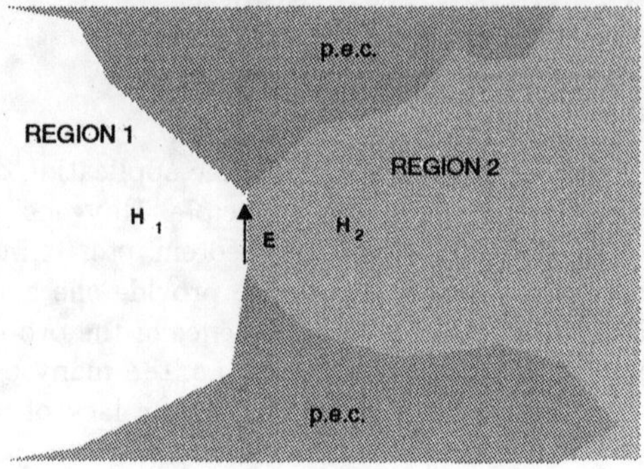

Figure 2.6 Application of the uniqueness theorem.

When the tangential components of **E** or **H** coincide on the boundary surface S, i.e. when either $\delta\mathbf{E}$ or $\delta\mathbf{H}$ are zero on S, we have

$$\oint_S (\delta\mathbf{E} \times \delta\mathbf{H}^* + \delta\mathbf{E}^* \times \delta\mathbf{H}) \cdot d\mathbf{S} = 0 \tag{2.74}$$

In this case, the right–hand side of (2.73) is zero only if $\delta\mathbf{E}$ and $\delta\mathbf{H}$ are identically zero in the region V. This proves the theorem.

As will be evident in the rest of the book, several problems are solved through application of the continuity of the tangential components of both the electric and the magnetic fields. As an example, the reader is referred to the solution of step discontinuity problems in planar dielectric waveguides in Chapter 5. As observed in [33] it is quite puzzling that we need to impose the continuity of both electric and magnetic fields while the uniqueness theorem states that it is sufficient to specify either the values of $\mathbf{n} \times \mathbf{E}$ or $\mathbf{n} \times \mathbf{H}$ on the boundary. The apparent discrepancy is resolved by referring to the example of Figure 2.6, where we have two different regions of space, regions 1 and 2, separated by an interface where a given electric field **E** is assumed. In this way, the fields in regions 1 and 2 are uniquely determined, being $\mathbf{H}_1$, $\mathbf{H}_2$ the magnetic fields in regions 1 and 2, respectively. However, such fields are not a solution of the entire problem unless $\mathbf{H}_1 = \mathbf{H}_2$ at the interface. Thus by imposing **E** at the interface we may determine in a unique manner the solution in the two separate regions. But, in order to calculate the solution in the entire structure, we have to impose the true electric field at the interface, i.e. the electric field for which also $\mathbf{H}_1(\mathbf{E}) = \mathbf{H}_2(\mathbf{E})$ holds.

As a final note, observe that for lossless structures, either closed or open, when we look for modal spectra we are seeking resonant solutions. In this case, the uniqueness theorem does not apply and an infinity of solutions is present.

2.3.2 Equivalence theorem

There are several forms in which the equivalence theorem may be conveniently stated [6, 19]. In view of its importance in the solution of electromagnetic field problems, especially when dealing with non–separable structures, it seems appropriate to examine it in a detailed manner.

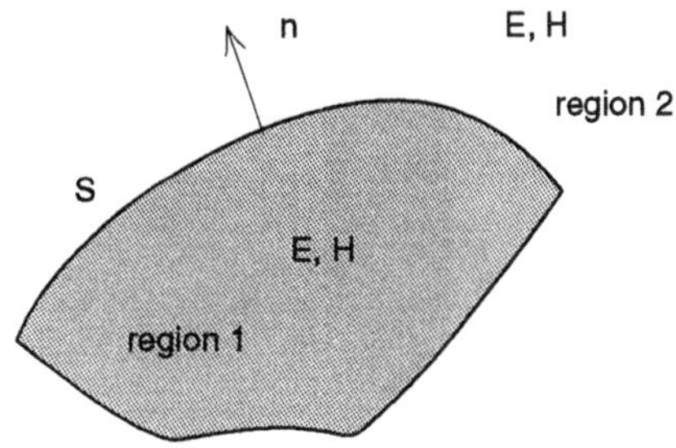

Figure 2.7 Original field.

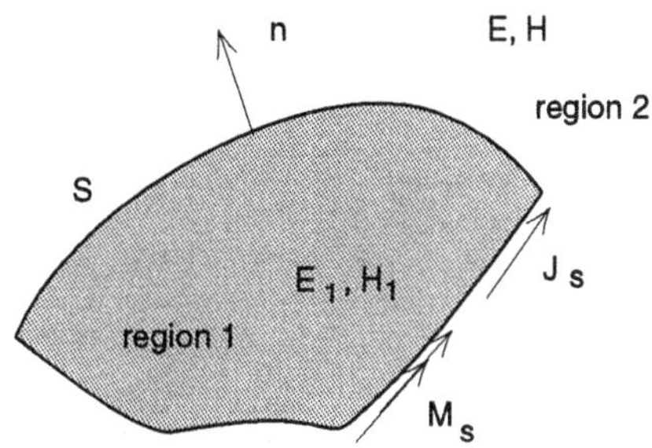

Figure 2.8 The field in region 1 has been modified. By inserting the equivalent electric and magnetic currents on the surface S the field in region 2 is unchanged.

Let us consider a volume V bounded by a surface S separating the internal region, denoted as region 1, from the external region, labelled as region 2. Our purpose in applying the equivalence theorem is to maintain the field in region 2 even when modifying the field in region 1. By so doing we obtain a modified field problem which, at least in region 2 and only in region 2, is equivalent to the original one.

We denote by $\mathbf{E}$ and $\mathbf{H}$ the original fields in regions 1 and 2, as shown in Figure 2.7. Now suppose the field in region 1 is altered, thus changing the field in this region from $\mathbf{E}, \mathbf{H}$ to $\mathbf{E}_1, \mathbf{H}_1$. In order to maintain the original field in region 2 we must insert equivalent magnetic and electric currents, $\mathbf{M}_s$ and $\mathbf{J}_s$ respectively, on the surface S such that

$$\begin{aligned} \mathbf{J}_s &= \mathbf{n} \times (\mathbf{H} - \mathbf{H}_1) \\ \mathbf{M}_s &= -\mathbf{n} \times (\mathbf{E} - \mathbf{E}_1) \end{aligned} \qquad (2.75)$$

as shown in Figure 2.8.

Love equivalence theorem. A particular case is to set the field in region 1 equal to zero. In this instance we get the case shown in Figure 2.9 where the currents are now given by

$$\mathbf{J}_s = \mathbf{n} \times \mathbf{H}$$

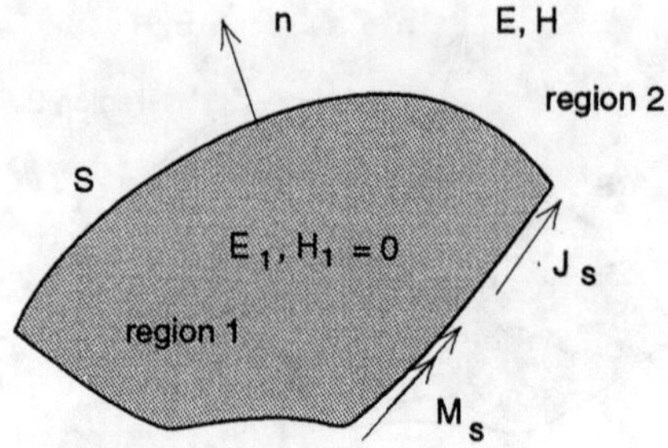

Figure 2.9 The field in region 1 has been set to zero. Equivalent currents maintain the original field in region 2.

$$\mathbf{M}_s = -\mathbf{n} \times \mathbf{E} \tag{2.76}$$

The Love equivalence theorem thus states that the field in region 2 produced by the given sources in region 1 is the same as that produced by a system of *virtual* sources on the surface S.

Perfect electric conductor. The Love theorem only specifies a zero field in region 1. Such a case may be obtained by filling region 1 with a perfect electric conductor (p.e.c.) as considered here, or by filling region 1 with a perfect magnetic conductor (p.m.c.) as considered next. It is easy to see that electric currents $\mathbf{J}_s$ on the p.e.c. do not radiate any field. In fact, near the perfect conductor the electric field is perpendicular to the surface S, while the magnetic field is parallel. The resulting Poynting vector is thus parallel to the surface of the conductor and no power is radiated into the space. A different proof of this fact may be obtained by using the Lorentz theorem. Thus, when region 1 is filled with a p.e.c., the resulting configuration is that shown in Figure 2.10. Note that this form of equivalence theorem is particularly used in practical applications where we have structures bounded by metallic walls. As an example, let us consider the case of a slotted coaxial cable that is a non–separable structure difficult to deal with. However, by applying the equivalence theorem, we may separate our structure into two regions, namely a closed coaxial cable and a wire in free space, both of which have a well defined modal spectrum. Several applications of this form of the equivalence theorem will be shown later in the book.

Perfect magnetic conductor. The other possibility of obtaining a null field in region 1, is to fill this region with a perfect magnetic conductor (p.m.c.). In this case, since magnetic currents do not radiate, we are left with the case of Figure 2.11. Note that in this case, as in the previous one, when calculating the field produced by sources present in region 2, we should take into account the presence of the p.m.c.; otherwise stated, the Green's function to be considered must satisfy the appropriate boundary conditions on the surface S. On the contrary, when applying the Love theorem without filling region 1 with either a p.m.c. or a p.e.c. the Green's function to be considered is that of free–space.

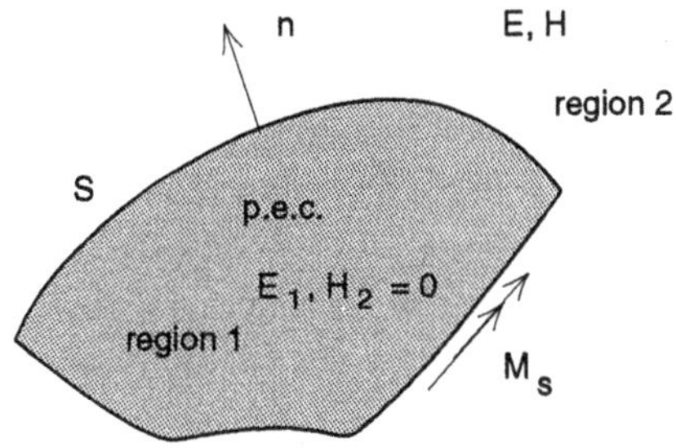

Figure 2.10 Region 1 has been filled with a p.e.c. Only magnetic currents should be considered since electric currents do not radiate.

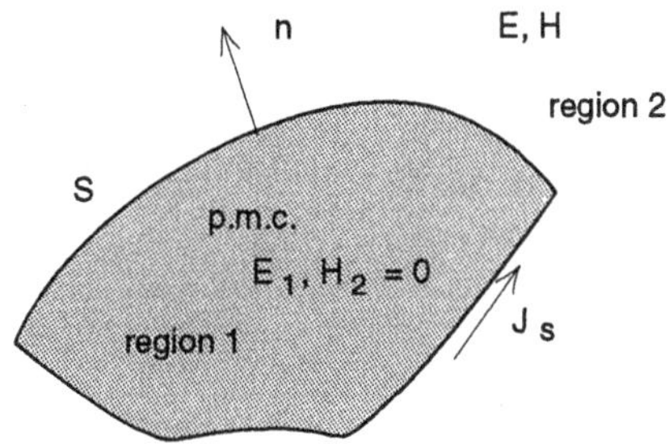

Figure 2.11 Region 1 has been filled with a p.m.c. Only electric currents are considered since magnetic currents do not radiate.

The circuit theory analogue of the equivalence theorem provides a simple and effective way to illustrate this theorem [13]. Let us consider region 2 without sources, represented by the passive network in Figure 2.12(a), while region 1 is represented by the source (active) network. We can set up an equivalent problem by

- switching off the sources in the active network leaving the source impedance connected;
- placing a shunt current generator I equal to the terminal current in the original problem;
- placing a series voltage generator V equal to the terminal voltages in the original problem.

In this way we have replaced the original sources in region 1, the active network, by the virtual sources at the interface as shown in Figure 2.12(b). From usual circuit concepts it is evident that there is no excitation of the source impedance from these equivalent sources whereas the excitation of the passive network is unchanged. This fact offers the possibility of replacing the source impedance by either a short or an open circuit. By considering a short circuit we obtain the case of Figure 2.12(c), equivalent to considering a p.e.c. when applying the Love theorem. When using an

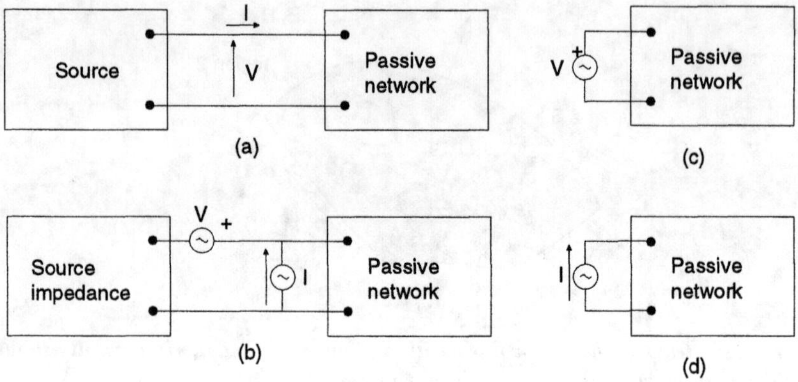

Figure 2.12 Circuit analogue of the equivalence theorem: (a) original problem; (b) source deactivated and equivalent (virtual) sources; (c) source impedance replaced by a short circuit; (d) source impedance replaced by an open circuit.

open circuit we obtain the case of Figure 2.12(d), equivalent to considering a p.m.c. in the Love theorem.

Induction theorem

Another theorem which finds several important applications is the induction theorem, sometimes referred to as *scattering superposition* [34]. With reference to Figure 2.13(a), let us consider some sources $\mathbf{J}, \mathbf{M}$ radiating in the presence of an obstacle enclosed in the volume V bounded by the surface S. We denote with $\mathbf{E}, \mathbf{H}$ the field produced by the sources in the presence of the obstacle. This field is assumed unknown. On the contrary, we assume known the field produced by the sources with the *obstacle removed*, i.e. the so called *incident field*; see also Figure 2.13(b). We define the *scattered field* $\mathbf{E}_s, \mathbf{H}_s$ as

$$\begin{aligned} \mathbf{E}_s &= \mathbf{E} - \mathbf{E}_i \\ \mathbf{H}_s &= \mathbf{H} - \mathbf{H}_i \end{aligned} \tag{2.77}$$

Let us now consider the case shown in Figure 2.13(c) where inside the obstacle we consider the original field $\mathbf{E}, \mathbf{H}$ whereas outside the obstacle we assume the scattered field $\mathbf{E}_s, \mathbf{H}_s$ to be present. Such a field calls for the presence of electric and magnetic currents $\mathbf{J}_s, \mathbf{M}_s$ given by

$$\begin{aligned} \mathbf{J}_s &= \mathbf{n} \times (\mathbf{H}_s - \mathbf{H}) \\ \mathbf{M}_s &= -\mathbf{n} \times (\mathbf{E}_s - \mathbf{E}) \end{aligned} \tag{2.78}$$

By substituting the previous equations into the above, we get

$$\begin{aligned} \mathbf{J}_s &= -\mathbf{n} \times \mathbf{H}_i \\ \mathbf{M}_s &= \mathbf{n} \times \mathbf{E}_i \end{aligned} \tag{2.79}$$

It is apparent that from the knowledge of the incident field we may compute the currents $\mathbf{J}_s, \mathbf{M}_s$. Moreover, by considering the latter currents radiating in free space

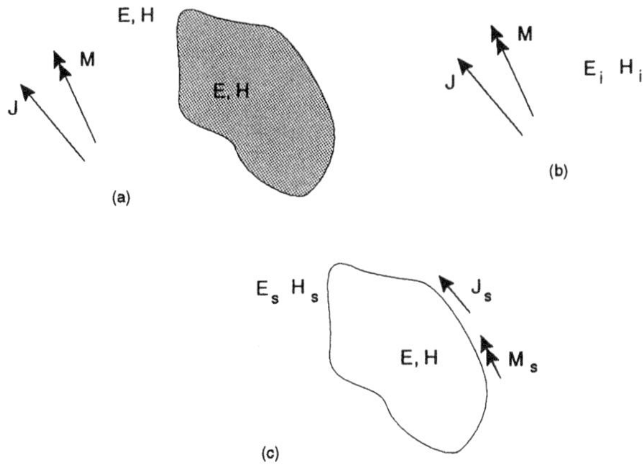

Figure 2.13 Application of the scattering superposition. (a) Original situation with sources radiating in presence of an obstacle (darker region). (b) Sources radiating in free space. (c) Equivalent sources radiating in free space with original sources deactivated. The superposition of cases (b) and (c) gives the same field as in (a) in the region outside the obstacle.

we obtain the scattered field outside the obstacle. Thus the total field, as given by the sum of the incident and scattered fields, may be computed as a superposition of two cases, namely:

- original sources $\mathbf{J}, \mathbf{M}$ radiating in free space;
- scattering sources $\mathbf{J}_s, \mathbf{M}_s$ radiating in free space.

Obviously, this approach does not provide the correct field inside the obstacle.

A significant simplification of the induction theorem occurs when the obstacle is a p.e.c. In this case the field inside the obstacle is zero, $\mathbf{E} = 0$, thus yielding $\mathbf{E}_s = -\mathbf{E}_i$. Since only magnetic currents radiate from a p.e.c. in order to obtain the total field it is sufficient to consider the superposition of the field $\mathbf{E}_i, \mathbf{H}_i$ and that produced by the magnetic current $\mathbf{M}_s = \mathbf{n} \times \mathbf{E}_i$.

Impressed and equivalent sources

It is interesting to consider a few cases, which have a relevance to applications as well as providing a certain insight into the use of the equivalence theorems. Let us start by considering the field produced by impressed magnetic, $\mathbf{M}$, and electric currents, $\mathbf{J}$, as shown in Figure 2.14. An impressed magnetic current produces two electric field components with opposite directions. However, since the magnetic field is the same for the two waves, it results in waves propagating in opposite directions. The effect of such an impressed magnetic current is the same as that of a series voltage generator. Similar considerations apply to the case of impressed electric currents. Also in this case, by impressing the currents, two waves are produced travelling in opposite directions. The impressed electric currents may be represented by a shunt current generator.

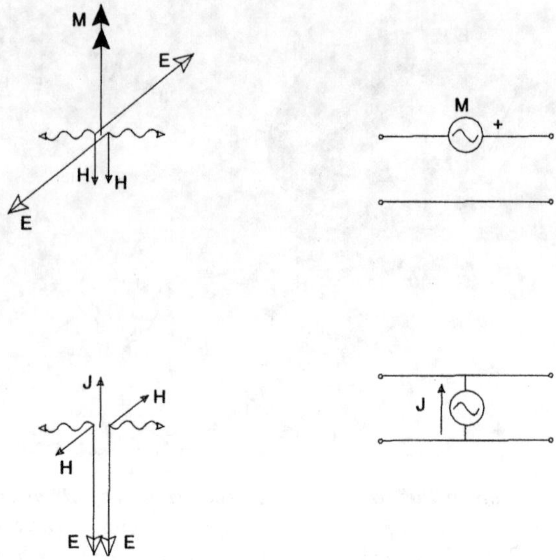

Figure 2.14 Impressed magnetic currents **M** *are represented by a series voltage generator; electric currents* **J** *are represented by a shunt current generator. The waving line shows wave propagation. The impressed sources produce two waves travelling in opposite directions.*

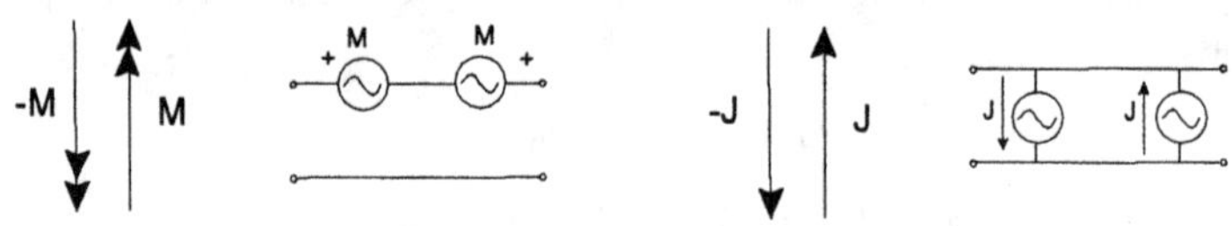

Figure 2.15 A double sheet of impressed magnetic or electric currents with the same strength but opposite directions does not produce any field.

We may also consider two impressed sources, of the same strength but of opposite directions, as shown in Figure 2.15. In this case no radiation is produced since the two currents annihilate.

Let us now insert in between the currents an infinitesimally thin layer of either a p.e.c. or a p.m.c., as shown in Figure 2.16. In the case of impressed magnetic currents on a p.e.c., the voltage generators corresponding to the impressed currents are short–circuited, thus creating the same voltage on the two lines, i.e. the same field to the left and to the right of the double sheet. This configuration is equivalent to having imposed a given electric field, while leaving the possibility of having two different magnetic fields on the two sides of the p.e.c. This case is perhaps the most important in practical applications since it allows complex regions to be separated into simpler ones, bounded by metallic walls. It is fairly obvious that when considering a p.e.c. between electric currents, no field is produced. The circuit interpretation is also shown in Figure 2.16.

Another interesting case is that of an impressed double sheet of electric currents with a thin layer of p.m.c. in between. In this case, the current generators represent-

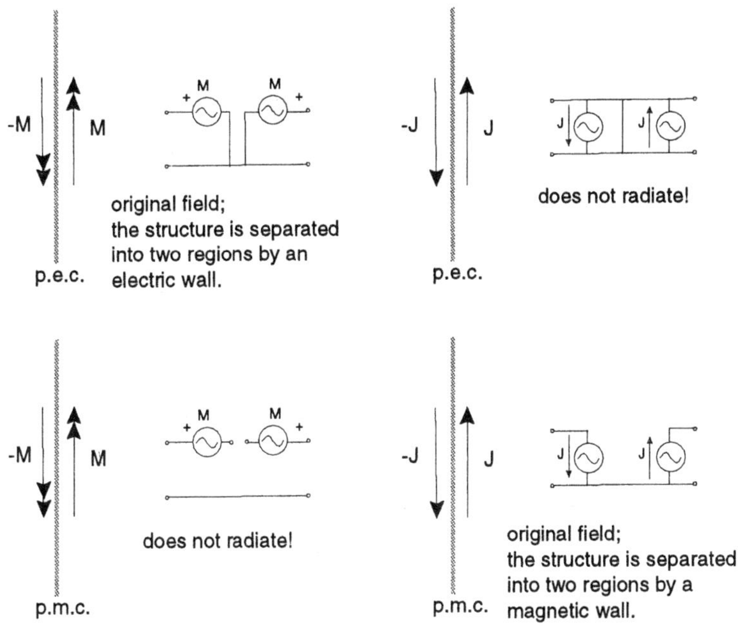

Figure 2.16 Impressed double sheet separated by an infinitesimally thin layer of either a p.e.c. or a p.m.c.

ing the electric currents are decoupled, since they are separated by an open circuit. We have imposed the value of the magnetic field but we have left the electric field free to assume different values on the two sides of the p.m.c. However, magnetic currents on a magnetic conductor do not radiate.

A further application of the equivalence theorem arises in conjunction with the use of image theory. Let us consider a system of sources radiating into free space, as shown in Figure 2.17 and producing an electromagnetic field $\mathbf{E}, \mathbf{H}$. We may consider the surface S and replace the region at the left of S by considering the equivalent sources $\mathbf{J}, \mathbf{M}$ on the surface S. In this way we maintain the original field in the region to the right of S, and a null field to the left of S. The zero field to left of S may be thought of as obtained by placing a p.e.c. in this region. In this case the electrical currents $\mathbf{J}$ on the surface of the p.e.c. do not radiate, leaving only the magnetic currents $\mathbf{M}$ producing the field to the right of S. However, by applying image theory we may remove the metallic screen and place *doubled* magnetic currents radiating into free space. Note that the last step, which requires the use of image theory, may only be applied to a flat surface S, while all the previous steps are always valid. A similar development may also be considered by placing a p.m.c. instead of the p.e.c. previously considered. By analogous considerations one finally obtains that the field in the right–hand region may be obtained by considering doubled electric currents radiating into free space.

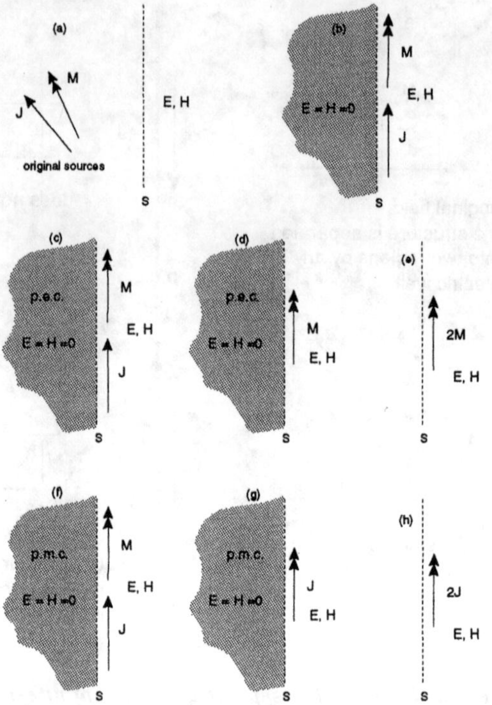

Figure 2.17 Application of the equivalence theorem and of the image principle. (a) Original situation. (b) The original sources have been replaced by equivalent sources that produce a null field to the left of S and the original field to the right of S. In (c) the region to the left of S has been filled with p.e.c. (d) Electric currents do not radiate on a p.e.c.; thus only the magnetic currents are considered. (e) Application of the image principle: the magnetic currents are doubled and the metallic screen is removed. (f) As in (c) but a magnetic screen is now used. (g) The magnetic currents are now ineffective. (h) Application of duality leads to doubled electric currents radiating into free space.

Duality

It is noted that after the following substitutions are carried out

$$\begin{array}{lll} \mathbf{E} \rightarrow \mathbf{H} & \varepsilon \rightarrow \mu & \mu \rightarrow \varepsilon \\ \mathbf{H} \rightarrow -\mathbf{E} & \mathbf{J} \rightarrow \mathbf{M} & \mathbf{M} \rightarrow -\mathbf{J} \end{array} \tag{2.80}$$

equation (2.31) becomes (2.32), and vice versa; this is called duality. However with the above substitutions we are considering a medium "dual" of free space, i.e. a medium with a permittivity of $4\pi \times 10^{-7}$ f/m and with permeability 8.854×10^{-12} h/m, which is an undesirable situation.

A form of duality which is more suitable for antenna and radiation problems is established by the following equalities [19]

$$\mathbf{E} \rightarrow \eta \mathbf{H} \quad \mathbf{H} \rightarrow -\frac{1}{\eta}\mathbf{E}$$

$$\mathbf{J} \rightarrow \frac{1}{\eta}\mathbf{M} \quad \mathbf{M} \rightarrow -\eta\mathbf{J} \tag{2.81}$$

with $\eta = \sqrt{\frac{\varepsilon}{\mu}}$ the free space impedance. With the above substitution equation (2.31) becomes (2.32) and vice versa without any need to replace free space with a different medium. However, the latter form of duality, (2.81), does not apply to anisotropic or bianisotropic media, while (2.80) is more general. In the next Section we introduce Babinet's principle also making use of duality, as expressed by (2.81).

Complementarity and the Babinet principle

Babinet's principle allows the reduction of several problems of slot waveguides to situations involving strip waveguides and vice versa.

In order to illustrate this principle it is convenient to refer to optics, where the principle is stated as follows [20]:

- *The field at any point behind a plane having a screen, if added to the field at the same point when the complementary screen is substituted, is equal to the field at the point when no screen is present.*

The principle is best illustrated with the help of Figure 2.18 where we consider a point source and two imaginary planes, i.e. the plane of the screen and a plane of observation. In Figure 2.18(a) we consider the screen with an aperture . This produces a shadow region and an illuminated region; we may denote the field on the plane B by F_s. We may now consider another case, see Figure 2.18(b), where the screen on the plane A has been replaced by a complementary screen, i.e. a screen extending over the region of the aperture. Also in this case a shadowed and an illuminated region are produced. We may denote by F_c this complementary field on the plane B. Finally let us consider the sources radiating in free space as shown in Figure 2.18(c) and let us denote by F_i the field produced in this case. Babinet's principle asserts that, at the same point, the fields satisfy the following relation

$$F_s + F_c = F_i \tag{2.82}$$

However, by superposition, the principle holds for arbitrary distribution of sources. While Babinet's principle seems quite obvious when considering simple optical cases, it also applies when diffraction is accounted for, as illustrated in the following.

Let us consider again the case of a system of sources radiating in the presence of a metallic screen, with an aperture, as shown in Figure 2.19(a). In this case we denote by

$$\begin{array}{ll} \mathbf{E}_1, \mathbf{H}_1 & \text{field with the screen present} \\ \mathbf{E}^i, \mathbf{H}^i & \text{field without the screen} \\ \mathbf{E}_1^s, \mathbf{H}_1^s & \text{scattered field} \end{array} \tag{2.83}$$

However, as noted when considering the induction theorem, the above fields are related in the following manner

$$\begin{aligned} \mathbf{E}_1 &= \mathbf{E}^i + \mathbf{E}_1^s \\ \mathbf{H}_1 &= \mathbf{H}^i + \mathbf{H}_1^s \end{aligned} \tag{2.84}$$

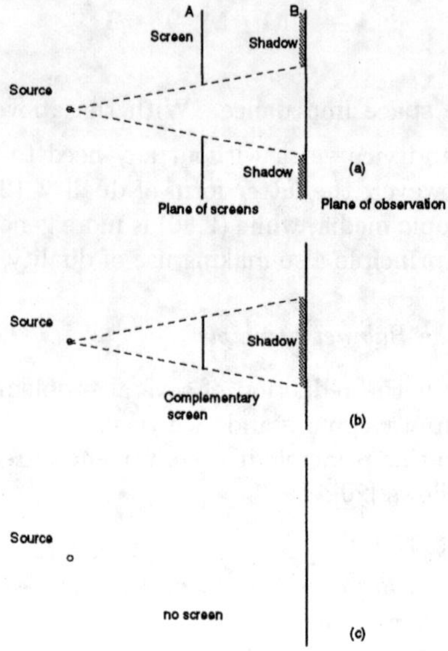

Figure 2.18 Babinet principle: the field at any point behind a plane having a screen, if added to the field at the same point when the complementary screen is substituted, is equal to the field at the point when no screen is present.

The field $\mathbf{E}_1, \mathbf{H}_1$ also satisfies Maxwell's curl equations as

$$\begin{aligned}\nabla \times \mathbf{E}_1 &= -j\omega\mu\mathbf{H}_1 \\ \nabla \times \mathbf{H}_1 &= -j\omega\varepsilon\mathbf{E}_1\end{aligned} \tag{2.85}$$

with the boundary conditions

$$\begin{aligned}\mathbf{n} \times \mathbf{E}_1 &= 0 \qquad \text{on } S_m \\ \mathbf{n} \times \mathbf{H}_1 &= \mathbf{n} \times \mathbf{H}^i \qquad \text{on } S_a\end{aligned} \tag{2.86}$$

Let us now consider Figure 2.19(b) where the aperture in the screen of Figure 2.19(a) has been replaced by a *magnetic* conductor and the screen eliminated. Let us denote by

$$\begin{aligned}&\mathbf{E}_2, \mathbf{H}_2 \quad \text{field with the p.m.c. strip present} \\ &\mathbf{E}_2^s, \mathbf{H}_2^s \qquad\qquad \text{scattered field}\end{aligned} \tag{2.87}$$

which are the field quantities relative to this example. Also in this case we have

$$\begin{aligned}\mathbf{E}_2 &= \mathbf{E}^i + \mathbf{E}_2^s \\ \mathbf{H}_2 &= \mathbf{H}^i + \mathbf{H}_2^s\end{aligned} \tag{2.88}$$

and the field $\mathbf{E}_2, \mathbf{H}_2$ also satisfies (2.85). However, in this case the boundary conditions are

$$\begin{aligned}\mathbf{n} \times \mathbf{E}^i &= \mathbf{n} \times \mathbf{E}_2 \qquad \text{on } S_m \\ \mathbf{n} \times \mathbf{H}_2 &= 0 \qquad \text{on } S_a\end{aligned} \tag{2.89}$$

Consider now the sum field $\mathbf{E}_1 + \mathbf{E}_2, \mathbf{H}_1 + \mathbf{H}_2$, which clearly satisfies (2.85) with the boundary conditions

$$\begin{aligned} \mathbf{n} \times (\mathbf{E}_1 + \mathbf{E}_2) &= \mathbf{n} \times \mathbf{E}^i \qquad \text{on } S_m \\ \mathbf{n} \times (\mathbf{H}_1 + \mathbf{H}_2) &= \mathbf{n} \times \mathbf{H}^i \qquad \text{on } S_a \end{aligned} \tag{2.90}$$

Since the tangential fields on S_m and S_a are identical to those of $\mathbf{E}^i$ and $\mathbf{H}^i$, by the uniqueness theorem we conclude that

$$\begin{aligned} \mathbf{E}_1 + \mathbf{E}_2 &= \mathbf{E}^i \\ \mathbf{H}_1 + \mathbf{H}_2 &= \mathbf{H}^i \end{aligned} \tag{2.91}$$

which is analogous to the result obtained for the optical case.

Note that in the above derivation of the Babinet principle we have made use of a p.m.c. which, however, has no physical counterpart. A more useful form of Babinet's principle may be obtained by considering a complementary *metallic* screen instead of the complementary *magnetic* screen. In this case we need to modify the system of sources in a suitable manner as detailed in the following. With reference to Figure 2.19(c) let us consider the modified sources producing the field

$$\begin{aligned} \mathbf{E}_3 &= \mathbf{E}_3^i + \mathbf{E}_3^s \\ \mathbf{H}_3 &= \mathbf{H}_3^i + \mathbf{H}_3^s \end{aligned} \tag{2.92}$$

with the symbols having analogous meaning to those defined previously. The field $\mathbf{E}_3, \mathbf{H}_3$ satisfies Maxwell's equations

$$\begin{aligned} \nabla \times \mathbf{E}_3^s &= -j\omega\mu \mathbf{H}_3^s \\ \nabla \times \mathbf{H}_3^s &= -j\omega\varepsilon \mathbf{E}_3^s \end{aligned} \tag{2.93}$$

and the following boundary conditions

$$\begin{aligned} \mathbf{n} \times \mathbf{E}_3^s &= -\mathbf{n} \times \mathbf{E}_3^i \qquad \text{on } S_a \\ \mathbf{n} \times \mathbf{H}_3^s &= 0 \qquad \text{on } S_m \end{aligned} \tag{2.94}$$

Considering the two problems (2.93), (2.85) and (2.94), (2.86) we find that they are mathematically dual if we set

$$\begin{aligned} \mathbf{E}_3^s &= -\eta \mathbf{H}_1 \\ \mathbf{H}_3^s &= \frac{1}{\eta} \mathbf{E}_1 \end{aligned} \tag{2.95}$$

If we also consider the sources related by duality as

$$\begin{aligned} \mathbf{E}_3^i &= \eta \mathbf{H}^i \\ \mathbf{H}_3^i &= -\frac{1}{\eta} \mathbf{E}^i \end{aligned} \tag{2.96}$$

we obtain the alternative form of Babinet's principle, i.e.

$$\begin{aligned} \mathbf{E}_3 &= \eta(\mathbf{H}^i - \mathbf{H}_1) \\ \mathbf{H}_3 &= -\frac{1}{\eta}(\mathbf{E}^i - \mathbf{E}_1) \end{aligned} \tag{2.97}$$

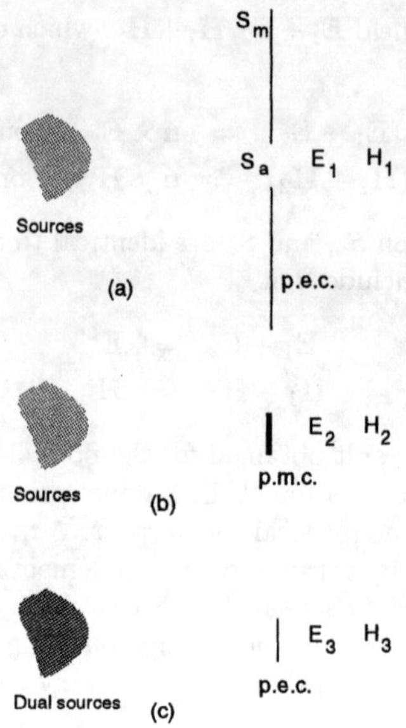

Figure 2.19 Babinet's principle. (a) A system of sources in the presence of a p.e.c. produces the field $\mathbf{E}_1, \mathbf{H}_1$*. (b) The same system of sources in the presence of the complementary p.m.c. screen produces the field* $\mathbf{E}_2, \mathbf{H}_2$*. (c) Dual sources in the presence of the complementary p.e.c. screen produce the field* $\mathbf{E}_3, \mathbf{H}_3$*.*

Two special cases are worth some consideration, e.g. when no screen is present and when no aperture is present. When no screen is present we have $\mathbf{E}_1 = \mathbf{E}^i$ and $\mathbf{H}_1 = \mathbf{H}^i$; in this case however $\mathbf{E}_3 = \mathbf{H}_3 = 0$. When no aperture is present, then $\mathbf{E}_1 = \mathbf{H}_1 = 0$; in this case we have $\mathbf{E}_3 = \eta \mathbf{H}^i$ and $\mathbf{H}_3 = -\frac{1}{\eta}\mathbf{E}^i$, which are the fields generated by the dual sources.

2.3.3 Huygens' principle

The equivalence principle is rigorously proved by introducing the mathematical version of Huygens' principle [34, 6, 33]. To this end, since use is made of the vector Green's theorem, we recall briefly [33] the scalar and vector Green's theorems.

Scalar Green's theorem

Let us consider a closed, regular, surface S bounding a volume V where the two scalar functions ϕ and ψ, continuous together with their first and second derivatives throughout V and on the surface S, are defined. By applying the divergence theorem to the vector $\psi\nabla\phi$ yields

$$\int_V \nabla \cdot (\psi\nabla\phi)\, dV = \int_S (\psi\nabla\phi) \cdot \mathbf{n}\, dS \tag{2.98}$$

The divergence on the left hand side may be expanded as,

$$\nabla \cdot (\psi\nabla\phi) = \nabla\psi \cdot \nabla\phi + \psi\nabla \cdot \nabla\phi = \nabla\psi \cdot \nabla\phi + \psi\nabla^2\phi \tag{2.99}$$

while, on the right hand side, we may replace the normal component of the gradient by the normal derivative, i.e.

$$\nabla\phi \cdot \mathbf{n} = \frac{\partial\phi}{\partial n} \tag{2.100}$$

Upon substituting (2.99) and (2.100) into (2.98) we obtain the Green's first identity

$$\int_V \nabla\psi \cdot \nabla\phi\, dV + \int_V \psi\nabla^2\phi\, dV = \int_S \psi\frac{\partial\phi}{\partial n}\, dS \tag{2.101}$$

The above identity holds also by interchanging the roles of the functions ϕ, ψ; by so doing we obtain

$$\int_V \nabla\psi \cdot \nabla\phi\, dV + \int_V \phi\nabla^2\psi\, dV = \int_S \phi\frac{\partial\psi}{\partial n}\, dS \tag{2.102}$$

By subtracting (2.102) from (2.101) we get another important identity, the Green's second identity, namely,

$$\int_V \left(\psi\nabla^2\phi - \phi\nabla^2\psi\right) dV = \int_S \psi\frac{\partial\phi}{\partial n}\, dS - \int_S \phi\frac{\partial\psi}{\partial n}\, dS \tag{2.103}$$

which is frequently referred to as Green's theorem.

Vector Green's theorem

Let us continue to consider the surface S and volume V as defined in the previous paragraph. Now consider two vector functions $\mathbf{P}$ and $\mathbf{Q}$ which, together with their first and second derivatives, are continuous throughout V and on the surface S. Then, by replacing the gradient with the curl, i.e. ∇ with $\nabla\times$, and ∇^2 with $\nabla\times\nabla\times$, we may obtain a vector analogue of the scalar Green's theorem. To this end, let us consider the divergence theorem applied to the vector $\mathbf{P} \times \nabla \times \mathbf{Q}$

$$\int_V \nabla \cdot (\mathbf{P} \times \nabla \times \mathbf{Q})\, dV = \int_S (\mathbf{P} \times \nabla \times \mathbf{Q}) \cdot \mathbf{n} dS \tag{2.104}$$

By expanding the divergence on the left hand side we get

$$\begin{aligned}\nabla\cdot(\mathbf{P}\times\nabla\times\mathbf{Q}) =\\ \nabla_P\cdot(\mathbf{P}\times\nabla\times\mathbf{Q})+\nabla_Q\cdot(\mathbf{P}\times\nabla\times\mathbf{Q}) =\\ \nabla\times\mathbf{P}\cdot\nabla\times\mathbf{Q}-\mathbf{P}\cdot\nabla\times\nabla\times\mathbf{Q}\end{aligned} \tag{2.105}$$

which, by introduction in (2.104), provides the vector analogue of the Green's first identity, i.e.

$$\int_V \nabla\times\mathbf{P}\cdot\nabla\times\mathbf{Q}-\mathbf{P}\cdot\nabla\times\nabla\times\mathbf{Q}\,dV=\int_S(\mathbf{P}\times\nabla\times\mathbf{Q})\cdot\mathbf{n}\,dS \tag{2.106}$$

Another form of the vector first identity may be obtained by interchanging the roles of $\mathbf{P}$ and $\mathbf{Q}$, namely

$$\int_V \nabla\times\mathbf{Q}\cdot\nabla\times\mathbf{P}-\mathbf{Q}\cdot\nabla\times\nabla\times\mathbf{P}dV=\int_S(\mathbf{Q}\times\nabla\times\mathbf{P})\cdot\mathbf{n}dS \tag{2.107}$$

By subtracting (2.107) from (2.106) we get the vector analogue of the Green's second identity,

$$\begin{aligned}\textstyle\int_V(\mathbf{Q}\cdot\nabla\times\nabla\times\mathbf{P}-\mathbf{P}\cdot\nabla\times\nabla\times\mathbf{Q})\,dV =\\ \textstyle\int_S(\mathbf{P}\times\nabla\times\mathbf{Q}\ -\mathbf{Q}\times\nabla\times\mathbf{P})\cdot\mathbf{n}dS\end{aligned} \tag{2.108}$$

The derivation of the dyadic form of Huygens' principle is obtained by considering in the above equation the vector $\mathbf{Q}$ replaced by the scalar product of a dyad with a vector, i.e. $\mathbf{Q}=\overline{\mathbf{G}}\cdot\mathbf{a}$.

Field equivalence principles

Let us now consider a volume V_1, containing all the sources, bounded by a smooth surface S, and the exterior volume V. The electric field in V is a solution of the source–free equation

$$\nabla\times\nabla\times\mathbf{E}-k^2\mathbf{E}=0 \tag{2.109}$$

Consider also a dyadic Green's function $\overline{\mathbf{G}}$ which is, in turn, a solution of

$$\nabla\times\nabla\times\overline{\mathbf{G}}_e-k^2\overline{\mathbf{G}}_e=\overline{\mathbf{I}}\delta(\mathbf{r}-\mathbf{r}') \tag{2.110}$$

and which satisfies the radiation condition at infinity. Forming the scalar products

$$\mathbf{E}\cdot\nabla\times\nabla\times\overline{\mathbf{G}}_e-\nabla\times\nabla\times\mathbf{E}\cdot\overline{\mathbf{G}}_e \tag{2.111}$$

and then applying the vector Green's second identity (2.108), yields

$$\begin{aligned}\textstyle\int_V\left(\mathbf{E}\cdot\nabla\times\nabla\times\overline{\mathbf{G}}_e-\nabla\times\nabla\times\mathbf{E}\cdot\overline{\mathbf{G}}_e\right)dV =\\ \textstyle\int_S\left(\overline{\mathbf{G}}_e\times\nabla\times\mathbf{E}-\mathbf{E}\times\nabla\times\overline{\mathbf{G}}_e\right)\cdot\mathbf{n}dS\end{aligned} \tag{2.112}$$

where the integral on the sphere at infinity has been set to zero by virtue of the fact that both $\mathbf{E}$ and $\overline{\mathbf{G}}_e$ satisfy the radiation condition. Hence, in consideration of (2.110) we have

$$\int_S\left(\mathbf{n}\times\mathbf{E}\cdot\nabla\times\overline{\mathbf{G}}_e+\mathbf{n}\times\nabla\times\mathbf{E}\cdot\overline{\mathbf{G}}_e\right)dS=\begin{cases}\mathbf{E}(\mathbf{r}') & \mathbf{r}'\text{ in }V\\ 0 & \mathbf{r}'\text{ in }V_1\end{cases} \tag{2.113}$$

and, by using Maxwell's curl equation

$$\nabla \times \mathbf{E} = -j\omega\mu\mathbf{H} \tag{2.114}$$

(2.113) may be written in terms of the currents flowing on S as

$$\mathbf{E}(\mathbf{r}') = \int_S \mathbf{n} \times \mathbf{E} \cdot \nabla \times \overline{\mathbf{G}}_e dS - j\omega\mu \int_S \mathbf{n} \times \mathbf{H} \cdot \overline{\mathbf{G}}_e dS \tag{2.115}$$

The above formula provides the electric field at each point of V in terms of the boundary fields on S and is the mathematical version of Huygens' principle (see [6, p.135], [34]). By following the same steps, or by using duality, it is possible to derive a formula analogous to (2.115) for the magnetic field, e.g.

$$\mathbf{H}(\mathbf{r}') = \int_S \mathbf{n} \times \mathbf{H} \cdot \nabla \times \overline{\mathbf{G}}_m dS + j\omega\varepsilon \int_S \mathbf{n} \times \mathbf{E} \cdot \overline{\mathbf{G}}_m dS \tag{2.116}$$

where $\overline{\mathbf{G}}_m$ satisfies (2.110) and magnetic field boundary conditions.

By recalling the equivalence theorem, it is clear that specification of the tangential components of the $\mathbf{E}, \mathbf{H}$ fields on S is the same as the specification of equivalent electric and magnetic currents. It is useful to write the two above equations (2.115) and (2.116) in the following way

$$\begin{aligned} \mathbf{E} &= \hat{Z}(\mathbf{J}) + \hat{T}_e(\mathbf{M}) \\ \mathbf{H} &= \hat{T}_m(\mathbf{J}) + \hat{Y}(\mathbf{M}) \end{aligned} \tag{2.117}$$

which express the electromagnetic field as obtained from the field on S in terms of operators as specified by (2.115) and (2.116). As may be readily proved by inserting the above equation into (2.31) and (2.32), and considering the arbitrariness of $\mathbf{J}, \mathbf{M}$, the above operators also satisfy the following equations

$$\begin{aligned} \nabla \times \hat{Z} &= -j\omega\mu\hat{T}_m \\ \nabla \times \hat{Y} &= j\omega\varepsilon\hat{T}_e \\ \nabla \times \hat{T}_e &= -j\omega\mu\hat{Y} - \mathbf{M} \\ \nabla \times \hat{T}_m &= j\omega\varepsilon\hat{Z} + \mathbf{J} \end{aligned} \tag{2.118}$$

from which one obtains

$$\begin{aligned} (\nabla \times \nabla \times -k^2)\hat{Z} &= -j\omega\mu\mathbf{J} \\ (\nabla \times \nabla \times -k^2)\hat{Y} &= -j\omega\varepsilon\mathbf{M} \\ (\nabla \times \nabla \times -k^2)\hat{T}_e &= -\nabla \times \mathbf{M} \\ (\nabla \times \nabla \times -k^2)\hat{T}_m &= \nabla \times \mathbf{J} \end{aligned} \tag{2.119}$$

However, the operators $\hat{Z}, \hat{T}_e$ satisfy the same boundary condition as the electric field, while the operators $\hat{Y}, \hat{T}_m$ satisfy the same boundary condition as the magnetic field does.

An interesting circuit analogy of (2.117) can be obtained by considering the equivalent sources $\mathbf{J}, \mathbf{M}$ on the surface S_1 and the observation point $\mathbf{r}'$ on the surface S_2. In this case it is apparent that (2.117) corresponds to an $ABCD$ representation of the region of space between the two surfaces. It is also apparent that, in order to

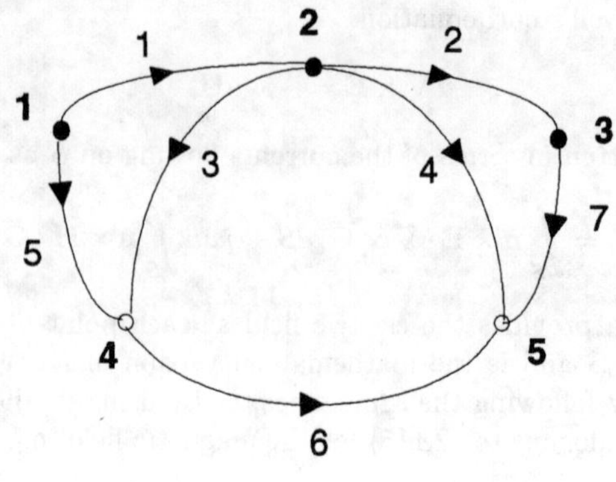

Figure 2.20 A typical network represented by a directed graph.

describe this region, we could have also chosen other representations, such as the Z or the Y representation. As an example, a Z representation is obtained by considering the two surfaces S_1, S_2 as magnetic walls. Accordingly only the electric currents, i.e. the magnetic fields, produce radiation away from the surfaces. By letting $\mathbf{E}_1, \mathbf{H}_1$ be the electric and magnetic fields on the surface S_1, and $\mathbf{E}_2, \mathbf{H}_2$ the electric and magnetic fields on the surface S_2, we may express the relationship between electric and magnetic fields on these surfaces as

$$\begin{aligned} \mathbf{E}_1 &= \hat{Z}_{11}(\mathbf{H}_1) + \hat{Z}_{12}(\mathbf{H}_2) \\ \mathbf{E}_2 &= \hat{Z}_{21}(\mathbf{H}_1) + \hat{Z}_{22}(\mathbf{H}_2) \end{aligned} \tag{2.120}$$

A similar relationship may be written for the admittance representation. Use of the impedance representation will be made in the following in the case of interacting step discontinuities in dielectric waveguides. Another example will be found in the computation of the mutual impedance of a system of dipoles radiating in the presence of an open waveguide.

Finally we would like to note that, when the operator is expressed in a diagonalised form, i.e. when the region we are dealing with is separable, we can pass from one representation, say the admittance representation, to another representation in a straightforward way [12].

2.3.4 Lorentz theorem

The Lorentz reciprocity theorem is introduced from and related to circuit concepts for time–harmonic signals [29]. We start by referring to network analysis in order to establish Tellegen's theorem and the reaction concept. From there, the reciprocity theorem is derived in terms of reactions, which are also used for deriving variational formulae.

Tellegen's theorem

For ease of reference let us refer to the network shown in Figure 2.20 representing a directed graph [18] where each terminal (or node) is numbered. Moreover, to each path (or branch) connecting terminals is assigned a direction, shown by arrows, and a number, as reported beside the arrow. We introduce the following quantities:

- V_k the voltage at the terminal k
- I_k the current supplied from external sources at terminal k
- v_j the voltage drop along the j-th path (or branch)
- i_j the current flowing in the direction of the arrow

which are obviously interrelated. As an example, in Figure 2.20, the current at terminal 2 is given by

$$I_2 = -i_1 + i_2 + i_3 + i_4$$

where currents exiting the terminal have been chosen as positive. Similar relationships hold at other terminals; hence, the vectors $\mathbf{I}$ and $\mathbf{i}$ with components I_k and i_j respectively, are related by the topological relationship

$$\mathbf{I} = A\mathbf{i} \tag{2.121}$$

where A, the topological matrix relating the branch currents to the nodal ones, has the following form for the graph of Figure 2.20

$$A = \begin{pmatrix} 1 & 0 & 0 & 0 & 1 & 0 & 0 \\ -1 & 1 & 1 & 1 & 0 & 0 & 0 \\ 0 & -1 & 0 & 0 & 0 & 0 & 1 \\ 0 & 0 & -1 & 0 & -1 & 1 & 0 \\ 0 & 0 & 0 & -1 & 0 & -1 & -1 \end{pmatrix}$$

Note that, if no current is supplied from outside at terminal k, i.e. if no external sources are present, Kirchhoff's law applies and I_k is equal to zero. The voltage drop v_j over branch j is related to the nodal voltages V_k in an analogous manner, e.g. we have for branch 3,

$$v_3 = V_2 - V_4$$

By introducing vectors $\mathbf{V}$ and $\mathbf{v}$ with components V_k and v_j respectively, it is evident that

$$\mathbf{v} = A^T\mathbf{V} \tag{2.122}$$

with superscript T denoting transposition. It is now useful to consider the scalar products of currents and voltages. In particular, two scalar products are meaningful,

$$\begin{aligned} \mathbf{I}^T \cdot \mathbf{V} &= (\mathbf{I}, \mathbf{V}) \qquad (a) \\ \mathbf{i}^T \cdot \mathbf{v} &= <\mathbf{i}, \mathbf{v}> \qquad (b) \end{aligned} \tag{2.123}$$

i.e. that between branch quantities as given by (2.123a), and that between nodal quantities as given by (2.123b). In (2.123) symbols $(,)$ and $<,>$ have been introduced in order to represent scalar products in the spaces of nodal and branch quantities, respectively. As stated before, when no currents are supplied from the outside, $\mathbf{I}$ is zero; thus also the product (2.123a) is zero. In addition, also the scalar product (2.123b) is zero since, by using (2.122) and (2.121),

$$\mathbf{i}^T\mathbf{v} = \mathbf{i}^T \cdot A^T\mathbf{V} = (A\mathbf{i})^T \cdot \mathbf{V} = \mathbf{I}^T \cdot \mathbf{V} = 0 \tag{2.124}$$

By using the notation introduced in (2.123), relationship (2.124) may also be written in the following way

$$<\mathbf{i},\mathbf{v}>=<\mathbf{i},A^T\mathbf{V}>=(A\mathbf{i},\mathbf{V})=(\mathbf{I},\mathbf{V})=0 \tag{2.125}$$

From (2.121), (2.122) and (2.125) it is apparent that A is an operator that transforms the branch currents $\mathbf{i}$ into the nodal currents $\mathbf{I}$; A^T is the adjoint operator transforming the nodal voltages $\mathbf{V}$ into branch voltages $\mathbf{v}$ and, as usual for real matrices, the adjoint operator is simply the transpose. This type of reasoning in terms of suitably defined scalar products is also of considerable help when applied to field quantities, as in the following parts of this section. Note that no complex conjugate appears in the definition of scalar products in (2.123), so that a quantity having the same dimension as power but different from complex power is obtained. This quantity, often referred to as *quasi–power* in network theory or *reaction* in electromagnetic parlance, expresses the coupling between voltages and currents. By using the concept of reaction, equation (2.124) may be stated as: the reaction between voltages and currents in a network is zero.

Let us now consider two different networks, N and N', but with identical topologies [7], i.e. with the same directed graph. By indicating with $\mathbf{v},\mathbf{i}$ the branch quantities of network N and with $\mathbf{v}',\mathbf{i}'$ the branch quantities of network N' it is easy to see that

$$\begin{aligned} &<\mathbf{i}',\mathbf{v}>=0 \qquad (a)\\ &<\mathbf{v}',\mathbf{i}>=0 \qquad (b) \end{aligned} \tag{2.126}$$

Equation (2.126a) is readily proved, since use of (2.125) yields

$$<\mathbf{i}',\mathbf{v}>=<\mathbf{i}',A^T\mathbf{V}>=(A\mathbf{i}',\mathbf{V})=(\mathbf{I}',\mathbf{V})=0 \tag{2.127}$$

The last equality in (2.127) comes from the Kirchhoff's current law for network N'. In a similar way, by using Kirchhoff's voltage laws in network N we can prove (2.126b). Equation (2.126) is Tellegen's theorem [35, 36] which states that the reaction between voltages and currents of two networks with the same topology is always zero. From Tellegen's theorem is readily obtained a weak or "difference" form of the type

$$<\mathbf{i}',\mathbf{v}>-<\mathbf{v}',\mathbf{i}>=0 \tag{2.128}$$

which is recognised as the network analogue of the Lorentz reciprocity theorem for electromagnetic fields.

Reaction in electromagnetic theory

The reaction concept in electromagnetic theory has been introduced in [31] in order to find a fundamental observable representing measurements which can practically be performed. In fact, if we want to measure the field radiated by some source of electromagnetic energy, we may use an antenna and observe the signal received at its terminals at the point of observation. Unfortunately, the latter measurement does not provide the field just at the observation point, but it measures the effect of the field over a small, but finite, region. To take into account this fact, it is convenient to define the reaction, i.e. the coupling between the field that we want to measure and the antenna (probe) that we are using.

Consider a monochromatic source of electromagnetic field, denoted by a, consisting of electric and magnetic currents $\mathbf{J}_a$ and $\mathbf{M}_a$ respectively, and producing the field $\mathbf{E}_a$, $\mathbf{H}_a$. Similarly, consider also a source b of electric and magnetic currents $\mathbf{J}_b$ and $\mathbf{M}_b$, generating the field $\mathbf{E}_b$, $\mathbf{H}_b$. The interaction of source a with field b may be characterised by the complex number $< a, b >$, defined as [19]

$$< a, b >= \int_V (\mathbf{J}_a \cdot \mathbf{E}_b - \mathbf{M}_a \cdot \mathbf{H}_b)\, dV \tag{2.129}$$

where the first entry, a, is associated with source (or probe), and the second entry, b, is associated with the observed field. The integration is extended over the volume V, i.e. the region containing the source a, which may contain both volume current densities and surface current densities. Note that, for an ideal electric field probe, $\mathbf{J}_a$ is a delta function which measures the field just at the observation point. As noted in the previous paragraph, also for electromagnetic field quantities the reaction is different from complex power since there is no complex–conjugate. Moreover, let S represent any scalar and Sa be the source a increased in strength by the factor S, then

$$< Sa, b >= S < a, b > \tag{2.130}$$

By considering another source c, radiating at the same frequency as a and b, we have

$$< a, (b + c) >=< a, b > + < a, c > \tag{2.131}$$

Having defined the reaction we can now derive the Lorentz reciprocity theorem.

Lorentz reciprocity theorem

A simple interpretation of the reciprocity theorem is that in isotropic media the response of a system to a source is unchanged when source and detector are interchanged [13]. In order to establish this theorem let us consider the two monochromatic sources a, b introduced in the previous paragraph and the field thereby produced. In each case Maxwell's equations are:

$$\begin{aligned} \nabla \times \mathbf{E}_a &= -j\omega\mu\mathbf{H}_a - \mathbf{M}_a \qquad (a) \\ \nabla \times \mathbf{H}_a &= j\omega\varepsilon\mathbf{E}_a + \mathbf{J}_a \qquad (b) \end{aligned} \tag{2.132}$$

and

$$\begin{aligned} \nabla \times \mathbf{E}_b &= -j\omega\mu\mathbf{H}_b - \mathbf{M}_b \qquad (a) \\ \nabla \times \mathbf{H}_b &= j\omega\varepsilon\mathbf{E}_b + \mathbf{J}_b \qquad (b) \end{aligned} \tag{2.133}$$

By scalar multiplying of (2.132b) by $\mathbf{E}_b$ and (2.133a) by $\mathbf{H}_a$, and by subtracting one from the other we get

$$\nabla \cdot (\mathbf{E}_b \times \mathbf{H}_a) = -j\omega\varepsilon\mathbf{E}_a \cdot \mathbf{E}_b - \mathbf{J}_a \cdot \mathbf{E}_b - j\omega\mu\mathbf{H}_a \cdot \mathbf{H}_b - \mathbf{M}_b \cdot \mathbf{H}_b \tag{2.134}$$

where use is made of the identity

$$\nabla \cdot (\mathbf{A} \times \mathbf{B}) = \mathbf{B} \cdot \nabla \times \mathbf{A} - \mathbf{A} \cdot \nabla \times \mathbf{B} \tag{2.135}$$

Similarly, by scalar multiplying of (2.133b) by $\mathbf{E}_a$, (2.132a) by $\mathbf{H}_b$, (2.132a) by $\mathbf{H}_b$ and (2.133b) by $\mathbf{E}_a$ and subtracting in pairs yields:

$$\nabla \cdot (\mathbf{E}_a \times \mathbf{H}_b) = -j\omega\varepsilon \mathbf{E}_a \cdot \mathbf{E}_b - \mathbf{J}_b \cdot \mathbf{E}_a - j\omega\mu \mathbf{H}_a \cdot \mathbf{H}_b - \mathbf{M}_a \cdot \mathbf{H}_b \qquad (2.136)$$

By subtracting (2.134) from (2.136), integrating throughout a source–free region and applying the divergence theorem we have

$$\oint_S (\mathbf{E}_a \times \mathbf{H}_b - \mathbf{E}_b \times \mathbf{H}_a)\, d\mathbf{S} = < a, b > - < b, a > \qquad (2.137)$$

By definition, isotropic media are reciprocal when

$$\oint_S (\mathbf{E}_a \times \mathbf{H}_b - \mathbf{E}_b \times \mathbf{H}_a)\, d\mathbf{S} = 0 \qquad (2.138)$$

In this case, the Lorentz reciprocity theorem can be stated as

$$< a, b > - < b, a > = 0 \qquad (2.139)$$

which is the analogue, for electromagnetic field quantities, of the Tellegen difference form (2.128). The surface integral on the left side of (2.137) vanishes also when the surface S encloses all the sources. In fact, in this case we can consider the source–free region of volume bounded by S and the surface S_∞ of a sphere of infinite radius. When the fields satisfy the radiation condition the integrand of the left side of (2.137) vanishes on S_∞ and (2.138) holds true. The Lorentz theorem is useful in many instances, since it allows one to derive in a direct way stationary formulas in variational problems. It is also useful to prove simple assertions, such as the fact that an electric current sheet impressed on the surface of a perfect conductor does not radiate [19]. This is a trivial result when the surface of the conductor is a plane, since image theory shows that no field is produced. In fact, by replacing the metallic plane by an image source, i.e. by an impressed current directed in the opposite direction, the two impressed currents annihilate producing no field. When the surface is not a plane, application of reciprocity shows the above assertion in the following way. With reference to Figure 2.21 let us consider source a on the perfect electric conductor. In order to measure the field $\mathbf{E}_a$, $\mathbf{H}_a$ produced by this source let us place a probe (source b) at the observation point and evaluate the reaction of source b on the field a, e.g. $< b, a >$. By the reciprocity theorem, the effect of source b on the field a is equal to the effect of source a on the field b, i.e.

$$< b, a > = < a, b > \qquad (2.140)$$

but the tangential component of the electric field produced by b is zero on the metallic surface where $\mathbf{J}_a$ is present, thus

$$< a, b > = 0 \qquad (2.141)$$

In view of the arbitrariness of source b it is proved that the impressed electric current sheets $\mathbf{J}_a$ on the surface of the perfect electric conductor do not produce any field.

As a further application of the concept of reaction, let us consider the example of Figure 2.22, where we have a field $\mathbf{E}_b$, $\mathbf{H}_b$ generated by source b. When a circuit

Figure 2.21 The impressed electric current sheets $\mathbf{J}_a$ *on the surface of a perfect electric conductor do not produce any field, as measured through probe b.*

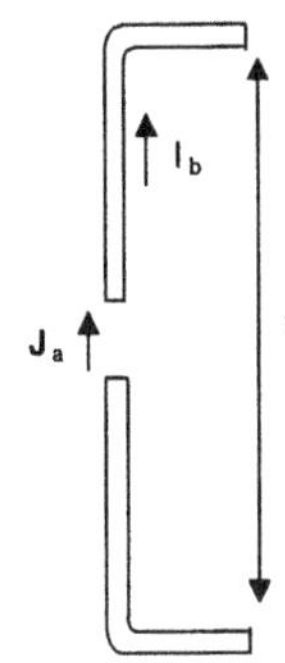

Figure 2.22 Reaction of a current source $\mathbf{J}_a$ *on the field* $\mathbf{E}_b, \mathbf{H}_b$ *generated by source b.*

(or an antenna) is present, this field induces current and voltages on the circuit elements and the reaction of the latter field with a current source $\mathbf{J}_a$ is given by

$$< a, b >= \int_V \mathbf{J}_a \cdot \mathbf{E}_b dV = I_a \int_{\ell_a} \mathbf{E}_b \cdot d\ell = -I_a V_b \tag{2.142}$$

Note that, by using a unit current source $I_a = 1$, we measure the voltage V_b at source a as generated by source b.

However, we may consider sources a and b as two antennas ([19, p.400], schematically depicted in Figure 2.23, which may be also represented by means of the linear network of Figure 2.24. The network describes the antennas as well as the region of space where transmission occurs. When reciprocity holds, the reaction of the field excited by b on source a is equal to the reaction of source b on the field generated by a, i.e. $< a, b >=< b, a >$. By considering only electric current sources this means that

$$\int_{V_a} \mathbf{J}_a \cdot \mathbf{E}_b dV = \int_{V_b} \mathbf{J}_b \cdot \mathbf{E}_a dV \tag{2.143}$$

For perfectly conducting antennas, the tangential components of the electric fields are zero on the antenna, while producing voltages at the terminal pairs. By denoting with

$$V_a^{oc} = -\int_{\ell_a} \mathbf{E}_b \cdot d\ell \tag{2.144}$$

the open circuit voltage at the terminal of antenna a due to the field produced by

Figure 2.23 Two antenna system.

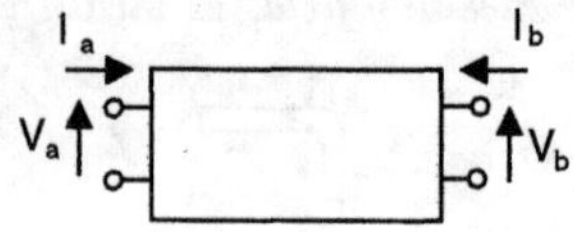

Figure 2.24 The representation of the two antenna system by a two–port network.

antenna b, and by

$$V_b^{oc} = -\int_{\ell_b} \mathbf{E}_a \cdot d\ell \tag{2.145}$$

the open circuit voltage at the terminal of antenna b due to the field produced by antenna a, (2.143) becomes

$$V_a^{oc} I_a = V_b^{oc} I_b \tag{2.146}$$

On the other hand, the two–port network representation of the antennas is

$$\begin{aligned} V_a &= Z_{aa} I_a + Z_{ab} I_b \\ V_b &= Z_{ba} I_a + Z_{bb} I_b \end{aligned} \tag{2.147}$$

which, under open circuit excitation, gives

$$\begin{aligned} V_a^{oc} &= Z_{ab} I_b \\ V_b^{oc} &= Z_{ba} I_a \end{aligned} \tag{2.148}$$

From (2.148) and (2.146) it follows that

$$Z_{ab} = Z_{ba} \tag{2.149}$$

which is a direct consequence of reciprocity. Since the two antennas may represent any network, it is also proved that any network made by linear isotropic matter has a symmetrical impedance matrix.

2.4 Method of moments and variational formulae

Electromagnetic field problems may be expressed in operator form as

$$Lf = g \tag{2.150}$$

where L is a linear operator, g is the known field and f the unknown field or source. As an example we may establish an equation of the type $Z(J) = E$ with Z an impedance operator, J the unknown current and E the known electric field; or we may deal with an equation of the type $Y(E) = H$ where Y is an admittance operator, E is the unknown electric field and H is the known magnetic field. Detailed expressions of the above immittance operators are found, for instance, in Chapter 5 when dealing with step discontinuities in planar dielectric waveguides.

In general, we may be interested in the field f itself, or in finding a functional ρ of the type

$$\rho = < h, f > \tag{2.151}$$

which may express immittances, scattering parameters or other quantities of interest. In (2.151) the function h is suitably selected and is a known continuous function, so that ρ is a continuous linear functional. The inner product appearing in (2.151) is a reaction as defined by (2.129). In addition to (2.150) it is also appropriate to consider the adjoint operator $L^\dagger$ and the adjoint equation

$$L^\dagger w = h \tag{2.152}$$

By using (2.152), (2.150) and the definition of adjoint, the following equivalences are obtained

$$< h, f > = < L^\dagger w, f > = < w, Lf > = < w, g > \tag{2.153}$$

which may be viewed as a reciprocity theorem of a sort (see [18, p.304]). The above equation also has a nice interpretation in terms of reaction. For example if the problem considered is $Z(J) = E$, we may introduce the adjoint equation $Z^\dagger(J_a) = E_a$. In this case, the inner product $< w, g >$ corresponds to $< J_a, E >$, which is the reaction (measure) of the field E by the probe J_a; similarly, the inner product $< h, f >$ corresponds to $< E_a, J >$, which is again a measure of the field E_a by the current J.

Moreover, by using (2.153), it is also possible to derive a variational formula for functional (2.151), that has the property that first–order errors in f give rise to second–order errors in the functional. In fact, on account of (2.153) we may consider the functional [14]

$$\rho = \frac{< h, f >< w, g >}{< w, Lf >} \tag{2.154}$$

with the following property. Let $f = \alpha\phi$ be a guess at our solution, with α representing an amplitude and ϕ being a trial function. With the above choice ρ becomes

$$\rho = \frac{< h, \phi >< w, g >}{< w, L\phi >} \tag{2.155}$$

which shows that our approximation to the value of the functional is independent of α. Let us now consider the variations $\delta\rho$ obtained by keeping w constant and by varying ϕ by $\delta\phi$

$$\delta\rho = \frac{< h, \delta\phi >< w, g >< w, L\phi > - < h, \phi >< w, g >< w, L\delta\phi >}{< w, L\phi >^2}$$

or, equivalently,

$$\delta\rho = \rho\left\{\frac{< h, \delta\phi >}{< h, \phi >} - \frac{< w, L\delta\phi >}{< w, L\phi >}\right\} = \rho\left\{\frac{< h, \delta\phi >}{< h, \phi >} - \frac{< L^\dagger w, \delta\phi >}{< L^\dagger w, \phi >}\right\} \tag{2.156}$$

It is apparent that if (2.152) is satisfied then $\delta\rho = 0$. By keeping ϕ constant and by varying w we obtain

$$\delta\rho = \rho\left\{\frac{<\delta w, g>}{<w, g>} - \frac{<\delta w, L\phi>}{<w, L\phi>}\right\} \qquad (2.157)$$

which is satisfied for

$$L\phi = g \qquad (2.158)$$

Hence, (2.155) is a variational formula for ρ with a stationary point when (2.152) and (2.158) are satisfied. In addition to (2.155) other variational formulae are also available; for instance we may use [14]

$$\rho = <h, f> + <w, g> - <w, Lf> \qquad (2.159)$$

which has a stationary point when (2.150) and (2.152) are satisfied. In several problems considered in the following Chapters, for example when dealing with step discontinuities as in Chapter 5, the operators are symmetrical since we are considering a reciprocal medium; they are often complex, since part of the energy is radiated away from the discontinuity and part remains confined near the discontinuity. Finally, they are often self–adjoint; in this case the variational formulas simplify since, for self–adjoint problems, $h = g$ and $f = w$.

2.4.1 Rayleigh–Ritz procedure

In order to obtain more systematic results it is convenient to apply the Rayleigh–Ritz procedure [14] and represent the unknown functions f and w by means of the following expansions

$$\phi = \sum_j \alpha_j f_j \qquad (a)$$

$$w = \sum_i \beta_i w_i \qquad (b) \qquad (2.160)$$

where the coefficients α_j, β_i are determined by imposing the functional ρ to be stationary with respect to variations of the coefficients, that is,

$$\frac{\partial\rho}{\partial\alpha_k} = \frac{\partial\rho}{\partial\beta_k} = 0 \qquad k = 1, 2, 3, \ldots \qquad (2.161)$$

Hence, by defining the following quantities,

$$\rho_f = \sum_j \alpha_j <h, f_j> \qquad (a)$$

$$\rho_w = \sum_i \beta_i <g, w_i> \qquad (b)$$

$$\rho = \frac{\rho_f \rho_w}{<\sum_i \beta_i w_i, \sum_j \alpha_j L f_j>} = \frac{\rho_f \rho_w}{D} \qquad (c) \qquad (2.162)$$

we have

$$\frac{\partial \rho}{\partial \alpha_k} = \rho \left\{ \frac{< h, f_k >}{\rho_f} - \frac{< \sum_i \beta_i w_i, L f_k >}{D} \right\} \tag{2.163}$$

which is equal to zero if the following two equations are satisfied:

$$\sum_i \beta_i < L^\dagger w_i, f_k > = < h, f_k > \qquad (a)$$

$$\rho_f = D \qquad (b) \tag{2.164}$$

Analogously, we also have

$$\frac{\partial \rho}{\partial \beta_k} = \rho \left\{ \frac{< w_k, g >}{\rho_w} - \frac{< w_k, \sum_j \alpha_j L f_j >}{D} \right\} \tag{2.165}$$

which, in turn, vanishes when the two following equations are satisfied:

$$\sum_j \alpha_j < w_k, L f_j > = < w_k, g > \qquad (a)$$

$$\rho_w = D \qquad (b) \tag{2.166}$$

Hence, by noting that if (2.164a) and (2.166a) are satisfied so are also (2.164b) and (2.166b), we see that the Rayleigh–Ritz solution is obtained by solving equations (2.164a) and (2.166a).

2.4.2 Method of moments

We have just seen that the solution of the operator equation (2.150) is obtained in the Rayleigh–Ritz approach by solving (2.166a). However, there is a much more direct approach which can be used to derive (2.166a). Let us consider (2.150) and apply to this equation the reaction, i.e. let us measure the field g by some properly selected sources w_k. In the context of the method of moments (MoM) approach the functions w_k are often referred to as testing functions or weighting functions. By taking N measurements we obtain the N equations

$$< w_k, L f > = < w_k, g > \qquad k = 1, 2, \ldots, N \tag{2.167}$$

Let us now look for an approximation of the function f as a linear combination of suitably selected basis functions, or expansion functions, f_j, with unknown amplitude coefficients α_j

$$f = \sum_j \alpha_j f_j \tag{2.168}$$

In the above formula it is common practice to use the same number of basis functions as the number of measurements, although the problem may be solved by a least squares approach when different numbers of basis and testing functions are selected. Now, inserting (2.168) into (2.167) yields

$$\sum_j \alpha_j < w_k, L f_j > = < w_k, g > \qquad k = 1, 2, \ldots, N \tag{2.169}$$

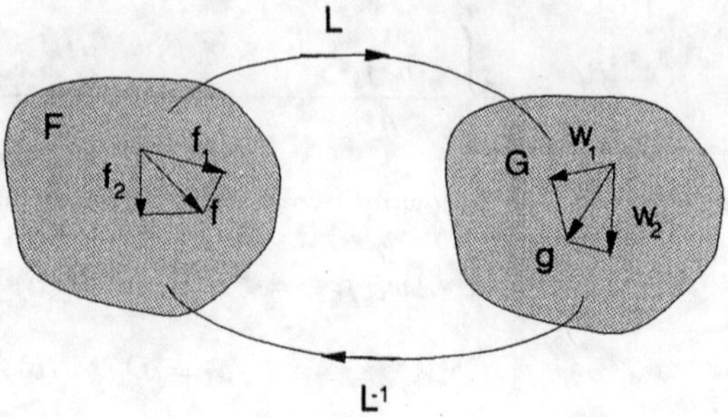

Figure 2.25 Space mapping for the method of moments.

i.e. equation (2.166a) as found by the Rayleigh–Ritz variational approach. Therefore, the MoM approach leads directly to the same equation as the Rayleigh–Ritz method. Note that, since the latter exhibits variational properties, so does also the MoM, notwithstanding the type of basis and testing function selected. Naturally, the great virtue of the Rayleigh–Ritz approach lies in showing explicitly the variational property of the solution. Equation (2.166a) may be obtained also considering other interpretations, one of which is often referred to as the method of *weighted residuals*. In the latter approach we start by considering the expansion (2.168) and we insert it into (2.150), thus obtaining

$$\sum_j \alpha_j L f_j - g = r \tag{2.170}$$

where r is the *residual*. Weighting the residual with the testing functions w_k and making all the weights equal to zero yields again equation (2.166a).

Each of the above versions of MoM has its geometrical interpretation, by considering that the reaction, i.e. the inner product, may be thought of as a projection of a vector over another one. Thus, with reference to Figure 2.25, we may consider the unknown function f in the space F, to be represented by the basis function terms f_j. Note that, as also shown in the figure, the basis functions need not be orthogonal; they only need to represent any function in F; thus only completeness is required. Analogously, we introduce a basis w_k in the data space G to which g belongs.

2.5 Potentials

The main difficulty posed by electromagnetic problems is the solution of Maxwell's equations with appropriate boundary conditions. In a homogeneous region the two curl equations

$$\begin{aligned} \nabla \times \mathbf{E} &= -j\omega\mu\mathbf{H} && (a) \\ \nabla \times \mathbf{H} &= j\omega\varepsilon\mathbf{E} + \mathbf{J} && (b) \end{aligned} \tag{2.171}$$

provide six scalar equations to be solved in order to find electric and magnetic fields. The divergence equations for the fields

$$\begin{aligned} \nabla \cdot \mathbf{D} &= \rho \\ \nabla \cdot \mathbf{B} &= 0 \end{aligned} \tag{2.172}$$

must be simultaneously satisfied while also the constitutive relations

$$\begin{aligned} \mathbf{D} &= \varepsilon \mathbf{E} \\ \mathbf{B} &= \mu \mathbf{H} \end{aligned} \tag{2.173}$$

hold. For the purpose of avoiding the solution of such a large set of equations, it is generally useful to introduce some potential functions in terms of which the electromagnetic fields can be expressed. The first potentials we consider are the vector and scalar potential functions $\mathbf{A}$ and ϕ, which represent an extension of the static magnetic vector potential and electrical scalar potential, respectively. Potential theory is generally developed considering the time–dependent form of Maxwell's equations [33, 17, 2]; since we are primarily concerned with time–harmonic fields, we will consider directly time–harmonic potentials. In this case, an $\exp(j\omega t)$ time–dependence is assumed and suppressed. By taking the divergence of (2.171a) we see that

$$\nabla \cdot \mathbf{H} = 0 \tag{2.174}$$

i.e. the divergence equation for $\mathbf{H}$ is automatically satisfied. Also, this fact makes it possible to express $\mathbf{H}$ as

$$\mathbf{H} = \frac{1}{\mu} \nabla \times \mathbf{A} \tag{2.175}$$

where $\mathbf{A}$ is called the magnetic vector potential. In general the vector $\mathbf{A}$ has both a lamellar part ($\nabla \cdot \mathbf{A} \neq 0$) and a solenoidal part ($\nabla \times \mathbf{A} \neq 0$). By using (2.175) we have imposed a condition on the solenoidal part of $\mathbf{A}$ but we have left unspecified its lamellar part. By substituting (2.175) into (2.171a) we obtain

$$\nabla \times (\mathbf{E} + j\omega \mathbf{A}) = 0 \tag{2.176}$$

and since

$$\nabla \times \nabla \phi = 0 \tag{2.177}$$

we can express $\mathbf{E}$ as

$$\mathbf{E} = -j\omega \mathbf{A} - \nabla \phi \tag{2.178}$$

where ϕ is an electric scalar potential. By substituting (2.175) and (2.178) into (2.171b) and by using the vector identity

$$\nabla \times \nabla \times \mathbf{A} = \nabla \nabla \cdot \mathbf{A} - \nabla^2 \mathbf{A} \tag{2.179}$$

we obtain

$$\nabla \nabla \cdot \mathbf{A} - \nabla^2 \mathbf{A} = k^2 \mathbf{A} - j\omega\mu\varepsilon \nabla \phi + \mu \mathbf{J} \tag{2.180}$$

The above equation can be simplified by forcing the as yet unspecified lamellar part of $\mathbf{A}$ to satisfy the Lorentz condition, or "gauge",

$$\nabla \cdot \mathbf{A} = -j\omega\mu\varepsilon\phi \tag{2.181}$$

With this choice (2.180) reduces to

$$\nabla^2\mathbf{A}+k^2\mathbf{A}=-\mu\mathbf{J} \tag{2.182}$$

i.e. to the vector Helmholtz equation. If we take the divergence of (2.178) and use the divergence equation for **E**, it is readily seen that the scalar potential ϕ satisfies the following scalar Helmholtz equation

$$\nabla^2\phi+k^2\phi=-\frac{\rho}{\varepsilon} \tag{2.183}$$

Using Lorentz condition (2.181), the fields may be written in terms of the vector potential as

$$\begin{aligned}\mathbf{H} &= \frac{1}{\mu}\nabla\times\mathbf{A}\\ \mathbf{E} &= -j\omega\mathbf{A}+\frac{\nabla\nabla\cdot\mathbf{A}}{j\omega\mu\varepsilon}\end{aligned} \tag{2.184}$$

However, in a region where only magnetic sources are present, the vector **E** has zero divergence. When magnetic sources are present, the same arguments that applied to deriving the magnetic potential **A** may also be applied in order to derive an electric vector potential **F** and a scalar potential ψ. By duality, we obtain (see [13, p.129])

$$\begin{aligned}\mathbf{E} &= -\frac{1}{\varepsilon}\nabla\times\mathbf{F}\\ \mathbf{H} &= -j\omega\mathbf{F}+\frac{\nabla\nabla\cdot\mathbf{F}}{j\omega\varepsilon\mu}\end{aligned} \tag{2.185}$$

with

$$\begin{aligned}\nabla^2\mathbf{F}+k^2\mathbf{F} &= -\varepsilon\mathbf{M}\\ \nabla^2\psi+k^2\psi &= -\frac{\rho_m}{\mu}\end{aligned} \tag{2.186}$$

M denotes magnetic currents and ρ_m magnetic charges. If electric and magnetic currents are both present in a linear system, we may employ superposition and express part of the field in terms of the magnetic vector potential **A** and part in terms of the electric vector potential **F**; hence we have

$$\begin{aligned}\mathbf{E} &= -j\omega\mathbf{A}+\frac{\nabla\nabla\cdot\mathbf{A}}{j\omega\mu\varepsilon}-\frac{1}{\varepsilon}\nabla\times\mathbf{F}\\ \mathbf{H} &= \frac{1}{\mu}\nabla\times\mathbf{A}-j\omega\mathbf{F}+\frac{\nabla\nabla\cdot\mathbf{F}}{j\omega\varepsilon\mu}\end{aligned} \tag{2.187}$$

In a source–free region Hertzian potentials are often preferred. In the time–domain these potentials are obtained from Lorentz potentials through a further integration with respect to time. As such, the Hertzian potentials are "less discontinuous" than the Lorentz potentials [2, 4]. When time–harmonic fields are considered the Hertzian potentials are recovered by setting [11, p.245]

$$\begin{aligned}\mathbf{A} &= j\omega\varepsilon\mu\mathbf{\Pi}_e\\ \mathbf{F} &= j\omega\varepsilon\mu\mathbf{\Pi}_h\end{aligned} \tag{2.188}$$

By using the Hertzian vector potentials the fields are expressed as

$$\begin{aligned}\mathbf{E} &= k^2\mathbf{\Pi}_e + \nabla\nabla\cdot\mathbf{\Pi}_e - j\omega\mu\nabla\times\mathbf{\Pi}_h \\ \mathbf{H} &= j\omega\varepsilon\nabla\times\mathbf{\Pi}_e + k^2\mathbf{\Pi}_h + \nabla\nabla\cdot\mathbf{\Pi}_h\end{aligned} \tag{2.189}$$

It is worthwhile to remark on the terminology used for the various potentials: the vector potential $\mathbf{A}$ is sometimes referred to as magnetic vector potential since it was introduced in magnetostatics; for similar reasons the vector potential $\mathbf{F}$ is called electric vector potential. The potential $\mathbf{\Pi}_e$ is often called the electric Hertzian potential since it is generated from electric currents and satisfies the same boundary conditions as the electric field. In the same manner $\mathbf{\Pi}_h$ is often referred to as the magnetic Hertzian potential.

Let us observe that we have justified the introduction of potentials by saying that we would like to solve fewer than the six equations provided by (2.171). However, since we have introduced two vector potentials such as $\mathbf{\Pi}_e, \mathbf{\Pi}_h$ we still have six unknowns. Luckily, not all the six scalar components of $\mathbf{\Pi}_e, \mathbf{\Pi}_h$ are necessary in order to represent an electromagnetic field. In fact, it is possible to prove [17, p.28] that, in a source–free region, only two scalars are needed to represent the field. However, there is still plenty of choice when selecting these two scalars out of the six appearing in (2.189). A common choice, often referred to as *Hertz–Debye potentials*, consists in forming the two vector potentials by multiplying two scalar functions by unit vectors in the longitudinal direction, as is illustrated in the next Section. When dealing with radiation from electric dipoles over stratified media, another common choice is that of Sommerfeld. In this case one chooses $F = 0$ while describing a vertical electrical dipole (VED), i.e. a dipole oriented along z, in terms of A_z. In this case a horizontal electrical dipole (HED) is described by introducing a second component parallel to the source. Another particular choice of potentials often used when dealing with stratified media, is that of taking one scalar component of A and one of F in the direction orthogonal to the stratification.

2.5.1 TE and TM potentials

In waveguiding structures it is possible to recognize a preferred direction, say z, along which the field is guided. It is advantageous, in this case, to simplify the field expressions (2.189) of the previous Section by separating the transverse dependence from the longitudinal one. The common choice is to derive the field from two potentials, an electric and a magnetic one, directed along the z–axis as

$$\begin{aligned}\mathbf{\Pi}_e &= \mathbf{z}_0\Pi_e(\mathbf{r}) \\ \mathbf{\Pi}_h &= \mathbf{z}_0\Pi_h(\mathbf{r})\end{aligned} \tag{2.190}$$

Obviously $\mathbf{\Pi}_e$ generates a magnetic field entirely contained in the transverse plane, therefore TM waves; while $\mathbf{\Pi}_e$ generates an electric field totally contained in the transverse plane, thus TE waves. For ease of notation we will consider separately the TE and TM cases. For TM modes we set

$$\mathbf{\Pi}_h = 0 \tag{2.191}$$

By decomposing the nabla operator into its transverse and longitudinal components

$$\nabla = \nabla_t + \mathbf{z}_0\partial_z \tag{2.192}$$

and by using (2.190, 2.192) in (2.189), we get

$$\begin{aligned} \mathbf{E} = \mathbf{E}_t + \mathbf{z}_0 E_z &= \nabla_t \left(\partial_z \Pi_e\right) + \mathbf{z}_0 \left(\partial_z^2 \Pi_e + k^2 \Pi_e\right) \qquad (a) \\ \mathbf{H} = \mathbf{H}_t &= -j\omega\varepsilon \nabla_t \Pi_e \qquad (b) \end{aligned} \tag{2.193}$$

It is convenient to introduce the following scalar quantities

$$\begin{aligned} I'(\mathbf{r}) &= j\omega\varepsilon \Pi_e(\mathbf{r}) \\ V'(\mathbf{r}) &= -\partial_z \Pi_e(\mathbf{r}) \end{aligned} \tag{2.194}$$

where in (2.194) and in the rest of this paragraph, the prime is used to denote TM type quantities.

If the structure under consideration, e.g. a uniform waveguide, allows separability of the longitudinal z–dependence from the transverse dependence ρ, i.e. $\Pi_e(\mathbf{r}) = \phi(\underline{\rho})\zeta(z)$, we can write the Helmholtz equation in a source–free locally or piecewise homogeneous region

$$\nabla^2 \Pi_e(\mathbf{r}) + k^2 \Pi_e(\mathbf{r}) = 0 \tag{2.195}$$

as a pair of equations

$$\begin{aligned} \nabla_t^2 \phi(\underline{\rho}; k_t) + k_t^2 \phi(\underline{\rho}; k_t) &= 0 \\ \frac{d^2\zeta(z)}{dz^2} + k_z^2 \zeta(z) &= 0 \end{aligned} \tag{2.196}$$

linked by wavenumber conservation

$$k_t^2 + k_z^2 = k^2 \tag{2.197}$$

In (2.196a), ϕ denotes a transverse eigenfunction, depending on the parameter k_t. If the waveguide transverse cross–section is bounded, then k_t only takes discrete values, as in closed metallic waveguides. On the other hand, if the waveguide cross–section extends to infinity, k_t varies with continuity and a continuous spectrum is present: a discrete spectrum may also be present in nonhomogeneous cross–sections. When a discrete spectrum is present, we can represent the current I' as a sum of all the eigenfunctions

$$I'(\mathbf{r}) = \sum_i I_i'(z)\phi_i(\underline{\rho}) \tag{2.198}$$

For a continuous spectrum we proceed in exactly the same way, but substituting the discrete sum with an integral over k_t

$$I'(\mathbf{r}) = \int I'(z; k_t)\phi(\underline{\rho}; k_t) dk_t \tag{2.199}$$

Since we are primarily concerned with open waveguides, a few discrete modes and a continuum will be present simultaneously in general. We will retain just the integral for short. Nonetheless, the reader can readily consider bounded domains simply by substituting the integral in (2.199) with the discrete summation in (2.198). By representing the quantity V' in a manner similar to (2.198)

$$V'(\mathbf{r}) = \int V'(z; k_t)\phi(\underline{\rho}; k_t) dk_t \tag{2.200}$$

and by making use of (2.194, 2.199, 2.200) we obtain the usual transmission line relationship

$$\begin{aligned} \frac{dI'(z)}{dz} &= -jk_z Y'(k_t) V'(z) \\ \frac{dV'(z)}{dz} &= -jk_z Z'(k_t) I'(z) \end{aligned} \tag{2.201}$$

with

$$Z'(k_t) = \frac{1}{Y'(k_t)} = \frac{k_z}{\omega\varepsilon} \tag{2.202}$$

for TM modes. It is now possible to express the transverse fields in (2.193) in a very convenient way, according to the classical representation of [22], also taking into account the presence of a continuous spectrum. By introducing the transverse vector eigenfunctions

$$\begin{aligned} \mathbf{e}'(\underline{\rho}; k_t) &= -\nabla_t \phi(\underline{\rho}; k_t) \\ \mathbf{h}'(\underline{\rho}; k_t) &= \mathbf{z}_0 \times \mathbf{e}'(\underline{\rho}; k_t) \end{aligned} \tag{2.203}$$

we write

$$\begin{aligned} \mathbf{E}_t(\mathbf{r}) &= \int V'(z; k_t) \mathbf{e}'(\underline{\rho}; k_t) dk_t \\ \mathbf{H}_t(\mathbf{r}) &= \int I'(z; k_t) \mathbf{h}'(\underline{\rho}; k_t) dk_t \end{aligned} \tag{2.204}$$

We proceed in a similar way for TE modes, where we label the corresponding quantities by a double prime

$$\begin{aligned} V''(\mathbf{r}) &= j\omega\mu\Pi_h(\mathbf{r}) \\ I''(\mathbf{r}) &= -\partial_z \Pi_h(\mathbf{r}) \end{aligned} \tag{2.205}$$

and we make use of the transverse eigenfunctions ψ satisfying the Helmholtz equation in the transverse plane,

$$\nabla_t^2 \psi(\underline{\rho}; k_t) + k_t^2 \psi(\underline{\rho}; k_t) = 0 \tag{2.206}$$

as well as the appropriate boundary conditions, i.e. the boundary condition of H_z, since ψ is proportional to this field. By introducing the transverse vector eigenfunctions

$$\begin{aligned} \mathbf{e}''(\rho; k_t) &= \mathbf{z}_0 \times \nabla_t \psi(\underline{\rho}; k_t) \\ \mathbf{h}''(\underline{\rho}; k_t) &= \mathbf{z}_0 \times \mathbf{e}''(\underline{\rho}; k_t) \end{aligned} \tag{2.207}$$

we write the transverse fields in a manner similar to (2.204) as

$$\begin{aligned} \mathbf{E}_t(\mathbf{r}) &= \int V''(z; k_t) \mathbf{e}''(\underline{\rho}; k_t) dk_t \\ \mathbf{H}_t(\mathbf{r}) &= \int I''(z; k_t) \mathbf{h}''(\underline{\rho}; k_t) dk_t \end{aligned} \tag{2.208}$$

with $V''(z)$ and $I''(z)$ satisfying the telegrapher's equations

$$\begin{aligned}\frac{dV''(z)}{dz} &= -jk_zZ''(k_t)I''(z)\\ \frac{dI''(z)}{dz} &= -jk_zY''(k_t)V''(z)\end{aligned} \tag{2.209}$$

with immittances now defined for the TE case as

$$Z''(k_t) = \frac{1}{Y''(k_t)} = \frac{\omega\mu}{k_z} \tag{2.210}$$

Returning to the general case note that, by following the above procedure, we have represented the fields in terms of the two scalars Π_e, Π_h. Since from (2.193) and from the analogous relationship for the TE case we have

$$\begin{aligned}E_z &= \partial_z^2\Pi_e + k^2\Pi_e = k_t^2\Pi_e\\ H_z &= \partial_z^2\Pi_h + k^2\Pi_h = k_t^2\Pi_h\end{aligned} \tag{2.211}$$

the transverse fields can also be expressed directly in terms of the longitudinal ones as [19]

$$\begin{aligned}\mathbf{E}_t &= \frac{1}{k_t^2}\left[\nabla_t\left(\partial_z E_z\right) + j\omega\mu\nabla_t \times H_z\right]\\ \mathbf{H}_t &= \frac{1}{k_t^2}\left[\nabla_t\left(\partial_z H_z\right) - j\omega\varepsilon\nabla_t \times E_z\right]\end{aligned} \tag{2.212}$$

Vice versa, by using the divergence equations it is also possible to eliminate the longitudinal fields and express the z–variation of the transverse fields in terms of the same transverse components and sources as

$$\begin{aligned}\frac{\partial\mathbf{E}_t}{\partial z} &= -j\omega\mu\left(\underline{\underline{\mathbf{I}}} + \frac{1}{k^2}\nabla_t\nabla_t\cdot\right)(\mathbf{H}_t \times \mathbf{z})\\ \frac{\partial\mathbf{H}_t}{\partial z} &= j\omega\varepsilon\left(\underline{\underline{\mathbf{I}}} + \frac{1}{k^2}\nabla_t\nabla_t\cdot\right)(\mathbf{E}_t \times \mathbf{z})\end{aligned} \tag{2.213}$$

where $\underline{\underline{\mathbf{I}}}$ is the unit dyadic.

2.5.2 LSE, LSM potentials

Let us now consider a region vertically stratified in a given direction, say y, as shown in Figure 2.26. As noted before, the field can be represented in terms of two scalars when no sources are present. The choice of these two potentials is somewhat arbitrary, as we can choose, for instance, an electric potential in the x–direction or a magnetic potential in the z–direction. However, it appears that the field obtained in terms of just one potential (for example the field obtained only from the z–directed magnetic potential) cannot satisfy all the boundary conditions at the dielectric interfaces. In general, it is necessary to consider both scalar potentials together in order to generate a field satisfying the boundary conditions at the various dielectric interfaces.

A favourable situation arises if we take potentials directed along the stratification (y–direction in Figure 2.26). In this case, fields generated by each individual

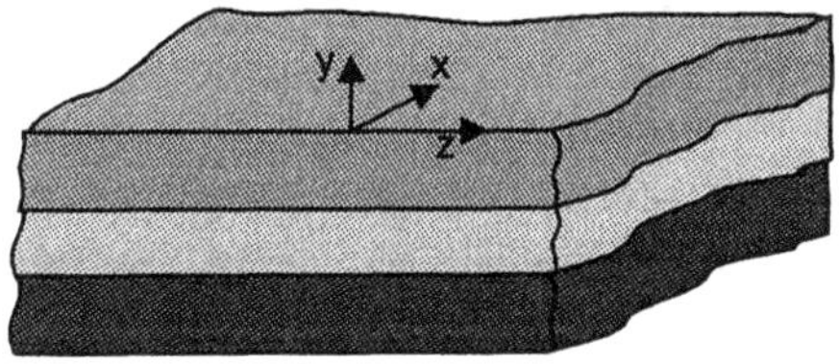

Figure 2.26 Geometry of a multi-layer planar dielectric structure.

potential fulfill the boundary conditions independently and are therefore separate solutions of Maxwell's equations. The possibility of using each potential separately greatly simplifies calculations, so that the above constitutes the common choice when dealing with a layered region. A time–harmonic dependence is assumed and suppressed. Moreover, since we are interested in modes propagating along z, a space dependence of the type $e^{-j\beta z}$ is assumed. With the above choices, with reference to Figure 2.26, the field may be represented in terms of the following potentials

$$\begin{aligned} \mathbf{\Pi}_e &= \hat{\mathbf{y}}\psi_e(x,y)e^{-j\beta z} \\ \mathbf{\Pi}_h &= \hat{\mathbf{y}}\psi_h(x,y)e^{-j\beta z} \end{aligned} \tag{2.214}$$

which satisfy the scalar wave equation

$$\nabla_t^2\psi_{e,h}(x,y) + \left(\varepsilon_r k_0^2 - \beta^2\right)\psi_{e,h}(x,y) = 0 \tag{2.215}$$

By introducing (2.214) into (2.189), we get in cartesian coordinates the following expression for the electric and magnetic fields

$$\begin{aligned} E_x &= \partial_x\partial_y\psi_e + \omega\mu_0\beta\psi_h \\ E_y &= \partial_y^2\psi_e + \varepsilon_r k_0^2\psi_e \\ E_z &= -j\beta\partial_y\psi_e - j\omega\mu_0\partial_x\psi_h \\ H_x &= -\omega\varepsilon\beta\psi_e + \partial_x\partial_y\psi_h \\ H_y &= \partial_y^2\psi_h + \varepsilon_r k_0^2\psi_h \\ H_z &= j\omega\varepsilon\partial_x\psi_e - j\beta\partial_y\psi_h \end{aligned} \tag{2.216}$$

In (2.216) and in the following the z–dependence $e^{-j\beta z}$ is understood but not shown for brevity. If we retain just the magnetic Hertzian potential ψ_h in (2.216), we obtain the following field

$$\begin{aligned} E_x &= \omega\mu_0\beta\psi_h \\ E_y &= 0 \\ E_z &= -j\omega\mu_0\partial_x\psi_h \\ H_x &= \partial_x\partial_y\psi_h \\ H_y &= \partial_y^2\psi_h + \varepsilon_r k_0^2\psi_h \\ H_z &= -j\beta\partial_y\psi_h \end{aligned} \tag{2.217}$$

In (2.217) the electric field has no component normal to the interface and lies therefore on a plane parallel to the latter. This solution is referred to as longitudinal section electric (LSE). Actually, there are infinite longitudinal sections; since this mode lies on a constant y–plane, or equivalently it has no component in the y–direction, it is completely identified by saying that it is of the LSE^y type. Such a field can also be referred to as a TE^y, since the electric field is transverse with respect to the y–direction. The admittance looking in the positive y–direction is given by the ratio of a pair of fields transverse to y, that is

$$Y_0^{LSE} = -\frac{H_z}{E_x} = \frac{j\beta\partial_y\psi_h}{\omega\mu_0\beta\psi_h} = -\frac{1}{j\omega\mu_0}\frac{\partial_y\psi_h}{\psi_h} \tag{2.218}$$

and, if the hypothesis of separability holds, that is, if

$$\partial_y\psi_h = -jk_y\psi_h \tag{2.219}$$

we get for TE^y modes

$$Y_0^{LSE} = \frac{k_y}{\omega\mu_0} \tag{2.220}$$

It is also interesting to note from (2.217) that, for slow changes in the x–direction, the terms containing the x–derivative are smaller than the other terms, thus giving

$$\begin{aligned} |E_z| &<< |E_x| \\ |H_x| &<< |E_x| \end{aligned} \tag{2.221}$$

We are left with a field with three components, namely E_x, H_y, H_z, that corresponds to a two–dimensional TE field.

In a similar manner, an LSM mode is obtained from (2.216) by setting

$$\psi_h = 0 \tag{2.222}$$

which, upon use in (2.216), gives

$$\begin{aligned} E_x &= \partial_x\partial_y\psi_e \\ E_y &= \partial_y^2\psi_e + \varepsilon_r k_0^2\psi_e \\ E_z &= -j\beta\partial_y\psi_e \\ H_x &= -\omega\varepsilon\beta\psi_e \\ H_y &= 0 \\ H_z &= j\omega\varepsilon\partial_x\psi_e \end{aligned} \tag{2.223}$$

In this case we have for the impedance in the positive y–direction

$$Z_0^{LSM} = \frac{E_z}{H_x} = \frac{-j\beta\partial_y\psi_e}{-\omega\varepsilon\beta\psi_h} = -\frac{1}{j\omega\varepsilon}\frac{\partial_y\psi_h}{\psi_h} \tag{2.224}$$

Also for the LSM case, if the hypothesis of separability holds, we have

$$\partial_y\psi_e = -jk_y\psi_e \tag{2.225}$$

so that we obtain,

$$Z_0^{LSM} = \frac{k_y}{\omega\varepsilon} \tag{2.226}$$

Similarly to the LSE case, when a slow variation in the x–direction is present, we recover a two-dimensional LSM field.

2.6 Scalar wave equation

The problem of finding the electromagnetic field in a source–free, homogeneous, region is solved once general solutions to the scalar Helmholtz equation are known. These solutions are generally obtained by the method called separation of variables. It has been shown by Eisenhart, [8], that the Helmholtz equation is separable in 11 three–dimensional orthogonal coordinate systems. In the next Sections we will use the separation of variables to solve the Helmholtz equation in the rectangular, circular and elliptical cylindrical, spherical coordinate systems; the reader is referred to [27] for solutions of the wave equations in other coordinate systems. Emphasis is placed on the fact that when the domain of a variable is finite we have a discrete spectrum of eigenvalues and eigenfunctions; while for an unbounded (infinite) domain the discrete spectrum coalesces into a continuum. Another feature which deserves some attention is that, depending on the chosen coordinate system, we can represent the field in our domain in several different ways.

2.6.1 *Plane wave expansion*

The Helmholtz equation in a rectangular coordinate system is [13],

$$\frac{\partial^2\psi}{\partial x^2}+\frac{\partial^2\psi}{\partial y^2}+\frac{\partial^2\psi}{\partial z^2}+k^2\psi=0 \tag{2.227}$$

The method of separation of variables seeks solutions of (2.227) in the form

$$\psi(x,y,z)=X(x)Y(y)Z(z) \tag{2.228}$$

By introducing (2.228) into (2.227) and by dividing by ψ we get

$$\frac{1}{X(x)}\frac{d^2X(x)}{dx^2}+\frac{1}{Y(y)}\frac{d^2Y(y)}{dy^2}+\frac{1}{Z(z)}\frac{d^2Z(z)}{dz^2}+k^2=0 \tag{2.229}$$

In (2.229) each term depends on only one coordinate. The various terms depending on the x, y, z coordinates are linked only via k^2. Thus, by separating k^2 into three constants k_x, k_y, k_z, such that

$$k^2=k_x^2+k_y^2+k_z^2 \tag{2.230}$$

we obtain the three equations

$$\begin{aligned}\frac{1}{X(x)}\frac{d^2X(x)}{dx^2}&=-k_x^2\\ \frac{1}{Y(y)}\frac{d^2Y(y)}{dy^2}&=-k_y^2\\ \frac{1}{Z(z)}\frac{d^2Z(z)}{dz^2}&=-k_z^2\end{aligned} \tag{2.231}$$

Equation (2.230) is often called the *separation equation.* The solutions of (2.231) are the harmonic functions, i.e. functions like

$$h(k_xx)\sim\sin k_xx,\ \cos k_xx,\ e^{-jk_xx},\ e^{jk_xx} \tag{2.232}$$

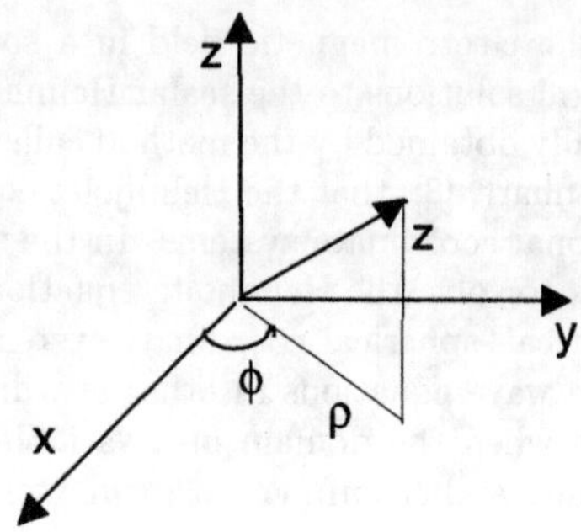

Figure 2.27 Cylindrical coordinate system.

Hence, the solutions of the Helmholtz equation are written in terms of the harmonic functions as

$$\psi(x, y, z) = h(k_x x)h(k_y y)h(k_z z) \tag{2.233}$$

and these solutions are often referred to as *elementary wave functions.*

It is noted that the vector $\mathbf{k} = (k_x, k_y, k_z)$ indicates the propagation direction and the type of field. In particular for $|\mathbf{k}| < k_0$ we have a field which is free to propagate in space, thus a radiative field, while for $|\mathbf{k}| > k_0$ the field is attenuated, at least in one direction, thus representing a localised, reactive field.

Finally, we observe that for spatial coordinates ranging over finite domains, such as in the case of closed waveguides and cavities, in order to satisfy the boundary conditions k_x, k_y, k_z may assume only discrete values; whenever the spatial domain becomes infinite, such as in the case of antennas, open waveguides, open resonators, etc., k_x, k_y, k_z vary over a continuous range.

2.6.2 *Cylindrical wave expansion*

When boundaries coincide with cylindrical coordinate surfaces, it is expedient to solve the Helmholtz equation in cylindrical coordinates. Even when no finite boundaries are present, if we have an a priori knowledge that the field has cylindrical wavefronts it is obviously advantageous to represent the field in this coordinate system. It is therefore important to investigate solutions of the Helmholtz equation in cylindrical coordinates. The latter solutions can be obtained by separation of variables [13, p.198]. The scalar Helmholtz equation in circular cylindrical coordinates, henceforth referred to as cylindrical coordinates, is

$$\frac{1}{\rho}\frac{\partial}{\partial \rho}\left(\rho\frac{\partial \psi}{\partial \rho}\right) + \frac{1}{\rho^2}\frac{\partial^2 \psi}{\partial \phi^2} + \frac{\partial^2 \psi}{\partial z^2} + k^2\psi = 0 \tag{2.234}$$

In order to solve (2.234) by the method of separation of variables we seek solutions in the form

$$\psi = R(\rho)\Phi(\phi)Z(z) \tag{2.235}$$

which, after introduction in (2.234) and division by ψ gives

$$\frac{1}{\rho R}\frac{d}{d\rho}\left(\rho\frac{dR}{d\rho}\right)+\frac{1}{\rho^2\Phi}\frac{d^2\Phi}{d\phi^2}+\frac{1}{Z}\frac{d^2Z}{dz^2}+k^2=0 \tag{2.236}$$

In the above equation, the first term will be seen to depend only on ρ, the second only on ϕ, and the third only on z. By separating the free–space wavenumber k^2 into its z–component, k_z^2, and a transverse part k_ρ^2 as

$$k_z^2+k_\rho^2=k^2 \tag{2.237}$$

we get, for the z–dependent coordinate, an equation similar to that obtained for the rectangular coordinate system

$$\frac{1}{Z}\frac{d^2Z}{dz^2}=-k_z^2 \tag{2.238}$$

After substitution of (2.238) into (2.236) we obtain

$$\frac{\rho}{R}\frac{d}{d\rho}\left(\rho\frac{dR}{d\rho}\right)+\frac{1}{\Phi}\frac{d^2\Phi}{d\phi^2}+\left(k^2-k_z^2\right)\rho^2=0 \tag{2.239}$$

where, as before, the first term depends only on ρ, and the second only on ϕ. By imposing

$$\frac{1}{\Phi}\frac{d^2\Phi}{d\phi^2}=-n^2 \tag{2.240}$$

with n denoting a constant, (2.239) becomes

$$\frac{\rho}{R}\frac{d}{d\rho}\left(\rho\frac{dR}{d\rho}\right)-n^2+\left(k^2-k_z^2\right)\rho^2=0 \tag{2.241}$$

which depends only on ρ. Thus, by using (2.235) we have separated (2.234) into the three following equations

$$\begin{aligned}\rho\frac{d}{d\rho}\left(\rho\frac{dR}{d\rho}\right)+\left[k_\rho^2\rho^2-n^2\right]R &= 0 \qquad (a)\\ \frac{d^2\Phi}{d\phi^2} &= -n^2\Phi \qquad (b)\\ \frac{d^2Z}{dz^2} &= -k_z^2Z \qquad (c)\end{aligned} \tag{2.242}$$

Equation (2.242a) is the Bessel equation whose solutions are

$$B_n(k_\rho\rho)\sim J_n(k_\rho\rho),N_n(k_\rho\rho),H_n^{(1)}(k_\rho\rho),H_n^{(2)}(k_\rho\rho)$$

with

- $J_n(k_\rho\rho)$ Bessel function of the first kind
- $N_n(k_\rho\rho)$ Bessel function of the second kind
- $H_n^{(1)}(k_\rho\rho)$ Hankel function of the first kind

- $H_n^{(2)}(k_\rho \rho)$ Hankel function of the second kind

The equations (2.242b and c) are harmonic equations that give rise to solutions of the type already seen for rectangular coordinates. But now a fundamental difference is introduced. Let us consider, for example, the free–space case, where z varies over the infinite domain $-\infty < z < \infty$, while ϕ varies over the bounded domain $0 < \phi < 2\pi$. Hence, while k_z varies over a continuum of values, n is forced to assume only discrete values. In particular if we require that $\psi(\phi) = \psi(\phi + 2\pi)$ and if we take solutions of the type $\sin(n\phi), \cos(n\phi)$, n is forced to take only integer values. Note that also ρ has unlimited support $0 \leq \rho < \infty$, thus possessing a continuous spectrum of eigenvalues k_ρ. To summarise the free–space situation, we now have eigenfunctions along z that are of the same type for rectangular and cylindrical coordinates. In the transverse plane, for a rectangular coordinate system we have two sets of eigenfunctions and eigenvalues (k_x, k_y) each varying over an unlimited support; while for cylindrical coordinates only the transverse wavenumber pertains to the continuous spectrum, since the angular variation is over a bounded domain. Note also that the transverse eigenfunctions, i.e. the Bessel functions, are now dependent on the angular solution. Hence, if we are considering a wave travelling along z, we can describe its modal spectrum in terms of either (k_x, k_y) or in terms of n, k_ρ.

Guided waves in cylindrical coordinates are often described by means of the TE and TM potentials. In particular, according to the formulas (2.189) on page 67, by setting

$$\mathbf{\Pi}_h = \mathbf{z}_0 \psi \tag{2.243}$$

we have the following expressions for TE fields

$$\begin{aligned} E_\rho &= -j\omega\mu \frac{1}{\rho}\frac{\partial \psi}{\partial \phi} & \quad \mathrm{H}_\rho &= \frac{\partial^2 \psi}{\partial \rho \partial z} \\ E_\phi &= j\omega\mu \frac{\partial \psi}{\partial \rho} & \quad H_\phi &= \frac{1}{\rho}\frac{\partial^2 \psi}{\partial \phi \partial z} \\ E_z &= 0 & \quad H_z &= \left(k^2 + \frac{\partial^2}{\partial z^2}\right)\psi \end{aligned} \tag{2.244}$$

In a similar manner for TM fields we set

$$\mathbf{\Pi}_e = \mathbf{z}_0 \psi \tag{2.245}$$

thus obtaining

$$\begin{aligned} E_\rho &= \frac{\partial^2 \psi}{\partial \rho \partial z} & \quad H_\rho &= j\omega\varepsilon \frac{1}{\rho}\frac{\partial \psi}{\partial \phi} \\ E_\phi &= \frac{1}{\rho}\frac{\partial^2 \psi}{\partial \phi \partial z} & \quad H_\phi &= -j\omega\varepsilon \frac{\partial \psi}{\partial \rho} \\ E_z &= \left(k^2 + \frac{\partial^2}{\partial z^2}\right)\psi & \quad H_z &= 0 \end{aligned} \tag{2.246}$$

2.6.3 Scalar wave equation in elliptical coordinates

Ridges in closed circular or elliptical waveguides are sometimes introduced in order to increase the bandwidth and/or to "straighten" electric field lines in a region

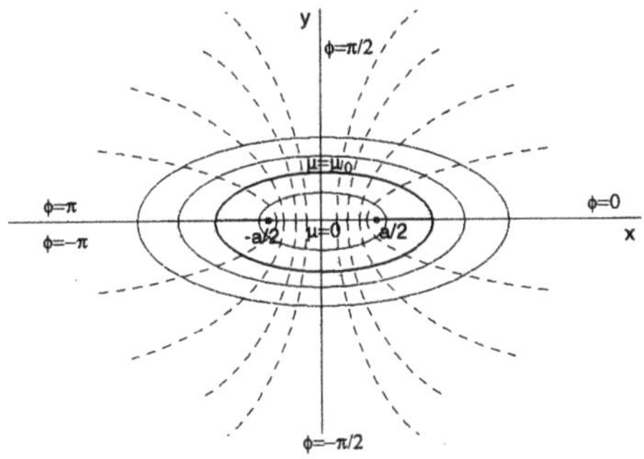

Figure 2.28 Elliptical coordinate system.

where field uniformity is required. The open region is a limiting case of the closed one as the radial distance goes to infinity.

Several important open waveguide cross–sections such as that of the slotted line, to be discussed in the following Sections, can be thought of as elliptical ridged cross–sections and become separable when described in that coordinate system. Its dual configuration, i.e. the metal strip in the open region is also solved as separable in the same system. Although inhomogeneity of the cross–section prevents separability even in this system, nonetheless a very compact description of the far field and of some features of the near field, e.g. the edge condition, is still possible.

We pass from the rectangular system (x, y) to the elliptical one (μ, ϕ) illustrated in Figure 2.28 by means of the transformation of variables

$$\begin{aligned} x &= \frac{1}{2}a\cosh\mu\cos\phi \\ y &= \frac{1}{2}a\sinh\mu\sin\phi \end{aligned} \tag{2.247}$$

The elliptical boundary of a closed guide coincides with the surface $\mu = \mu_0$, $\mu_0 \to \infty$ in the open case, $-\pi < \phi \leq \pi$, whereas the ridge surfaces are defined as

$$\begin{aligned} \frac{a}{2} \leq x, \quad y = 0 \quad &: \quad \phi = 0, \quad 0 \leq \mu \leq \mu_0 \\ x \leq -\frac{a}{2}, \quad y = 0 \quad &: \quad \phi = \pm\pi, \quad 0 \leq \mu \leq \mu_0 \end{aligned} \tag{2.248}$$

The major and minor axes of the ellipse are given by

$$A = a\cosh\mu_0 \qquad B = a\sinh\mu_0 \tag{2.249}$$

A scalar potential ψ in this cross–section is a solution of the Helmholtz equation

$$\left(\nabla_t^2 + k_t^2\right)\psi = 0 \tag{2.250}$$

with boundary conditions of the type

$$\begin{aligned} \partial\psi/\partial n = 0 \quad &\text{on the metal surfaces} \quad TE\ modes \\ \psi = 0 \quad &\text{on the metal surfaces} \quad TM\ modes \end{aligned} \tag{2.251}$$

By utilizing the Jacobian coordinate transformation rule

$$\frac{\partial\psi}{\partial x} = \frac{\frac{\partial(\psi,y)}{\partial(\mu,\phi)}}{J} \qquad \frac{\partial\psi}{\partial y} = \frac{\frac{\partial(x,\psi)}{\partial(\mu,\phi)}}{J} \tag{2.252}$$

$$J = \begin{vmatrix} \frac{\partial x}{\partial\mu} & \frac{\partial x}{\partial\phi} \\ \frac{\partial y}{\partial\mu} & \frac{\partial y}{\partial\phi} \end{vmatrix} = \left(\frac{a}{2}\right)^2 (\sinh^2\mu\cos^2\phi + \cosh^2\mu\sin^2\phi) \tag{2.253}$$

$$\frac{\partial(\psi,y)}{\partial(\mu,\phi)} = \begin{vmatrix} \frac{\partial\psi}{\partial\mu} & \frac{\partial\psi}{\partial\phi} \\ \frac{\partial y}{\partial\mu} & \frac{\partial y}{\partial\phi} \end{vmatrix} \qquad \frac{\partial(x,\psi)}{\partial(\mu,\phi)} = \begin{vmatrix} \frac{\partial x}{\partial\mu} & \frac{\partial x}{\partial\phi} \\ \frac{\partial\psi}{\partial\mu} & \frac{\partial\psi}{\partial\phi} \end{vmatrix} \tag{2.254}$$

(2.250) reduces to the following scalar transverse wave equation in elliptical coordinates:

$$\left[\frac{\partial^2}{\partial\mu^2} + \frac{\partial^2}{\partial\phi^2} + \left(\frac{ak_t}{2}\right)^2 (\cosh^2\mu - \cos^2\phi)\right]\psi = 0 \tag{2.255}$$

whose solutions separate out as products of a radial function $J(\mu)$ and an angular one $S(\phi)$

$$\psi(\mu,\phi) = J(\mu)S(\phi) \tag{2.256}$$

The latter functions are solutions of the Mathieu ordinary differential equation [27]

$$\begin{aligned} S'' + (b^2 - h^2\cos^2\phi)S &= 0 \qquad (a) \\ -J'' + (b^2 - h^2\cosh^2\mu)J &= 0 \qquad (b) \end{aligned} \tag{2.257}$$

where we have posed

$$h = \frac{ak_t}{2}$$

and b is the eigenvalue. Angular solutions are required to possess periodicity of $\pi, 2\pi$ so that Φ is the Mathieu function of odd, even type denoted by $S_{on}(h,\phi)$; $S_{en}(h,\phi)$, respectively. It is noted that in the case of vanishing eccentricity ($h \to 0$), $S_{on}(h,\phi)$ reduces to the ordinary $\sin\phi$, $S_{en}(h,\phi)$ to $\cos\phi$ and they satisfy the same parity, symmetry and boundary conditions. An ample discussion of these functions can be found in [27, vol. 1, pp.550-557]. Given h, b and n two types of radial function satisfy the "modified" Mathieu equation (2.257b): these are denoted by $J_{on}(h,\mu)$ and $N_{on}(h,\mu)$, or $J_{en}(h,\mu)$ and $N_{en}(h,\mu)$, which in the limit of vanishing excentricity, ($h \to 0$), reduce to the ordinary Bessel and Neumann functions respectively, for cylindrical coordinates. They too satisfy the same parity, symmetry and boundary conditions.

Transverse Electric Modes

For even TE modes, ψ is symmetrical about the plane $y = 0$ or $\mu = 0$ and $\phi = 0$ for $\mu > 0$, where it also satisfies the Neumann boundary condition:

$$\left.\frac{\partial\psi}{\partial y}\right|_{y=0} = 0 \tag{2.258}$$

Hence, we form ψ by the product

$$\psi = S_{en}(h, \phi)Q_{en}(h, \mu) \tag{2.259}$$

where

$$Q_{en}(h, \mu) = \frac{J_{en}(h, \mu)}{J_{en}(h, 0)} \tag{2.260}$$

so that $Q_{en}(h, 0) = 1$; $Q'_{en}(h, 0) = 0$ and noting that

$$\left.\frac{\partial}{\partial y}\right|_{\mu=0} = \frac{2}{a}\frac{1}{\sin\phi}\frac{\partial}{\partial\mu} \qquad \left.\frac{\partial}{\partial x}\right|_{\mu=0} = -\frac{2}{a}\frac{1}{\sin\phi}\frac{\partial}{\partial\phi} \tag{2.261}$$

(2.258) is satisfied for any x by the choice (2.259) that is regular with its first derivative anywhere in the cross–section. By recalling that

$$E_x = -\frac{\partial\psi}{\partial y} \qquad E_y = \frac{\partial\psi}{\partial x} \tag{2.262}$$

and the first expression in (2.261), it is seen that the electric field is purely normal to the plane $y = 0$. Hence, this kind of mode does not feel the presence of the ridge and they constitute acceptable solutions of the elliptical cross–section in the absence of a ridge. For finite μ_0 in fact their dispersion equation is determined by the boundary condition on the guide walls

$$J'_{en}(h, \mu_0) = 0 \tag{2.263}$$

whereas no such condition is present for the open case. For *odd* TE modes ψ ought to vanish on the plane of symmetry $y = 0$ on the slot between the ridges, i.e. for $\mu = 0$ the Dirichlet condition is fulfilled whereas the Neumann condition is to be satisfied for $\mu > 0$, $\phi = 0, \pi$, i.e. on the ridges. The latter constitutes an interesting discontinuous boundary problem typical of slotted line/strip line configurations where a singular edge condition prevails at the slot/strip edges.

It is possible to construct a solution by appropriately modifying the even TE case in order to cater for both constraints, as follows:

$$\begin{aligned} \psi &= S_{en}(h, \phi)P_{en}(h, \mu) \quad &: \quad 0 < \phi < \pi \\ \psi &= -S_{en}(h, \phi)P_{en}(h, \mu) \quad &: \quad -\pi < \phi < 0 \end{aligned} \tag{2.264}$$

where

$$P_{en}(h, \mu) = J_{en}(h, 0)N_{en}(h, \mu) - N_{en}(h, 0)J_{en}(h, \mu) \tag{2.265}$$

so that $P_{en}(h, 0) = 0$; $P'_{en}(h, 0) = 1$. The angular variation in (2.264) satisfies Neumann's condition on the ridge ($\phi = 0, \mu > 0$), whereas the chosen combination of radial functions ensures fulfilment of Dirichlet's condition on the slot ($\mu = 0$). Also ψ is continuous with its first derivative across the slot and discontinuous across the ridge, due to the electric currents flowing on it. For this type of mode, the electric field at the plane of the slot is tangential and it becomes singular as prescribed by the $r^{-1/2}$ edge condition at the tips. Clearly no equivalent is present in a regular elliptical waveguide cross–section. The above statements are verified by considering (2.262); hence, by recalling the first part of (2.264), E_y is seen to vanish for $\mu = 0$, whereas the term $1/\sin\phi$ in the second part of (2.264) produces the well known square root type of singularity

$$\left(1-\left[\frac{2x}{a}\right]^2\right)^{-1/2}$$

Transverse Magnetic Modes

In view of the boundary condition $\psi = 0$ at $\phi = 0, \pi, -\pi$ the appropriate angular functions are now the Mathieu odd functions $S_{on}(h,\phi)$. *Odd* solutions are constructed by employing odd radial functions of the type

$$Q_{on}(h,\mu) = \frac{J_{on}(h,\mu)}{J'_{on}(h,\mu)} \tag{2.266}$$

so that $Q'_{on}(h,0) = 1$; $Q_{on}(h,0) = 0$. Consequently

$$\psi = S_{on}(h,\phi)Q_{on}(h,\mu) \tag{2.267}$$

These are seen to be regular with their first derivatives over the cross–section and are therefore acceptable solutions without a ridge. In particular, modes of a closed elliptical waveguide correspond to the roots of the dispersion equation

$$J_{on}(h,\mu_0) = 0 \tag{2.268}$$

$\mu = \mu_0$ being the elliptical guide boundary.

Even solutions are dual to the odd TE solutions previously seen, satisfying a mixed boundary condition on the plane $y = 0$. The condition $\psi = 0$ on the ridge $(\phi = 0, \pi, -\pi)$ is catered for by assuming, as in the odd case, an odd angular dependence, whereas that of the slot, namely $\frac{\partial\psi}{\partial y} = 0$, is achieved by taking a radial dependence of the type

$$P_{on}(h,\mu) = N'_{on}(h,0)J_{on}(h,\mu) - J'_{on}(h,0)N_{on}(h,\mu) \tag{2.269}$$

so that $P_{on}(h,0) = 0$; $P'_{on}(h,0) = 1$. We assume therefore

$$\begin{aligned} \psi &= S_{on}(h,\phi)P_{on}(h,\mu) \quad : \quad 0 < \phi < \pi \\ \psi &= -S_{on}(h,-\phi)P_{on}(h,\mu) \quad : \quad -\pi < \phi < 0 \end{aligned} \tag{2.270}$$

As for the TE–case, it is noted that the electric field deriving from (2.270) is purely tangential to the slot, whereas that deriving from the odd solution (2.267) is normal to it.

The strip problem

The dual problem of an infinitesimal thin conducting strip in an unbounded medium also admits the closed form solutions discussed above. Regular solutions of the TE and TM type are described by potentials (2.259) and (2.267), respectively. Singular fields at the edges of the strip are produced instead by the following potentials:

For TE modes :

$$\psi = P_{on}(h,\mu)S_{on}(h,\phi) \quad : \quad n = 1, 2, \ldots \tag{2.271}$$

where

$$P_{on}(h,\mu) = J_{on}(h,\mu)N'_{on}(h,0) - J'_{on}(h,0)N_{on}(h,\mu) \tag{2.272}$$

with the boundary conditions $P_{on}(h,0) = 0$; $P'_{on}(h,0) = 1$. ψ is an odd function of ϕ about $\phi = 0$.

For TM modes :

$$\psi = P_{en}(h,\mu)S_{en}(h,\phi) \quad : \quad n = 0,1,2,\ldots \tag{2.273}$$

where P_{en} is defined by (2.265). ψ is an even function of ϕ about $\phi = 0$ and it satisfies the condition $\psi = 0$ for $\mu = 0$.

Connection between Mathieu's functions and Chebyshev polynomials

The Chebyshev polynomials $T_n(x) = \cos n\left(\text{arc cos}\,\frac{2x}{a}\right)$ are orthonormal over the slot with weight function $\left(1 - \left(\frac{2x}{a}\right)^2\right)^{-1/2}$ and have been extensively used as expanding functions in solutions of the slot/strip problem. As the above weight function is the same as appears in the fields generated by the Mathieu function potentials, it is interesting to explore the connection between the sets.

Let us carry out in the Mathieu equation, i.e. in (2.25b), the following change of variables, according to (2.247), in the domain of the slot (strip), $\mu = 0$

$$\cos\phi = \frac{2x}{a} = \bar{x} \tag{2.274}$$

Recalling the second part of (2.261), we find

$$\frac{d^2S}{d\phi^2} = \left(1 - \bar{x}^2\right)\frac{d^2S}{d\bar{x}^2} - \bar{x}\frac{dS}{d\bar{x}} \tag{2.275}$$

and Mathieu's equation (2.257) becomes

$$\left(1 - \bar{x}^2\right)\frac{d^2S}{d\bar{x}^2} - \bar{x}\frac{dS}{d\bar{x}} + \left(b^2 - h^2\bar{x}^2\right)S(\bar{x}) = 0 \tag{2.276}$$

In the limit $h^2 \to 0$, we have $b^2 \to n^2$ and (2.276) becomes

$$\left(1 - \bar{x}^2\right)\frac{d^2S}{d\bar{x}^2} - \bar{x}\frac{dS}{d\bar{x}} + n^2S(\bar{x}) = 0 \tag{2.277}$$

i.e., Chebyshev's equation [1] and $S(\bar{x}) = T_n(\bar{x})$. The physical interpretation of the above limit is that the Chebyshev polynomials approximate very well the true solution for "small" slot apertures and/or for $k_t \to 0$, i.e. for $\beta \sim k$, that is the quasi–TEM limit.

2.6.4 Spherical wave expansion

The spherical coordinate system is represented in Figure 2.29 and is the natural coordinate system to adopt when dealing with spherical waves, or with boundaries conformal to such geometry. Solutions of the Helmholtz equation in the spherical coordinate system may be obtained by separation of variables by setting

$$\psi = R(r)\Theta(\theta)\Phi(\phi) \tag{2.278}$$

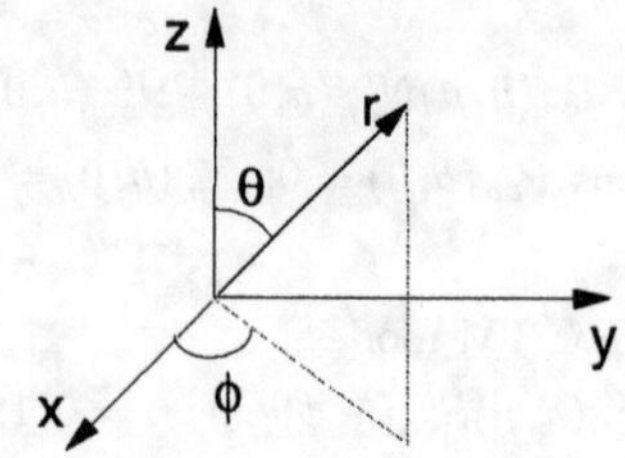

Figure 2.29 Spherical coordinate system.

which is used in the Helmholtz equation expressed in spherical coordinates, i.e. in

$$\frac{1}{r^2}\frac{\partial}{\partial r}\left(r^2\frac{\partial\psi}{\partial r}\right)+\frac{1}{r^2\sin\theta}\frac{\partial}{\partial\theta}\left(\sin\theta\frac{\partial\psi}{\partial\theta}\right)+\frac{1}{r^2\sin^2\theta}\frac{\partial^2\psi}{\partial\phi^2}+k^2\psi=0 \qquad (2.279)$$

By substituting (2.278) into (2.279), dividing by ψ, and multiplying by $r^2\sin^2\theta$, we obtain

$$\frac{\sin^2\theta}{R}\frac{d}{dr}\left(r^2\frac{dR}{dr}\right)+\frac{\sin\theta}{\Theta}\frac{d}{d\theta}\left(\sin\theta\frac{d\Theta}{d\theta}\right)+\frac{1}{\Phi}\frac{d^2\Phi}{d\phi^2}+k^2r^2\sin^2\theta=0 \qquad (2.280)$$

from which the ϕ–dependence can be separated out by requiring the third term in the above equation to be equal to a constant $(-m^2)$ as

$$\frac{d^2\Phi}{d\phi^2}+m^2\Phi=0 \qquad (2.281)$$

After use of (2.281) in (2.280) and division by $\sin^2\theta$ we can separate (2.280) into

$$\frac{1}{\Theta\sin\theta}\frac{d}{d\theta}\left(\sin\theta\frac{d\Theta}{d\theta}\right)-\frac{m^2}{\sin^2\theta}=-n(n+1) \qquad (2.282)$$

and

$$\frac{1}{R}\frac{d}{dr}\left(r^2\frac{dR}{dr}\right)-n(n+1)+k^2r^2=0 \qquad (2.283)$$

To summarise we now have the set of equations (2.281, 2.282 and 2.283) which represent the separated Helmholtz equation in spherical coordinates. The solutions of (2.281) are the well–known harmonic functions; they extend over a limited support, thus corresponding to discrete eigenvalues and eigenfunctions. The solutions (2.282) are called associated Legendre functions since (2.282) is related to Legendre's equation. Such solutions are referred to as

$$L_n^m(\cos\theta)\sim P_n^m(\cos\theta), Q_n^m(\cos\theta) \qquad (2.284)$$

and they also represent a discrete eigensystem. The solutions of equations (2.283) are the spherical Bessel functions since (2.283) is closely related to the Bessel equation. The various different situations arising when describing the free–space with rectangular, cylindrical and spherical coordinate systems are summarised in Table 2.5.

Table 2.5 Types of spectra arising from the solution of the free-space Helmholtz equation in different coordinate systems.

Coordinate system	space domain	spectrum
rectangular	$-\infty < x < \infty$	continuous
	$-\infty < y < \infty$	continuous
	$-\infty < z < \infty$	continuous
cylindrical	$0 \leq \rho < \infty$	continuous
	$0 \leq \phi < 2\pi$	discrete
	$-\infty < z < \infty$	continuous
spherical	$0 \leq r < \infty$	continuous
	$0 \leq \phi < 2\pi$	discrete
	$0 \leq \theta < \pi$	discrete

2.6.5 Connections between different wave expansions

The wave expansions introduced previously are not independent, as it is possible to represent a spherical or a cylindrical wave in terms of plane waves. These representations have particular interest in practical applications. The expansion of a spherical wave in terms of plane waves is known as the *Weyl identity* [13, p.188] and may be written as

$$\frac{e^{-jkr}}{r} = -\frac{j}{2\pi}\int_{-\infty}^{\infty}\int_{-\infty}^{\infty}\frac{e^{-j(k_x x+k_y y)-jk_z|z|}}{k_z}\,dk_x\,dk_y \tag{2.285}$$

where in (2.285)

$$k_z = \sqrt{k^2-k_x^2-k_y^2} \qquad r = \sqrt{x^2+y^2+z^2} \tag{2.286}$$

Another useful relationship, called the *Sommerfeld identity* [19, p.312], relates a spherical wave to the cylindrical wave expansion and it states

$$\frac{e^{-jkr}}{r} = -\frac{j}{2}\int_{-\infty}^{\infty}\frac{k_\rho}{k_z}H_0^{(2)}(k_\rho\rho)e^{-jk_z|z|}dk_\rho \tag{2.287}$$

or, in terms of Bessel functions

$$\frac{e^{-jkr}}{r} = -j\int_{0}^{\infty}\frac{k_\rho}{k_z}J_0(k_\rho\rho)e^{-jk_z|z|}dk_\rho \tag{2.288}$$

where, in (2.287) and (2.288) we have set

$$k_z = \sqrt{k^2-k_\rho^2} \tag{2.289}$$

The above identities (2.285, 2.287 and 2.288) are important not only in order to express a field in different coordinate systems, but also for quick evaluation of the Sommerfeld–Weyl–type integrals. In fact, in several applications such as microstrip antennas, frequency selective surfaces, etc., one needs to evaluate this type of integral in the radiation zone or the far–field region. As we will see later, solutions for this class of problem are often obtained in the spectral domain via Fourier or Hankel transforms. Naturally, when transforming back to the real space a Sommerfeld–Weyl–type integral is involved. Such integrals become difficult to evaluate by numerical integration routines, since when the observation point is far away from the source, the integrand becomes rapidly oscillatory: in this case the saddle–point method of integration has to be used. Nonetheless, as cleverly observed in [5], by using the above identities it is possible to bypass most of the algebra involved in the steepest–descent calculations. As an example, consider a Weyl–type integral of the form commonly encountered in microwave integrated circuits, i.e.

$$I = \int_{-\infty}^{\infty}\int_{-\infty}^{\infty} g(k_x, k_y) e^{-j(k_x x + k_y y) - jk_z|z|} dk_x\, dk_y \tag{2.290}$$

The integral (2.290) can be easily factorised in a form that resembles the Weyl identity; in fact we can write

$$I = \int_{-\infty}^{\infty}\int_{-\infty}^{\infty} \{g(k_x, k_y)k_z\} \left\{\frac{e^{-j(k_x x + k_y y) - jk_z|z|}}{k_z}\right\} dk_x dk_y \tag{2.291}$$

If x, y or z are large in the above integral, the integrand oscillates rapidly and contributions arise only from the neighbourhood of the point where the phase of the exponential function is stationary, i.e. when

$$\begin{aligned} \frac{\partial}{\partial k_x}(k_x x + k_y y + k_z|z|) &= 0 \\ \frac{\partial}{\partial k_y}(k_x x + k_y y + k_z|z|) &= 0 \end{aligned} \tag{2.292}$$

which are easily solved to give the stationary phase point at

$$k_{x0} = \frac{k}{r}x \qquad k_{y0} = \frac{k}{r}y \tag{2.293}$$

The physical interpretation of (2.293) is that only the plane wave whose **k**–vector points from the source to the observation point is important when evaluating the far field at the observation point. By replacing the slowly varying part of the integrand with its value at the points of stationary phase we obtain

$$I \approx \{g(k_{x0}, k_{y0})k_{z0}\} \int_{-\infty}^{\infty}\int_{-\infty}^{\infty} \left\{\frac{e^{-j(k_x x + k_y y) - jk_z|z|}}{k_z}\right\} dk_x dk_y \tag{2.294}$$

which after use of the Weyl identity gives the sought approximation

$$I \approx j2\pi g(k_{x0}, k_{y0})k_{z0}\frac{e^{-jkr}}{r} \tag{2.295}$$

The location of the stationary phase point can be easily found by (2.293), which is easily remembered through its physical interpretation. In fact a source radiates plane waves in all directions, with wavenumbers given by $k_x^2 + k_y^2 + k_z^2 = k^2$.

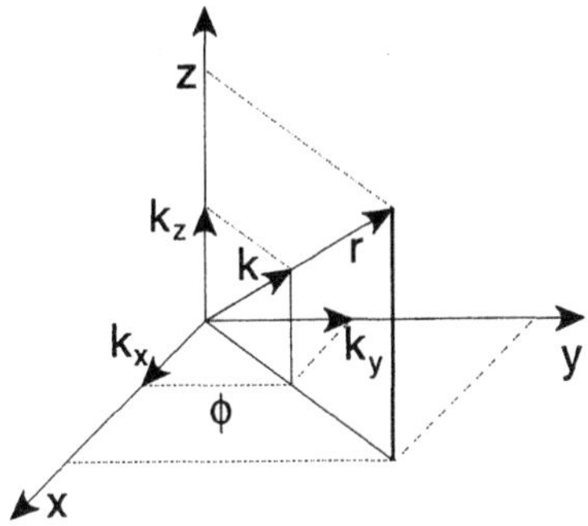

Figure 2.30 Only the wave travelling from the origin to the observation point placed at (x, y, z) contributes to the asymptotic evaluation of the radiation integral.

However, as is clear from the geometrical construction of Figure 2.30, only the wave which satisfies the condition

$$\frac{k_x}{x} = \frac{k_y}{y} = \frac{k_z}{z} = \frac{k}{r} \tag{2.296}$$

reaches the observation point; it is then apparent that the evaluation of (2.290) is achieved by simple inspection. A similar argument is also applicable to the Sommerfeld integral [5].

2.7 Plane waves at interfaces

2.7.1 Snell's law

Consider a plane wave impinging on a surface between two media with different dielectric constants, as in Figure 2.31. In general, part of the incident power will be transmitted to the other region, while part will be reflected. Regardless of how the wave is polarised, it is easy to see that the tangential components of all wave vectors are continuous. In fact, by denoting with $\mathbf{k}_1^i, \mathbf{k}_1^r, \mathbf{k}_2^t$ the incident, reflected and transmitted wave vectors respectively, in rectangular coordinates we have

$$\begin{aligned} \mathbf{k}_1^i &= \mathbf{x}_0 k_{1x}^i + \mathbf{y}_0 k_{1y}^i + \mathbf{z}_0 k_{1z}^i \\ \mathbf{k}_1^r &= \mathbf{x}_0 k_{1x}^r + \mathbf{y}_0 k_{1y}^r + \mathbf{z}_0 k_{1z}^r \\ \mathbf{k}_2^t &= \mathbf{x}_0 k_{2x}^t + \mathbf{y}_0 k_{2y}^t + \mathbf{z}_0 k_{2z}^t \end{aligned} \tag{2.297}$$

with the spatial dependence of incident, reflected and transmitted fields respectively given by $\exp(-j\mathbf{k}_1^i \cdot \mathbf{r})$, $\exp(-j\mathbf{k}_1^r \cdot \mathbf{r})$ and $\exp(-j\mathbf{k}_2^t \cdot \mathbf{r})$.

Boundary conditions at $x = 0$ require the continuity of the tangential components of the electromagnetic field. By denoting by E_1^i, E_1^r, E_2^t the magnitudes of the tangential components of the electric fields, the boundary conditions at $x = 0$ imply

$$E_1^i e^{-jk_{1y}^i y - jk_{1z}^i z} + E_1^r e^{-jk_{1y}^r y - jk_{1z}^r z} = E_2^t e^{-jk_{2y}^t y - jk_{2z}^t z} \tag{2.298}$$

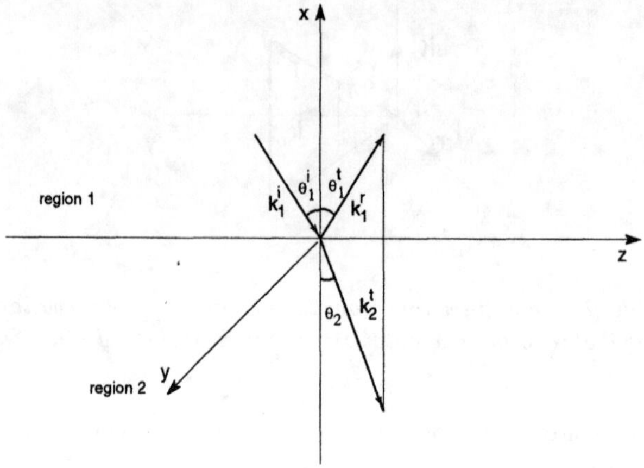

Figure 2.31 A wave coming from region 1 ($x > 0$) is partially reflected and partially transmitted into region 2. The incident, reflected and transmitted waves all lie in the plane of incidence, i.e. the plane defined by the incident wave vector and the normal to the interface between the two media.

which must hold for all y and z. This can only occur if

$$E_1^i + E_1^r = E_2^t \tag{2.299}$$

and

$$\begin{aligned} k_{1y}^i &= k_{1y}^r = k_{2y}^t = k_y \\ k_{1z}^i &= k_{1z}^r = k_{2z}^t = k_z \end{aligned} \tag{2.300}$$

Hence, satisfaction of the boundary conditions requires the *phase–matching conditions* (2.300) to hold, i.e. the continuity of the tangential components of all wave vectors [19]. Another way of stating the phase–matching condition is that the incident, reflected and transmitted wave vectors must all lie in the same plane, the *plane of incidence*, i.e. the plane determined by the incident vector $\mathbf{k}_1^i$ and the normal to the boundary surface. According to this result, it is expedient to develop the subsequent part of the theory assuming the plane $y = cost$ as the plane of incidence, thus considering $k_y = 0$. In order to find out the angle of reflection, θ_1^r, and the angle of transmission, θ_2^t, as a function of the angle of incidence θ_1^i let us apply (2.300). With reference to Figure 2.31 we get

$$\begin{aligned} k_1^i \sin\theta_1^i &= k_1^r \sin\theta_1^r \\ k_1^i \sin\theta_1^i &= k_2^t \sin\theta_2^t \end{aligned} \tag{2.301}$$

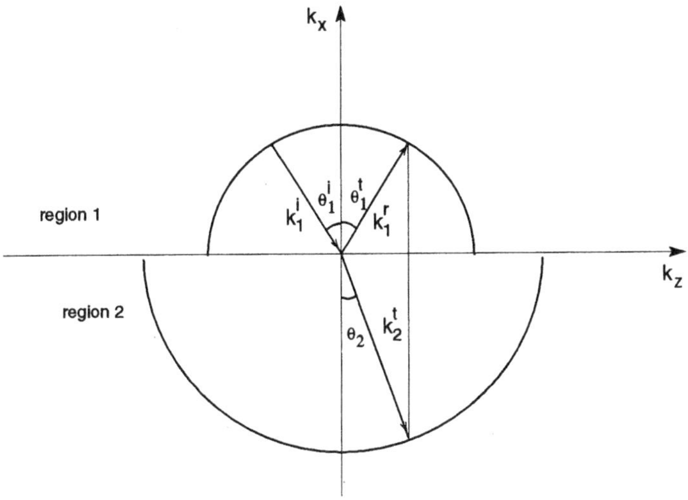

Figure 2.32 Phase matching using wavenumber surfaces.

Since the magnitude of the wave vectors is

$$\begin{aligned} k_1^i = k_1^r &= \omega\sqrt{\mu_1\varepsilon_1} \\ k_2^t &= \omega\sqrt{\mu_2\varepsilon_2} \end{aligned} \tag{2.302}$$

we see that the angle of reflection is equal to the angle of incidence, i.e.

$$\theta_1^r = \theta_1^i \tag{2.303}$$

and that Snell's law holds, namely

$$\sqrt{\mu_1\varepsilon_1}\sin\theta_1^i = \sqrt{\mu_2\varepsilon_2}\sin\theta_2^t \tag{2.304}$$

which has the simple geometrical interpretation shown in Figure 2.32. In fact, we may plot the k–locus for the two regions on the k_x, k_z plane. Let us assume that the medium filling region 2 is denser than the medium filling region 1, so that $k_2 > k_1$. From $\mathbf{k}_1^i$ we determine k_z, that is the projection of the incident wave vector on the z–axis. The reflected and transmitted waves must have the same k_z, while k_x is determined from the intersection with the k–locus, as shown in Figure 2.32. This technique may also be used for media which are not isotropic, as exemplified in [19].

Total internal reflection

It is apparent that, if the wave is incident from the region with smaller k, then it is always possible to find the wave vectors of the reflected and transmitted waves.

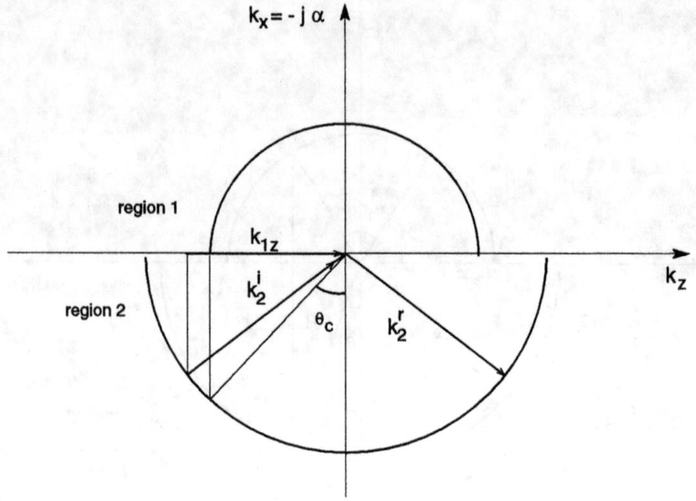

Figure 2.33 Total internal reflection arises for incidence from an optically denser medium at an angle greater than the critical angle θ_c.

A different case arises when the incidence is from the denser medium, as in Figure 2.33. In this case, when the z–component of the incident wave vector exceeds the wave vector of region 1, it is impossible to satisfy the phase–matching conditions unless we take a complex wave vector in region 1. By making this choice, the tangential components of the propagation constants are real and matched, while the x–directed component is given by

$$k_{1x}^{t} = -j\alpha = \sqrt{\left(k_{2z}^{i}\right)^{2} - k_1^2} \tag{2.305}$$

The above choice corresponds to a transmitted wave exponentially decaying in region 1, as we have $\exp(-jk_{1x}x) = \exp(-\alpha x)$. Consequently we have a plane wave, with constant phase front perpendicular to the boundary surface, which propagates along the z–direction with phase velocity ω/k_z; the amplitude of this wave is maximum at the interface and decays exponentially away from the surface. This type of wave is generated when radiation impinges from a denser medium at an angle greater than θ_c, the critical angle of incidence given by

$$k_2 \sin\theta_c = k_1$$

Since there is no time–average power penetrating into medium 1, the phenomenon is also referred to as total internal reflection.

As we have seen, Snell's law provides the propagation direction for reflected and transmitted waves, when present, but does not tell us the amplitude of these waves. In order to obtain their magnitude it is necessary to examine separately the TE and TM cases. In fact, an incident wave with any polarisation can be decomposed into TE and TM wave components. The TE wave is linearly polarised with the electric field vector perpendicular to the plane of incidence and is often called perpendicularly polarised or horizontally polarised. The TM wave is linearly polarised with the electric vector parallel to the plane of incidence and is referred to as parallel polarised or vertically polarised. The two wave components can be studied separately.

2.7.2 TE incidence

Let us consider a plane wave, incident from region 1, with the electric field given by

$$\mathbf{E}_1^i = \mathbf{y}_0 E_1^i e^{-j\mathbf{k}_1 \cdot \mathbf{r}} \tag{2.306}$$

The electric field in region 1 is the superposition of the incident and reflected field, given by

$$E_{1y} = \left[E_1^i e^{jk_{1x}x} + E_1^r e^{-jk_{1x}x}\right] e^{-jk_z z} \tag{2.307}$$

In the above we have used (2.300), stating that the tangential component of the propagation vector of incident, reflected and transmitted waves is the same. The z–component of the magnetic field can be obtained from Faraday's law

$$\nabla \times \mathbf{E} = -j\omega\mu\mathbf{H} \tag{2.308}$$

By inserting (2.307) into (2.308) we obtain the magnetic field in region 1 as

$$H_{1z} = -\frac{k_{1x}}{\omega\mu_1}\left[E_1^i e^{jk_{1x}x} + E_1^r e^{-jk_{1x}x}\right] e^{-jk_z z} \tag{2.309}$$

Similarly for region 2, where only a transmitted wave exists we have

$$\begin{aligned} E_{2y} &= E_2^t e^{+jk_{2x}x - jk_z z} \\ H_{2z} &= -\frac{k_{2x}}{\omega\mu_2} E_2^t e^{+jk_{2x}x - jk_z z} \end{aligned} \tag{2.310}$$

It is expedient to define the *characteristic admittance* in the transverse direction, Y_1, given by the ratio between H_z and E_y in region 1 as

$$\frac{k_{1x}}{\omega\mu_1} = \sqrt{\frac{\varepsilon_1}{\mu_1}}\cos\theta_1 = Y_1 = \frac{1}{Z_1} \tag{2.311}$$

In the above, we have also introduced the *characteristic impedance*, given by the inverse of the characteristic admittance. Similarly, for region 2 we define:

$$\frac{k_{2x}}{\omega\mu_2} = \sqrt{\frac{\varepsilon_2}{\mu_2}}\cos\theta_2 = Y_2 = \frac{1}{Z_2} \tag{2.312}$$

Continuity of the tangential components of **E** and **H** requires that the ratio

$$Z = -\frac{E_y}{H_z}$$

be continuous. As a consequence of (2.307), (2.309) and (2.310) at $x = 0$, we have

$$Z_2 = Z_1 \frac{E_1^i + E_1^r}{E_1^i - E_1^r} \tag{2.313}$$

It is convenient to introduce the reflection coefficient, defined by

$$\Gamma = \frac{E_1^r}{E_1^i} \tag{2.314}$$

and the transmission coefficient, defined as

$$T = \frac{E_2^t}{E_1^i} \tag{2.315}$$

which are related by

$$T = 1 + \Gamma \tag{2.316}$$

By solving (2.313) we obtain the reflection coefficient

$$\Gamma = \frac{Z_2 - Z_1}{Z_2 + Z_1} \tag{2.317}$$

which can be expressed more explicitly by using Snell's law as

$$\Gamma = \frac{\sqrt{1-\sin^2\theta_1} - \sqrt{1-\frac{\varepsilon_1\mu_1}{\varepsilon_2\mu_2}\sin^2\theta_1}\sqrt{\frac{\varepsilon_2\mu_1}{\varepsilon_1\mu_2}}}{\sqrt{1-\sin^2\theta_1} + \sqrt{1-\frac{\varepsilon_1\mu_1}{\varepsilon_2\mu_2}\sin^2\theta_1}\sqrt{\frac{\varepsilon_2\mu_1}{\varepsilon_1\mu_2}}} \tag{2.318}$$

2.7.3 TM incidence

For TM incidence, since it is the magnetic field which is parallel to the boundary between the two media, it is convenient to describe reflection and transmission in terms of the y–directed magnetic field. The latter, in region 1, is the superposition of the incident and reflected waves, as

$$H_{1y} = \left[H_1^i e^{jk_{1x}x} + H_1^r e^{-jk_{1x}x}\right] e^{-jk_z z} \tag{2.319}$$

As for the TE incidence we have used (2.300), stating that the tangential component of the propagation vector of incident, reflected and transmitted waves is the same. The z–component of the electric field may be obtained from Ampere's law

$$\nabla \times \mathbf{H} = j\omega\varepsilon\mathbf{E} \tag{2.320}$$

By inserting (2.319) into (2.320) we obtain the z–component of the electric field in region 1 as

$$E_{1z} = -\frac{k_{1x}}{\omega\varepsilon_1}\left[H_1^i e^{jk_{1x}x} + H_1^r e^{-jk_{1x}x}\right] e^{-jk_z z} \tag{2.321}$$

while the transmitted field is given by

$$\begin{aligned} H_{2y} &= H_2^t e^{+jk_{2x}x - jk_z z} \\ E_{2z} &= \frac{k_{2x}}{\omega\varepsilon_2} H_2^t e^{+jk_{2x}x - jk_z z} \end{aligned} \tag{2.322}$$

In this case, the characteristic admittance in the transverse direction, given by the ratio between H_y and E_z, is

$$Y_1 = \frac{1}{Z_1} = \frac{\omega\varepsilon_1}{k_{1x}} = \sqrt{\frac{\varepsilon_1}{\mu_1}}\frac{1}{\cos\theta_1} \tag{2.323}$$

in region 1, while in region 2 we have:

$$Y_2 = \frac{1}{Z_2} = \frac{\omega\varepsilon_2}{k_{2x}} = \sqrt{\frac{\varepsilon_2}{\mu_2}}\frac{1}{\cos\theta_2} \tag{2.324}$$

Observe that, in this TM case, the cosine appears in the denominator, leading to an infinite admittance for an incidence with $\theta = \pi/2$. Continuity of the tangential components of **E** and **H** requires that the ratio

$$Z = \frac{E_z}{H_y}$$

be continuous at $x = 0$, which yields

$$Z_2 = Z_1 \frac{H_1^i - H_1^r}{H_1^i + H_1^r} \tag{2.325}$$

Also in this case we may introduce the transmission and reflection coefficients, this time defined by

$$\Gamma = -\frac{H_1^r}{H_1^i} \tag{2.326}$$

which can be expressed, by using Snell's law, as

$$\Gamma = \frac{\sqrt{1-\sin^2\theta_1} - \sqrt{1-\frac{\varepsilon_1\mu_1}{\varepsilon_2\mu_2}\sin^2\theta_1}\sqrt{\frac{\varepsilon_1\mu_2}{\varepsilon_2\mu_1}}}{\sqrt{1-\sin^2\theta_1} + \sqrt{1-\frac{\varepsilon_1\mu_1}{\varepsilon_2\mu_2}\sin^2\theta_1}\sqrt{\frac{\varepsilon_1\mu_2}{\varepsilon_2\mu_1}}} \tag{2.327}$$

Brewster angle

From (2.327) it is noted that, for $\mu_1 = \mu_2 = \mu_0$, by choosing

$$\theta_B = \tan^{-1}\sqrt{\frac{\varepsilon_2}{\varepsilon_1}} \tag{2.328}$$

the reflection coefficient vanishes. This means that there exists a real angle of incidence $\theta_1 = \theta_B$ for which all the incident power is transmitted to region 2 and no reflection arises. No such case is present for TE incidence.

2.7.4 Goos–Hänschen shift

We have already considered the case of total internal reflection, arising when a medium is optically denser. It is now interesting to look at the shape of the fields in region 1 and 2 when total internal reflection takes place. At optical frequencies $\mu \sim \mu_0$; moreover, let us consider the medium in region 1 denser than the medium in region 2, i.e. $\varepsilon_1 > \varepsilon_2$. We have seen that, for an angle of incidence greater than the critical angle, the x–directed propagation constant becomes imaginary, thus implying an exponentially decaying field in region 2. Accordingly, we may express the field in region 2 as

$$\begin{aligned} E_{2y} &= E_2^t e^{\alpha x - jk_z z} \\ H_{2z} &= \frac{j\alpha}{\omega\mu_0} E_2^t e^{\alpha x - jk_z z} \end{aligned} \tag{2.329}$$

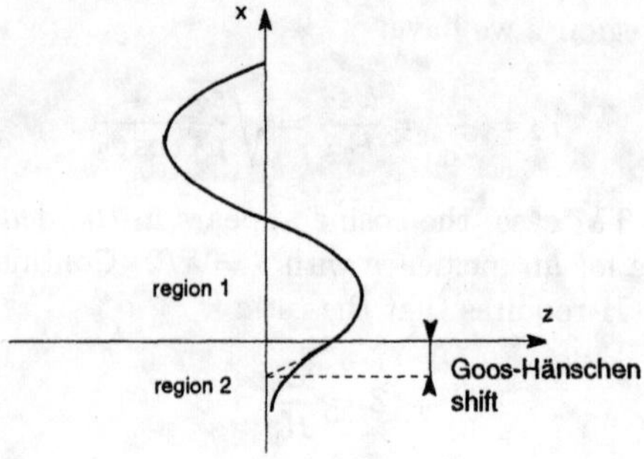

Figure 2.34 Critical internal reflection of a wave at the interface between two media illustrating the Goos–Hänschen shift.

It is apparent that the wave impedance $-E_y/H_z$ is now imaginary in region 2, and thus we may set $Z_2 = jX_2$ with X_2 real. The amplitude of fields in regions 1 and 2 must satisfy (2.313) from which we obtain the reflection coefficient as

$$\Gamma = \frac{jX_2 - Z_1}{jX_2 + Z_1} \tag{2.330}$$

The above shows that $|\Gamma| = 1$; thus the magnitude of incident and reflected waves is equal and a standing wave in the x–direction is present in region 1, while no power is transmitted to region 2. The electric field in region 1 is given by

$$E_{1y} = E_1^i \left[e^{jk_{1x}x} + \Gamma e^{-jk_{1x}x} \right] e^{-jk_z z} = e^{-j\phi} 2E_1^i \cos(k_{1x}x - \phi) e^{-jk_z z} \tag{2.331}$$

with

$$\phi = -\frac{1}{2} \arg \Gamma$$

This situation is thus similar to that arising when a perfect conductor is present, i.e. the incident wave is completely reflected. The interesting fact is that the total reflection does not take place at the physical boundary but at a small distance inside region 2, as shown by Figure 2.34; this distance takes the name of the Goos–Hänschen shift. A similar analysis may be carried out for TM polarisation.

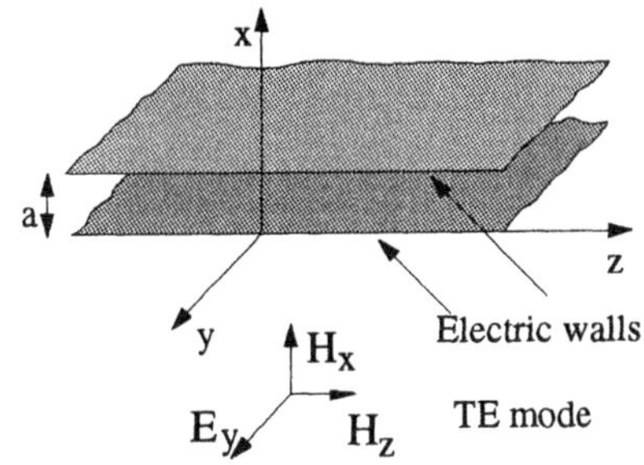

Figure 2.35 Parallel plate waveguide, infinite in the y−, z−directions. Propagation is assumed in the z−direction.

2.8 Spectral representations for closed and open regions

We have seen in the previous Chapter several examples of open waveguides and of their applications. It has been noted that a peculiar characteristic of open waveguides is to possess, apart from a few bound modes, a continuous spectrum. While, generally, the reader is quite familiar with the discrete spectrum of closed waveguides, as studied from basic e.m. courses [30], the notion of a continuous spectrum is relatively uncommon. In order to introduce the latter in a simple way, we refer in the following to the case of a parallel plate waveguide where we move one of the plates to infinity. By considering this example the properties of completeness and orthogonality for the continuous spectrum are introduced, also showing the analogy between discrete and continuous cases.

Let us consider a parallel plate region such as that shown in Figure 2.35 with a TE mode present. In this region $E_y = \phi$ satisfies the equation

$$\frac{d^2\phi}{dx^2} + k_x^2\phi = 0 \tag{2.332}$$

with the boundary conditions at $x = 0, a$ (see Figure 2.35), requiring the vanishing of the tangential component of the electric field on the metallic planes, thus $\phi = 0$. The solution of the above eigenvalue problem is given by

$$\phi_n(x) = \sqrt{\frac{2}{a}}\sin(\frac{n\pi}{a}x) \tag{2.333}$$

The normalisation constant in (2.333) has been chosen so that the mode set is normalised to unity; moreover the orthogonality is readily proved so that we have

$$\int_0^a \phi_m(x)\phi_n(x)dx = \delta_{mn} = \begin{Bmatrix} 1 & m = n \\ 0 & m \neq n \end{Bmatrix} \tag{2.334}$$

In order to ascertain formally the completeness of the eigenfunction set we expand the delta function in terms of the above eigenfunctions as

$$\delta(x - x') = \sum_{n=1}^{\infty} f_n\phi_n(x)$$

By multiplying both sides by $\phi_m(x)$ and by integrating over the interval $[0, a]$ we obtain

$$f_m = \int_0^a \delta(x - x')\phi_m(x)dx = \phi_m(x')$$

which inserted in the delta representation gives

$$\delta(x - x') = \sum_{n=1}^{\infty} \phi_n(x)\phi_n(x') \tag{2.335}$$

i.e. a concise expression of completeness and orthonormality.

When the plate separation becomes infinite, the parallel plate guide turns into a half-space region over a ground plane and the set of discrete eigenvalues

$$k_n = \frac{n\pi}{a}$$

become a continuous range [12, 10]. By introducing

$$\Delta k_n = k_{n+1} - k_n = \frac{\pi}{a}$$

and rewriting (2.335) as

$$\delta(x - x') = \frac{2}{\pi} \sum_{k_n = \Delta k_n}^{\infty} \sin(k_n x)\sin(k_n x')\Delta k_n$$

while performing the limit for $\Delta k_n \to 0$, we obtain the completeness relationship for the open region as

$$\delta(x - x') = \frac{2}{\pi} \int_0^{\infty} \sin(kx)\sin(kx')dk \tag{2.336}$$

It is also apparent that the continuous eigenfunction is given by

$$\phi(x; k) = \sqrt{\frac{2}{\pi}} \sin(kx) \tag{2.337}$$

Note that the eigenfunctions appearing in (2.335) have infinite energy on the interval $[0, \infty)$; for this reason they are sometimes referred to as improper [12]. Note also that, while in the parallel plate case the eigenfunctions have finite energy, in the half space only the integral of the energy is finite. The orthogonality between eigenfunctions may be obtained by noting that in (2.336) after replacing

$$x \to k; x' \to k'; k \to x$$

we obtain the orthogonality relationship

$$\delta(k - k') = \frac{2}{\pi} \int_0^{\infty} \sin(kx)\sin(k'x)dx \tag{2.338}$$

Table 2.6 compares some examples of eigenfunctions, completeness and orthogonality relationships for discrete and corresponding continuous cases. From Table 2.6, the following observations are in order:

Table 2.6 Summary of the completeness and orthogonality relationship for the discrete and continuous eigenvalue cases.

	discrete case	**continuous case**
eigenfunction	$\phi_n(x) = \sqrt{\frac{2}{a}} \sin\left(\frac{n\pi x}{a}\right)$	$\phi(x;k) = \sqrt{\frac{2}{\pi}} \sin(kx)$
completeness (δ representation in the space domain) $\delta(x-x') =$	$\sum_{n=1}^{\infty} \phi_n(x)\phi_n(x')$	$\int_0^{\infty} \phi(x;k)\phi(x';k)dk$
orthogonality (δ representation in the spectral domain)	$\delta_{mn} = \begin{Bmatrix} 1 & m=n \\ 0 & m \neq n \end{Bmatrix} =$ $\int_0^a \phi_n(x)\phi_m(x)dx$	$\delta(k-k') =$ $\int_0^{\infty} \phi(x;k)\phi(x;k')dx$

1. in both cases, discrete and continuous, the eigenfunction set provides a basis which is both orthogonal and complete;

2. the eigenfunctions of the continuous spectrum satisfy the boundary condition at $x = 0$, but only finiteness is required at infinity;

3. from a physical point of view the continuous eigenfunction corresponds to a field that originates from a generator placed at infinity;

4. as such, the continuous eigenfunctions taken individually do not satisfy the radiation condition, i.e. that the field vanish at infinity; the only condition satisfied is that they remain finite at infinity.

However, on physical grounds it is quite obvious that the total field, which is given by a superposition of eigenfunctions each with its own amplitude, should vanish at infinity, thus satisfying the radiation condition. This is not in contrast to the above point 4 stating that each eigenfunction has a finite amplitude at infinity. It is only their sum, i.e. the wave packet formed by the superposition of all eigenfunctions, which vanishes at infinity. As we will see in the following discussion, it is exactly this point as elucidated by [32] which allows us to determine the spectrum of open waveguides.

For reference purposes, it is advantageous to tabulate the spectral representation for some cases which occur frequently in applications [12]. In the first column of Table 2.7 are specified the boundary conditions, while in the second and third columns are given the eigenfunction expansion and the spectral representation for the operator

$$L = -\frac{d^2}{dx^2}$$

over $[0, a]$.

When the domain of the operator becomes infinite the above eigenfunction expansion is modified as in Table 2.8 in the interval $[0, \infty)$. Note that in the lower right–hand box of Table 2.8, apart from the usual superposition integral, another

Table 2.7 Spectral representation over the finite domain $[0, a]$.

Boundary conditions	Eigenfunction	$\delta(x-\xi)$
$u(0)=u(a)=0$	$\sqrt{\frac{2}{a}}\sin\left(\frac{n\pi}{a}x\right)$	$\frac{2}{a}\sum_{n=1}^{\infty}\sin\left(\frac{n\pi x}{a}\right)\sin\left(\frac{n\pi\xi}{a}\right)$
$u'(0)=u'(a)=0$	$\sqrt{\frac{1}{a}}$ $n=0$ $\sqrt{\frac{2}{a}}\cos\left(\frac{n\pi}{a}x\right)$ $n\geq 1$	$\frac{1}{a}+\frac{2}{a}\sum_{n=1}^{\infty}\cos\left(\frac{n\pi x}{a}\right)\cos\left(\frac{n\pi\xi}{a}\right)$
$u(0)=u'(a)=0$	$\sqrt{\frac{2}{a}}\sin\left(\frac{[n+1/2]\pi x}{a}\right)$	$\frac{2}{a}\sum_{n=1}^{\infty}\sin\left(\frac{[n+1/2]\pi x}{a}\right)\sin\left(\frac{[n+1/2]\pi\xi}{a}\right)$
$u(0)=u(a)$, $u'(0)=u'(a)$		$\frac{1}{a}+$ $+\frac{2}{a}\sum_{n=1}^{\infty}\cos\left(\frac{2n\pi x}{a}\right)\cos\left(\frac{2n\pi\xi}{a}\right)+\sin\left(\frac{2n\pi x}{a}\right)\sin\left(\frac{2n\pi\xi}{a}\right)$

Table 2.8 Spectral representation on the infinite domain $[0, \infty)$.

Boundary conditions	Eigenfunction	$\delta(x-\xi)=$
$u(0)=0$	$\sqrt{\frac{2}{\pi}}\sin(kx)$	$\frac{2}{\pi}\int_0^{\infty}\sin(kx)\sin(k\xi)\,dk$
$u'(0)=0$	$\sqrt{\frac{2}{\pi}}\cos(kx)$	$\frac{2}{\pi}\int_0^{\infty}\cos(kx)\cos(k\xi)\,dk$
$u'(0)=\alpha u(0)$		$-2\alpha e^{\alpha(x+\xi)}+$ $\frac{2}{\pi}\int_0^{\infty}\left[\cos(kx)+\frac{\alpha}{k}\sin(kx)\right]\left[\cos(k\xi)+\frac{\alpha}{k}\sin(k\xi)\right]\frac{k^2}{k^2+\alpha^2}\,dk$ The first term is missing if $Re(\alpha)>0$

term also appears. The latter corresponds to a surface wave, which will be investigated in greater detail in Chapter 3, referring to the case of planar dielectric waveguides. For the moment it is enough to note that, since for 2D TE modes we have

$$\frac{dE_y}{dx} = -j\omega\mu H_z \tag{2.339}$$

then identifying E_y with u and using 2.339 yields

$$\frac{E_y}{H_z} = -\frac{j\omega\mu}{\alpha} \tag{2.340}$$

where the latter quantity may be interpreted as the impedance of a wave travelling in the x-direction. By varying α between 0 and infinity we recover the two cases of a magnetic and metallic plate, respectively. Hence, when the plate has a finite, negative impedance, in addition to the continuous spectrum of modes, also a surface wave exists. The reason why we have a surface wave contribution only for positive values of $Re(\alpha)$ will become clear in Chapter 3, where we discuss surface waves as obtained from transverse resonance.

Finally, we observe that when the interval is unbounded in both directions, e.g. by removing both plates, the delta representation over $(-\infty, \infty)$ is given by

$$\delta(x-\xi) = \frac{1}{2\pi} \int_{-\infty}^{\infty} e^{-jk(x-\xi)} dk \tag{2.341}$$

Consequently, the eigenfunctions, representing a plane wave spectrum, are:

$$\phi(x,k) = \frac{e^{-jkx}}{\sqrt{2\pi}} \tag{2.342}$$

2.9 Excitation by a source

In the next Chapter we consider the problem of a source distribution exciting a dielectric slab waveguide and we solve it directly as we deal with a simple two–dimensional field. In a three–dimensional case, where TE and TM waves may coexist, it is advantageous to investigate first the method of solution. To this end, it is expedient to formulate the problem in a somewhat abstract manner. Let us consider a source f, a *forcing field*, which creates a certain *unknown field* u. The latter must satisfy some linear relationships, i.e. Maxwell's equations and appropriate boundary conditions, which we can briefly summarise as a linear operator L. Therefore, our problem is to find a solution to the following equation

$$Lu = f \tag{2.343}$$

A numerical solution of the above problem may be obtained by applying the Rayleigh-Ritz approach, or the method of moments, as described previously in Section 2.4. However, whenever possible, it is often advantageous to find an analytical solution of our problem. To this end we may follow two different approaches [12, p.134]: we could make use of the spectral representation of our operator, or we could use directly the Green's function pertaining to our problem. Let us consider first the solution by means of a spectral representation.

2.9.1 Solution by spectral representation

In this case we imagine we know the eigenvalues (λ) and eigenfunctions (ϕ) of our operator L, satisfying the homogeneous equation

$$L\phi = \lambda\phi \tag{2.344}$$

As always, due to completeness of the eigenfunctions, we may represent both our unknown field u and our source f in terms of an eigenfunction expansion. If the eigenfunctions form a discrete set, then the expansion is represented by a discrete summation, whereas an integral is used when the eigenfunctions form a continuous set. Accordingly, we write

$$\begin{aligned} f(r) &= \textstyle\int_k \tilde{f}(k)\phi(r;k)dk \\ u(r) &= \textstyle\int_k \tilde{u}(k)\phi(r;k)dk \end{aligned} \tag{2.345}$$

where we have also stressed the dependence of source f and generated field u on the spatial coordinate r. By inserting the above spectral representation into (2.343), and by using (2.344), neglecting for the moment the problem of interchanging the integral with the operator L, we find

$$\int_k \tilde{u}(k)\lambda(k)\phi(r;k)dk = \int_k \tilde{f}(k)\phi(r;k)dk \tag{2.346}$$

from which the transform $\tilde{u}(k)$ of the field is obtained

$$\tilde{u}(k) = \frac{\tilde{f}(k)}{\lambda(k)} \tag{2.347}$$

In fact, it is now sufficient to insert the above expression into (2.345) to obtain the solution of our problem. It is important to note that, by using the eigenfunction expansion, we have transformed the operator equation (2.343) into an algebraic one, for the solution is simply obtained through (2.347) by means of algebraic operations.

Green's function by spectral representation

Once the spectral representation (2.344) is known, the Green's function can also be computed in a straightforward way. The latter is, in fact, a solution of the problem

$$LG(r,r') = \delta(r-r') \tag{2.348}$$

We can represent both the Green's function and the delta function in terms of the eigenfunction set as

$$G(r,r') = \int_k \tilde{g}(r';k)\phi(r;k)dk \tag{2.349}$$

$$\delta(r-r') = \int_k \phi(r';k)\phi(r;k)dk \tag{2.350}$$

By introducing the above expressions into (2.348) and by making use of (2.344) we readily obtain

$$\tilde{g}(r';k) = \frac{\phi(r';k)}{\lambda(k)} \tag{2.351}$$

and therefore,

$$G(r,r') = \int_k \frac{\phi(r';k)\phi(r;k)}{\lambda(k)}dk \tag{2.352}$$

which represents the sought expression for the Green's function.

2.9.2 Direct calculation of the Green's function

The previous approach, although very useful, is not the only possible one; in several cases, the Green's function can be calculated almost immediately, at least for one dimensional cases, and without any previous knowledge of the eigensystem. The procedure for directly calculating the Green's function is here illustrated by an example [12, p.160]. Let us consider an operator

$$L = -\frac{d^2}{dx^2} \tag{2.353}$$

with the boundary conditions

$$u(0) = 0 \qquad u(1) = 0 \tag{2.354}$$

In order to find the Green's function we must solve

$$\frac{d^2 G(x, x')}{dx^2} = -\delta(x - x') \tag{2.355}$$

Let us now consider two solutions of the homogeneous equation corresponding to (2.355) such that they satisfy the boundary conditions for $x < x'$ and $x > x'$, respectively. One such solution can evidently be

$$\phi_0(x) = c_0 x \tag{2.356}$$

which satisfies the boundary condition at $x = 0$, and where c_0 is a constant still to be determined; another solution is

$$\phi_1(x) = c_1(1 - x) \tag{2.357}$$

where c_1 is another suitable constant; (2.357) satisfies the prescribed condition for $x > x'$. However, the solution of (2.355) must be continuous at $x = x'$, which is obtained by setting

$$c_0 = (1 - x') \qquad c_1 = x' \tag{2.358}$$

Therefore a solution of (2.355) of the form

$$G(x, x') = \begin{cases} (1 - x')x & x < x' \\ (1 - x)x' & x' < x \end{cases} \tag{2.359}$$

which satisfies both boundary conditions and continuity requirements. It is also apparent that the jump in the first derivative at $x = x'$ equals -1 as stated by (2.355).

For the same problem, the spectral expansion of the Green's function would have required first the determination of the eigenfunctions and eigenvalues, then the application of (2.352). Since in this simple case the eigenfunctions are

$$\sqrt{2}\sin(n\pi x) \tag{2.360}$$

while the eigenvalues are $n\pi$, the Green's function assumes the form

$$G(x, x') = \frac{2}{\pi^2} \sum_{n=1}^{\infty} \frac{\sin(n\pi x)\sin(n\pi x')}{n^2} \tag{2.361}$$

which looks quite different from that calculated with the direct approach (2.359). Naturally, if we are dealing with one–dimensional problems we may apply either the method of the eigenfunction expansion or the method of the direct Green's function evaluation. When facing higher dimensional problems (2D or 3D), however, we are free to use a mixture of the two methods. Before looking in some detail into the above statement, let us first consider the eigenfunction expansion in more dimensions for problems allowing separation of variables.

2.9.3 Eigenfunction expansion in more dimensions

We have seen in the previous subsection that, given the problem (2.343), once we know the eigensystem (2.344), the solution can be found in an algebraic fashion by using (2.347). In this Section we investigate the problem (2.343), considering now functions and operators in a higher dimensional domain. As a simple example, consider the operator L, satisfying the inhomogeneous equation (2.343), as given by

$$Lu = -(u_{xx} + u_{yy}) \tag{2.362}$$

where the domain of L is the set of square–integrable functions $u(x, y)$, which possess square–integrable second derivatives (here denoted with the subscript) and satisfy appropriate boundary conditions. L is the sum of two commuting operators, say L_1 and L_2, so that according to (2.362),

$$L = L_1 + L_2 \tag{2.363}$$

We can find the inverse of L by considering either one as a "constant" with respect to the other. In our example we have

$$L_1 u = -u_{xx} \qquad L_2 u = -u_{yy} \tag{2.364}$$

and L_1, L_2, commute since

$$L_1 L_2 u = L_2 L_1 u \tag{2.365}$$

In order to find solutions of (2.362), we consider the two eigensystems pertaining to the two operators L_1, L_2 as

$$\begin{aligned} L_1 \phi_k^{(1)} &= \lambda_k^{(1)} \phi_k^{(1)} \\ L_2 \phi_k^{(2)} &= \lambda_k^{(2)} \phi_k^{(2)} \end{aligned} \tag{2.366}$$

By expressing the unknown function u and the source function f in terms of the second set, namely

$$\begin{aligned} u &= \sum_k u_k \phi_k^{(2)} \\ f &= \sum_k f_k \phi_k^{(2)} \end{aligned} \tag{2.367}$$

and by introducing (2.367) into the equation $Lu = f$, i.e. into

$$(L_1 + L_2) \sum_k u_k \phi_k^{(2)} = \sum_k f_k \phi_k^{(2)} \tag{2.368}$$

the problem reduces to the solution of the one–dimensional expression

$$L_1 u_k + \lambda_k^{(2)} u_k = f_k \tag{2.369}$$

which can be solved by using either of the two methods described in the previous Section. In particular, if we apply again the method of eigenfunction expansion, thus writing

$$\begin{aligned} u_k &= \sum_j u_{kj} \phi_j^{(1)} \\ f_k &= \sum_j f_{kj} \phi_j^{(1)} \end{aligned} \tag{2.370}$$

and if we substitute the above expansions in (2.369) we obtain a fully algebraic form of our problem, leading to a solution of the type

$$u = \sum_k \sum_j \frac{f_{kj}}{\lambda_j^{(1)} + \lambda_k^{(2)}} \phi_j^{(1)} \phi_k^{(2)} \tag{2.371}$$

An alternative expression to the latter formula, sometimes more convenient for numerical computation, is obtained by solving the last one–dimensional problem by the direct Green's function method. The procedure just described naturally also applies to three–dimensional problems; actually, this procedure is the operator version of the classical separation of variables already seen in Section 2.6. However, in this way it is emphasised that by expanding the fields in terms of eigenfunctions we achieve algebraic forms of the solution, thus reducing complexity. We now proceed with a couple of examples showing the application of this procedure to determining the scalar Green's function for free–space and for a planar stratified medium.

2.9.4 Free–space scalar Green's function

In free–space we look for solutions of

$$\nabla^2 g(\mathbf{r};\mathbf{r}') + k_0^2 g(\mathbf{r};\mathbf{r}') = -\delta(\mathbf{r}-\mathbf{r}') \tag{2.372}$$

where we have denoted by $\mathbf{r}'$ the source point and by $\mathbf{r}$ the observation point, while k_0 is the free space wavenumber. Solutions of (2.372) can be found by using expansions in plane waves, i.e. an eigenfunction expansion in Cartesian coordinates. In this case we write

$$g(\mathbf{r};\mathbf{r}') = \frac{1}{(2\pi)^3} \int_{-\infty}^{\infty} \int_{-\infty}^{\infty} \int_{-\infty}^{\infty} \tilde{g}(\mathbf{k}) e^{-j\mathbf{k}\cdot(\mathbf{r}-\mathbf{r}')} d\mathbf{k} \tag{2.373}$$

where $\mathbf{k}$ stands for the vector (k_x, k_y, k_z) and $\tilde{g}(\mathbf{k})$ is the Fourier transform of $g(\mathbf{r};\mathbf{r}')$. The integral appearing in (2.373) is a triple integral, extending from $-\infty$ to ∞, over the three wavenumbers (k_x, k_y, k_z). By taking advantage of the completeness of the plane waves, we express the delta function as a superposition of eigenfunctions in the following manner

$$\delta(\mathbf{r}-\mathbf{r}') = \frac{1}{(2\pi)^3} \int_{-\infty}^{\infty} \int_{-\infty}^{\infty} \int_{-\infty}^{\infty} e^{-j\mathbf{k}\cdot(\mathbf{r}-\mathbf{r}')} d\mathbf{k} \tag{2.374}$$

After introducing (2.374) and (2.373) into (2.372) and by recalling that nabla operation corresponds to multiplication times $-j\mathbf{k}$

$$\nabla \Rightarrow -j\mathbf{k} \tag{2.375}$$

we get the following expression for the Fourier transformed Green's function

$$\tilde{g}(\mathbf{k}) = \frac{1}{k^2 - k_0^2} \tag{2.376}$$

which, after substitution in (2.373), yields the free–space scalar Green's function in rectangular coordinates as

$$g(\mathbf{r};\mathbf{r}') = \frac{1}{(2\pi)^3} \int_{-\infty}^{\infty} \int_{-\infty}^{\infty} \int_{-\infty}^{\infty} \frac{e^{-jk_x(x-x')-jk_y(y-y')}}{k_x^2 + k_y^2 + k_z^2 - k_0^2} e^{-jk_z(z-z')} d\mathbf{k} \tag{2.377}$$

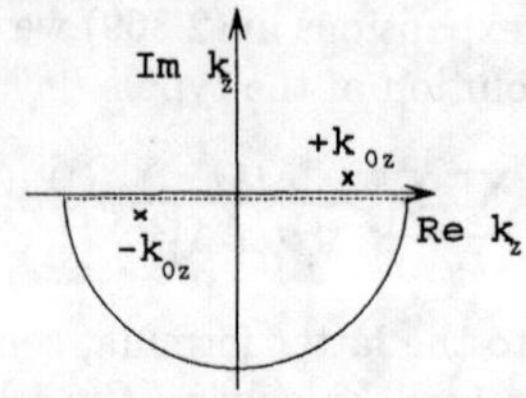

Figure 2.36 Contour of integration for $(z - z') > 0$.

As apparent from (2.377) a simplification is possible by contour evaluation of one of the integrals. In particular, with reference to the integral in k_z, and by assuming $(z - z') > 0$, the contour of integration can be enclosed in the lower half space as shown in Figure 2.36. Since (2.377) has poles for

$$k_z^2 = k_0^2 - k_x^2 - k_y^2 \tag{2.378}$$

we may apply the residue theorem, stating that

$$\oint_C f(\alpha) d\alpha = -j2\pi \sum_n f_n \tag{2.379}$$

where in (2.379) the integral is over the closed contour C; f_n are the residues of the function f obtained, in the case of simple poles at α_n, as

$$f_n = \lim_{\alpha \to \alpha_n} \{(\alpha - \alpha_n) f(\alpha)\} \tag{2.380}$$

Upon use of (2.379) we get the following expression for the free–space Green's function:

$$g(\mathbf{r};\mathbf{r}') = -\frac{j}{(2\pi)^2} \int_{-\infty}^{\infty} \int_{-\infty}^{\infty} \frac{e^{-jk_x(x-x')-jk_y(y-y')-jk_{0z}(z-z')}}{2k_z} dk_x dk_y \tag{2.381}$$

valid for $(z - z') > 0$ and where

$$k_{0z}^2 = k_0^2 - k_x^2 - k_y^2 \tag{2.382}$$

Equation (2.381) deserves some comment: it can be compared with the analogous formula resulting from modal expansion in closed waveguides; formulae (2.377) and (2.381) can be read by considering the various Green's function representations introduced in the previous Section; finally, it is also interesting to compare (2.381) with the Green's function in spherical and cylindrical coordinates as reported in the paragraph on the connections between different wave expansions.

2.9.5 Free–space vector Green's function

We have just described a way of obtaining the free–space scalar Green's function. It is now interesting to look at the problem of finding the Green's function for the vector wave equation, i.e. for the equation

$$\nabla \times \nabla \times \mathbf{G}(\mathbf{r},\mathbf{r}') - k^2 \mathbf{G}(\mathbf{r},\mathbf{r}') = \mathbf{I}\delta(\mathbf{r} - \mathbf{r}') \tag{2.383}$$

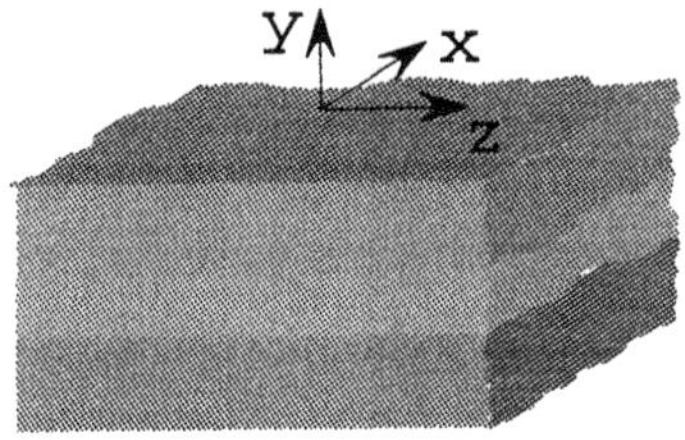

Figure 2.37 A multi–layer planar dielectric structure.

By taking the divergence of the above equation the first term on the left hand side disappears, since $\nabla \cdot \nabla\times = 0$, thus yielding

$$\nabla \cdot \mathbf{G}(\mathbf{r},\mathbf{r}') = -\frac{1}{k^2}\nabla \cdot \mathbf{I}\delta(\mathbf{r}-\mathbf{r}') = -\frac{1}{k^2}\nabla\delta(\mathbf{r}-\mathbf{r}') \tag{2.384}$$

On the other hand we may use in (2.383) the equivalence $\nabla \times \nabla\times = \nabla\nabla\cdot - \nabla^2$, leading to

$$-\nabla^2\mathbf{G}(\mathbf{r},\mathbf{r}') + \nabla\nabla\cdot\mathbf{G}(\mathbf{r},\mathbf{r}') - k^2\mathbf{G}(\mathbf{r},\mathbf{r}') = \mathbf{I}\delta(\mathbf{r}-\mathbf{r}') \tag{2.385}$$

By substituting (2.384) into the latter equation and rearranging terms, one obtains

$$\nabla^2\mathbf{G}(\mathbf{r},\mathbf{r}') + k^2\mathbf{G}(\mathbf{r},\mathbf{r}') = -\left(\mathbf{I} + \frac{\nabla\nabla}{k^2}\right)\delta(\mathbf{r}-\mathbf{r}') \tag{2.386}$$

Now, let us consider a solution $g(\mathbf{r},\mathbf{r}')$ of the *scalar* wave equation (2.372); by letting

$$\mathbf{G}(\mathbf{r},\mathbf{r}') = -\left(\mathbf{I} + \frac{\nabla\nabla}{k^2}\right)g(\mathbf{r},\mathbf{r}') \tag{2.387}$$

we get a solution of the *vector* wave equation (2.386). As an example, substituting (2.381) into (2.387)yields the free–space vector Green's function.

This short and elegant method for obtaining the vector Green's function was introduced by Levine and Schwinger [21] and it is particularly useful for free–space problems. Note that when boundaries are present the above equation is valid in each piecewise homogeneous region; where boundaries are located along planes of symmetry, further simplification is also possible.

2.10 Immitance approach

The immittance approach was introduced in [16] in order to simplify the quite tedious procedure of field matching at the interfaces in multilayer structures. It is easily described by following the approach suggested in [16, 26]. It essentially consists in an eigenfunction expansion plus a suitable choice of coordinate. As such, this is another interesting application of the procedure described in eigenfunction expansion in more dimensions, applied to the case of a multi–layer planar dielectric structure, as shown in Figure 2.37. We have already found in Section 2.5.2 that we can analyse the problem of a multilayer in the y–direction in terms of Hertzian LSE and LSM potentials with z–dependence of the type $\exp(-j\beta z)$. On that occasion, this choice was motivated by the fact that we were considering modes propagating along the dielectric interface in the z–direction. However, by using the previous considerations on the eigenfunction expansion, and by noting that $\exp(-j\beta z)$ is the eigenfunction in the z–direction, we can look at this as the first step in finding a solution, by eigenfunction expansion, of the Helmholtz equation for the structure of Figure 2.37.

Let us thus consider the LSE field, here repeated for convenience

$$\begin{aligned} E_x &= \omega\mu_0\beta\psi_h \\ E_y &= 0 \\ E_z &= -j\omega\mu_0\partial_x\psi_h \\ H_x &= \partial_x\partial_y\psi_h \\ H_y &= \partial_y^2\psi_h + \varepsilon_r k_0^2\psi_h \\ H_z &= -j\beta\partial_y\psi_h \end{aligned} \tag{2.388}$$

In a similar manner, we may also consider an eigenfunction expansion in the x–direction. In this case eigenfunctions are of the type $\exp(-j\alpha x)$. Upon inserting this dependence in (2.388) one obtains the following equations, where the tilde denotes Fourier transforms in x, z

$$\begin{aligned} \tilde{E}_x &= \omega\mu_0\beta\tilde{\psi}_h \\ \tilde{E}_y &= 0 \\ \tilde{E}_z &= -\omega\mu_0\alpha\tilde{\psi}_h \\ \tilde{H}_x &= -j\alpha\partial_y\tilde{\psi}_h \\ \tilde{H}_y &= (\alpha^2+\beta^2)\tilde{\psi}_h \\ \tilde{H}_z &= -j\beta\partial_y\tilde{\psi}_h \end{aligned} \tag{2.389}$$

Therefore, by taking advantage of the eigenfunction expansion in the x and z–directions we have achieved an algebraic form of the dependence on these coordinates by eliminating differentiation along x or z; as apparent from (2.389) the only differential is along the y–direction. The quantities in the spectral domain α, β, and in the space domain, are related by two–dimensional Fourier transform pairs as

$$\begin{aligned} g(x,z;y) &= \int_{-\infty}^{\infty}\int_{-\infty}^{\infty} \tilde{g}(\alpha,\beta;y)e^{-j(\alpha x+\beta z)}\,d\alpha\,d\beta \\ \tilde{g}(\alpha,\beta;y) &= \left(\frac{1}{2\pi}\right)^2 \int_{-\infty}^{\infty}\int_{-\infty}^{\infty} g(x,z;y)e^{j(\alpha x+\beta z)}\,dx\,dz \end{aligned} \tag{2.390}$$

It is now useful to look graphically at the $\mathbf{E}, \mathbf{H}_t$ vectors, $\mathbf{H}_t$ being the projection of $\mathbf{H}$ on the x, z plane; since E_x is proportional to β and E_z is proportional to α, we have that $\mathbf{E}$ is directed as in Figure 2.38 (left side). Analogously, for the magnetic field in the x, z plane, since H_x is proportional to $-\alpha$ and H_z is proportional to $-\beta$, $\mathbf{H}_t$ is directed as depicted in Figure 2.38 (right side). It is apparent that the components of $\mathbf{E}$ and $\mathbf{H}$ transverse to y are orthogonal. It is therefore convenient to introduce a coordinate system u, v so that $\mathbf{E}$ is directed along u and $\mathbf{H}_t$ is directed along v. The new coordinates u, v depend on α, β as

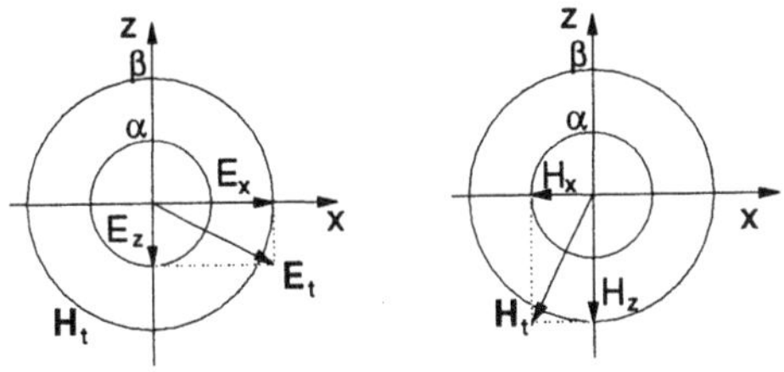

Figure 2.38 Representation of the components of $\mathbf{E}, \mathbf{H}$ *on the* x, z *plane for an LSE field.*

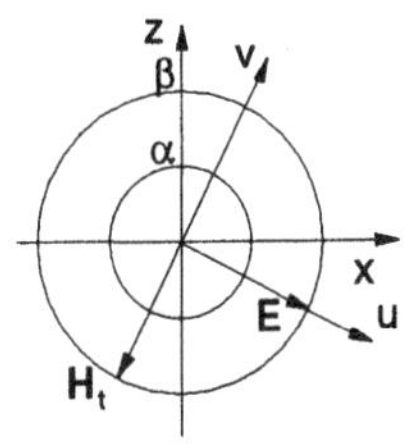

Figure 2.39 Spectral waves TE^y *in the (u,v) coordinate system.*

$$\begin{aligned} u &= x\cos\theta - z\sin\theta \\ v &= x\sin\theta + z\cos\theta \end{aligned} \tag{2.391}$$

with

$$\begin{aligned} \sin\theta &= \frac{\alpha}{\sqrt{\alpha^2+\beta^2}} \\ \cos\theta &= \frac{\beta}{\sqrt{\alpha^2+\beta^2}} \end{aligned} \tag{2.392}$$

Consequently, we have found a coordinate system which reduces our field, having five scalar components in a general system, to just three, namely E_u, H_v, H_y. This is therefore a field TE^y but also TE^v. Since, from (2.389),

$$\begin{aligned} E_u &= \omega\mu\sqrt{\alpha^2+\beta^2}\tilde{\psi}_h \\ H_v &= j\sqrt{\alpha^2+\beta^2}\partial_y\tilde{\psi}_h \end{aligned} \tag{2.393}$$

the impedance, in the positive y–direction, is given by

$$\frac{E_u}{H_v} = -j\omega\mu\frac{\tilde{\psi}_h}{\partial_y\tilde{\psi}_h} \tag{2.394}$$

In the LSM case, by proceeding in the same manner as for the LSE case, we obtain

$$\begin{aligned} \tilde{E}_x &= -j\alpha\partial_y\tilde{\psi}_e \\ \tilde{E}_y &= \partial_y^2\tilde{\psi}_e + \varepsilon_r k_0^2\tilde{\psi}_e \end{aligned}$$

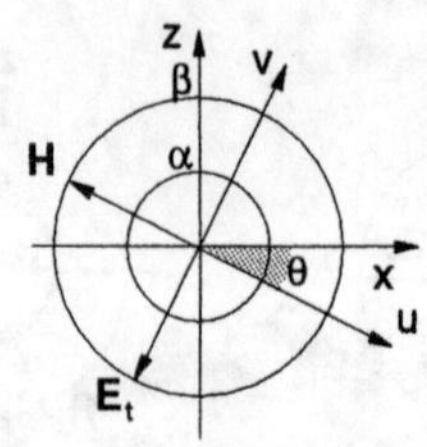

Figure 2.40 Spectral waves TM^y in the (u,v) coordinate system.

$$
\begin{aligned}
\tilde{E}_z &= -j\beta\partial_y\tilde{\psi}_e \\
\tilde{H}_x &= -\omega\varepsilon\beta\tilde{\psi}_e \\
\tilde{H}_y &= 0 \\
\tilde{H}_z &= \omega\varepsilon\alpha\tilde{\psi}_e
\end{aligned} \tag{2.395}
$$

Graphically $\mathbf{E}_t, \mathbf{H}$ lie now in the x, z plane as shown in Figure 2.40, i.e. $\mathbf{E}_t$ is now directed along v while $\mathbf{H}$ is now u–directed. In this case we have therefore a TM^y, TM^v field. Therefore, by using the coordinate transformation (2.391) we have achieved that for LSE modes the magnetic field in the x, z plane lies along v, while for LSM modes the magnetic field lies along u. This is particularly important when considering excitation of our structure by planar currents, i.e. currents in the x, z plane. In fact, by virtue of (2.391) and (2.392), when exciting with currents J_x, J_z we have

$$
\begin{pmatrix} J_u \\ J_v \end{pmatrix} = \begin{pmatrix} \cos\theta & -\sin\theta \\ \sin\theta & \cos\theta \end{pmatrix} \begin{pmatrix} J_x \\ J_z \end{pmatrix} \tag{2.396}
$$

On the u, v plane we have $\mathbf{J} = \mathbf{y}\times\mathbf{H}$; hence the current J_u excites only the magnetic field H_v (LSE field) and the current J_v only produces effects on H_u (LSM field). Equivalent circuits for the TM and TE fields are obtained from the transmission line analogy as shown in Figure 2.41. Then the electric field of LSE and LSM modes may be obtained from the magnetic field once the immittance is known. In fact for TE^y modes with reference to Figure 2.41 we have

$$
\tilde{E}_{u1} = \tilde{E}_{u2} = \tilde{E}_u \tag{2.397}
$$

$$
-\frac{\tilde{H}_{v1}}{\tilde{E}_u} = Y_1^h \qquad \frac{\tilde{H}_{v2}}{\tilde{E}_u} = Y_2^h \tag{2.398}
$$

$$
J_u = \tilde{H}_{v1} - \tilde{H}_{v2} \tag{2.399}
$$

from which follows

$$
\tilde{E}_u = -\frac{J_u}{Y_1^h + Y_2^h} \tag{2.400}
$$

while similarly for TM^y modes

$$
\tilde{E}_{v1} = \tilde{E}_{v2} = \tilde{E}_v \tag{2.401}
$$

$$
-\frac{\tilde{H}_{u1}}{\tilde{E}_v} = Y_1^e \qquad \frac{\tilde{H}_{u2}}{\tilde{E}_v} = Y_2^e \tag{2.402}
$$

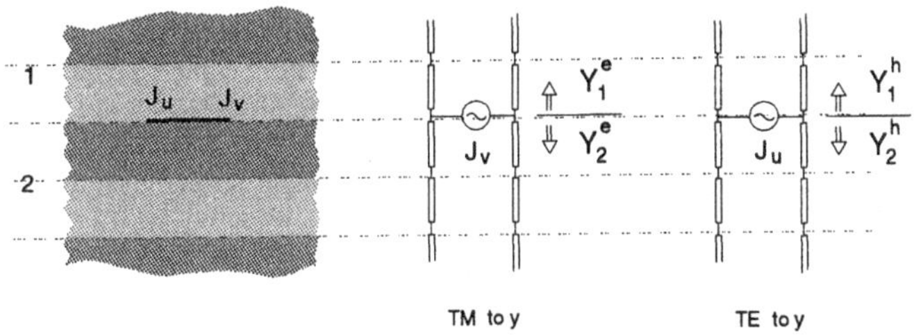

Figure 2.41 Equivalent transmission lines for the TM and TE fields.

$$J_v = \tilde{H}_{u1} - \tilde{H}_{u2} \tag{2.403}$$

$$\tilde{E}_v = -\frac{J_v}{Y_1^e + Y_2^e} \tag{2.404}$$

By introducing the following impedances

$$Z^h = \frac{1}{Y_1^h + Y_2^h} \qquad Z^e = \frac{1}{Y_1^e + Y_2^e} \tag{2.405}$$

we may write the relationship between currents and fields as

$$\begin{pmatrix} E_u \\ E_v \end{pmatrix} = -\begin{pmatrix} Z^h & 0 \\ 0 & Z^e \end{pmatrix} \begin{pmatrix} J_u \\ J_v \end{pmatrix} \tag{2.406}$$

By substituting (2.396) in the above equation and by considering that the fields in cartesian coordinates are obtained as

$$\begin{pmatrix} E_x \\ E_z \end{pmatrix} = \begin{pmatrix} \cos\theta & \sin\theta \\ -\sin\theta & \cos\theta \end{pmatrix} \begin{pmatrix} E_u \\ E_v \end{pmatrix} \tag{2.407}$$

we obtain

$$\begin{pmatrix} E_x \\ E_z \end{pmatrix} = -\frac{1}{\alpha^2 + \beta^2} \begin{pmatrix} Z^e\alpha^2 & (Z^e - Z^h)\alpha\beta \\ (Z^e - Z^h)\alpha\beta & Z^h\beta^2 \end{pmatrix} \begin{pmatrix} J_x \\ J_z \end{pmatrix} \tag{2.408}$$

This relationship provides the basis for the application of the spectral domain method of analysis of microstrips, slotlines, coplanar waveguides, etc.

Some comments on the above procedure are now appropriate. First of all it is noted that the first part of the procedure followed is an eigenfunction expansion which, since our domain is infinite along x and z, corresponds to a Fourier integral transform. However, the second part of the described procedure is the so called *immittance approach*, due to a clever idea of T. Itoh [16]. This latter part allows us to express an LSE or LSM field in terms of just three components instead of five. On the other hand this is quite apparent by considering that our structure is essentially a 2D structure. Accordingly, by properly arranging the axis orientation we should be able to describe the field by just three components.

From the point of view of the Green's function representation, we have expressed the x, z dependences by means of an eigenfunction expansion, while the problem in the y-direction has been solved by a circuit approach, equivalent to the direct method introduced on page 98. It is also appropriate to observe that the eigenfunction expansion in terms of plane waves is not the only physically meaningful one. In fact, in several applications, an eigenfunction expansion in cylindrical waves may be a more appropriate choice [28].

Bibliography

[1] M. Abramowitz and I. Stegun. *Handbook of mathematical functions.* Dover Publications, New York, 1965.

[2] J. Van Bladel. *Electromagnetic Fields.* McGraw-Hill, New York, 1964.

[3] C. J. Bouwkamp. A note on singularities occurring at sharp edges in electromagnetic diffraction theory. *Physica*, 12:467–474, 1946.

[4] J. J. Bowman, T. B. A. Senior, and P. L. E. Uslenghi. *Electromagnetic and Acoustic Scattering by Simple Shapes.* Hemisphere Publishing Corp., New York, 1987.

[5] W.C. Chew. A quick way to approximate Sommerfeld-Weyl-type integrals. *IEEE Trans. Antennas Propagat.*, 36(11):1654–1657, November 1988.

[6] R. E. Collin. *Field Theory of Guided Waves.* IEEE Press, New York, 1991.

[7] J. A. Dobrowolski. *Introduction to Computer Methods for Microwave Circuit Analysis and Design.* Artech House, London, 1991.

[8] L. P. Eisenhart. Separable systems of staeckel. *Ann. Math.*, 135:284, 1934.

[9] R. S. Elliott. *Electromagnetics.* IEEE Press, New York, 1993.

[10] L. Felsen and N. Marcuvitz. *Radiation and Scattering of Waves.* Prentice Hall, Englewood Cliffs, NJ, 1973.

[11] G. Franceschetti. *Campi Elettromagnetici.* Boringhieri, Torino, 1983.

[12] B. Friedman. *Principles and Techniques of Applied Mathematics.* John Wiley & Sons Inc., New York, 1956.

[13] R. F. Harrington. *Time Harmonic Electromagnetic Fields.* McGraw-Hill, New York, 1961.

[14] R. F. Harrington. Origin and development of the method of moments for field computations. In *Computational Electromagnetics.* IEEE Press, 1992.

[15] R. A. Hurd. The edge condition in electromagnetics. *IEEE Trans. Antennas Propagat.*, 24(1):70–73, January 1976.

[16] T. Itoh. Spectral domain immittance approach for dispersion characteristics of generalized printed transmission lines. *IEEE Trans. Microwave Theory Tech.*, 28:733–736, 1980.

[17] D. S. Jones. *Acoustic and Electromagnetic Waves.* Clarendon Press, Oxford, England, 1986.

[18] D. S. Jones. *Methods in Electromagnetic Wave Propagation.* Clarendon Press, Oxford, England, 1987.

[19] J. A. Kong. *Electromagnetic Wave Theory.* John Wiley & Sons, Singapore, 1986.

[20] J. D. Kraus. *Antennas.* McGraw-Hill, New York, 1988.

[21] H. Levine and J. Schwinger. On the theory of electromagnetic wave diffraction by an aperture in an infinite plane conducting screen. *Comm. Pure and Appl. Math.*, 3:355–391, 1950.

[22] N. Marcuvitz. *Waveguide Handbook.* McGraw-Hill, New York, 1951.

[23] A.W. Maue. Zur formulierung eines allgemeinen beugungsproblems durch eine integralleichung. *Z. Phys.*, 126:601–618, 1949.

[24] J. Meixner. Die kantenbedingung in der theorie der beugung electromagnetischer wellen an vollkommen leitenden ebenen schirmen. *Ann. Phys.*, 6:1–9, 1949.

[25] J. Meixner. The behaviour of electromagnetic fields at edges. *IEEE Trans. Antennas Propagat.*, 20(4):442–446, July 1972.

[26] W. Menzel. A new interpretation of the spectral domain immittance matrix approach. *IEEE Microwave and Guided Wave Letters*, pages 305–306, September 1993.

[27] P. M. Morse and H. Feshback. *Methods of Theoretical Physics.* McGraw-Hill, New York, 1950.

[28] J. R. Mosig. Integral equation technique. In T. Itoh, editor, *Numerical Techniques for Microwave and Millimeter-wave Passive Sructures.* John Wiley & Sons, New York, 1989.

[29] P. Penfield, R. Spence, and S. Duinker. *Tellegen's theorem and electrical networks.* MIT Press, Cambridge, Massachusetts, 1970.

[30] S. Ramo, J. R. Whinnery, and T. Van Duzer. *Fields and Waves in Communication Electronics.* John Wiley & Sons Inc., New York, 1965.

[31] V. H. Rumsey. Reaction concept in electromagnetic theory. *Phys. Rev., ser. 2*, 94(6):1483–1491, 1954.

[32] V. V. Shevchenko. *Continuous Transition in Open Waveguides.* Golem Press, Boulder, Colorado, 1971.

[33] J. A. Stratton. *Electromagnetic Theory.* McGraw-Hill, New York, NY, 1941.

[34] C. T. Tai. *Dyadic Green's Functions in Electromagnetic Theory.* Intext Educational Publishers, Scranton, PA, 1971.

[35] B. D. H. Tellegen. A general network theorem with applications. *Philips Research Reports*, 7:259–269, 1952.

[36] B. D. H. Tellegen. A general network theorem with applications. *Proc. Inst. Radio Engineers*, 14:265–270, 1953.

Chapter 3

Planar dielectric waveguides

This Chapter deals with planar dielectric waveguides, also called slab waveguides, which are possibly the most simple and yet practically useful open waveguides [6, 7, 8, 10]. In this introduction we briefly recall how a bound mode can occur; we also discuss why it is natural to look for the complete spectrum of slab waveguides and what features it should possess. These concepts are then considered in a more rigorous form later in the Chapter in Sections 3.2 and 3.3. In addition, in Section 3.4, the bound modes of slab waveguides are also introduced from a transverse resonance approach; whereas the continuous spectrum is derived from the relative transverse network.

A more complete understanding of the use of the modal spectra is obtained in Section 3.5 by studying the case of a field excited by a delta source. Finally, in Section 3.6, we consider the asymmetric three–layer waveguide which is a basic structure for practical integrated optics circuits.

In order to introduce dielectric waveguides let us consider first the interface between two different media of refractive indices n_1, n_2 from a simple ray optics point of view. We know from Snell's law that a ray impinging from region 1 on the interface between the two media (see Figure 3.1), is transmitted in region 2 with an angle θ_2 such that

$$n_1 \sin\theta_1 = n_2 \sin\theta_2 \tag{3.1}$$

By assuming the medium 1 denser than the medium 2 ($n_1 > n_2$), then $\theta_2 > \theta_1$ and there will be an angle of incidence $\theta_1 = \theta_c$ for a ray coming from 1 such that

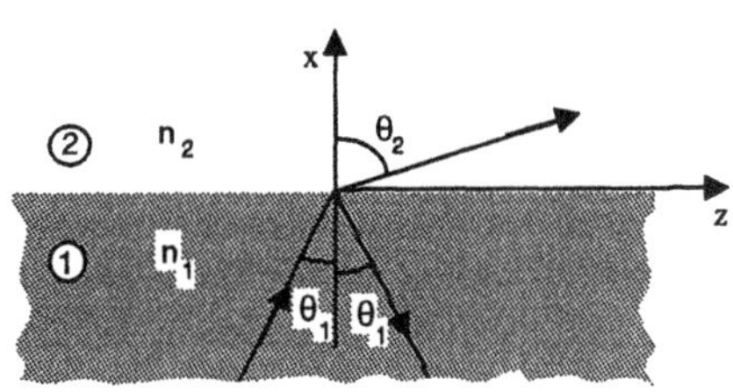

Figure 3.1 Interface between two different media with refractive indices ($n_1 > n_2$).

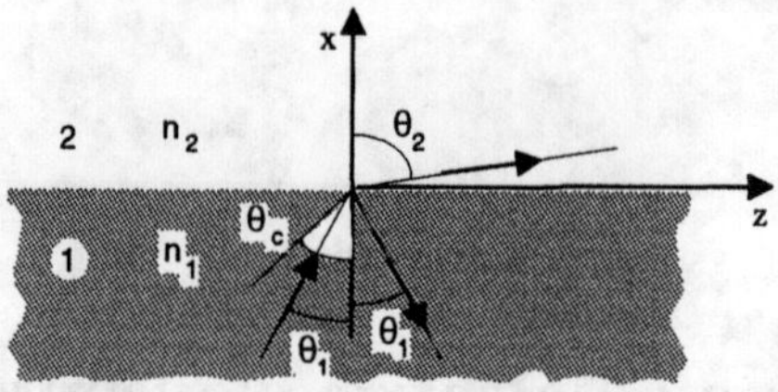

Figure 3.2 A ray impinging from region 1 at an angle $\theta_1 < \theta_c$ will emerge in region 2 after undergoing refraction.

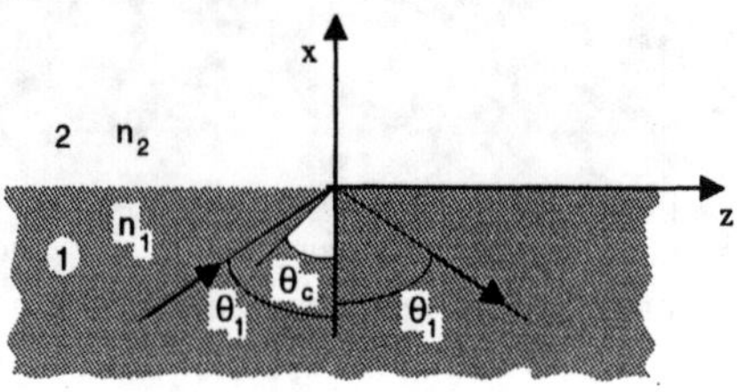

Figure 3.3 A ray impinging at any angle $\theta_1 > \theta_c$ will undergo total reflection and stay in the denser medium.

$\theta_2 = \pi/2$, i.e.

$$\frac{n_1}{n_2}\sin\theta_c = 1$$

$$\theta_c = \arcsin\frac{n_2}{n_1} \qquad (3.2)$$

This angle θ_c is called the critical angle. A ray impinging from region 1 at an angle θ_1 less than the critical angle θ_c will emerge in region 2 after undergoing refraction as shown in Figure 3.2. A ray impinging at any angle $\theta_1 > \theta_c$ will undergo total reflection and stay in the denser medium as illustrated in Figure 3.3.

Suppose now we have a slab of thickness $2d$ as in Figure 3.4. A ray undergoing total reflection at the upper interface will hit the lower interface again at the same angle and in principle propagate down the slab. This seems identical to the situation of the parallel plate waveguide. In order for the transverse distribution of the mode to be the same at A (i.e. $e^{-jk_x x}$) and at A' ($e^{-jk_x(x+4d)}$) we require that the phase of a round trip in the transverse direction be a multiple of 2π, i.e.

$$4k_x d = 4k_0 d\cos\theta = 2n\pi$$

implying that only discrete values θ_n of the angle are allowed, e.g.

$$\theta_n = \arccos\frac{n\pi}{2k_0 d} \qquad (3.3)$$

If $n_1 \gg n_2$ the e.m. field is totally confined to the denser medium. If n_1 is only slightly larger than n_2, as in many practical cases, internal reflection is not total;

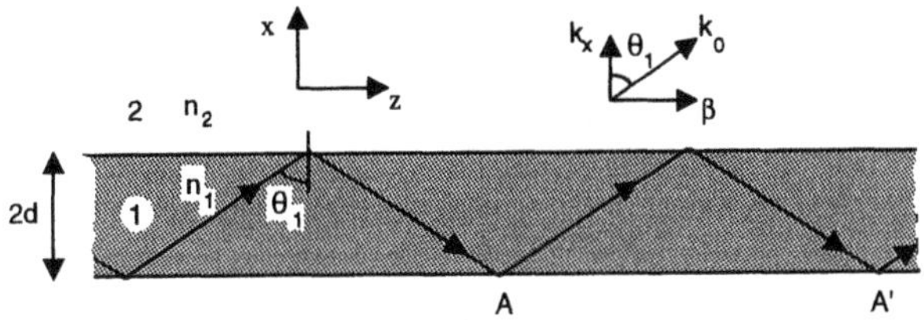

Figure 3.4 Ray propagation along a dielectric slab.

some of the field escapes from the inner region, the core, into the outer region, say air for short even if the outside medium is not air. Still the field remains fairly well localised to the neighbourhood of the interface and waveguiding is possible.

Such a mode of propagation is called a "surface wave", indicating a wave guided by the dielectric interface. The guiding of a plane wave by a plane interface between two different dielectric media was investigated initially by Uller in 1903 and later by Zenneck in 1907. This surface wave is associated with the Brewster angle phenomenon of total transmission across a dielectric interface and it is often referred to as a Zenneck wave. A detailed description of the bound modes of slab waveguides is given in Section 3.2 of this Chapter.

Once we have found the bound modes, it is natural to generalise a number of ideas used in the theory of closed waveguides so that they could be applied also to open waveguides. Of fundamental importance for closed waveguides are modal expansions, which allow us to represent any field in any cross–section as a sum of modes. It is obvious that, for open waveguides, it is necessary to replace the discrete summation by an integral, for in open waveguides the number of eigenmodes (bound modes) is finite. In fact, not more than one mode can propagate along a finitely conducting plane; the number of modes propagating along a dielectric cylinder is always finite, etc. In conclusion, in open waveguides we do not have that innumerable set of eigenmodes present in closed waveguides. In fact, in open waveguides, there are only a few modes whose fields satisfy the radiation condition far from the waveguide (at infinity in the radial direction). Since the total field must satisfy this condition it would seem reasonable at first to require that the modes in which the total field is to be expanded also satisfy this condition. This requirement, however, is excessive [9]: the radiation condition at infinity in the radial direction is not necessary for the individual modes in terms of which the total field is expanded. For the condition at infinity it is only necessary to require that each modal field is bounded, this condition being sufficient for constructing a complete system of orthogonal eigenmodes.

In order to state the previous ideas in a more rigorous way we now introduce the equations for the two–dimensional guided fields and we derive the bound modes and complete spectrum of open waveguides.

3.1 Two–dimensional guided fields

In this Chapter we deal with structures uniform in one direction, hence making it feasible to separate Maxwell's equations into two independent sets of three–

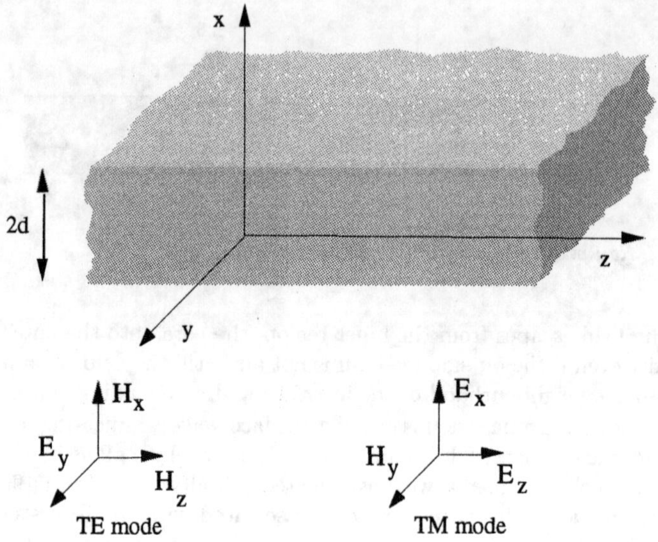

Figure 3.5 Slab waveguide, infinite in the y–, z–directions. Propagation is assumed in the z–direction, whereas no variation occurs along the y–direction.

component TE and TM fields. As customary, the field is considered to have a time–harmonic dependence of the type $e^{j\omega t}$, suppressed in the following, which allows any time–varying fields to be obtained by Fourier analysis. Maxwell's curl equations for time–harmonic signals in a source–free region are

$$\begin{aligned} \nabla \times \mathbf{E} &= -j\omega\mu\mathbf{H} \\ \nabla \times \mathbf{H} &= j\omega\varepsilon\mathbf{E} \end{aligned} \tag{3.4}$$

where μ is the magnetic permeability (H/m) and ε is the dielectric permittivity (F/m). For a wave travelling in the z–direction we assume the fields to be of the form

$$\begin{aligned} \mathbf{E}(x,y,z,t) &= \mathbf{E}(x,y)\, e^{j\omega t-\gamma_z z} \\ \mathbf{H}(x,y,z,t) &= \mathbf{H}(x,y)\, e^{j\omega t-\gamma_z z} \end{aligned} \tag{3.5}$$

where $\gamma_z = \alpha_z + j\beta_z$ is the propagation constant of the guide in the z–direction. Equation (3.4) may be expanded in component form in rectangular coordinates as

$$\begin{aligned} \frac{\partial E_z}{\partial y} + \gamma_z E_y &= -j\omega\mu H_x & (a) \\ -\frac{\partial E_z}{\partial x} - \gamma_z E_x &= -j\omega\mu H_y & (b) \\ \frac{\partial E_y}{\partial x} - \frac{\partial E_x}{\partial y} &= -j\omega\mu H_z & (c) \\ \frac{\partial H_z}{\partial y} + \gamma_z H_y &= j\omega\varepsilon E_x & (d) \end{aligned}$$

$$\begin{aligned} -\frac{\partial H_z}{\partial x} - \gamma_z H_x &= j\omega\varepsilon E_y \qquad (e) \\ \frac{\partial H_y}{\partial x} - \frac{\partial H_x}{\partial y} &= j\omega\varepsilon E_z \qquad (f) \end{aligned} \tag{3.6}$$

In (3.6) by letting $E_z = 0$ we have a Transverse Electric (TE) mode; similarly, by setting $H_z = 0$, we obtain a Transverse Magnetic (TM) mode. By considering a dielectric slab waveguide uniform in the y–direction as in Figure 3.5, we further simplify (3.6) by using

$$\frac{\partial}{\partial y} = 0 \tag{3.7}$$

Hence, from (3.6a, 3.6c, 3.6e) we obtain the
TE modes ($E_z = 0$)

$$\begin{aligned} \gamma_z E_y &= -j\omega\mu H_x \qquad (a) \\ \frac{dE_y}{dx} &= -j\omega\mu H_z \qquad (b) \\ \frac{dH_z}{dx} &= \left(\frac{\gamma_z^2}{j\omega\mu} - j\omega\varepsilon\right) E_y \qquad (c) \end{aligned} \tag{3.8}$$

while the remaining equations (3.6b, 3.6d, 3.6f) provide the
TM modes ($H_z = 0$)

$$\begin{aligned} \gamma_z H_y &= j\omega\varepsilon E_x \qquad (a) \\ \frac{dH_y}{dx} &= j\omega\varepsilon E_z \qquad (b) \\ \frac{dE_z}{dx} &= \left(-\frac{\gamma_z^2}{j\omega\varepsilon} + j\omega\mu\right) H_y \qquad (c) \end{aligned} \tag{3.9}$$

The field in (3.8) is also a TE wave with respect to the x–direction; similarly, the field in (3.9) is a TM wave with respect to x. As a consequence, (3.8b, 3.8c) and (3.9b, 3.9c) may be recognised as the equations for a transmission line in the x–direction for TE and TM modes, respectively.

3.1.1 Transmission lines for two–dimensional fields

By introducing the propagation constant in the x–direction, defined as

$$k_x^2 = \omega^2\mu\varepsilon + \gamma_z^2 \tag{3.10}$$

and relative impedances,

$$Z(x) = \frac{1}{Y(x)} = \begin{cases} \omega\mu/k_x & \text{for TE} \\ \\ k_x/\omega\varepsilon & \text{for TM} \end{cases} \tag{3.11}$$

for either TM or TE modes we may write

$$
\begin{aligned}
-\frac{dV(x)}{dx} &= jk_x\, Z(x)\, I(x) \\
-\frac{dI(x)}{dx} &= jk_x\, Y(x)\, V(x)
\end{aligned}
\tag{3.12}
$$

In (3.12) $V(x)$ and $I(x)$ are the equivalent voltage and current corresponding respectively to electric and magnetic fields. By substituting one equation into the other in (3.12), we obtain the Sturm–Liouville differential form, hence

$$
\begin{aligned}
\left\{\frac{d}{dx}\frac{1}{k_x\, Z(x)}\frac{d}{dx} + k_x Y(x)\right\} V(x) &= 0 \\
\left\{\frac{d}{dx}\frac{1}{k_x\, Y(x)}\frac{d}{dx} + k_x Z(x)\right\} I(x) &= 0
\end{aligned}
\tag{3.13}
$$

The solutions to the above equations in the homogeneous case, when both ε and μ are piecewise constants, may be written in the following form

$$
\begin{aligned}
V(x) &= V(x')\,[\cos k_x(x-x') - jY(x')\sin k_x(x-x')] \\
I(x) &= I(x')\,[\cos k_x(x-x') - jZ(x')\sin k_x(x-x')]
\end{aligned}
\tag{3.14}
$$

where we have introduced the x–dependent impedance

$$
Z'(x) = \frac{1}{Y'(x)} = \frac{V(x)}{Z\, I(x)}
\tag{3.15}
$$

i.e. the impedance normalised with respect to Z, (the characteristic impedance given by (3.11)). In (3.14) $V(x')$ and $I(x')$ are voltage and current values at the point x' on the transmission line.

3.2 Bound modes of slab waveguides

3.2.1 Even TE mode

Let us consider an even TE wave in the slab, with components as shown in Figure 3.6. Due to symmetry, the structure is the same as in Figure 3.5, with the lower half replaced by a magnetic wall.

The tangential component of $\mathbf{E}$, E_y, is continuous across the interface $x = d$ and, from (3.8a), so is also the normal component of $\mathbf{H}$, H_x. For symmetric modes the distribution of E_y is given, apart for an amplitude factor, by

$$
E_y(x) = \begin{cases} \cos q_s x & : d \geq x \geq 0 \\ \cos q_s d\, e^{-\gamma_x (x-d)} & : x > d \end{cases}
\tag{3.16}
$$

where we have denoted with q_s the wavenumber of the surface wave. Equation (3.16) represents a field which is symmetric with respect to $x = 0$, continuous at $x = d$ and exponentially decaying outside the core. If the field is totally confined then $q_s = \pi/(2d)$; otherwise the wavenumbers q_s and γ_x are as yet undetermined. We shall determine them presently by imposing the continuity of H_z. Since H_z is

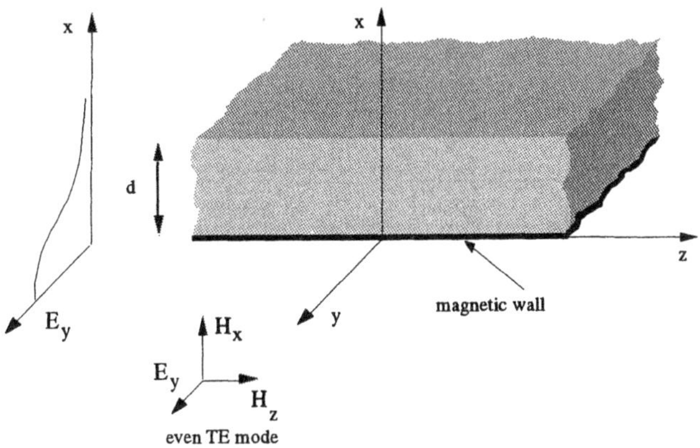

Figure 3.6 Even TE modes of a dielectric slab. By using symmetry, the field in the half-space $x > 0$ is obtained by considering just one half of the slab, placed above a magnetic wall.

proportional to dE_y/dx (3.8b), not only has E_y to be continuous at $x = d$ but also its derivative has to be continuous. By assuming the same magnetic permeability in the slab as in the air we find from the continuity of dE_y/dx,

$$-q_s \sin q_s d = -\gamma_x \cos q_s d$$

or, equivalently,

$$q_s \tan q_s d = \gamma_x \tag{3.17}$$

This constraint on the transverse wavenumbers, arising from continuity, is called the *dispersion relation* or eigenvalue equation of the waveguide. However, wavenumbers also satisfy another constraint, namely:

$$\begin{aligned} -\gamma_x^2 + \beta^2 &= n_2^2 k_0^2 \qquad \text{in air} \\ q_s^2 + \beta^2 &= n_1^2 k_0^2 \qquad \text{in waveguide} \end{aligned} \tag{3.18}$$

By subtracting one equation from the other, one obtains

$$\gamma_x^2 + q_s^2 = (n_1^2 - n_2^2)\, k_0^2$$

which, together with the dispersion equation, forms a system of two transcendental equations for the two unknowns, q_s and γ_x. It is convenient to normalise all quantities to the half-thickness of the slab, d, so as to make them dimensionless. We define

$$\begin{aligned} u &= q_s d \\ v &= \sqrt{(n_1^2 - n_2^2)}\, k_0 d \\ w &= \gamma_x d \end{aligned} \tag{3.19}$$

where v is called the normalised or effective frequency. Accordingly,

$$\begin{aligned} u \tan u &= w \\ u^2 + w^2 &= v^2 \end{aligned} \tag{3.20}$$

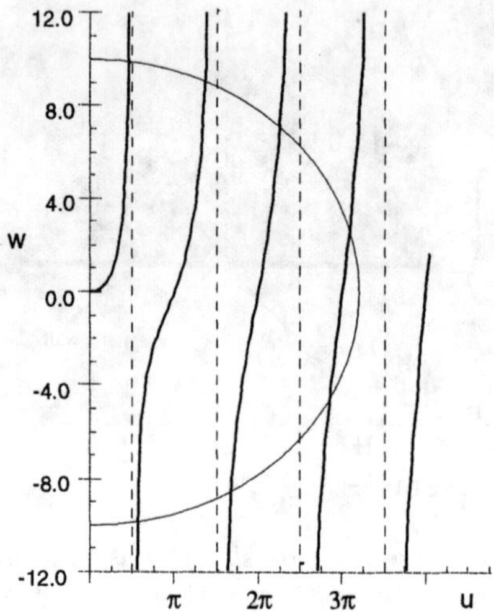

Figure 3.7 Numerical solution of the normalised dispersion equation. Intersections between circle and curves $u \tan u$ provide the solutions (surface waves).

Equations (3.20) may be solved either numerically or graphically; the latter approach is rather illustrative. On the $u-w$ plane, see Figure 3.7, the first equation represents a modified tan function, which is zero at $u = 0, \pi, 2\pi, \ldots$, and is infinity at $\pi/2, \frac{3\pi}{2}, \ldots$, etc. The second equation represents a circle of radius v. A crossing of the two curves in the upper half of the plane is a solution, i.e. a surface wave mode. The system (3.20) is even with respect to u, so that just $u \geq 0$ need be considered. Observe moreover that for $v < \pi$ only one solution exists for positive values of w, i.e. for waves decaying away from the slab. This is the TE_1 mode of the slab; as $v \geq \pi$, we have two modes, TE_1 and TE_2. Hence, we say that $v = \pi$ is the cut–off frequency of the higher order modes. We are not considering solutions for $w < 0$ since their behaviour for large x is $e^{(-wx)}$, implying that the wave grows exponentially away from the interface, thus a field not physically acceptable.

However, equations (3.20) also admit solutions for complex values of u, w. Even if these solutions are not acceptable, since they do not satisfy boundary conditions at infinity, nonetheless they have a physical interpretation. These non–modal solutions have complex propagation constant, as opposed to modal solutions of a lossless system which are either purely imaginary, for surface modes, or purely real, for radiation modes. The intensity pattern of these non–modal solutions is illustrated in Figure 3.8, where closer lines correspond to greater intensity of the Poynting vector. Due to the complex nature of the transverse propagation constant, the direction of power flow is not along the guide but at an angle from it. As apparent from the figure, the field grows exponentially as we move away from the slab in the x–direction. Moreover, since the field is leaking out of the guide, it rapidly diminishes in magnitude as it travels along the guide. Hence the term leaky wave is used to denote this type of wave. Leaky waves imply real power flowing away from the guide, which feature makes them useful for a qualitative description of

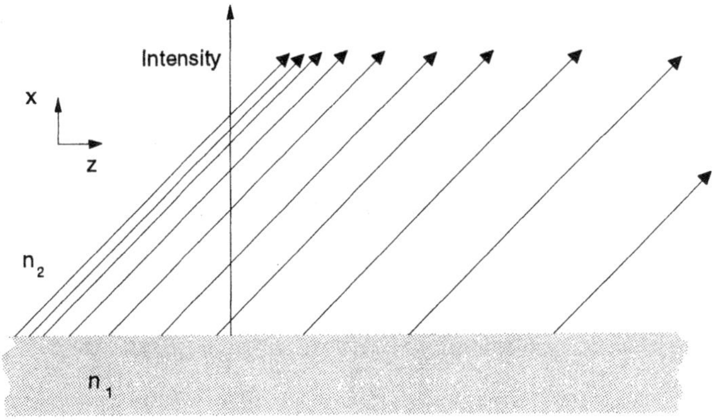

Figure 3.8 Intensity patterns of the leaky modes. Closer lines correspond to higher intensity. Field intensity increases with x and decreases along the propagation direction z.

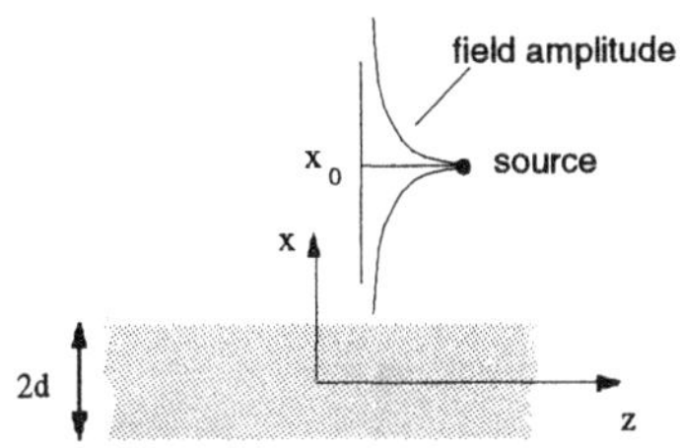

Figure 3.9 The field produced by a source external to the dielectric slab cannot, in general, be represented just in terms of surface wave modes, since they are exponentially decaying away from the dielectric air interface.

the radiation fields caused by sources or some discontinuities, their contribution being sometimes met in a steepest descent approximation. For the above reason, leaky waves are an interesting and stimulating subject which has been the object of considerable attention in the literature.

If a source is located in the air region above the guide, see e.g. Figure 3.9, this type of solution describes the field in the finite range between the slab waveguide and the source, since in this case the field amplitude decays away from the source towards the guide. However, leaky waves are *not* waveguide modes, since the latter must be solutions of the wave equation, satisfying boundary and radiation conditions. Hence, when searching for the spectrum of the slab waveguide we must reject the solutions $w < 0$ and the complex ones, which could not be sustained in a source–free region.

The actual shape of the field depends on the combination of refractive index, frequency and thickness of the slab considered. As thickness increases or, equivalently, as frequency increases, the field becomes more and more confined in the slab. These effects are shown in Figure 3.10, while Figure 3.11 displays the field contour lines of the fundamental even TE mode.

Table 3.1 u, w values for an He–Ne laser operating at λ_0 = 0.63 μm, a slab of thickness d and refractive index n_1 =1.5, n_2 = 1.

d (in μm)	u	w
0.25	1.15	2.54
0.5	3.93	3.95
0.75	6.89	4.74
1.	7.16	8.55
1.25	10.18	9.52
1.5	10.33	13.15
2	13.49	17.76

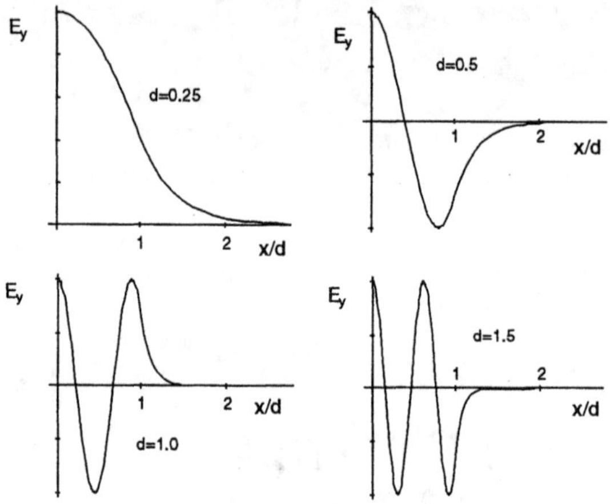

Figure 3.10 An He–Ne laser operating at λ_0 = 0.63 μm produces the above field distribution in a slab of thickness d and refractive index n_1 = 1.5, n_2 = 1. The slab thickness reported in the inset is in μm.

3.2.2 Odd TE mode

Let us consider an odd TE wave in the slab, with components as shown in Figure 3.12. As for even modes, by using symmetry, the lower half of the structure of Figure 3.5 is replaced by an electric wall as shown in Figure 3.12.

Also in this case, **E** tangential, E_y, is continuous across the interface $x = d$ and, from (3.8a), so is also **H** normal, H_x. For anti–symmetric modes the distribution of the tangential components is

$$E_y(x) = \begin{cases} \sin q_s x & : d \le x \le 0 \\ \sin q_s d\, e^{-\gamma_x(x-d)} & : x > d \end{cases} \tag{3.21}$$

Equation 3.21 represents a field which is anti–symmetric with respect to $x = 0$, continuous at $x = d$ and decays exponentially in air, therefore satisfying the wave equation. If the field is totally confined then $q_s = \pi/2d$; otherwise the wavenumbers

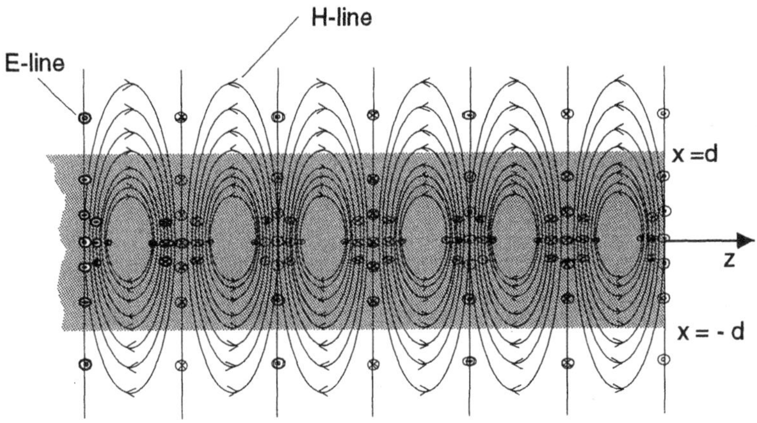

Figure 3.11 Dominant TE mode of dielectric slab.

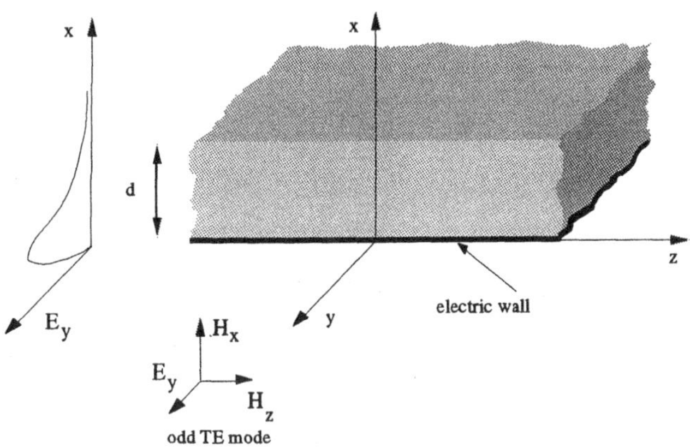

Figure 3.12 Odd TE modes of a dielectric slab. By using symmetry, the field is obtained by considering just one half of the slab, placed above an electric wall.

q_s and γ_x are as yet undetermined and will be determined by imposing the continuity of H_z. Since the latter is proportional to E_y, (3.8b), not only has E_y to be continuous at $x = d$ but also its derivative has to be continuous. By assuming the same permeability in the slab and in air we find from the continuity of dE_y/dx ,

$$q_s \cos q_s d = -\gamma_x \sin q_s d$$

or, equivalently

$$q_s \cot q_s d = \gamma_x \tag{3.22}$$

By following the same procedure and normalisation adopted for the even case we get

$$\begin{aligned} -u \cot u &= w \\ u^2 + w^2 &= v^2 \end{aligned} \tag{3.23}$$

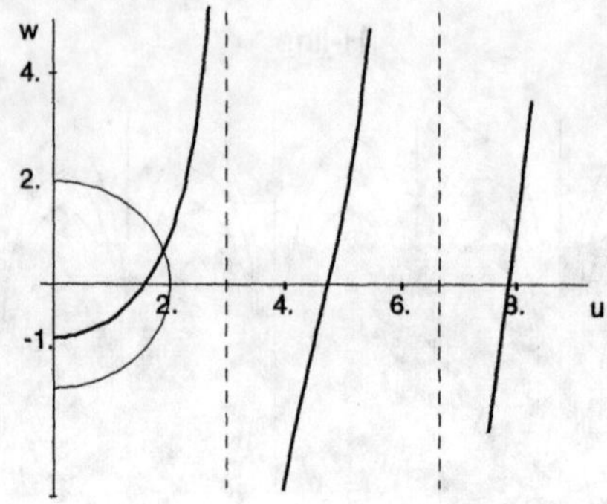

Figure 3.13 Numerical solution of the transcendental equations. The intersections represent solutions (surface waves), on the $u - w$ plane.

Also in this case (3.23) may be solved graphically as shown in Figure 3.13. The only acceptable solutions are those for $w > 0$, but a substantial difference is now present with respect to the previous even case; in fact, no solution occurs for $v < \pi/2$:

$0 < v < \pi/2$	no solutions	no propagating mode
$\pi/2 < v < 3\pi/2$	1 solution	1 propagating mode
$3\pi/2 < v < 5\pi/2$	2 solutions	2 propagating modes, etc.

It is interesting to consider what is happening to a dipole source radiating inside the guide when $0 < v < \pi/2$. For this combination of n_1, n_2, k_0 and d there is no guided mode. If we are considering a metallic guide below cut–off, the power radiated by the dipole is completely reflected toward the source. By contrast, if we measure the reflection coefficient in a dielectric guide we notice that the source is fairly well matched even if we are below cut–off. This means that the power radiated from the dipole at frequencies such that $v < \pi/2$ is radiated away from the slab guide in a manner that is clearly not describable in terms of surface waves only.

3.2.3 Even TM mode

Let us consider an even TM wave in the slab, with components as shown in Figure 3.14. The derivation of the TM modes is essentially the same followed for the TE field. Consequently, only the most important differences are outlined. In the TM case the field components are E_x,, E_z, H_y, and all components can be derived from H_y (3.9). Therefore, the even case refers to an even H_y field and this leads to the replacement of half the structure by an electric wall as shown in Figure 3.14.

Naturally, H_y must be continuous across the interface ($x = d$) and also E_z must be so. For symmetric modes the distribution of the tangential components is

$$H_y(x) = \begin{cases} \cos q_s x & : d \le x \le 0 \\ \cos q_s d \, e^{-\gamma_x (x-d)} & : x > d \end{cases} \tag{3.24}$$

The wavenumbers q_s and γ_x are determined by imposing the continuity of E_z. Since E_z is proportional to the derivative of H_y (3.9b) we see that not only has H_y to be continuous at $x = d$ but also its derivative has to be continuous. By assuming the

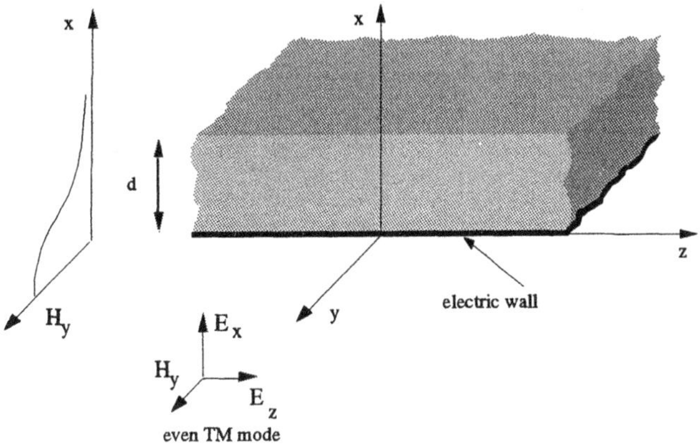

Figure 3.14 Even TM modes of a dielectric slab. By using symmetry, the field in the half space $x > 0$ is obtained by considering just one half of the slab, placed above an electric wall.

same permeability in the slab and in air we find from the continuity of dH_y/dx ,

$$-q_s \frac{\sin q_s d}{\varepsilon_1} = -\gamma_x \frac{\cos q_s d}{\varepsilon_2}$$

or, equivalently,

$$q_s \tan q_s d = \gamma_x \frac{\varepsilon_1}{\varepsilon_2} \tag{3.25}$$

where $\varepsilon_1, \varepsilon_2$ are the dielectric constants inside and outside the slab, respectively. By proceeding in the same way as for the TE case, and by adopting the same normalisation of all quantities to d, the half–thickness of the slab, we obtain:

$$\begin{aligned} u \tan u &= \tfrac{\varepsilon_1}{\varepsilon_2} w \\ u^2 + w^2 &= v^2 \end{aligned} \tag{3.26}$$

Note that in this case, similarly to the even TE case, *no cut–off* frequency is present. Moreover, the dielectric slab over a ground plane is a quite common configuration in practical microwave and millimetre wave transmission lines such as microstrips, etc. This means that when discontinuities occur in a microstrip circuit, there is a natural tendency to excite this type of surface wave. Note also that the power of the surface wave decays rather slowly, only with the square root of the distance from the source; hence, coupling phenomena can take place affecting the performance of the circuit. This subject will be treated with more detail in following Chapters.

3.2.4 Odd TM mode

Let us consider an odd TM wave in the slab, with the components as shown in Figure 3.15. As before, use is made of symmetry.

In this case, for anti–symmetric modes the distribution of the tangential components is

$$H_y(x) = \begin{cases} \sin q_s x & : d \le x \le 0 \\ \sin q_s d \, e^{-\gamma_x (x-d)} & : x > d \end{cases} \tag{3.27}$$

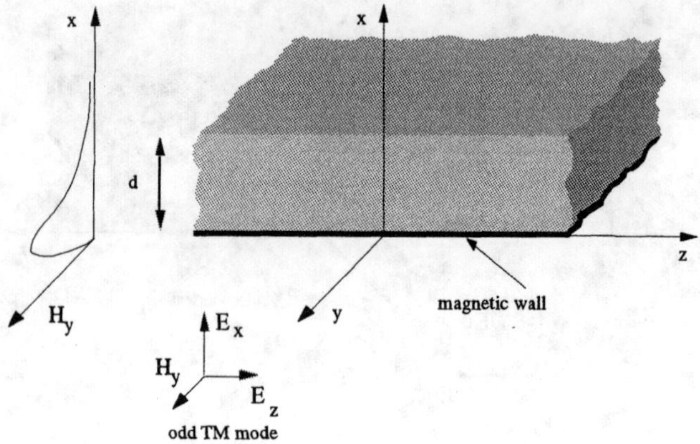

Figure 3.15 Odd TM modes of a dielectric slab. By using symmetry, the field in the half space $x > 0$ is obtained by considering just one half of the slab, placed above a magnetic wall.

The wavenumbers q_s and γ_x are determined by imposing the continuity of H_y, E_z. Since the latter is proportional to the derivative of H_y, (3.8b), we have that H_y together with its derivative has to be continuous at $x = d$. By assuming the same permeability in the slab and in the air we find from the continuity relationship

$$q_s \cot q_s d = -\gamma_x \frac{\varepsilon_1}{\varepsilon_2} \tag{3.28}$$

By following the same procedure and normalisation adopted for the previous cases we get

$$\begin{aligned} -u \cot u &= \tfrac{\varepsilon_1}{\varepsilon_2} w \\ u^2 + w^2 &= v^2 \end{aligned} \tag{3.29}$$

Also in this case the only acceptable solutions are those for $w > 0$ and no solution is present for $v < \pi/2$. The various cases possible and the relative equations are summarised in Table 3.2.

3.2.5 Experimental observations of surface waves

It is relatively simple to verify experimentally some of the main features of a surface mode of an open dielectric slab guide [1]. The two parameters of a surface mode which can be measured are the propagating constant, β, and the decay constant transverse to the slab, α. The propagation constant is given by $\beta = 2\pi/\lambda_g$ where λ_g is the guide wavelength of the slab guide. The experimental setup for measuring the guide wavelength of a dielectric slab in the 10 GHz region is shown schematically in Fig 3.16. The dielectric slab may be, e.g., a 1 cm thick perspex sheet, 60 cm wide and 100 cm long. The mode converters/launchers shown in the schematic are specially built horns with an inset perspex lens designed to give a parallel wavefront at 10 GHz. The lens, besides giving a parallel wavefront (which is essential for our assumption of an infinite slab), also reduces the reflection from the side of the finite width slab guide. The converters/launchers are further tuned by an adjustable

Table 3.2 Summary of TE, TM fields in a slab waveguide.

	Even	Odd
TE	Even E_y no cut–off $u\tan u = w$ $u^2 + w^2 = v^2$	Odd E_y cut–off present $-u\cot u = w$ $u^2 + w^2 = v^2$
TM	Even H_y no cut–off $u\tan u = \frac{\epsilon_1}{\epsilon_2} w$ $u^2 + w^2 = v^2$	Odd H_y cut–off present $-u\cot u = \frac{\epsilon_1}{\epsilon_2} w$ $u^2 + w^2 = v^2$

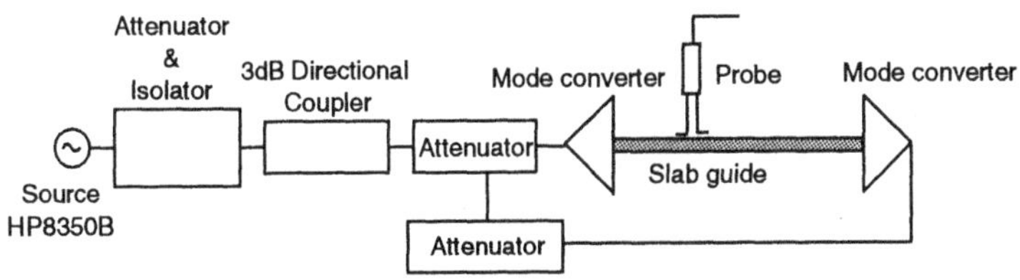

Figure 3.16 Experimental setup for the measurement of guide wavelength.

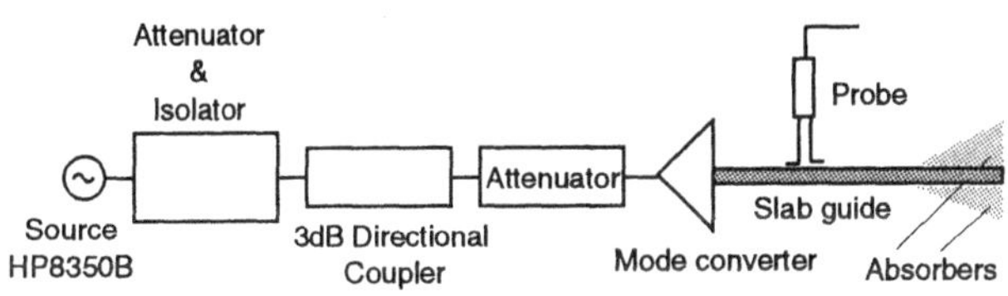

Figure 3.17 Experimental setup for the measurement of field attenuation away from a slab guide.

inductive post at the input flange, in order to give an input VSWR of less than 1.2 across the whole x–band. Adjustable attenuators are present at the input of each converter/launcher for fine tuning in order to achieve a large VSWR on the slab guide. The standing wave pattern in the slab is detected by a small dipole probe on a semi–rigid cable connected to a stub tuner and diode detector. The surface mode launched into the slab is predominantly of the fundamental even TE mode and the dipole probe is aligned so as to pick up the E_y field component of this mode.

The surface wave field attenuation constant α is measured by replacing one of the launchers by a matched load, e.g. two pieces of tapered absorbing material stuck onto both sides of the dielectric slab as shown in Figure 3.17.

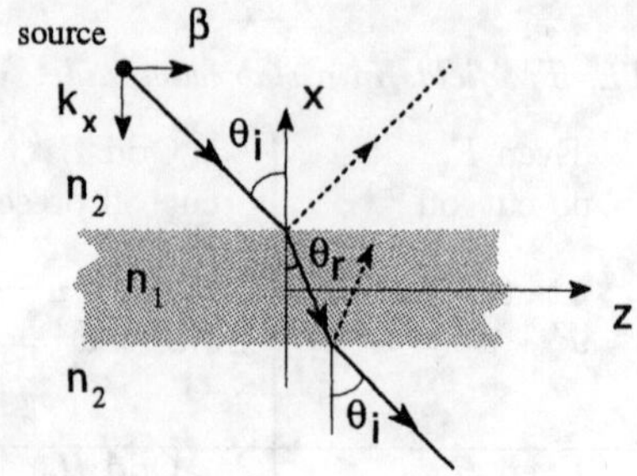

Figure 3.18 A source illuminates a slab producing a reflected and a refracted wave. This type of situation, impossible to describe just in terms of bound modes, is explained in terms of the complete spectrum (surface waves plus continuous spectrum) of the slab waveguide.

3.3 Complete spectrum of slab waveguides

So far we have considered only the surface waves of the slab. The question now arises as to whether these are the only type of waves that may exist in the slab. In order to answer this question consider a simple experiment where a symmetric slab of refractive index n_1 is illuminated by an external source as depicted in Figure 3.18. We know that a refraction phenomenon takes place; each ray produced by the source, incident at an angle θ_i onto the slab, in addition to being reflected, travels along the slab with an angle θ_r and escapes from the lower part of the slab at angle θ_i. Obviously a surface wave, or a combination of surface waves, cannot represent a situation of this kind. To describe the above phenomenon in terms of waves and modes, we may separate the incident ray into two waves: one propagating in the z direction of the type $e^{(-j\beta z)}$ and the other propagating in the $-x$ direction and, as such, of the type $e^{(jk_x x)}$. In the two media the wavenumber conservation must hold and is expressed as

$$\begin{aligned}(n_2 k_0)^2 &= \beta^2 + k_x^2 \\ (n_1 k_0)^2 &= \beta^2 + q^2\end{aligned} \tag{3.30}$$

where we have indicated with k_x and q the wavenumbers in the x–direction in air and inside the slab, respectively. Geometrically, from Figure 3.18 we have

$$\begin{aligned}k_x &= n_2 k_0 \cos\theta_i \\ q &= n_1 k_0 \cos\theta_r\end{aligned} \tag{3.31}$$

and since the two waves must propagate in phase along z with the same wavenumber β,

$$\beta = n_2 k_0 \sin\theta_i = n_1 k_0 \sin\theta_r$$

from which follows Snell's law

$$n_2 \sin\theta_i = n_1 \sin\theta_r \tag{3.32}$$

It is expedient to work in terms of even and odd modes as it is obvious that, for this geometry, any case can be solved as a superposition of even and odd excitations. In particular, let us now consider the even TE case, as shown in Figure 3.19 where

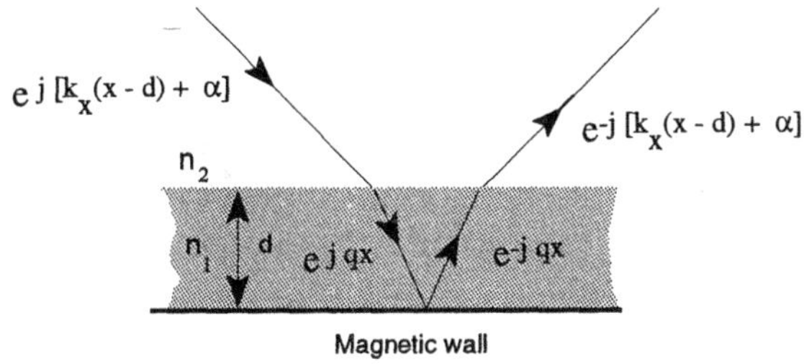

Figure 3.19 Reflection of a plane wave impinging on the slab. α is the phase shift due to propagation inside the slab.

a magnetic wall has been introduced to enforce symmetry of the electric field. The incident ray undergoes refraction at the air–dielectric interface; it also undergoes a total reflection at the magnetic wall and finally escapes with the same angle as that of incidence. This situation is equivalent to the superposition of a ray generated from the actual source and a ray generated from the image source. By choosing as reference plane the magnetic wall, the propagation in the x–direction is of the type

$$\begin{array}{cc} e^{\pm jqx} & \text{in the slab} \\ e^{\pm j[k_x(x-d)+\alpha]} & \text{in air} \end{array} \tag{3.33}$$

By adding the incident and reflected waves in the two dielectrics and by considering the electric field continuity at the interface we obtain

$$\varphi(x) = \begin{cases} \dfrac{\cos qx}{\cos qd} & \text{in the slab } (d \geq x \geq 0) \\ \dfrac{\cos[k_x(x-d)+\alpha]}{\cos\alpha} & \text{in air } (x \geq d) \end{cases} \tag{3.34}$$

From the continuity of H_z at $x = d$, since

$$\varphi'(x) = \begin{cases} -\dfrac{q \sin qx}{\cos qd} \\ -\dfrac{k_x \sin[k_x(x-d)+\alpha]}{\cos\alpha} \end{cases} \tag{3.35}$$

we get

$$\tan\alpha = \frac{q}{k_x}\tan qd \tag{3.36}$$

From the above equation it is possible to find α; moreover (3.30) yields

$$\left(n_1^2 - n_2^2\right) k_0^2 = q^2 - k_x^2 \tag{3.37}$$

It is possible to note the following fundamental differences with respect to the case of surface waves (guided modes):

1. for each value of k_x there exists one value of q ($q > 0$),
2. for real k_x the wave does not vanish at infinity (a single wave does not satisfy a radiation condition).

Let us first discuss point 1. With reference to Figure 3.20 note that each value of k_x between 0 and $n_2 k_0$ is acceptable and corresponds to an angle of incidence

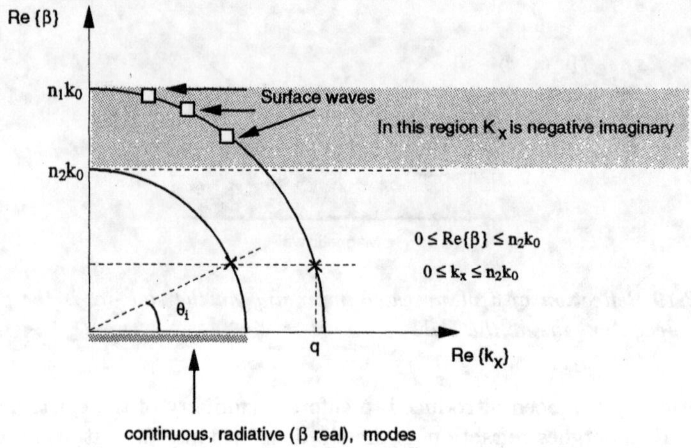

Figure 3.20 Relationship between β and k_x. $0 < k_x < n_2 k_0$ corresponds to a real propagation constant β also varying between 0 and $n_2 k_0$. In this range q is given by the abscissa of the intercept of the horizontal line with the circle $n_1 k_0$. When $n_2 k_0 < \beta < n_1 k_0$ we have only a few discrete solutions with real q and imaginary k_x.

between 90° and 0° respectively. In this range the propagation constant β is also real and contained between 0 and $n_2 k_0$. The corresponding value of q is given by (3.30) or graphically by the abscissa of the crossing point with the circle $n_1 k_0$. When β is in the range between $n_2 k_0$ and $n_1 k_0$ we have only a few discrete solutions: here we still have a real propagation constant q, but k_x is now imaginary so that k_x^2 is negative. All the modes described up to now can propagate in the z–direction without any attenuation.

Some of the previous considerations are simply summarised by using a dispersion diagram relating β to ω. Let us assume that the slab is in free space, so that the normalised propagation constant for surface waves is:

$$1 \leq \frac{\beta}{k_0} \leq \sqrt{\frac{\varepsilon}{\varepsilon_0}}$$

and the phase velocity of these waves, v_{ph}, is less than the speed of light c, since

$$v_{ph} = \frac{\omega}{\beta} = \frac{\omega}{k_0 \frac{\beta}{k_0}} = \frac{c}{\frac{\beta}{k_0}} \leq c \tag{3.38}$$

This type of wave, present in a grounded slab, is often referred to as a slow wave, as illustrated in Figure 3.21 [8].

Apart from bound and radiative modes, other modes are also possible corresponding to the modes below cut–off (non propagating, reactive modes) of closed waveguides. In this case, by considering a propagation of the type $e^{-\gamma z}$, β is purely imaginary and, in order to have attenuated waves at infinity

$$\begin{gathered} \beta = -j\gamma \\ \gamma = \sqrt{k_x^2 - n_2^2 k_0^2} \end{gathered} \tag{3.39}$$

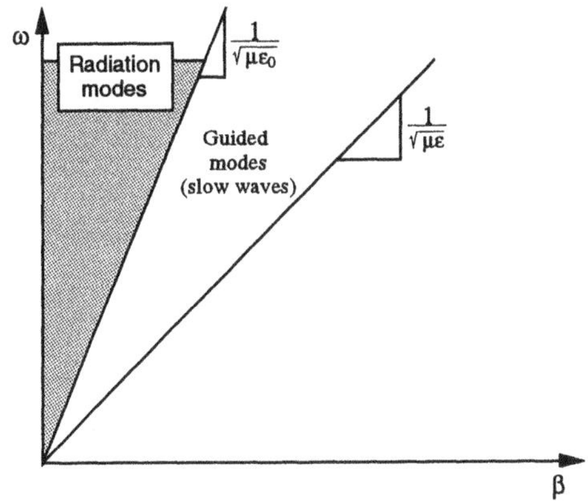

Figure 3.21 Dispersion diagram illustrating the location of radiative and guided modes.

This case corresponds to an imaginary incidence angle. With reference to Figure 3.22 we have therefore the following possibilities: when $\mathrm{Re}\{k_x\} < n_2k_0$ the wave propagates in the z–direction and has a standing wave behaviour in the x–direction; this is a wave generated by the source and radiated away. On the other hand, the case $\mathrm{Re}\{k_x\} > n_2k_0$ corresponds to attenuation in the z–direction and a standing wave behaviour in the x–direction. Clearly, the latter field is localised near the source, since these waves cannot travel in the z–direction, corresponding to modes below cut–off in closed waveguides.

To visualise the whole situation it is helpful to refer to a representation in terms of k_x^2 as in Figure 3.23. Both radiative and reactive modes have real k_x forming a continuous spectrum, while bound modes have imaginary k_x thus forming a finite discrete spectrum on the negative k_x^2 axis. The various possibilities occurring in the slab waveguide are summarised in Table 3.3.

Let us now discuss point 2, i.e. the behaviour at infinity of the continuous spectrum. As we have seen previously, in the slab waveguide there are only a few bound modes (surface waves) whose fields satisfy the radiation condition far from the waveguide (at infinity in the radial direction). Since the total field must satisfy this condition, it was generally believed that the modes in which the total field is expanded must also satisfy this condition.This statement is erroneous [9], since the radiation condition at infinity in the radial direction is not required of each individual field in terms of which the total field is expanded. Therefore, in order to construct a complete set of eigenmodes or eigen–solutions for an open waveguide, the boundary condition at infinity must be relaxed. This means that instead of the normal requirement that each modal field should tend to zero at infinity, the weaker condition that the field only be finite at infinity is imposed [9]. The system of eigenmodes thus derived proves to be complete and suitable for the representation of any physically realisable field in an open waveguide. The complete set of eigenmodes will consist of a finite number of discrete modes together with a continuous spectrum. The discrete modes are often termed [4] proper modes as they decay exponentially away from the guide. The continuous modes, also called improper modes, do not

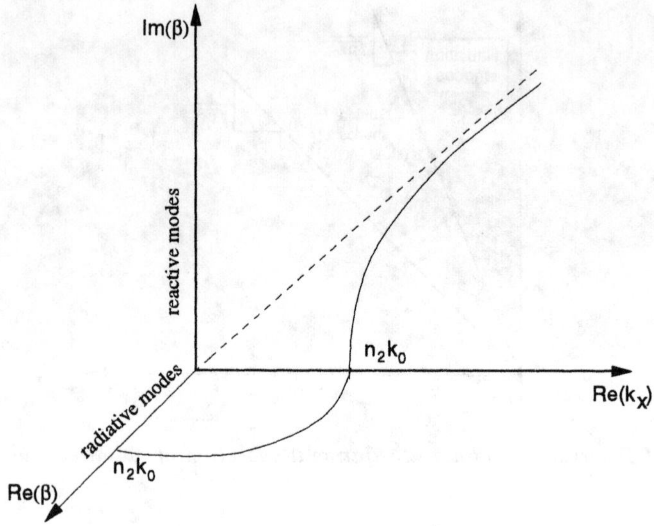

Figure 3.22 For $k_x > n_2k_0$ the modes are attenuated in the propagation direction z.

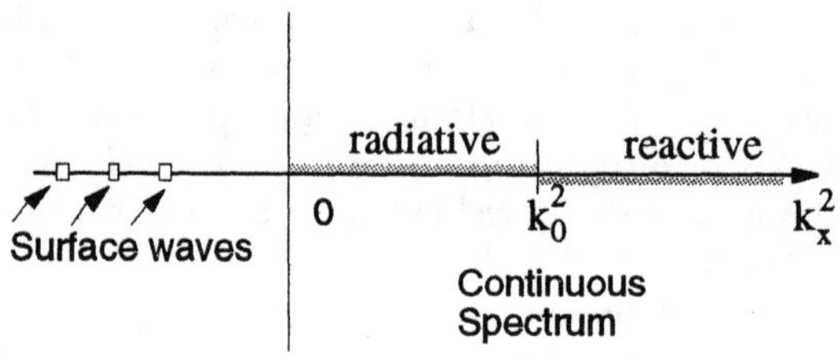

Figure 3.23 Positive values of k_x^2 correspond to the continuous spectrum while negative values represent surface waves.

Table 3.3 Spatial behaviour of modal fields.

	positive x direction	positive z direction
surface wave	attenuation	propagation
radiative spectrum	standing wave	propagation
reactive spectrum	standing wave	attenuation

individually decay to zero at infinity but remain finite. This behaviour is possible only because such modes can never be excited individually by any source of finite energy but they always occur as a continuum and under an integral sign.

3.3.1 Completeness and normalisation of the discrete spectrum

So far we have found modal solutions of Maxwell's equations for a dielectric slab, but we have not discussed orthogonality, normalisation and completeness of these modal solutions. Let us start with normalisation and orthogonality of the discrete eigenfunctions. In this case, each eigenfunction must satisfy the normalisation condition

$$\int_0^\infty \varphi^2(x)dx = 1 \tag{3.40}$$

where

$$\varphi(x) = \begin{cases} A\cos q_s x & : x \le d \\ A\cos q_s d e^{-\gamma_x(x-d)} & : x > d \end{cases} \tag{3.41}$$

Due to the exponential decay in air the above integral is easily evaluated giving the following value for the normalisation constant A

$$A = \sqrt{\frac{2}{d + \frac{1}{\gamma_x}}} \tag{3.42}$$

where, as before,

$$q_s \tan q_s d = \gamma_x \tag{3.43}$$

The orthogonality of discrete, bound, modes is readily proved by using the standard approach,since the field is zero at infinity (see [8, pp.212-218], [5, p.389]).For future reference, it is worthwhile to provide the proof in two dimensions. Two different eigenfunctions satisfy the transverse wave equations

$$\begin{aligned} \nabla_t^2\varphi + k_t^2\varphi &= 0 \qquad (a) \\ \nabla_t^2\overline{\varphi} + \overline{k}_t^2\overline{\varphi} &= 0 \qquad (b) \end{aligned} \tag{3.44}$$

where

$$\nabla_t^2 = \frac{\partial^2}{\partial x^2} + \frac{\partial^2}{\partial y^2}$$

and

$$k_t^2 = q_s^2 + k_y^2$$

Moreover they vanish, together with their derivatives for large $r = \sqrt{x^2+y^2}$; they also satisfy boundary conditions on the wall such that the function or its normal derivative is equal to zero on the wall, i.e.

$$\begin{aligned} \varphi = 0 \quad &\text{or} \quad \frac{\partial\varphi}{\partial n} = 0 \\ \overline{\varphi} = 0 \quad &\text{or} \quad \frac{\partial\overline{\varphi}}{\partial n} = 0 \end{aligned} \tag{3.45}$$

By multiplying (3.44a) by $\overline{\varphi}$ and (3.44b) by φ and by subtracting the resulting (3.44b) from (3.44a) we get

$$\frac{1}{k_t^2 - \overline{k}_t^2}\int\!\!\int \left(\overline{\varphi}\nabla_t^2\varphi - \varphi\nabla_t^2\overline{\varphi}\right) ds = \int\!\!\int \overline{\varphi}\varphi\, ds \tag{3.46}$$

In order to evaluate the above integral on the surface $z = const.$ it is convenient to employ the Green's second identity in two dimensions

$$\int\!\!\int \left(\overline{\varphi}\nabla_t^2\varphi - \varphi\nabla_t^2\overline{\varphi}\right) ds = \oint \left(\overline{\varphi}\frac{\partial\varphi}{\partial n} - \varphi\frac{\partial\overline{\varphi}}{\partial n}\right) d\ell \tag{3.47}$$

Owing to the exponential decay for large radial distance from the guide the contour integral vanishes. As a consequence we obtain that for $k_t \neq \overline{k}_t$

$$\int\int \overline{\varphi}\varphi \, ds = 0 \tag{3.48}$$

The above argument can also be used to prove the orthogonality between a surface wave and a wave pertaining to the continuous spectrum. In this case, far away from the waveguide core both the function describing the surface wave and its derivative vanish, whilst that of the continuous mode remains finite, therefore giving a zero contribution to the integral.

3.3.2 Orthonormalisation of the continuous spectrum of the symmetric slab

For finite intervals the mode spectrum is discrete and normalisation of the eigenfunctions is a straightforward task. On the contrary, for open intervals, where a continuous spectrum is present, the normalisation of the modal set may become a formidable problem [3]. Nevertheless, a general direct procedure for the determination of both discrete and continuous complete sets of modal representations can be devised [3]. Unfortunately, this procedure is quite cumbersome and requires several concepts. For our present purposes, we can circumvent the problem since the normalised continuous modal spectra of open configurations can be derived by direct integration. For the slab waveguide the eigenfunctions, not yet normalised, pertaining to the continuous spectrum may be written in the following form [4]

$$\varphi(x;k_x) = \begin{cases} \sqrt{\frac{2}{\pi}}\left[\cos(k_x x) + \frac{b}{k_x}\sin(k_x x)\right] & \text{in air} \quad (x \geq 0) \\ \sqrt{\frac{2}{\pi}}\frac{\cos[q(x+d)]}{\cos[qd]} & \text{in the slab} \quad (0 \geq x \geq -d) \end{cases} \tag{3.49}$$

Note that the origin is now placed at the air dielectric interface. In the above expression we have used the general form for an eigenfunction in a half space which satisfies the following boundary condition at $x = 0$,

$$\frac{\frac{d\varphi(x;k_x)}{dx}}{\varphi(x;k_x)} = b \tag{3.50}$$

From the continuity of the function and of its derivative at the air dielectric interface we also get

$$b = -q\tan qd \tag{3.51}$$

while wavenumber conservation provides

$$q^2 - k_x^2 = \left(n_1^2 - n_2^2\right)k_0^2 = \nu^2 \tag{3.52}$$

By using the following results,

$$\frac{2}{\pi}\int_0^\infty \sin(k_x x)\sin(\overline{k}_x x)dx = \frac{2}{\pi}\int_0^\infty \cos(k_x x)\cos(\overline{k}_x x)dx = \delta(k_x - \overline{k}_x) \tag{3.53}$$

and

$$\frac{2}{\pi}\int_0^\infty \sin(k_x x)\cos(\overline{k}_x x)dx = \frac{2}{\pi}\frac{k_x}{k_x^2 - \overline{k}_x^2} \tag{3.54}$$

it is readily seen that

$$I_1 = \int_0^\infty \varphi(x;k_x)\overline{\varphi}(x;\overline{k}_x)dx = \delta(k_x - \overline{k}_x)\left[1 + \left(\frac{b}{k_x}\right)^2\right] + \frac{2}{\pi}\frac{b - \overline{b}}{k_x^2 - \overline{k}_x^2} \tag{3.55}$$

In the slab, we have by direct integration

$$I_2 = \int_{-d}^0 \varphi^2(x;q)dx = \frac{\frac{d}{\pi} + \frac{\sin(2qd)}{2q\pi}}{\cos^2(qd)} \tag{3.56}$$

The normalisation is then accomplished by multiplying (3.49) by the constant A given by

$$A = \frac{1}{\sqrt{I_1 + I_2}} = \frac{\cos(qd)}{\sqrt{\frac{d}{\pi} + \frac{\sin(2qd)}{2q\pi} + 1 + \left[\frac{\nu}{k_x}\sin(qd)\right]^2}} \tag{3.57}$$

where use has been made of the following identity

$$\left[1 + \left(\frac{b}{k_x}\right)^2\right] = \frac{\left[1 + \left(\frac{\nu}{k_x}\sin(qd)\right)^2\right]}{\cos^2(qd)} \tag{3.58}$$

In order to prove the orthogonality of the continuous spectrum it is necessary to evaluate

$$\int_{-d}^0 \overline{\varphi}\varphi\, dx \tag{3.59}$$

in the dielectric layer. This can be easily done by using the second Green's identity, since

$$\int_{-d}^0 \overline{\varphi}\varphi\, dx = \frac{1}{q^2 - \overline{q}^2}\int_{-d}^0 \left(\overline{\varphi}\partial_x^2\varphi - \varphi\partial_x^2\overline{\varphi}\right) ds = \frac{1}{q^2 - \overline{q}^2}\left[\left(\overline{\varphi}\frac{\partial\varphi}{\partial x} - \varphi\frac{\partial\overline{\varphi}}{\partial x}\right)\right]_{-d}^0 = \frac{2}{\pi}\frac{\overline{b} - b}{q^2 - \overline{q}^2} \tag{3.60}$$

By summing the above expression with (3.55), and by noting that

$$k_x^2 - \overline{k}_x^2 = q^2 - \overline{q}^2$$

we obtain the orthonormalised eigenfunctions over the whole space, i.e.

$$A\overline{A}\int_{-d}^\infty \varphi(x;k_x)\overline{\varphi}(x;\overline{k}_x)dx = \delta(k_x - \overline{k}_x) \tag{3.61}$$

3.4 Transverse resonance

When considering a dielectric slab we can easily match fields at the interfaces, as already shown in Sections 3.2 and 3.3. In the case where our waveguide is a multilayer structure however, this procedure becomes too involved and inadequate. A different way of looking at the problem, particularly useful in a multilayer situation,

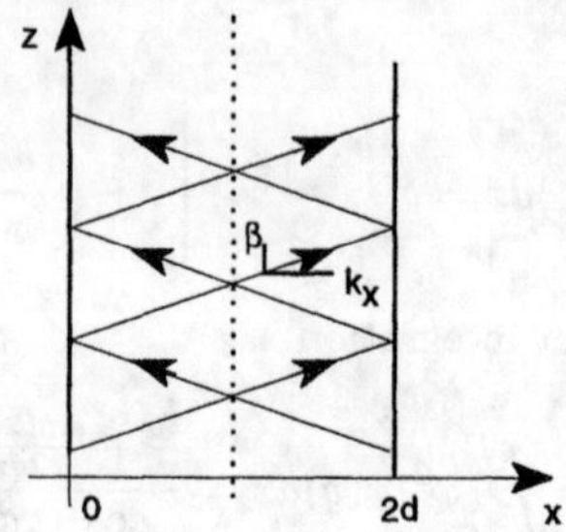

Figure 3.24 Top view of wave propagation in a closed rectangular waveguide.

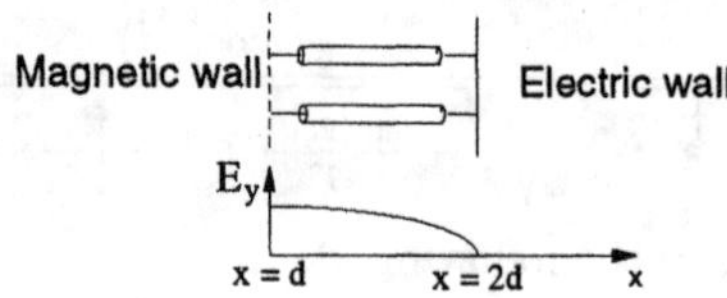

Figure 3.25 Transverse equivalent circuit for an even mode in a rectangular waveguide and corresponding electric field for the fundamental mode.

is based on network considerations. This is the so called transverse resonance technique, [11], which is easier to illustrate by starting with a closed waveguide example. Consider the cross–section of the rectangular waveguide excited by its fundamental TE_{10} mode as shown in Figure 3.24. We know that wave propagation in the guide can be broken down into a wave propagating in the longitudinal z–direction, with propagation constant β, and a plane wave bouncing between the walls at $x = 0, 2d$ with propagation constant k_x. Wavenumber conservation yields

$$k_x^2 + \beta^2 = k_0^2 \tag{3.62}$$

Naturally, k_x is restricted by the round trip condition to be

$$k_x = \frac{n\pi}{2d} \tag{3.63}$$

Let us look at the round trip condition in circuit terms. By symmetry we may place a magnetic wall, i.e. an open circuit, in the mid–plane of the guide, as shown in Figure 3.25. In circuit terms a short circuit is located at $x = 2d$ and an open circuit at $x = d$, the plane of even symmetry. The transverse transmission line between the two has propagation constant k_x. A TE wave propagating in the positive z–direction has a characteristic impedance given by the ratio between the fields transverse to the propagation direction, i.e.

$$Z_0^z = \frac{E_y}{-H_x} = \frac{\omega\mu_0}{\gamma_z} \tag{3.64}$$

This wave is also TE if we consider propagation in the x–direction; the characteristic impedance of the line, in this case, is given by the ratio

$$Z_0^x = \frac{E_y}{H_z} = \frac{\omega\mu_0}{k_x} \tag{3.65}$$

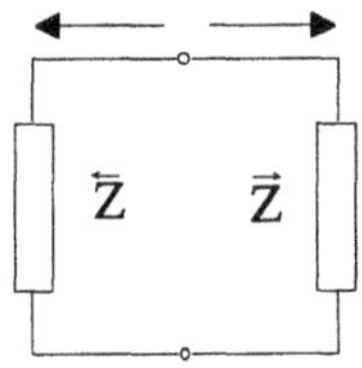

Figure 3.26 Equivalent network used to evaluate the impedance at a certain point along the line.

In the following, unless otherwise stated, we are considering the propagation in the x–direction and we omit the x–dependence in the corresponding impedance. Resonance implies that a finite response ensues from zero excitation, that is, either Z or Y must be infinite. For instance, if we write

$$I = YV \tag{3.66}$$

at resonance I is finite for V tending to zero; hence Y is infinite and Z is zero. In the circuit of Figure 3.26, the total impedance is

$$\overleftrightarrow{Z} = \overrightarrow{Z}(d) + \overleftarrow{Z}(d) \tag{3.67}$$

The arrow points according to whether we take the impedance to the left or right of the observation point. It does not matter at which point we are taking the impedance, as the total $\overleftrightarrow{Z}$ is not affected by this choice. With reference to Figure 3.25, placing the observation point at $x = 2d$ yields

$$\overrightarrow{Z}(2d) = 0 \tag{3.68}$$

because of the short circuit at $x = 2d$ while on the left

$$\overleftarrow{Z}(2d) = -\frac{j\omega\mu_0}{k_x}\cot k_x d \tag{3.69}$$

because of the open circuit at $x = d$. Hence $\overleftrightarrow{Z} = 0$ implies $\cot k_x d = 0$ or

$$k_x d = \frac{n\pi}{2} \tag{3.70}$$

which is the same condition we found from the round trip phase difference.

3.4.1 Surface waves from transverse resonance

Let us consider the case of the slab waveguide. The equivalent network appropriate to this structure is shown in Figure 3.27 for the even TE case. When looking from $x = d$ (the air dielectric interface), the transverse resonance condition is given by

$$\overleftarrow{Z} + \overrightarrow{Z} = -j\frac{\omega\mu_0}{q_s}\cot q_s d + j\frac{\omega\mu_0}{\gamma_x} = 0 \tag{3.71}$$

Hence, we recover the dispersion relation

$$q_s \tan q_s d = \gamma_x \tag{3.72}$$

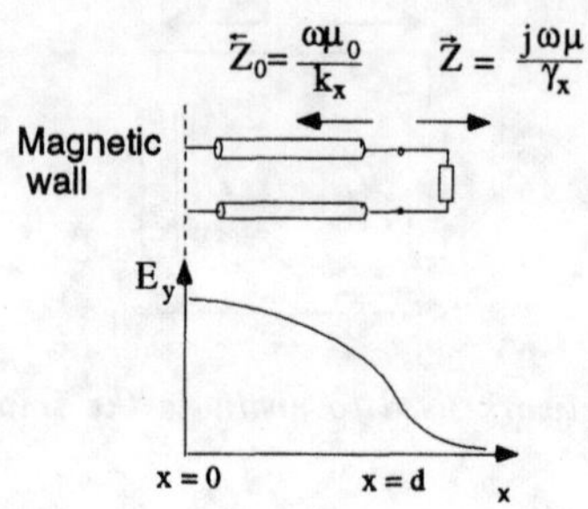

Figure 3.27 Equivalent network of a TE even field in a slab waveguide.

The above equation is, however, identical to 3.17 obtained by matching fields.

Consider odd modes next. Their x–dependence in the slab will be given by a sine function inside the slab due to the electric wall at $x = 0$, and an exponential outside. As a consequence, the total impedance obtained is

$$\overleftrightarrow{Z} = j\frac{\omega\mu_0}{q_s}\tan q_s d + j\frac{\omega\mu_0}{\gamma_x} = 0 \tag{3.73}$$

yielding the dispersion relation

$$q_s \cot q_s d = -\gamma_x \tag{3.74}$$

As already pointed out, we have a solution only if $v > \pi/2$, which corresponds to the TE cut–off.

A question now arises: what is the network interpretation of the cut–off? In the even case, for $q_s d < \pi/2$ the transverse equivalent circuit looks like a parallel LC. In fact, from the impedance transformation formula we have

$$Z'_L = Z_0 \frac{Z_L + jZ_0 \tan \beta\ell}{Z_0 + jZ_L \tan \beta\ell} \tag{3.75}$$

and if Z_L is an open circuit at the plane of symmetry $x = 0$, looking down from the interface $x = d$ we see a capacitive load

$$Z'_L = -j\frac{Z_0}{\tan q_s d} \tag{3.76}$$

On the other hand, the impedance seen from $x = d$ looking towards the air half–space, is the TE characteristic impedance below cut–off, that is an inductance L, as shown in Figure 3.28. We can always resonate the L with the C so that the total susceptance vanishes.

In the odd case for $q_s d < \pi/2$ the transmission line terminated by a short circuit looks like an inductor and we cannot resonate the transverse equivalent circuit until $q_s d > \pi/2$, when the distributed susceptance changes its nature from inductive to capacitive: hence the cut–off effect. The various cases are summarised in Figure 3.29.

Transverse resonance is particularly convenient in practical cases when a multilayer structure, such as an optical amplifier, or a laser, with five layers at least,

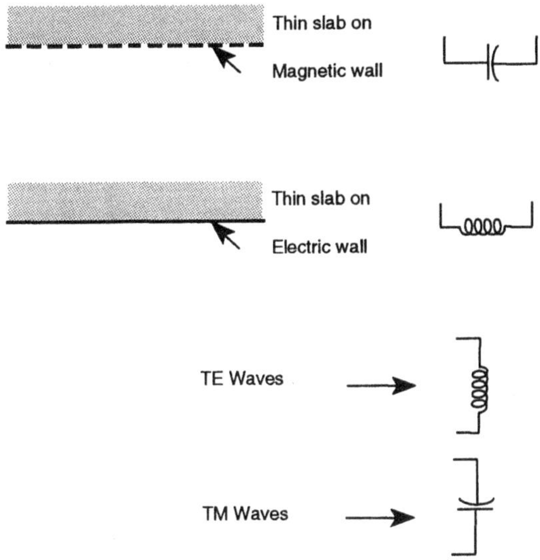

Figure 3.28 A thin dielectric slab on a magnetic wall behaves like a capacitor, whereas a thin dielectric slab backed by an electric wall behaves like an inductor. TE waves possess an inductive behaviour, while TM waves are capacitive.

is considered. It is also very useful to employ circuit concepts in order to calculate the dependence on x of the eigenfunctions by using $ABCD$ matrices. Moreover, the transverse resonance approach provides further insight into the nature of the continuous part of the spectrum. The next Section will therefore treat the continuous modes as derived from the transverse network approach.

3.4.2 Continuous spectra from the transverse network

As in the previous discussion of the bound modes of the dielectric slab, we consider even TE propagation and break it down into a wave propagating in the longitudinal z–direction and a wave propagating in the transverse x–direction. The circuit corresponding to the transverse propagation in the slab is represented by a transmission line, also TE in x, of length $\theta = qd$ and admittance

$$Y_0 = \frac{q}{\omega\mu_0} \tag{3.77}$$

This line is closed by a magnetic wall, that is, it is terminated by an open circuit. At $x = d$, if we look in the positive x–direction and if we assume that the field decays in free–space, i.e. if we are considering a surface wave, we observe a purely reactive load. This circuit is lossless and will exhibit resonances; that is a field can be present even for vanishing excitation, and therefore no source appears in this equivalent circuit.

The situation is quite different for continuous modes, on two main counts. First, the presence of a source is now required. In fact, since waves are travelling in the air half–space, energy can leave the guide. In order to compensate for this loss or,

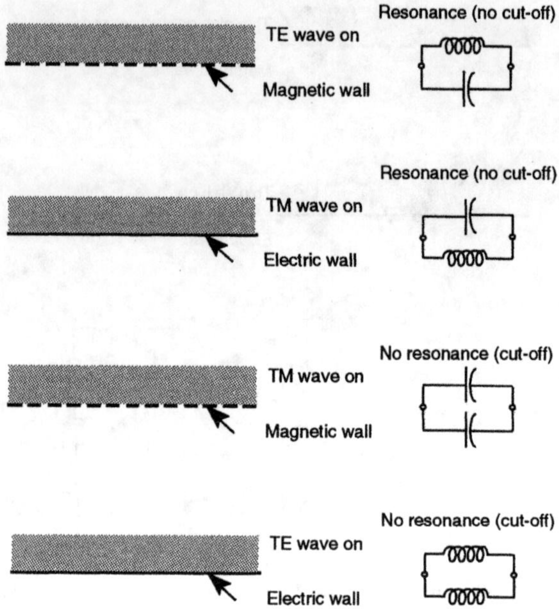

Figure 3.29 Equivalent networks of TE/TM even/odd fields in a slab waveguide.

equivalently, in order to maintain the standing wave pattern, a source is necessary. The second major difference is that the admittance of the air half–space is now resistive, since in this half–space the wave propagates. Therefore the admittance of the air half–space is the characteristic admittance of a TE wave propagating in the semi–infinite medium, that is,

$$Y_0 = \frac{k_x}{\omega\mu_0} \tag{3.78}$$

It is now evident that, in this case, no resonance condition has to be fulfilled.

We can also determine the field pattern by simple circuit considerations. Let us consider the equivalent circuit corresponding to a mode sinusoidally varying in air. Since we associate the field E_y with the voltage on the transmission line, we find the shape of E_y simply by calculating the voltage distribution on the equivalent circuit. In the same manner, by associating the amplitude of the magnetic field H_z with the current along the transmission line, we also determine the H_z field distribution. Now let the voltage at $x = d$ assume unit amplitude. It is easy to calculate the voltage amplitude at $x = 0$. In fact, by using the $ABCD$ description of the transmission line and by starting from the magnetic wall we have

$$\begin{pmatrix} V(x) \\ I(x) \end{pmatrix} = \begin{pmatrix} \cos qx & j\frac{\omega\mu_0}{q}\sin qx \\ j\frac{q}{\omega\mu_0}\sin qx & \cos qx \end{pmatrix} \begin{pmatrix} V(0) \\ 0 \end{pmatrix} \tag{3.79}$$

That is, voltage and current along the line are determined by the voltage on the magnetic wall $V(0)$, since the current on the magnetic wall is obviously zero. In particular, if we have imposed that at $x = d$, $V(d) = 1$ we have

$$V(0) = \frac{1}{\cos qd} \tag{3.80}$$

It is also straightforward to calculate the current, e.g. at $x = d$ we have

$$I(d) = j\frac{q}{\omega\mu_0}\tan qd \tag{3.81}$$

In order to calculate the fields in the air half space, we proceed along similar lines only noting that the characteristic admittance is now that pertaining to air, i.e.

$$\begin{pmatrix} V(x) \\ I(x) \end{pmatrix} = \begin{pmatrix} \cos k_x(x-d) & j\frac{\omega\mu_0}{k_x}\sin k_x(x-d) \\ j\frac{k_x}{\omega\mu_0}\sin k_x(x-d) & \cos k_x(x-d) \end{pmatrix} \begin{pmatrix} 1 \\ j\frac{q}{\omega\mu_0}\tan qd \end{pmatrix} \tag{3.82}$$

The transverse circuit approach also attaches a physical meaning to several quantities already used. As an example, let us recall (3.36) which can be rewritten as

$$j\frac{k_x}{\omega\mu_0}\tan\alpha = j\frac{q}{\omega\mu_0}\tan qd \tag{3.83}$$

It is apparent that the above formula allows a circuit interpretation: in fact, the right hand side is the admittance seen from the air–dielectric interface looking into the dielectric layer; while the left hand side is the admittance of a magnetic wall placed in air and at an electrical distance α from the interface. However, this latter configuration is completely equivalent to the previous one from the electromagnetic point of view. As a further remark, note that the effect of the dielectric and of the magnetic wall is completely summarised by α. This property holds just as well when several dielectric layers are present. In this case, by starting from the impedance of the last layer and by using the *ABCD* matrices of the individual layers, we calculate the input admittance of the multi–layer stack and represent it synthetically by means of the equivalent phase shift α.

3.5 Field excited by a delta source

In Sections 3.2 and 3.3 we have examined the trapped (surface) waves and the continuous spectrum of two–dimensional structures. We have then answered questions concerning orthogonality and completeness of the modal representations. However, up to now, the field was considered as being already present in our structure without any care about its excitation. In this paragraph we investigate how these fields are actually excited. Since we are dealing with two–dimensional structures independent of y, our excitation must also be independent of y. This means that we consider either a current filament in the y–direction, with currents flowing along y; or a current sheet with currents directed in the x or in the z–direction but uniform along y. Moreover, due to the field equivalence theorems, it is useful to consider as possible excitations both electric (**J**) and magnetic (**M**) currents. Figure 3.30 summarises the possible types of excitations and the fields generated thereby.

Let us consider, as an example, the field produced by a source placed at $(x', 0)$ and constant along y, as shown in the upper left–hand corner of Figure 3.30, that is

$$\mathbf{J} = \mathbf{y}_0\delta(x - x')\delta(z) \tag{3.84}$$

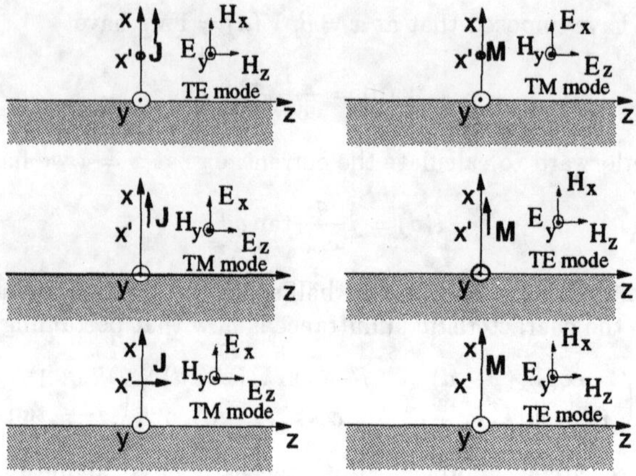

Figure 3.30 Possible types of two–dimensional excitations and the type of fields generated.

It is our present purpose to determine what type of field is generated by this source and in order to do so, it is convenient to express the field in terms of modes. The main advantage of using modes is that they provide the field distribution over the whole space, once their amplitudes are known on a given section. In fact, once modal amplitudes are known, it is sufficient to multiply each of them by the corresponding propagation factor to obtain the field at a given distance. By doing so however, we have also chosen a particular propagation direction for our problem. This choice is by no means unique and its consequences are particularly relevant from the computational point of view. By introducing potential Ψ we can write the field as

$$\begin{aligned} E_y &= -j\omega\mu_0\Psi \\ \mu_0 H_x &= -\frac{\partial\Psi}{\partial z} \\ \mu_0 H_z &= \frac{\partial\Psi}{\partial x} \end{aligned} \tag{3.85}$$

where the potential Ψ satisfies the following equation

$$\frac{\partial^2\Psi}{\partial x^2} + \frac{\partial^2\Psi}{\partial z^2} + k^2\Psi = -\delta(x - x')\delta(z) \tag{3.86}$$

In order to solve the above equation, we may equally well choose:

i) modes propagating in the z–direction (this is equivalent to performing an eigenfunction expansion in terms of eigenfunctions in x);

ii) modes propagating in the x–direction, i.e. eigenfunctions depending on z.

In the next two paragraphs we will follow both approaches highlighting the relationships between the two solution procedures.

3.5.1 Modes propagating along the dielectric interface

If we follow choice i) the solution of the above equation can be written as a superposition of modes, each with its own amplitude, as

$$\Psi(x,z) = \int_0^{\infty} V(\xi)\phi(x;\xi)f(z;\xi)d\xi \tag{3.87}$$

where ξ is the wavenumber in the x–direction, and $\phi(x;\xi)$ are the eigenfunctions along x. $V(\xi)$ represents the distribution of modal amplitudes generated by our source while the function f takes into account the z–dependence. Actually, in (3.87) only the continuous part of the spectrum is explicitly written. In addition to the continuous spectrum, a few bound modes may also be present: then the latter ought to be included in (3.87) as a finite sum. With a view to minimising notation, we will not consider explicitly this part of the spectrum in the rest of this Section. Since the orthogonality between the continuous and discrete spectra has already been proven, their inclusion does not introduce any difficulty from a theoretical point of view. Note that in the above expansion we have inserted the eigenfunction in the x–direction (in fact the integration variable is the spectral wavenumber in x); thus we have considered z as the propagation direction. The latter assumption is particularly useful if we consider discontinuities in the z–direction.

In order to find the modal amplitudes it is convenient to insert the expansion (3.87) in (3.86) and to use the eigenfunction representation of the delta function

$$\delta(x-x') = \int_0^{\infty} \phi(x;\xi)\phi(x';\xi)d\xi \tag{3.88}$$

thus obtaining

$$\begin{aligned}\textstyle\int_0^{\infty} V(\xi)\left[\phi''(x;\xi)f(z;\xi) + \phi(x;\xi)f''(z;\xi) + k^2\phi(x;\xi)f(z;\xi)\right]d\xi = \\ \delta(z)\textstyle\int_0^{\infty}\phi(x;\xi)\phi(x';\xi)d\xi\end{aligned} \tag{3.89}$$

We multiply the above expression by $\phi(x;\xi')$ and integrate in x, while denoting this inner product as

$$\langle\phi(x;\xi),\phi(x;\xi')\rangle = \int_0^{\infty}\phi(x;\xi)\phi(x;\xi')dx = \delta(\xi-\xi') \tag{3.90}$$

Also recalling that

$$\phi''(x;\xi) + \xi^2\phi(x;\xi) = 0$$

and that, consequently,

$$\langle\phi(x;\xi),\phi''(x;\xi')\rangle = -\xi^2\delta(\xi-\xi') \tag{3.91}$$

we obtain

$$V(\xi)\left[\beta^2 f(z;\xi) + f''(z;\xi)\right] = -\delta(z)\phi(x';\xi) \tag{3.92}$$

with

$$\beta^2 = k^2 - \xi^2 \tag{3.93}$$

The eigensolutions for the z–dependence are of the type

$$f(z;\xi) = \begin{cases} e^{-j\beta z} & z>0 \\ e^{j\beta z} & z<0 \end{cases} \tag{3.94}$$

Near $z = 0$, f is continuous, while its first derivative has a jump

$$\left[\frac{df}{dz}\right]_{0^-}^{0^+} = -2j\beta \tag{3.95}$$

By integration of (3.92) around $z = 0$ we obtain

$$V(\xi) = \frac{\phi(x', \xi)}{2j\beta} \tag{3.96}$$

Our solution is therefore

$$\Psi(x, z) = \int_0^\infty \frac{\phi(x', \xi)}{2j\beta} \phi(x; \xi) \left\{ \begin{array}{c} e^{-j\beta z} \\ e^{j\beta z} \end{array} \right\} d\xi \tag{3.97}$$

The above formula provides, apart from a factor $-j\omega\mu_0$, the electric field generated by a current distribution concentrated at $(x', 0)$. Thus the potential Ψ can be regarded as the response, or Green's function, of our structure. Owing to linearity, if our source extends over an interval of x', we simply have to integrate over x' to calculate the field generated by this source distribution.

3.5.2 Modes propagating normally to the dielectric interface

Let us now consider the type of solution which can be obtained by following approach ii): this is particularly suited to situations where the structure is uniform along the z–direction, whereas cascaded discontinuities are present along the x–direction. In this case, we also seek a solution of (3.86) but in terms of eigenfunctions in the z–direction. However, since the domain in the z–direction is unbounded and without any discontinuity, the eigenfunctions in the z–direction are simply provided by the Fourier basis $e^{-j\beta z}$, with β and ξ mutually related through (3.93). We represent our potential function as a superposition of eigenfunctions, each with a certain amplitude given by $g(x; \beta)$, as

$$\psi(x, z) = \frac{1}{2\pi} \int_{-\infty}^{\infty} g(x; \beta) e^{-j\beta z} d\beta \tag{3.98}$$

The above expression can be regarded as a Fourier transform and, consequently, we also have

$$g(x; \beta) = \int_{-\infty}^{\infty} \psi(x, z) e^{+j\beta z} dz \tag{3.99}$$

By noting that

$$\frac{\partial^2 \psi(x, z)}{\partial z^2} = -\beta^2 \psi(x, z) \tag{3.100}$$

by multiplying (3.86) by $e^{j\beta z}$ and integrating in z from $-\infty$ to $+\infty$, after making use of (3.99), we get the following equations for g:

$$\begin{aligned} \frac{d^2 g(x; \beta)}{dx^2} + (n_2^2 k_0^2 - \beta^2) g(x; \beta) &= -\delta(x - x') \qquad x \geq 0 \\ \frac{d^2 g(x; \beta)}{dx^2} + (n_1^2 k_0^2 - \beta^2) g(x; \beta) &= 0 \qquad 0 \geq x \geq -d \end{aligned} \tag{3.101}$$

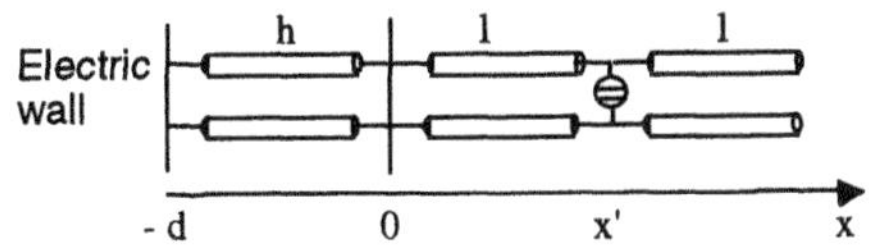

Figure 3.31 Equivalent circuit of the current source over a dielectric slab bounded by a metallic wall.

Since g is proportional to the electric field, it may be found by using the transmission line analogy and by considering the voltage of the equivalent circuit shown in Figure 3.31.

However, by taking advantage of the knowledge of the eigenfunctions along x, it is also possible to write g as a superposition of eigenfunctions with different amplitudes of the type

$$g(x, x') = \int_{-\infty}^{\infty} H(x'; \xi')\phi(x; \xi')d\xi' \tag{3.102}$$

with H to be determined in the following. By making use of the eigenfunction expansion of the delta function (3.88), we substitute the latter equation into (3.101), obtaining

$$\int_{-\infty}^{\infty} H(x'; \xi') \left[\xi^2 - \xi'^2\right] \phi(x; \xi')d\xi' = -\int_{0}^{\infty} \phi(x; \xi')\phi(x'; \xi')d\xi' \tag{3.103}$$

From the last equation it is readily seen that

$$H(x'; \xi') = -\frac{\phi(x'; \xi')}{\xi^2 - \xi'^2} \tag{3.104}$$

and we obtain the function g as

$$g(x, x'; \xi) = -\frac{1}{2}\int_{-\infty}^{\infty} \frac{\phi(x'; \xi')\phi(x; \xi')}{\xi^2 - \xi'^2} d\xi' \tag{3.105}$$

We can now evaluate the above expression by contour integration closed in the upper half–plane, yielding for g

$$g(x, x'; \xi) = -j\frac{\pi}{2}\frac{\phi(x; \xi)\phi(x'; \xi)}{\xi} \tag{3.106}$$

In conclusion, we have the following result for the potential

$$\psi(x, z) = -\frac{j}{4}\int_{-\infty}^{\infty} \frac{\phi(x; \xi)\phi(x'; \xi)}{\xi} e^{-j\beta z} d\beta \tag{3.107}$$

Note that in the above integral the sum is performed on the variable β and not on x as in (3.97). Naturally, both representations (3.97, 3.107) represent the Green's function of our structure and are equally valid from a theoretical point of view; they are both obtained in terms of the eigenfunctions in the z– and x–directions respectively. It is therefore apparent that the complete characterisation

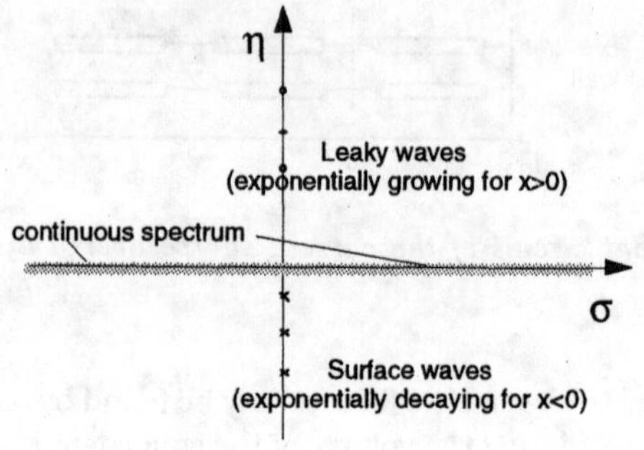

Figure 3.32 Complex ξ -plane ($\xi = \sigma + j\eta$). The integration path is along the σ axis.

of the structure is obtained when the eigenfunction expansion in either direction is known. It is also interesting to note that it is almost straightforward to go from (3.97) to (3.107) and vice versa. In fact by taking into account the relation (3.93) and by changing the variable of integration in (3.97) from x to β we obtain (3.107) The latter equivalence, as well as the various alternative representations of the Green's function are, however, well known [3] and can be proven in a general way. For the moment it is sufficient to be aware of the different possibilities at hand and of their use in practical computation when treating discontinuity problems. In fact, as will be shown for the dielectric step, some problems find a straightforward solution by suitably choosing the proper Green's function representation, while becoming quite involved when using the other representation.

3.5.3 Far field by the steepest descent

The field generated by a certain source distribution has been determined in the previous Section. In several applications, however, the main interest is to know the shape of the field quite far from the source or, equivalently stated, the high frequency asymptotic behaviour of the field. The far field, or the radiative part, of the E_y component of the electric field is given by (3.97) that, for $z > 0$, can be conveniently rewritten as

$$E_y^r = \int_0^\infty b(\xi)\phi(x;\xi)e^{-j\beta z}d\xi \tag{3.108}$$

After use of the modal form pertaining to ϕ in the air half–space,

$$\phi(x;\xi) = \sqrt{\frac{2}{\pi}}\cos\left[\xi\left(x-d\right)+\alpha\right] \tag{3.109}$$

we obtain the far field as

$$E_y^r = \frac{1}{\sqrt{2\pi}}\int_{-\infty}^{\infty} b(\xi)e^{-j[\xi(x-d)+\alpha+\beta z]}d\xi \tag{3.110}$$

Integral (3.110) can be evaluated by using the saddle point method of approximate integration. The saddle point, or steepest descents method, was first introduced by Debye in 1909 for obtaining the well known asymptotic expansion of the Hankel function [12, p.118]. Successively, it was improved by Ott (1942), van de Waerden (1951) and generalised by Oberhettinger (1959). A rather exhaustive discussion of the saddle point method can be found in [3, ch. 4] . This method applies when the exponential appearing in the integrand contains a large parameter in terms of which the integral can be approximated. In our case such a large parameter is the distance of the observation point from the origin. In order to facilitate the application of the above method it is convenient to express the space coordinates in a cylindrical coordinate system as

$$x - d = r\sin\theta \qquad z = r\cos\theta \tag{3.111}$$

where r is the distance from the origin and θ is the angle of the observation point with respect to the z–axis. This type of transformation can also be applied to the spectral coordinates ξ, β, namely we set

$$\xi = n_2 k_0 \sin w \qquad \beta = n_2 k_0 \cos w \tag{3.112}$$

where w is the complex variable $u + jv$. The above substitution simplifies the details of the saddle–point integration to be discussed in the following, reducing the integral to:

$$E_y^r = \frac{1}{\sqrt{2\pi}} \int_C f(w) e^{-jrn_2 k_0 \cos(w-\theta)} dw \tag{3.113}$$

The contour of integration C is shown in Figure 3.33; $f(w)$ is given by

$$f(w) = n_2 k_0 \cos w \; b(n_2 k_0 \sin w) e^{-j\alpha(w)} \tag{3.114}$$

The transformation (3.112) represents a mapping from the complex ξ–plane into strips of the complex w–plane as shown by Figures 3.32 and 3.33. The original integration path along the real axis of the ξ–plane is now mapped onto the path C in the w–plane. The shaded region is the area in which $\text{Im}\{\beta\} < 0$ i.e. that of decaying fields in the z–direction. By introducing the function

$$g(w) = -jn_2 k_0 \cos(w - \theta) \tag{3.115}$$

in the exponent of the integrand (3.113) becomes

$$E_y^r = \frac{1}{\sqrt{2\pi}} \int_C f(w) e^{g(w) r} dw \tag{3.116}$$

where r now plays the role of the large parameter in terms of which it is possible to find approximate expressions of the integral. In order to accomplish this task we proceed as follows: let us suppose we find a point w_s such that $\text{Re}\{g(w_s)\}$ has the maximum value on the path. When r is very large the main contribution to the integral comes from an interval around w_s, since contributions from the remainder of the path will be exponentially smaller. If $f(w)$ is regular and slowly varying in the vicinity of w_s, we can take the function f outside the integral. Subsequently, we can expand $g(w)$ near its maximum value w_s in a Taylor series and integrate analytically the first few terms of the series. However, the accuracy of the result obtained in this way will be strongly dependent on the path followed for the integration. In order to provide an accurate evaluation, this path should possess the following requisites:

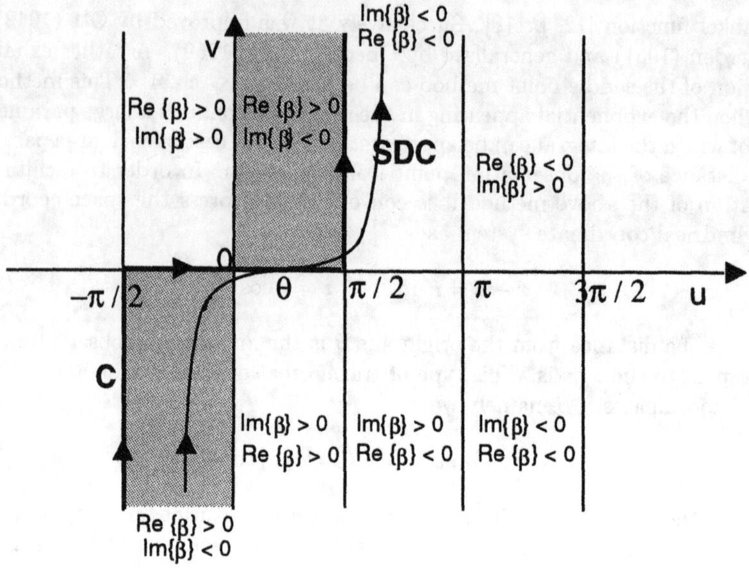

Figure 3.33 Complex w-plane, $w = u + jv$. The integration path corresponding to the σ-axis is represented by the contour C. The steepest descent path is indicated by the curve SDC.

i) $\mathrm{Re}\{g\}$ ought to vary as rapidly as possible, so that just the region near w_s contributes;

ii) the phase ought to remain constant along the path, so that it can be factored out of the integral.

The above requirements also explain the name of this asymptotic evaluation technique referred to as the steepest descent method, or the method of stationary phase. In order to ascertain if it is feasible to find an integration path which satisfies both requirements, it is necessary to look more carefully at the function $g(w)$. Since w_s is a stationary point we have

$$\frac{dg(w)}{dw} = 0 \quad \text{at} \quad w = w_s \tag{3.117}$$

Since $w = u + jv$ we also have

$$\frac{\partial \mathrm{Re}\{g\}}{\partial u} = \frac{\partial \mathrm{Re}\{g\}}{\partial v} = \frac{\partial \mathrm{Im}\{g\}}{\partial u} = \frac{\partial \mathrm{Im}\{g\}}{\partial v} = 0 \quad \text{at} \quad w = w_s \tag{3.118}$$

Hence, the function $g(w)$ satisfies the Cauchy–Riemann equations, i.e.,

$$\frac{\partial \mathrm{Re}\{g\}}{\partial u} = \frac{\partial \mathrm{Im}\{g\}}{\partial v} \qquad \frac{\partial \mathrm{Re}\{g\}}{\partial v} = -\frac{\partial \mathrm{Im}\{g\}}{\partial u} \tag{3.119}$$

and from the latter we have at the stationary point w_s

$$\frac{\partial^2 \mathrm{Re}\{g\}}{\partial u^2} = -\frac{\partial^2 \mathrm{Re}\{g\}}{\partial v^2} \qquad \frac{\partial^2 \mathrm{Im}\{g\}}{\partial u^2} = -\frac{\partial^2 \mathrm{Im}\{g\}}{\partial v^2} \tag{3.120}$$

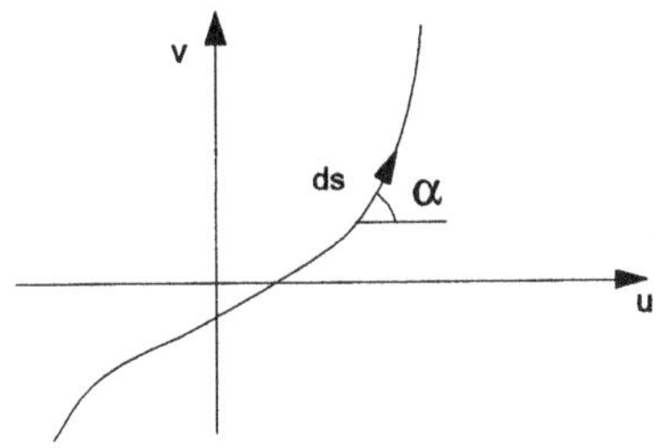

Figure 3.34 Steepest descent path in the w plane.

which are the conditions for a saddle point. In fact, if at $w = w_s$ the curvature of the surfaces $\mathrm{Re}\{g\}, \mathrm{Im}\{g\}$ is positive along the u–direction, this curvature is negative along the perpendicular v–direction. It is also important to see what happens along the path of maximum change for $\mathrm{Re}\{g\}$. With reference to Figure 3.34 we have that the variation along a path is given by

$$\frac{d\mathrm{Re}\{g\}}{ds} = \frac{\partial \mathrm{Re}\{g\}}{\partial u}\frac{\partial u}{\partial s} + \frac{\partial \mathrm{Re}\{g\}}{\partial v}\frac{\partial v}{\partial s} = \frac{\partial \mathrm{Re}\{g\}}{\partial u}\cos\alpha + \frac{\partial \mathrm{Re}\{g\}}{\partial v}\sin\alpha \tag{3.121}$$

where we have denoted

$$\frac{\partial u}{\partial s} = \cos\alpha \qquad \frac{\partial v}{\partial s} = \sin\alpha \tag{3.122}$$

The maximum rate of change is thus obtained when

$$\frac{\partial^2 \mathrm{Re}\{g\}}{\partial\alpha\partial s} = -\frac{\partial \mathrm{Re}\{g\}}{\partial u}\sin\alpha + \frac{\partial \mathrm{Re}\{g\}}{\partial v}\cos\alpha = 0 \tag{3.123}$$

In this case, the variation of $\mathrm{Im}\{g\}$ is readily seen to be zero too, by using the Cauchy–Riemann equations,

$$\frac{d\mathrm{Im}\{g\}}{ds} = \frac{\partial \mathrm{Im}\{g\}}{\partial u}\frac{\partial u}{\partial s} + \frac{\partial \mathrm{Im}\{g\}}{\partial v}\frac{\partial v}{\partial s} = \frac{\partial \mathrm{Im}\{g\}}{\partial u}\cos\alpha + \frac{\partial \mathrm{Im}\{g\}}{\partial v}\sin\alpha = 0 \tag{3.124}$$

In conclusion, the path of maximum change for $\mathrm{Re}\{g\}$ is a constant amplitude path for $\mathrm{Im}\{g\}$ and vice versa. The above concepts are visualised in Figures 3.35 and 3.37 by means of a 3–D plot and in Figures 3.36 and 3.38 by means of a "level map" of the behaviour of the function $g(w)$ over the complex w plane. It is seen that the two surfaces $\mathrm{Re}\{g\}$ and $\mathrm{Im}\{g\}$ provide a picture of a terrain with mountains, valleys and interconnecting passes. The path of integration to be followed is a constant level path in the $\mathrm{Im}\{g\}$ surface, passing for the saddle point w_s . However, while moving on this path, we are also moving on the steepest descent path on the $\mathrm{Re}\{g\}$ surface. This contour is clearly depicted in Figure 3.39 which reports at the same time the constant level contours for both $\mathrm{Re}\{g\}$ and $\mathrm{Im}\{g\}$.

The original integration path along the real axis of the ξ–plane has been mapped into the path C, and this path has been subsequently deformed into the steepest descent path SDC of Figures 3.33. However, this change from the path C to the path SDC does not change the value of the integral as long as no poles in the w–plane are swept across. When poles are crossed their residues should be included.

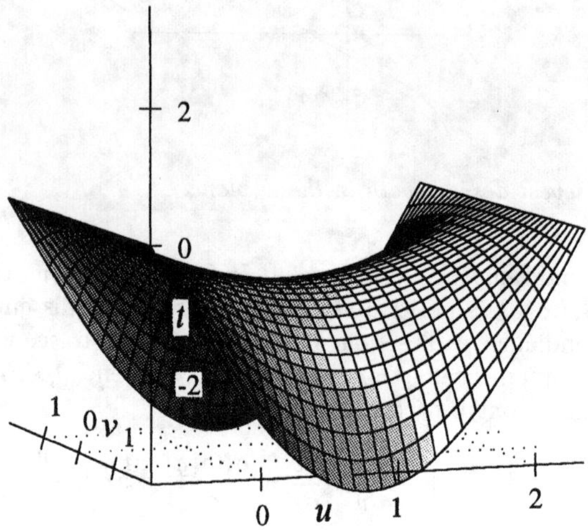

Figure 3.35 3–D plot of the imaginary part of the function g. The stationary phase path is the path along the line of amplitude -1 from the lower left corner to the top right corner in the contour plot.

The saddle points of the function $g(w)$ are readily found by considering that

$$\frac{dg(w)}{dw} = jn_2k_0 \sin(w - \theta) \tag{3.125}$$

is equal to zero at $w = \theta$. Furthermore, the SDC path is chosen along the path of constant phase, such that

$$\mathrm{Im}\{g(w) - g(\theta)\} = 0 \tag{3.126}$$

It is therefore apparent that the path of constant phase, and therefore the steepest descent path for $\mathrm{Re}\{g\}$, is given by the equation

$$\cos(u - \theta)\cosh v = 1 \tag{3.127}$$

Note that when the SDC is made to go through the saddle point and r is made arbitrarily large, the value of the integrand in (3.116) is negligible away from the saddle point only if the real part of $g(w)$ is negative. This requires that

$$\begin{array}{lcc} \text{when } v > 0 & \rightarrow & u < \theta + \pi \\ \text{when } v < 0 & \rightarrow & u < \theta \end{array} \tag{3.128}$$

and both the above conditions are satisfied by the SDC path. After contour deformation from C to SDC our integral is of the type

$$E_y^r = \frac{1}{\sqrt{2\pi}} \int_{SDC} f(w) e^{g(w)r} \, dw \tag{3.129}$$

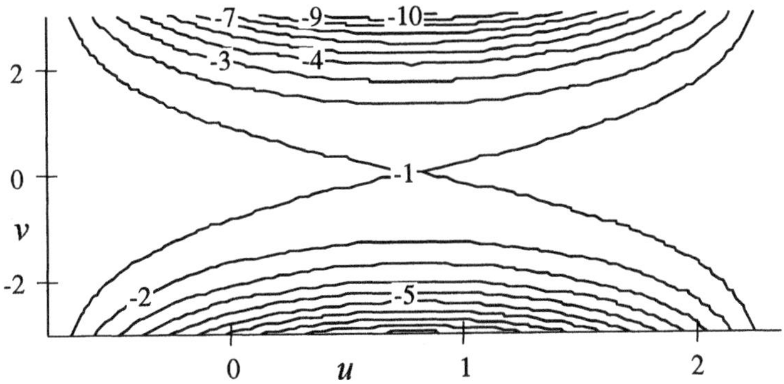

Figure 3.36 Contour plot of the imaginary part of the function g. The stationary phase path is the path along the line of amplitude -1 from the lower left corner to the top right corner in the contour plot.

By expanding $g(w)$ as a Taylor series around the saddle point w_s as

$$g(w) = g(w_s) + \frac{g''(w_s)}{2!}(w - w_s)^2 + \ldots \tag{3.130}$$

and using this expansion in the integral 3.129 we obtain

$$E_y^r = \frac{f(w_s)}{\sqrt{2\pi}} e^{rg(w_s)} \int_{SDC} e^{r\frac{g''(w_s)}{2}(w-w_s)^2} dw \tag{3.131}$$

It is expedient to introduce in the above integral the variable

$$s = -\frac{g''(w_s)}{2}(w - w_s)^2 \tag{3.132}$$

from which follows

$$dw = \sqrt{-\frac{2}{g''}} ds \tag{3.133}$$

and since

$$\int_{-\infty}^{\infty} e^{-rs^2} ds = \sqrt{\frac{\pi}{r}} \tag{3.134}$$

we get a second order approximation of the integral for arbitrarily large r in the form

$$E_y^r = \sqrt{-\frac{1}{rg''}} f(w_s) e^{rg(w_s)} \tag{3.135}$$

Finally, by inserting (3.114, 3.115) into the above we obtain the far–field representation as

$$E_y^r \approx \frac{f(\theta)}{\sqrt{rn_2k_0}} e^{-j(rn_2k_0 - \frac{\pi}{4})} \tag{3.136}$$

From the above expression we can immediately compute the radiated power from our source as a function of the angle θ and of the modal amplitude of the continuous spectrum as given by $b(w)$.

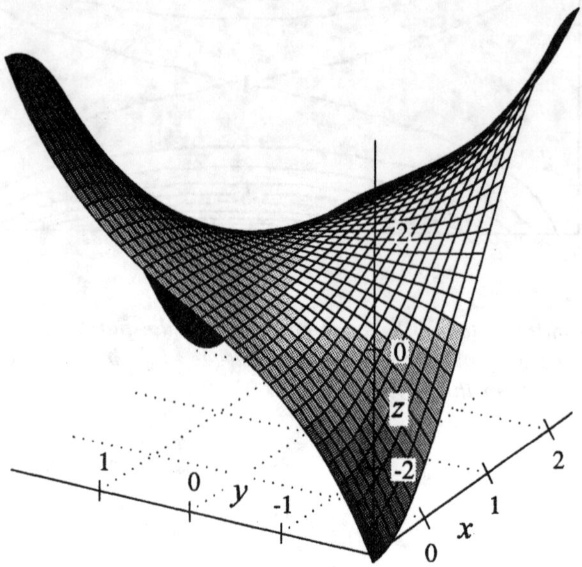

Figure 3.37 3–D plot of the real part of the function g. The steepest descent path is the one that crosses the level line s in the most rapid way.

Note that when passing from (3.129) to (3.131) we have considered just the term $f(w_s)$ to represent $f(w)$. However, the method of steepest descent permits us to find higher order terms by expanding $f(w)$ in a Taylor series about the saddle point. This is not done here, while the interested reader can find the complete steepest descent asymptotic expansion in [3, p.384], [2, p.830], [12, p.118]. A final word concerns the differences between steepest descent and the closely related method of the stationary phase. The latter was formulated late in the nineteenth century and was used by Kelvin in hydrodynamics and by Stokes in order to extract asymptotic values from Airy integrals. Essentially, the method of stationary phase is based on the argument that the integrand fluctuates rapidly except at saddle points. These fluctuations make the contributions to the integral cancel each other. Thus the latter integral can be approximated by its value in the neighbourhood of the saddle point. From a practical point of view there is no difference in using the steepest descent with a first order approximation, as above, or in using the stationary phase approach. However, from a theoretical point of view, the steepest descent is derived by following the rigorous procedure of contour deformation; moreover it also allows higher order asymptotic expansions.

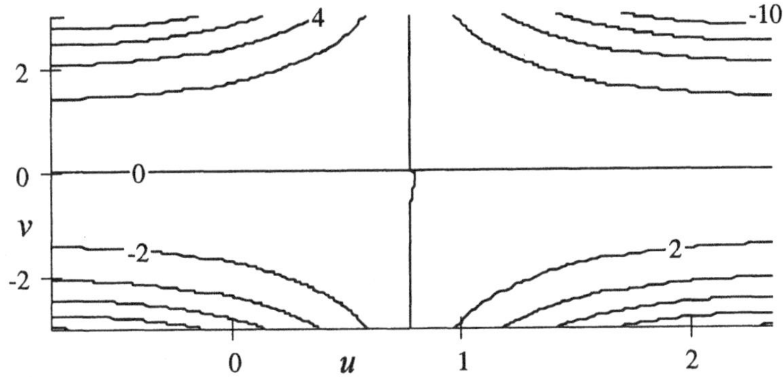

Figure 3.38 Contour plot of the real part of the function g. The steepest descent path is the one that crosses the level line s in the most rapid way.

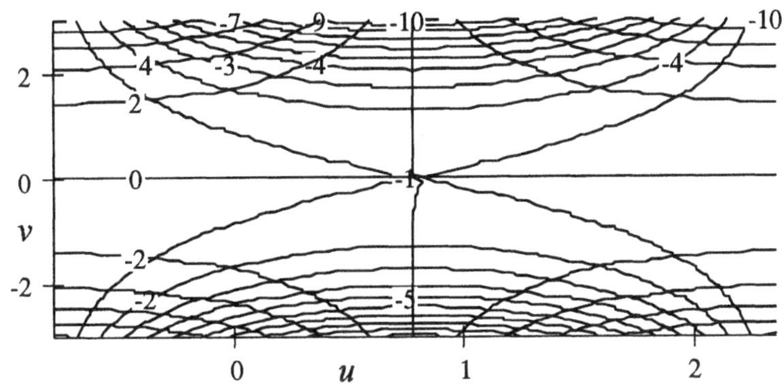

Figure 3.39 Contour plot of both the real and imaginary part of the function g.

3.6 Asymmetric three layer slab waveguide

Up to now we have considered symmetric structures; this geometry, although useful for theoretical considerations, is quite rare in integrated optics. In fact, in the optic frequency range, and sometimes also at millimetre waves, it is not possible to use metallic conductors due to the very high losses. As a consequence, we cannot use a metallic surface to support the dielectric waveguides; however, since the waveguide must be supported in some way, a dielectric substrate is generally used. In practice we have therefore, a substrate, used as a physical support, with a thin dielectric film on it acting as a waveguide, and air above the waveguide. This is an asymmetric planar waveguiding structure, as shown in Figure 3.40, with three layers with refractive indices such that $n_1 > n_2 > n_3$. In a typical configuration the upper layer is air ($n_3 = 1$), the core is GaAs ($n_1 = 3.6$) and the substrate is GaAlAs ($n_2 = 3.4$). As before, the layers are assumed to be of infinite extent in both the y– and z–directions, with waves propagating in the z–direction, without any variation with y.

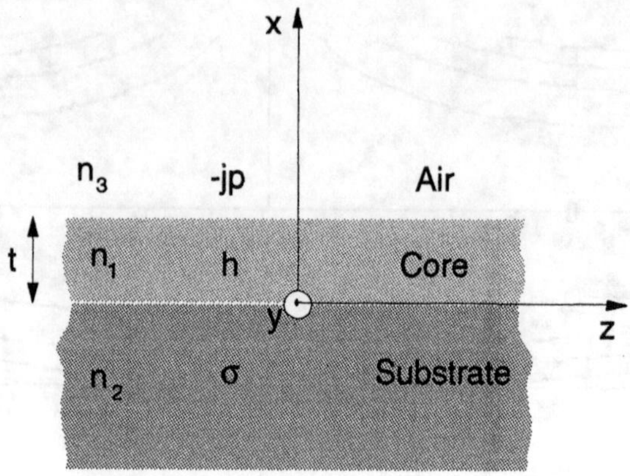

Figure 3.40 Geometry of the asymmetric three layer planar waveguide.

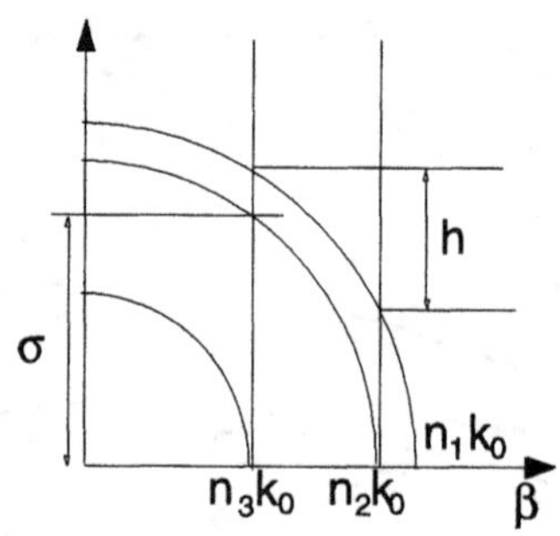

Figure 3.41 Transverse propagation constants in the core, h, and in the substrate, σ, for substrate modes.

For symmetric waveguides we have seen two types of modes; namely surface waves and radiative modes. Now a third possibility arises, since we may have total reflection at the air GaAs interface, but simply refraction at the interface between GaAs and GaAlAs. In this way, a source placed in the substrate produces rays which undergo refraction at the interface between substrate and core, while they are totally reflected at the interface between core and air. This wave decays exponentially in air while it varies sinusoidally both in the core and in the substrate. This is a mode which is guided at one side of the interface and radiated at the other side. In conclusion, three types of propagating modes can now exist:

Bound modes exponentially decaying along x both in the substrate and in air. These are discrete, finite in number, and their field shape and propagation constant can be obtained by the solution of an eigenvalue equation. Since they propagate in the core, and they are attenuated in the substrate, their propagation constant is in the range:

$$n_2 k_0 \leq \beta \leq n_1 k_0 \tag{3.137}$$

This type of mode can be thought of as being generated from a source placed in the core at $z = -\infty$; all transients have died out and the field is guided by the total reflection arising at the two interfaces. Therefore we have a zig–zag propagation in the core with reflection at the interfaces. The reflection does not quite take place at the interfaces, but the fields extend just a little into the semi infinite regions (Goos–Hänschen shift). Note also that there is no real power either in the substrate or in air; the Poynting vector is real only in the z–direction, while it is imaginary in the transverse direction.

Substrate modes corresponding to modes which, along x, vary sinusoidally in the core as well as in the substrate, but are exponentially damped in air. Substrate modes, in order to have an imaginary transverse wavenumber in air and a real one in the core and in the substrate must satisfy the condition

$$n_3 k_0 \leq \beta \leq n_2 k_0 \tag{3.138}$$

As evident from Figure 3.41, when (3.138) holds the transverse propagation constants in regions 1, 2 and 3 are respectively $h, \sigma, -jp$ and are in the range

$$\begin{aligned} 0 \leq \sigma \leq \sqrt{n_2^2 - n_3^2}\, k_0 \\ \sqrt{n_1^2 - n_2^2}\, k_0 \leq h \leq \sqrt{n_1^2 - n_3^2}\, k_0 \end{aligned} \tag{3.139}$$

This type of mode is continually losing energy from the core to the substrate, corresponding to the damping of the mode as it propagates. However, these modes are not useful for communication purposes since they decay rapidly. Nevertheless, they find use in coupler applications. This mode can be thought of as generated from a source in the substrate, but this time the source has to be placed not too far from the observation point due to the decay of the mode.

Air modes i.e. fields vary sinusoidally in air, in the core, and in the substrate so that:

$$0 \leq \beta \leq n_3 k_0 \tag{3.140}$$

In this case, the energy is free to spread out both in air and in the substrate. A graphical representation of this type of mode is provided in Figure 3.42 where h, σ, p are the propagation constants in the x–direction, in the core, substrate and air respectively, corresponding to a given propagation constant in the z–direction.

When this type of field is excited we have propagation constants as those shown in Figure 3.43, while the field is sinusoidally varying in all the three regions as shown in Figure 3.45. It is also clear that, when (3.140) holds we have the following continuous range of variations

$$\begin{aligned} 0 \leq p \leq n_3 k_0 \\ \sqrt{n_2^2 - n_3^2}\, k_0 \leq \sigma \leq n_2 k_0 \\ \sqrt{n_1^2 - n_3^2}\, k_0 \leq h \leq n_1 k_0 \end{aligned} \tag{3.141}$$

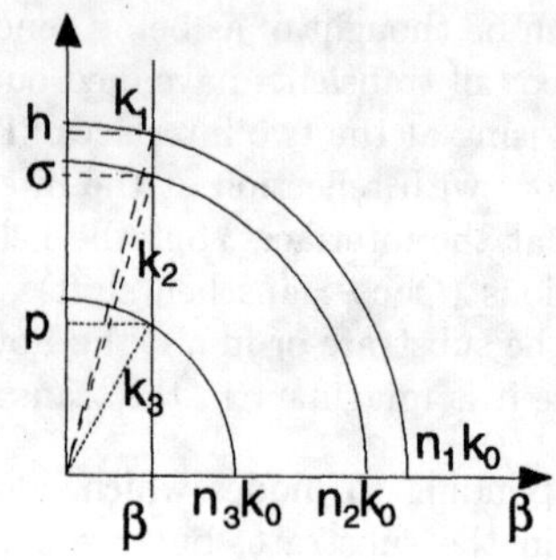

Figure 3.42 Transverse propagation constants for air modes.

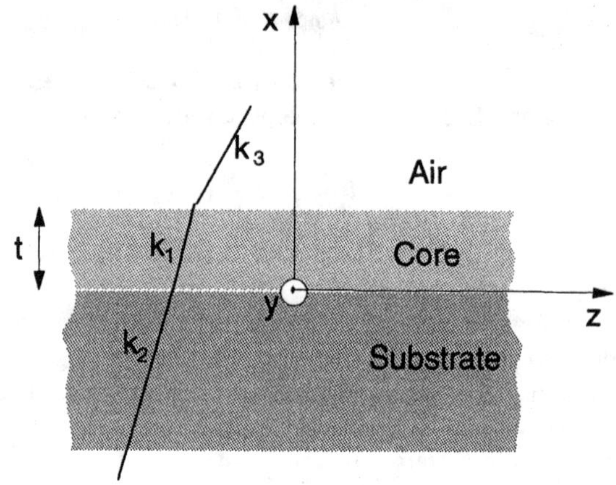

Figure 3.43 Direction of the propagation vectors for air modes.

which are graphically shown in Figure 3.44. For the latter type of modes it is noted that as the transverse variation becomes more and more rapid β becomes imaginary. Thus, the modes are exponentially attenuated in the propagation direction and this fact corresponds to energy localised near their excitation region. These type of modes are also part of the continuous spectrum and are necessary to describe the field arising near a discontinuity.

As a last possibility, should the propagation constant become

$$\beta \geq n_3 k_0 \tag{3.142}$$

we would get a mode of the exponential type in all three regions, namely in air, in the substrate and in the core. This mode should also satisfy the continuity condition at the interface for both the field and its transverse derivative. Clearly, this mode is not physically realisable, since it increases toward infinity. The modal field distributions which can actually take place in the three–layer asymmetric slab waveguide are shown in Figure 3.45.

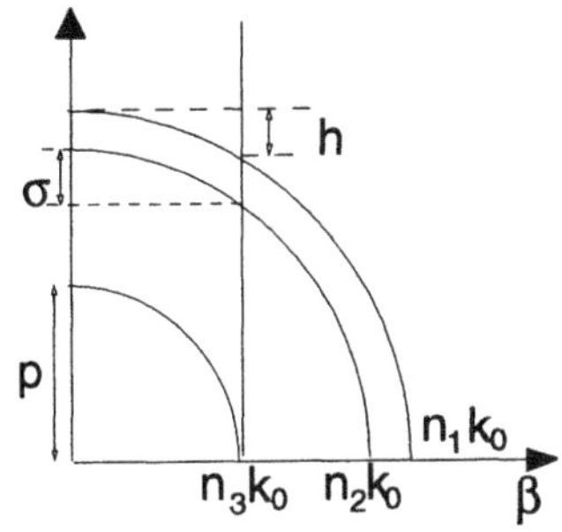

Figure 3.44 Range of values for the transverse propagation constants of air modes.

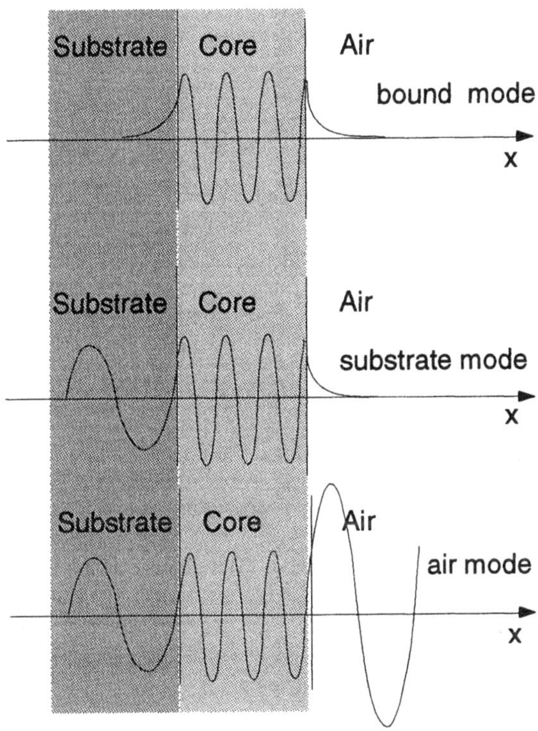

Figure 3.45 The three types of modes possible in an asymmetric three-layer slab waveguide.

3.6.1 TE surface waves

The electromagnetic study of the three–layer planar waveguide follows essentially the same procedure outlined in the TE case for the dielectric slab in homogeneous dielectric (Sections 3.2 and 3.3). Naturally, since the transverse resonance formulation has been introduced (Section 3.4), we will use this type of approach to determine the eigenvalue equation; once the latter has been established we derive the corresponding field distribution E_y. In this case, since the field is confined to the core, we have only reactive energy in the substrate and in air. Thus, the terminations corresponding to these regions are purely reactive in the equivalent network, as shown in Figure 3.46. By denoting the transverse propagation constants in the core, substrate and air as h, q and p respectively, we have

$$\begin{aligned} \beta^2 + h^2 &= (n_1 k_0)^2 \\ \beta^2 - q^2 &= (n_2 k_0)^2 \\ \beta^2 - p^2 &= (n_3 k_0)^2 \end{aligned} \tag{3.143}$$

or, briefly,

$$\beta^2 = (n_1 k_0)^2 + p^2 = (n_2 k_0)^2 + q^2 = (n_3 k_0)^2 - h^2 \tag{3.144}$$

It is also convenient to normalise the various admittances to $1/\omega\mu_0$, thus obtaining

$$Y_{01} = h; \quad Y_{02} = -jq; \quad Y_{03} = -jp \tag{3.145}$$

By choosing the reference point at $x = t$ we have looking towards $x > t$

$$\overrightarrow{Y}(t) = -jp \tag{3.146}$$

whereas for $x < t$,

$$\overleftarrow{Y}(t) = h\frac{[-jq + jh\tan(ht)]}{[h + j(-jq)\tan(ht)]} = jh\tan(ht - \overline{\phi}) \tag{3.147}$$

where we have set

$$\frac{q}{h} = \tan\overline{\phi} \tag{3.148}$$

The resonance condition is obtained as

$$\overleftrightarrow{Y}(t) = \overleftarrow{Y}(t) + \overrightarrow{Y}(t) = 0 \tag{3.149}$$

which is equivalent to say

$$h\tan(ht - \overline{\phi}) = p \tag{3.150}$$

The latter equation when solved together with

$$\begin{aligned} q^2 &= \left(n_3^2 - n_2^2\right) k_0^2 - h^2 \\ p^2 &= \left(n_3^2 - n_1^2\right) k_0^2 - h^2 \end{aligned} \tag{3.151}$$

provides the discrete solution corresponding to surface waves.

Once the transverse wavenumbers have been calculated it is a relatively simple task to compute the fields corresponding to this type of wave. In fact the field in the air region is of the type

$$\phi(x) = e^{-p(x-t)} \qquad x \geq t \tag{3.152}$$

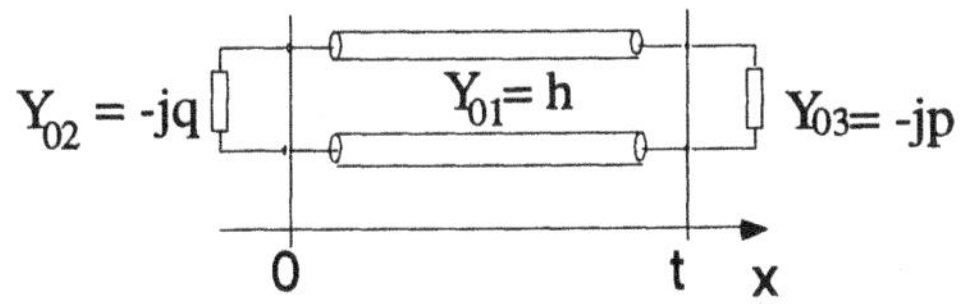

Figure 3.46 Equivalent network used to establish the eigenvalue equation for surface modes. The impedances relative to air and substrate are both imaginary, thus implying that no energy is radiated in these regions.

By using an $ABCD$ representation, and by referring to the equivalent circuit of Figure 3.46, it is also possible to calculate the field in the core as

$$\begin{aligned}\phi(x) &= \left[\cos\left[h\left(x-t\right)\right] \quad \tfrac{j}{h}\sin\left[h\left(t-x\right)\right]\right]\begin{bmatrix}1\\ -jp\end{bmatrix}\\ &= \cos\left[h\left(x-t\right)\right] + \frac{p}{h}\sin\left[h\left(t-x\right)\right]\end{aligned} \tag{3.153}$$

After inserting in the above equation the ratio p/h as found from (3.150), we obtain the field in the core region given by

$$\phi(x) = \cos\left[h\left(x-t\right)\right] - \tan\left(ht-\overline{\phi}\right)\sin\left[h\left(t-x\right)\right] \tag{3.154}$$

and after straightforward trigonometric manipulations we get

$$\phi(x) = \frac{\cos\left(hx-\overline{\phi}\right)}{\cos\left(ht-\overline{\phi}\right)} \qquad 0 \le x \le t \tag{3.155}$$

It is apparent that the above field and its derivative are continuous at $x = t$. This field when calculated at $x = 0$ also provides the factor which multiplies the exponential function in the substrate region, thus giving for the substrate

$$\phi(x) = \frac{\cos\overline{\phi}}{\cos\left(ht-\overline{\phi}\right)}e^{qx} \qquad x \le 0 \tag{3.156}$$

In conclusion, after introducing a factor A for normalisation purposes, we get the following field representation

$$\phi(x) = \begin{cases} Ae^{-p(x-t)} & x \ge t \\ A\dfrac{\cos\left(hx-\overline{\phi}\right)}{\cos\left(ht-\overline{\phi}\right)} & 0 \le x \le t \\ A\dfrac{\cos\left(\overline{\phi}\right)}{\cos\left(ht-\overline{\phi}\right)}e^{qx} & x \le 0 \end{cases} \tag{3.157}$$

with the constant A obtained through direct integration as,

$$A = \sqrt{\frac{2}{1/p+t+1/q}}\,\cos(ht-\overline{\phi}) \tag{3.158}$$

3.6.2 TE substrate modes

For substrate modes the termination on the substrate side is resistive, while the termination on the air side remains reactive. Accordingly, by referring to Figure 3.40, where in the inset the propagation constants are also reported, the equivalent network representing our structure is given by Figure 3.47. Propagation constants in the different regions are determined by

$$\begin{aligned}\beta^2 + h^2 &= (n_1 k_0)^2 \\ \beta^2 + \sigma^2 &= (n_2 k_0)^2 \\ \beta^2 - p^2 &= (n_3 k_0)^2\end{aligned} \tag{3.159}$$

Also in this case it is convenient to normalise the various admittances to $1/\omega\mu_0$, thus obtaining

$$Y_{01} = h; \quad Y_{02} = \sigma; \quad Y_{03} = -jp \tag{3.160}$$

By choosing the reference point at the top of the substrate ($x = 0$), with reference to the equivalent network of Figure 3.47, we have

$$\begin{aligned}\overleftarrow{Y} &= \sigma \\ \overrightarrow{Y}(0) &= jh\frac{-p + h\tan(ht)}{h + p\tan(ht)} = jh\tan\phi_s\end{aligned} \tag{3.161}$$

where we have introduced the phase shift ϕ_s which satisfies

$$\tan\phi_s = \frac{-p + h\tan(ht)}{h + p\tan(ht)} \tag{3.162}$$

This phase shift corresponds to considering an equivalent transmission line, of propagation constant h, closed on a magnetic wall at an electrical distance ϕ_s from the core–substrate interface. Total reflection of the equivalent ray takes place at this wall, as predicted by the Goos–Hänschen shift. The total admittance seen from $x = 0$ is given by

$$\overleftrightarrow{Y}(0) = \sigma + jh\tan\phi_s = \sigma\left(1 + \frac{jh}{\sigma}\tan\phi_s\right) \tag{3.163}$$

which, after introduction of the phase shift α_s

$$\tan\alpha_s = \frac{h}{\sigma}\tan\phi_s \tag{3.164}$$

gives

$$\overleftrightarrow{Y}(0) = \sigma(1 + j\tan\alpha_s) \tag{3.165}$$

Also in this case, it is possible to consider α_s as arising from the Goos–Hänschen effect, for α_s represents the electrical length of a line with characteristic impedance σ closed on a magnetic wall. After determining the propagation constants, it is a relatively simple matter to find the eigenfunctions normalised so that

$$\int_{-\infty}^{\infty} \phi(x;\sigma)\phi(x;\sigma')dx = \delta(\sigma - \sigma') \tag{3.166}$$

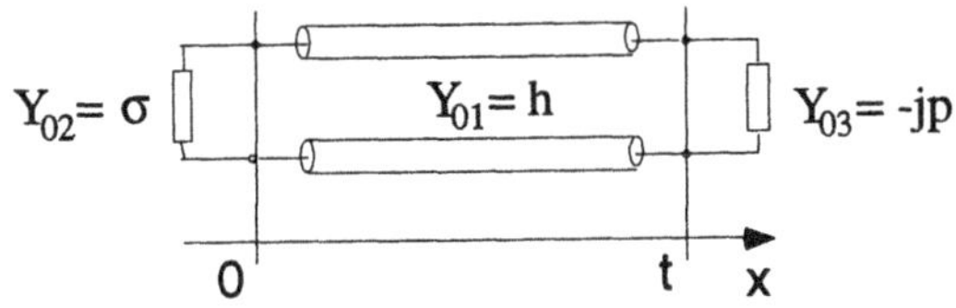

Figure 3.47 Equivalent circuit for the determination of the substrate mode. The termination corresponding to air is purely reactive, while the substrate termination is resistive.

which are given by

$$\phi(x;\sigma) = \begin{cases} \sqrt{\frac{2}{\pi}}\cos(\sigma x - \alpha_s) & x \le 0 \\ \sqrt{\frac{2}{\pi}\frac{\cos\alpha_s}{\cos\phi_s}}\cos(hx - \phi_s) & 0 \le x \le t \\ \sqrt{\frac{2}{\pi}\frac{\cos\alpha_s}{\cos\phi_s}}\cos(ht - \phi_s)e^{-p(x-t)} & x \ge t \end{cases} \tag{3.167}$$

3.6.3 TE air modes

Air modes are fields propagating in air as well as in the substrate region. The transverse equivalent network representing this situation is shown in Figure 3.48; this network is a cascade of transmission lines closed at both ends by resistive terminations (i.e. $q = j\sigma, p = j\rho$). Since the air modes pertain to the continuous part of the spectrum, their source is placed at $x = \pm\infty$. However, if the structure is symmetric, i.e. when substrate and air region have the same dielectric constants, by placing equal excitations at $x = \pm\infty$, a magnetic wall occurs at the mid plane. Similarly, if opposite excitations are applied at $x = \pm\infty$, then an electric wall will occur at the plane of symmetry. By proceeding in this way we obtain even and odd modes with real field distribution of the form of a stationary wave in the substrate, rather than a complex exponential denoting travelling wave propagation through the dielectric slab. A similar procedure can also be applied to asymmetric structures; the magnetic/electric wall has now to be placed in a suitable position to compensate the refractive index difference between air and substrate. In order to simplify the task of finding the fields in an asymmetric slab, it is convenient to locate this symmetry plane and then to use it as a reference plane. This will result in the two halves of the transverse equivalent network being decoupled (see Figure 3.48), so that the fields at the left hand side and right hand side of the magnetic/electric wall can be found separately. We will now determine the position of the magnetic wall for the three layer asymmetric waveguide.

The transverse equivalent network for the three layer slab representing air modes is shown in Figure 3.48; if a magnetic wall is placed at $x = x_0$, the total current $I(x_0)$ equals zero. Let a_1, b_1 be the incident and reflected wave amplitude at $x = x_0$, excited by the generator E_1 placed at $x = -\infty$, and a_2, b_2 be the incident and reflected wave amplitudes at $x = x_0$ excited by the generator E_2 placed at $x = +\infty$. The currents I_1, I_2 due to either generator are proportional to

$$\begin{aligned} I_1 &\quad \propto a_1 - b_1 \\ I_2 &\quad \propto a_2 - b_2 \end{aligned} \tag{3.168}$$

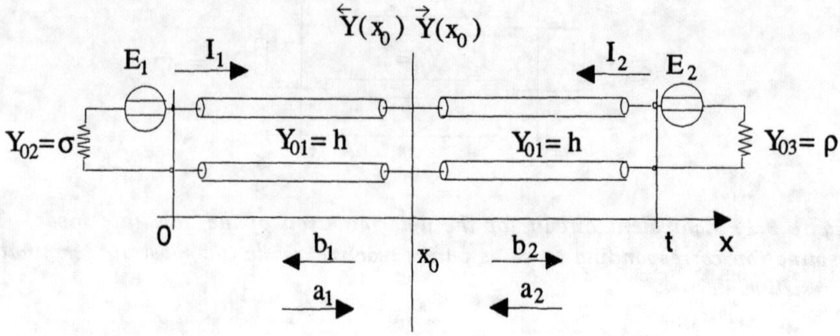

Figure 3.48 Transverse equivalent network representing air modes.

At $x = x_0$ the total current flowing is equal to zero (this is the condition for the existence of a magnetic wall) implying $I_1 = I_2$ and, consequently,

$$1 - \frac{b_1}{a_1} = \left(1 - \frac{b_2}{a_2}\right)\frac{a_2}{a_1} \tag{3.169}$$

Let us now define two phase shifts along the transmission lines as

$$\overline{\phi} = hx_0 \tag{3.170}$$
$$\chi = h(t - x_0) \tag{3.171}$$

From elementary network theory, with reference to Figure 3.48, wave amplitudes are given by

$$\begin{aligned} a_1 &= \frac{E_1}{1+\frac{h}{\sigma}}e^{-j\overline{\phi}} \\ b_1 &= \frac{h-\rho}{h+\rho}e^{-j2\chi} \end{aligned} \tag{3.172}$$

and similarly,

$$\begin{aligned} a_2 &= \frac{E_2}{1+\frac{h}{\rho}}e^{-j\chi} \\ b_2 &= \frac{h-\sigma}{h+\sigma}e^{-j2\overline{\phi}} \end{aligned} \tag{3.173}$$

By inserting the above expressions into (3.169) we get

$$1 - \frac{(h-\rho)}{h+\rho}e^{-j2\chi} = \left[1 - \frac{(h-\sigma)}{h+\sigma}e^{-j2\overline{\phi}}\right] E\frac{\rho}{\sigma}\frac{h+\sigma}{h+\rho}e^{-j(\chi-\overline{\phi})} \tag{3.174}$$

where we have introduced $E = E_1/E_2$. In the above expression the ratio E is arbitrary and describes a family of possible solutions. By equating the magnitudes of left hand side and right hand side of the above expression, and by also equating the phases of (3.174), while setting

$$E = \sqrt{\frac{\sigma}{\rho}}e^{j(\chi-\overline{\phi})} \tag{3.175}$$

we obtain the equation

$$\frac{(\rho^2 - h^2)\sin 2\chi}{\rho} = \frac{(\sigma^2 - h^2)\sin 2\overline{\phi}}{\sigma} \tag{3.176}$$

that represents the condition to be fulfilled in order to recover a magnetic wall at $x = x_0$. We are now in a position to calculate the even TE air modes.

Even TE air modes

The equivalent circuit to be used is that with a magnetic wall placed at $x = x_0$. We will first consider the fields at the left hand side of the magnetic wall and the correspondent equivalent network. By taking $x = 0$ as reference plane we have

$$\begin{aligned} \overleftarrow{Y}(0) &= \sigma \\ \overrightarrow{Y}(0) &= jh\tan\overline{\phi} \\ \overleftrightarrow{Y} &= \overleftarrow{Y}(0) + \overrightarrow{Y}(0) = \sigma + jh\tan\overline{\phi} \end{aligned} \tag{3.177}$$

and by defining

$$\tan\alpha_0 = \frac{h}{\sigma}\tan\overline{\phi} \tag{3.178}$$

we have

$$\overleftrightarrow{Y} = \frac{\sigma}{\cos\alpha_0} e^{j\alpha_0} \tag{3.179}$$

From the above it is possible to find the even air modes normalised between $-\infty$ and $x = x_0$ as given by

$$\phi_e(x) = \begin{cases} \sqrt{\frac{2\rho}{\pi\sigma}} \frac{\cos\alpha_0}{\cos\overline{\phi}} \cos(hx - \overline{\phi}) & 0 \leq x \leq x_0 \\ \sqrt{\frac{2\rho}{\pi\sigma}} \cos(\sigma x - \alpha_0) & x \leq 0 \end{cases} \tag{3.180}$$

Following a similar procedure at $x = t$ we can find the fields at the other side of the magnetic wall, where we obtain

$$\begin{aligned} \overrightarrow{Y}(t) &= \rho \\ \overleftarrow{Y}(t) &= jh\tan(ht - \overline{\phi}) \end{aligned} \tag{3.181}$$

The total admittance becomes

$$\overleftrightarrow{Y} = \overleftarrow{Y}(t) + \overrightarrow{Y}(t) = \rho + jh\tan(ht - \overline{\phi}) \tag{3.182}$$

and by defining

$$\tan\alpha_t = \frac{h}{\rho}\tan(ht - \overline{\phi}) \tag{3.183}$$

we have

$$\overleftrightarrow{Y}(t) = \frac{\rho}{\cos\alpha_t} e^{j\alpha_t} \tag{3.184}$$

From the above we find the even air modes normalised between $x = x_0$ and $+\infty$ as

$$\phi_e(x) = \begin{cases} \sqrt{\frac{2}{\pi}} \frac{\cos\alpha_t}{\cos[h(t - x_0)]} \cos[h(x - x_0)] & x_0 \leq x \leq t \\ \sqrt{\frac{2}{\pi}} \cos[\rho(x - x_0) + \alpha_t] & x \geq t \end{cases} \tag{3.185}$$

Normalising now the fields between $-\infty$ and $+\infty$ so that

$$\int_{-\infty}^{\infty} \phi_e(x;\rho)\phi_e(x;\rho')dx = \delta(\rho-\rho')$$

we obtain

$$\phi_e(x) = \begin{cases} \sqrt{\frac{2}{\pi}}A_e \frac{\cos\left[ht-\overline{\phi}\right]}{\cos\alpha_t}\cos\left[\rho(x-t)+\alpha_t\right] & x \geq t \\ \sqrt{\frac{2}{\pi}}A_e \cos\left[hx-\overline{\phi}\right] & 0 \leq x \leq t \\ \sqrt{\frac{2}{\pi}}A_e \frac{\cos\overline{\phi}}{\cos\alpha_t}\cos\left[\sigma x-\alpha_t\right] & x \leq 0 \end{cases} \tag{3.186}$$

where the normalisation constant is given by

$$A_e = \left[\frac{\sigma}{\rho}\left(\cos^2\overline{\phi} + \left(\frac{h}{\sigma}\right)^2 \sin^2\overline{\phi}\right) + \cos^2\left(ht-\overline{\phi}\right) + \left(\frac{h}{\rho}\right)^2 \sin^2\left(ht-\overline{\phi}\right)\right]^{-1} \tag{3.187}$$

The odd modes are obtained from the previous formulae by replacing all cosine functions by sine and tangent by cotangent.

3.6.4 Summary of TM modes

The TM mode spectrum of an asymmetric three–layer waveguide can be found by following the procedure outlined for the TE case. For reference purposes, the normalised eigenfunctions pertaining to this case are reported in the following. The following expressions hold for surface, substrate and air modes.

Surface modes

$$\phi_{se}(x) = \begin{cases} Ae^{-p(x-t)} & x \geq t \\ A\frac{\cos(hx-\overline{\phi})}{\cos(ht-\overline{\phi})} & t \geq x \geq 0 \\ A\frac{\cos\overline{\phi}}{\cos(ht-\overline{\phi})}e^{qx} & x \leq 0 \end{cases} \tag{3.188}$$

with the normalisation constant A given by

$$A = \left(\frac{\frac{p}{h^2} + \frac{1}{p} + \left(\frac{t}{\varepsilon_1} + \frac{\varepsilon_2(h^2+q^2)}{q(\varepsilon_2^2h^2+\varepsilon_1^2q^2)}\frac{1}{\cos^2(ht-\overline{\phi})}\right)}{2}\right)^{-1/2} \tag{3.189}$$

and

$$\overline{\phi} = \tan^{-1}\left(\frac{q\varepsilon_1}{h\varepsilon_2}\right) \tag{3.190}$$

Naturally, the wavenumbers in the three regions are related in the usual manner as

$$\begin{aligned} \varepsilon_1 k_0^2 &= h^2+\beta^2 \\ \varepsilon_2 k_0^2 &= -q^2+\beta^2 \\ \varepsilon_3 k_0^2 &= -p^2+\beta^2 \end{aligned} \tag{3.191}$$

Substrate modes

In this case a standing wave is present in the substrate, so that the propagation constant in this layer is now $q = -j\sigma$. With this substitution, the normalised eigenfunctions are given by

$$\phi_{sb}(x) = \begin{cases} \sqrt{\frac{2}{\pi}\varepsilon_2} A_{sb} \cos(ht - \overline{\phi}_{sb}) e^{-p(x-t)} & x \geq t \\ \sqrt{\frac{2}{\pi}\varepsilon_2} A_{sb} \cos(hx - \overline{\phi}_{sb}) & t \geq x \geq 0 \\ \sqrt{\frac{2}{\pi}\varepsilon_2} \cos(\sigma x - \overline{\alpha}_{sb}) & x \leq 0 \end{cases} \tag{3.192}$$

with the various constants defined by

$$\begin{aligned} \alpha_{sb} &= \tan^{-1}\left(\tfrac{h\varepsilon_2}{\sigma\varepsilon_1}\tan(\overline{\phi}_{sb})\right) \\ \overline{\phi}_{sb} &= \tan^{-1}\left(\frac{\frac{h}{\varepsilon_1}\tan(ht) - p}{\frac{h}{\varepsilon_1} + p\tan(ht)}\right) \\ A_{sb} &= \frac{\cos\alpha_{sb}}{\cos\overline{\phi}_{sb}} \end{aligned} \tag{3.193}$$

Wavenumber conservation is expressed as before, apart from the substrate, where now transverse propagation takes place

$$\begin{aligned} \varepsilon_1 k_0^2 &= h^2 + \beta^2 \\ \varepsilon_2 k_0^2 &= \sigma^2 + \beta^2 \\ \varepsilon_3 k_0^2 &= -p^2 + \beta^2 \end{aligned} \tag{3.194}$$

Air modes

In this case we have transverse propagation in all three layers and consequently $p = -j\rho$. The wavenumbers are related from the following relationships

$$\begin{aligned} \varepsilon_1 k_0^2 &= h^2 + \beta^2 \\ \varepsilon_2 k_0^2 &= \sigma^2 + \beta^2 \\ \varepsilon_3 k_0^2 &= \rho^2 + \beta^2 \end{aligned} \tag{3.195}$$

By following the same arguments applied to the TE case, it is convenient to consider separately the even and odd cases. For the even case we get

$$\phi_e(x) = \begin{cases} \sqrt{\frac{2}{\pi}} A_e \frac{\cos(ht - \overline{\phi}_e)}{\cos(\alpha_e)} \cos(\rho(x-t) + \alpha_e) & x \geq t \\ \sqrt{\frac{2}{\pi}} A_e \cos(hx - \overline{\phi}_e) & t \geq x \geq 0 \\ \sqrt{\frac{2}{\pi}} A_e \frac{\cos(\overline{\phi}_e)}{\cos(\overline{\alpha}_e)} \cos(\sigma x - \overline{\alpha}_e) & x \leq 0 \end{cases} \tag{3.196}$$

where the constants appearing in the above expression are given by

$$\begin{aligned} \alpha_e &= \tan^{-1}\left(\tfrac{h}{\rho\varepsilon_1}\tan(ht - \overline{\phi}_e)\right) \\ \overline{\alpha}_e &= \tan^{-1}\left(\tfrac{h\varepsilon_2}{\rho\varepsilon_1}\tan(\overline{\phi}_e)\right) \end{aligned} \tag{3.197}$$

and the angle $\overline{\phi}_e$ is given by the solution of the equation

$$\frac{\sin 2\overline{\phi}_e}{\sin\left[2(ht - \overline{\phi}_e)\right]} = \frac{\sigma}{\rho\varepsilon_2}\left[\frac{\rho^2 - \frac{h^2}{\varepsilon_1^2}}{\frac{\sigma^2}{\varepsilon_2^2} - \frac{h^2}{\varepsilon_1^2}}\right] \tag{3.198}$$

while the normalisation constant is obtained as

$$A_e = \left\{ \frac{\sigma}{\rho\varepsilon_2} \left(\cos^2\overline{\phi}_e + \frac{h^2\varepsilon_2^2}{\sigma^2\varepsilon_1^2} \sin^2\overline{\phi}_e \right) + \frac{h^2}{\rho^2\varepsilon_1^2} \sin^2(ht - \overline{\phi}_e) + \cos^2(ht - \overline{\phi}_e) \right\}^{-1/2} \tag{3.199}$$

The odd air modes are obtained from the previous formulae by replacing all cosine functions by sine and tangent by cotangent.

Bibliography

[1] C. S. Chang. *A Rigorous analysis of cascaded step discontinuities in open waveguides.* PhD thesis, University of Bath, Bath, U.K., 1993.

[2] R. E. Collin. *Field Theory of Guided Waves.* IEEE Press, New York, 1991.

[3] L. Felsen and N. Marcuvitz. *Radiation and Scattering of Waves.* Prentice Hall, Englewood Cliffs, NJ, 1973.

[4] B. Friedman. *Principles and Techniques of Applied Mathematics.* John Wiley & Sons Inc., New York, 1956.

[5] R. F. Harrington. *Time Harmonic Electromagnetic Fields.* McGraw-Hill, New York, 1961.

[6] H. A. Haus. *Waves and Fields in Optoelectronics.* Prentice-Hall, Englewood Cliffs, New Jersey, 1984.

[7] R. G. Hunsperger. *Integrated Optics: Theory and Technology.* Springer-Verlag, Berlin, 1991.

[8] D. L. Lee. *Electromagnetic Principles of Integrated Optics.* John Wiley & Sons, Singapore, 1986.

[9] V. V. Shevchenko. *Continuous Transition in Open Waveguides.* Golem Press, Boulder, Colorado, 1971.

[10] E. W. Snyder and J. D. Love. *Optical Waveguide Theory.* Chapmann and Hall, 1983.

[11] R. Sorrentino. Transverse resonance technique. In T. Itoh, editor, *Numerical Techniques for Microwave and Millimeter-wave Passive Sructures.* John Wiley & Sons, New York, 1989.

[12] G. Tyras. *Radiation and propagation of electromagnetic waves.* Academic Press, New York, 1969.

Chapter 4

Bound modes by approximate methods

Bound modes of open waveguides may be found in several different ways. In particular, in Chapter 6, a rigorous, full–wave approach will be presented. However, in several instances, bound modes may also be found by relatively simple approximate methods, such as the effective dielectric constant (EDC) method and conformal mapping. Such methods have been devised both for optical waveguides and for microwave guiding structures.

The method used for optical waveguides is the EDC and will be illustrated in the first Section of this Chapter. Not surprisingly the accuracy of the results obtained by this method improves as the frequency increases. It will be observed also that the results relative to the propagation constant are fairly accurate. This is valid for all methods since, as discussed in the last part of Section 4.1, the propagation constant is a variational quantity. An example of application of the EDC is then provided in Section 4.2, where a waveguide commonly used in integrated optics is investigated.

Open waveguides operating in the microwave frequency range are often investigated by a different approximate method, i.e. by conformal mapping. The latter is introduced by referring to the example of a microstrip line in the last Section of this Chapter.

4.1 Bound modes by the effective dielectric constant method

In the previous Chapter we have seen the analysis of waveguiding structures which are uniform along one direction, namely the y–direction. These structures, although providing confinement of the optical fields in only one dimension, are important for theoretical analysis and useful in many integrated optics applications [3]. However, most applications require field confinement in two–dimensions and lead to the open, nonseparable, waveguiding structures, used in optic and in millimetre–wave frequency ranges, which are shown in Figure 4.1. The rigorous analysis of these structures is complicated by the fact that the geometry is non–uniform in the y–

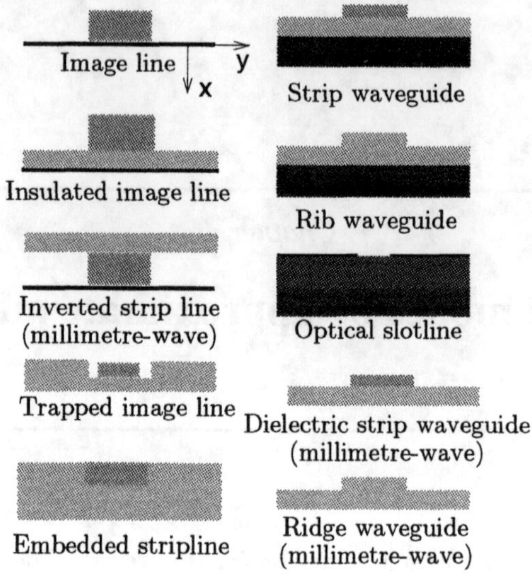

Figure 4.1 Some commonly used waveguiding structures for integrated optics and millimetre–wave integrated circuits.

direction; this produces diffraction of the electromagnetic field. In other words, the presence of corners and edges has an effect on the mode coupling of the fields, so that solutions for the above structures cannot be obtained in a separable form. As a consequence, the modes pertaining to the guides of Figure 4.1 are of hybrid nature, in the sense that a mode has all six field components ($E_x, E_y, E_z, H_x, H_y, H_z$). It is still possible to derive rigorously the modes of these structures by using the transverse resonance diffraction method, or other numerical methods, as described in Chapter 6.

However, in many practical cases, designers are content to know propagation constants and impedances of the transmission lines within a certain accuracy, so that suitable approximations can be effected in order to simplify circuit synthesis. Historically, the first semi–analytic treatment of rectangular dielectric guides was presented by Marcatili in his pioneering work [7]. This method, illustrated in the next Section, is based on the assumption that modes are strongly guided; i.e. fields are mainly concentrated in the core region with little or no penetration in the cladding regions or in corners. In addition, it is also assumed that fields are separable in the various regions of interest. As such, this type of analysis is only applicable for modes well above cut–off.

A more accurate method of analysis was introduced by Knox and Toulios [5] and is known as effective dielectric constant (EDC). This method, sometimes also referred to as the effective index method (EIM) [6] has been used e.g. in [9] for the analysis of the image line. Since then, the EDC technique has been widely used, also in modified formulations, for the purpose of analyzing dielectric waveguides in integrated optics and in millimetrics [4]. We shall introduce the EDC technique with reference to the image guide as was done in the original formulation [5]. In Section 4.2 we shall also use the EDC technique to analyse rib waveguides.

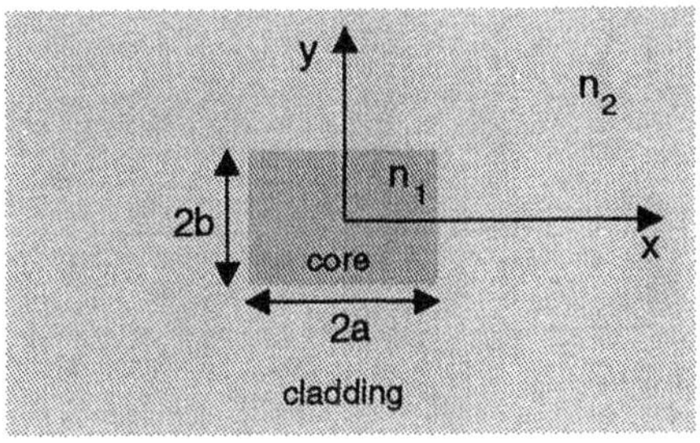

Figure 4.2 Cross–section of rectangular core dielectric waveguide. The refractive index of the core is n_1, while that of the cladding is n_2.

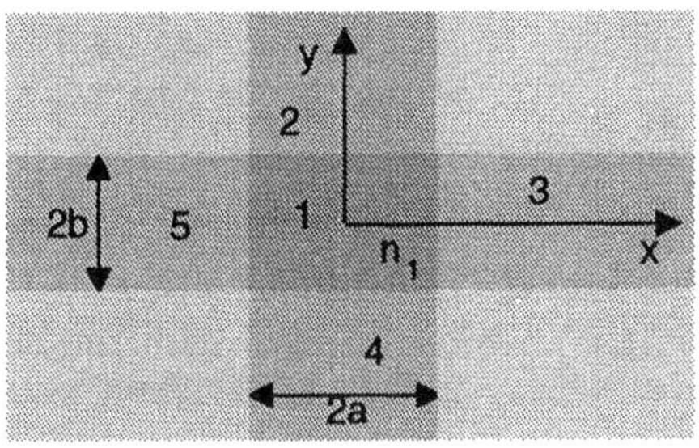

Figure 4.3 Cross–section of rectangular dielectric waveguide also showing the subdivisions used for analysis purposes.

4.1.1 Marcatili's method

Let us consider a waveguide with rectangular core cross–section with refractive indices as shown in Figure 4.2. We assume fields well above cut–off and strongly concentrated in the core. This guided mode is supported by total internal reflection which takes place at each interface. As a consequence, fields in the cladding region are exponentially decaying when moving away from the side walls. Moreover, in the cladding region, the field in the four corners (Figure 4.3) is supposed to be significantly smaller than the field above the walls, an assumption that is now seen as physically incorrect. Ignoring the effect of corners, one neglects completely the fields in these regions. However, one does not impose the presence of either a magnetic or electric wall at the boundary of the corner regions, leaving therefore an arbitrary boundary condition at the interface between, say, region 2 and the confining corner regions. On the other hand, the fields are matched at the four interfaces between regions 1 and regions 2, 3, 4, 5 (see Figure 4.3). Moreover, to further simplify matters, the field in the central region is obtained by separation of variables. The latter assumption, although clearly inaccurate, does again significantly facilitate the application of the method. As a final simplification we note that, while in general the field in the guide has to be described in terms of both potentials (electric and magnetic), we may attempt to use just one. In fact, if the core were infinite in the y–direction we would recover the slab waveguide considered in the previous Chapter; the field in the slab waveguide can be either TE or TM. If a TE field is

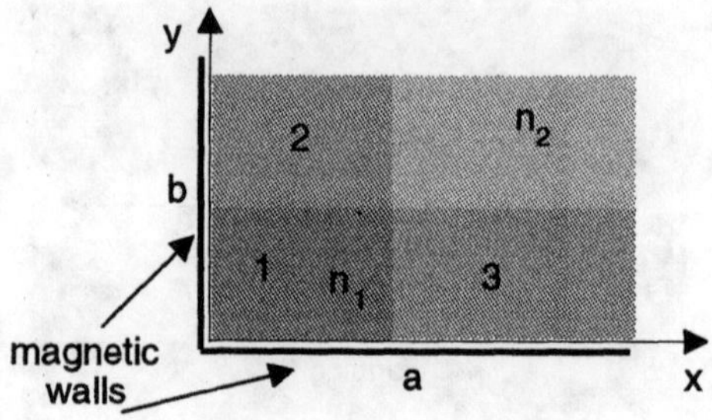

Figure 4.4 Reduced structure used for the analysis after taking symmetry into account.

present it can be described just in terms of the potential ψ_e. If we now introduce a small variation with y, the field remains TE but with five field components. It is the rapid field variation near the corner of the core which excites TM modes. By supposing that only a small amount of the field is present in this region, we also say that only a small fraction of the field is represented by the magnetic potential. The latter is then neglected in this method.

In order to simplify the analysis it is also convenient to consider symmetry. When taking even modes we may place a magnetic wall on both the x and y–axis thereby obtaining the structure of Figure 4.4. By suppressing the time–harmonic dependence as well as the z–dependence, we write for the electric potential

$$\begin{aligned} \phi^e &= A_1 \cos(k_x x)\cos(k_y y) \qquad \text{region 1} \\ \phi^e &= A_2 \cos(k_x x)\exp(-\eta y) \qquad \text{region 2} \\ \phi^e &= A_3 \cos(k_y y)\exp(-\xi x) \qquad \text{region 3} \end{aligned} \tag{4.1}$$

where k_x, k_y represent the propagation constant inside the core, while ξ, η describe the field decay in the cladding region. By imposing the continuity of tangential components of the fields, i.e. by requiring the potential and its normal derivative to be continuous at the interfaces, we obtain at the interface between regions 1 and 2

$$\begin{aligned} \eta &= k_y \tan k_y b \\ k_y^2 + \eta^2 &= (n_1^2 - n_2^2)k_0^2 \end{aligned} \tag{4.2}$$

while at the interface between regions 1 and 3 we get

$$\begin{aligned} \xi &= k_x \tan k_x a \\ k_x^2 + \xi^2 &= (n_1^2 - n_2^2)k_0^2 \end{aligned} \tag{4.3}$$

The solution of the two eigenvalue equations (4.2, 4.3) provides the value for the propagation constant since

$$k_z = \sqrt{n_1^2 k_0^2 - k_x^2 - k_y^2} \tag{4.4}$$

Note that (4.2, 4.3) are the same eigenvalue equations already obtained for even TE modes of a slab waveguide in Chapter 3. Also note that there is no link between the solution of (4.2) and that of (4.3), implying a complete decoupling in the two transverse directions: this is obviously a consequence of the hypothesis assumed. With slight modification, Marcatili's method can also be applied to other guiding structures such as, for example, the image guide.

Generally, the results obtained by the Marcatili method are inaccurate at lower frequencies and near cut–off, i.e. when the field is not tightly bound to the waveguide. On the other hand, in the high frequency range, when the mode is strongly bound to the guide, the method produces quite accurate results for the propagation constant, albeit the actual field distribution is quite inaccurate near the corners. An idea of the accuracy of the method can be obtained by looking at [5] where Marcatili's method is compared with a point matching approach and with the EDC theory.

4.1.2 Effective dielectric constant

The main deficiencies of Marcatili's method occur at low frequency and near cut–off of a mode, since we neglect the fields in the corner region, where the E–field on the $x-y$ plane becomes singular. A simple way to improve this situation was suggested in [5]. It consists in solving equation (4.2) first. This is equivalent to finding the eigenvalue of the slab waveguide shown in Figure 4.5. Once this eigenvalue has been found, we can say that the phase velocity in the structure of Figure 4.5 is given by

$$k_z^{slab} = n_{eff} k_0 = \sqrt{n_1^2 k_0^2 - k_y^2} \tag{4.5}$$

where we have introduced an effective dielectric constant, or equivalently the effective refractive index, of a hypothetical medium having the same phase velocity as the surface wave in the slab. Now, let us use this EDC as the dielectric constant of the core region for the problem in the x–direction. Therefore instead of (4.3) we solve the following eigenvalue problem

$$\begin{aligned} \xi &= k_x \tan k_x a \\ k_x^2 + \xi^2 &= (n_{eff}^2 - n_2^2) k_0^2 \end{aligned} \tag{4.6}$$

The above is an eigenvalue equation for the slab waveguide shown in Figure 4.6. Note that now n_{eff} contains the information obtained from the solution of the structure of Figure 4.5. Consequently, the two problems represented by (4.2) and (4.6) are now coupled. The new phase constant is obtained as

$$k_z = \sqrt{n_{eff}^2 k_0^2 - k_x^2} \tag{4.7}$$

Results obtained by the EDC method agree well with those obtainable from Marcatili's method in the high frequency range. In addition the EDC method also provides more accurate results near cut–off and at the lower frequencies. In fact, in the EDC method, the fields in the corner regions are not completely neglected. Even if the solution process is still an approximate one, and we are still describing the phenomena in terms of just one potential, we can now obtain quite accurate results.

Further improvement can be obtained if this technique is applied iteratively, that is, if the value of k_x from (4.6) is now entered in (4.5), i.e.:

$$k'^{slab}_z = n'_{eff} k_0 = \sqrt{n_1^2 k_0^2 - k_x^2 - k_y^2} \tag{4.8}$$

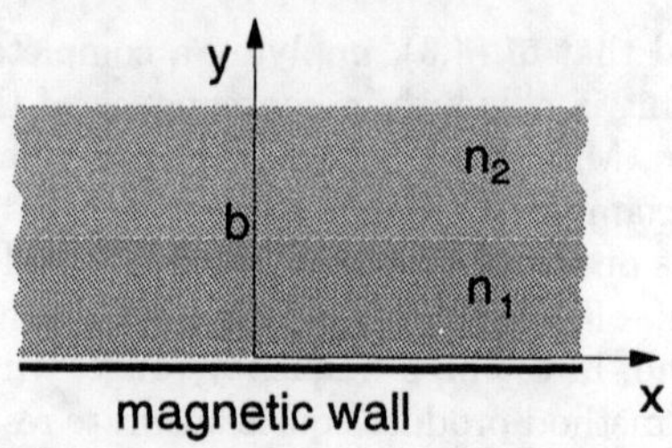

Figure 4.5 Subdivision used for the effective dielectric constant method.

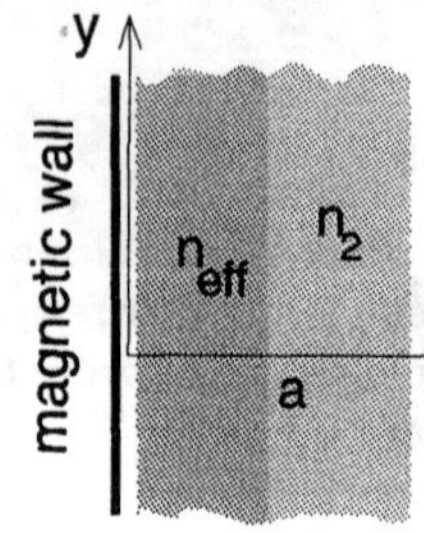

Figure 4.6 Subdivision used for the effective dielectric constant method. Note that now the refractive index of the core region has been replaced by n_{eff}.

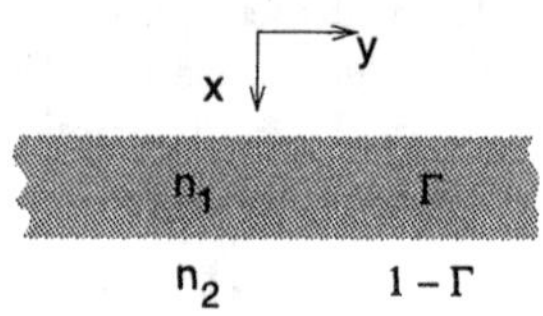

Figure 4.7 Slab waveguide uniform along the y–direction.

The solution in the y–direction is then repeated, yielding a new value of the effective permittivity n'_{eff} that is then employed in (4.6), yielding a better estimate, k'_x, of k_x. This process converges after very few iterations.

The EDC method has been the subject of several investigations and has also been extended to more general waveguide structures. An excellent overview of the EDC method and its application to the inverted strip dielectric waveguide can be found in [4]. In the following, we will consider some applications of the EDC method.

4.1.3 Variational method for the propagation constant

Let us consider a slab waveguide, uniform along the y–direction, as shown in Figure 4.7. The field in this waveguide is found from the wave equation in the transverse (x) direction. By solving this equation for any component of the electromagnetic field, say ϕ, we can calculate the other components as shown in the previous Chapter.

With reference to Figure 4.7 we have

$$\frac{d^2\phi(x)}{dx^2} + \left[\varepsilon_r(x)k_0^2 - \beta^2\right]\phi(x) = 0 \tag{4.9}$$

By multiplying the above equation by ϕ, and by integrating over the entire cross–section (which, in general, extends from $-\infty$ to $+\infty$) equation (4.9) becomes

$$\int_{-\infty}^{+\infty} \phi(x)\frac{d^2\phi(x)}{dx^2}dx + \int_{-\infty}^{+\infty} \left[\varepsilon_r(x)k_0^2 - \beta^2\right]\phi^2(x)dx = 0 \tag{4.10}$$

where $\varepsilon_r(x)$ will change according to the x position, while β is a constant so that

$$\int_{-\infty}^{+\infty} \phi(x)\frac{d^2\phi(x)}{dx^2}dx + k_0^2\int_{-\infty}^{+\infty} \varepsilon_r(x)\phi^2(x)dx = \beta^2\int_{-\infty}^{+\infty} \phi^2(x)dx \tag{4.11}$$

From the latter, by dividing both members of (4.11) by the quantity

$$\int_{-\infty}^{+\infty} \phi^2(x)dx \tag{4.12}$$

we get the following expression for the propagation constant

$$\beta^2 = \frac{\int_{-\infty}^{+\infty} \phi(x)\frac{d^2\phi(x)}{dx^2}dx + k_0^2\int_{-\infty}^{+\infty} \varepsilon_r(x)\phi^2(x)dx}{\int_{-\infty}^{+\infty} \phi^2(x)dx} \tag{4.13}$$

and consequently, we define the EDC ε_e as

$$\varepsilon_e = \left(\frac{\beta}{k_0}\right)^2 \tag{4.14}$$

The above is a *variational* or *stationary* expression for the propagation constant. We see, at a glance, that its value does not depend upon the magnitude of ϕ, but rather just on its distribution. Hence, if we normalise ϕ so that

$$\int_{-\infty}^{+\infty} \phi^2(x)dx = 1 \tag{4.15}$$

the denominator of (4.13) becomes unity. Moreover, let us suppose we obtain in some way a reasonable estimate of ϕ, different from the true distribution, say ϕ_0, by a certain "small" amount $\delta\phi$, that is

$$\phi = \phi_0 + \delta\phi \tag{4.16}$$

Strictly, $\delta\phi$ is a function of x, but in the following we will identify it with either its peak or mean square magnitude. By inserting (4.16) in (4.13) we cannot expect to get the true value β_0; but rather a somewhat incorrect value

$$\beta^2 = \beta_0^2 + \delta\left(\beta^2\right) \tag{4.17}$$

The interesting point is that, by introducing (4.16) in (4.13) and neglecting terms like $(\delta\phi)^2$, we see that the error arising from approximating β^2 is of the order of the *square* of the error arising from approximating the field:

$$\delta\left(\beta^2\right) = O\left((\delta\phi)^2\right) \tag{4.18}$$

As a consequence, an error of 10% made when calculating the field becomes an error of only 1% when calculating β^2.

By making some further assumptions, (4.13) can be further simplified leading to a slightly different expression for the EDC. Let us consider the integration by parts of

$$\int_{-\infty}^{+\infty} \phi(x) \frac{d^2\phi(x)}{dx^2} dx = \left[\frac{d^2\phi(x)}{dx^2} \phi(x) \right]_{-\infty}^{+\infty} - \int_{-\infty}^{+\infty} \left(\frac{d\phi(x)}{dx} \right)^2 dx \tag{4.19}$$

The bracketed term gives zero contribution in (4.19) when bound modes are considered, since both the field and its derivative are exponentially evanescent at infinity. In addition, let us suppose that the field changes smoothly: this is the case, for instance, if the refractive index of the slab varies smoothly or, anyway, is not very different from that of the surrounding cladding; such a situation is quite well verified in the dielectric fibre where in the core one has $n = 1.5$ while in the cladding $n = 1.4$. Under the above hypothesis we may neglect the square of the derivative of the field as compared to the field itself, so that (4.13) reduces to

$$\beta^2 \approx \frac{k_0^2 \int_{-\infty}^{+\infty} \varepsilon_r(x)\phi^2(x)dx}{\int_{-\infty}^{+\infty} \phi^2(x)dx} \tag{4.20}$$

If we now apply the above expression to the symmetrical slab waveguide of Figure 4.7, by separating the integral in the numerator into two contributions, namely the contribution from inside the slab and that from outside, we get

$$\beta^2 \approx \frac{2k_0^2 \int_d^{+\infty} n_2^2\phi^2(x)dx + k_0^2 \int_{-d}^{+d} n_1^2\phi^2(x)dx}{\int_{-\infty}^{+\infty} \phi^2(x)dx} \tag{4.21}$$

n_1 and n_2 are now constant and can be taken out of the integrals. The ratio

$$\Gamma = \frac{\int_{-d}^{+d} \phi^2(x)dx}{\int_{-\infty}^{+\infty} \phi^2(x)dx} \tag{4.22}$$

gives the fraction of the power present in the slab with respect to the total power in the guide. Similarly, the ratio

$$1 - \Gamma = \frac{2\int_d^{+\infty} \phi^2(x)dx}{\int_{-\infty}^{+\infty} \phi^2(x)dx} \tag{4.23}$$

gives the fraction of power present outside the guide. By using the above two equations, we may write the propagation constant as

$$\beta^2 \approx \Gamma n_1^2 k_0^2 + (1 - \Gamma)\, n_2^2 k_0^2 \tag{4.24}$$

or, equivalently

$$\varepsilon_e = \frac{\beta^2}{k_0^2} \approx \Gamma n_1^2 + (1 - \Gamma)\, n_2^2 \tag{4.25}$$

In (4.25) the square of the ratio between β and the wavenumber gives the effective refractive index. We have therefore found an approximate expression for the EDC in terms of the fractional power contained inside the slab; the range of validity of the above expression, however, is limited by the hypothesis of smooth variation. In particular, since the field variation gives some contribution to (4.20) we have to expect (4.25) to provide an excess estimate for β, since we have neglected a negative term in (4.20) as is actually found when comparing EDC calculations with exact ones. It is also noted that the approximation (4.20) improves as the frequency increases.

4.2 Analysis of rib waveguide by EDC

Let us consider a rib waveguide with dimensions and attached coordinate system as shown in Figure 4.8. We apply the EDC procedure by considering two different regions, the central one, that is the region of the rib (of width $2s$ in Figure 4.8), and the outer region, that is the region $|x| > s$. We also assume that just one value of k_x, k_y, is present in each section. Moreover, we consider the field as obtainable in terms of just one potential, e.g. in terms of Π_h directed along y. In order to apply the EDC to the central region, we consider a structure like the one shown in Figure 4.9, i.e. a structure which is uniform in the x–direction. From the wavenumber conservation we have

$$\begin{aligned} \varepsilon_1 k_0^2 &= \beta^2 + k_y^2 + k_x^2 \\ \varepsilon_2 k_0^2 &= \beta^2 - \Gamma_2^2 + k_x^2 \\ \varepsilon_3 k_0^2 &= \beta^2 - \Gamma_3^2 + k_x^2 \end{aligned} \tag{4.26}$$

where Γ_2, Γ_3 denote the decay constants in the regions with refractive indices n_2, n_3, respectively. By subtracting and setting $\varepsilon_3 = 1$, we get the following

$$\begin{aligned} (\varepsilon_1 - 1)\, k_0^2 &= \Gamma_3^2 + k_y^2 \\ (\varepsilon_1 - \varepsilon_2)\, k_0^2 &= \Gamma_2^2 + k_y^2 \end{aligned} \tag{4.27}$$

i.e. two equations in the three unknowns k_y, Γ_2, Γ_3. As pointed out in Chapter 3 we can find the other equation in two different ways. One possibility is to apply the continuity of the transverse components of the electromagnetic field; the other is to solve the same problem by using transverse resonance. In order to provide further examples we will illustrate both approaches. Let us start from the field continuity,

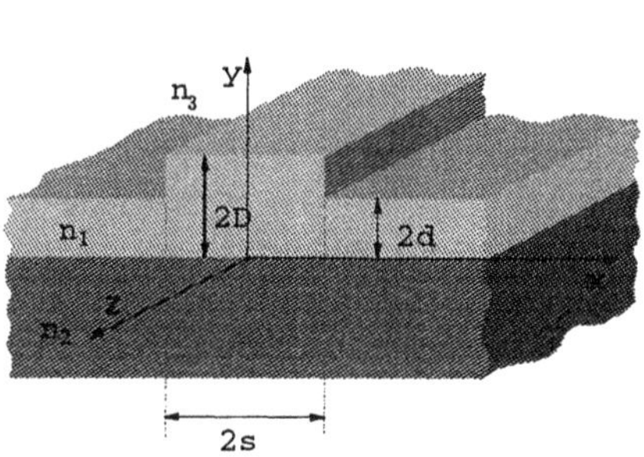

Figure 4.8 Cross-section of a rib waveguide.

where we write the electric field in the three regions as

$$E_x = \begin{cases} \cos(k_y 2D - \alpha)\, e^{-\Gamma_3(y-2D)} & y \geq 2D \\ \cos(k_y y - \alpha) & 0 \leq y \leq 2D \\ \cos\alpha\, e^{\Gamma_2 y} & y \leq 0 \end{cases} \tag{4.28}$$

Since we are considering a field constant with x, the only components of the electromagnetic field different from zero are (compare with Chapter 3, noting that x and y are now interchanged) E_x, H_y, H_z. Since E_x as given from (4.28) is already continuous at $y = 0, y = 2D$, we have only to impose the continuity of H_z at the dielectric interface. Moreover, H_z is proportional to $\partial_y E_x$

$$H_z \propto \begin{cases} -\Gamma_3 \cos(k_y 2D - \alpha)\, e^{-\Gamma_3(y-2D)} & y \geq 2D \\ -k_y \sin(k_y y - \alpha) & 0 \leq y \leq 2D \\ \Gamma_2 \cos\alpha\, e^{\Gamma_2 y} & y \leq 0 \end{cases} \tag{4.29}$$

Imposing the continuity condition at $y = 0$ yields

$$\tan\alpha = \frac{\Gamma_2}{k_y} \tag{4.30}$$

while from the continuity condition at $y = 2D$ we obtain

$$\tan(k_y 2D - \alpha) = \frac{\Gamma_3}{k_y} \tag{4.31}$$

By expanding the tangent in (4.31) and taking into account (4.30) to eliminate α we get

$$\Gamma_3 = k_y \frac{k_y \tan k_y 2D - \Gamma_2}{k_y + \tan k_y 2D} \tag{4.32}$$

which, together with the two equations (4.27), is the third equation necessary to solve for k_y, Γ_2, Γ_3. It is convenient to set

$$\begin{gathered} u = k_y D \quad w_2 = \Gamma_2 D \quad w_3 = \Gamma_3 D \\ (\varepsilon_1 - 1)\, k_0^2 D^2 = v_3^2 \quad (\varepsilon_1 - \varepsilon_2)\, k_0^2 D^2 = v_2^2 \end{gathered} \tag{4.33}$$

which allows us to write the system of three equations in the three unknowns in the following manner

$$\begin{aligned} u^2 + w_2^2 &= v_2^2 \\ u^2 + w_3^2 &= v_3^2 \\ w_3 &= u \frac{u \tan 2u - w_2}{u + w_2 \tan 2u} \end{aligned} \tag{4.34}$$

It is noted that if we take $v_2 = v_3, w_2 = w_3$ we recover the equations for even and odd TE modes as already found in the previous Chapter.

The same equation can also be obtained by using the transverse resonance approach. The admittances for the TE mode under consideration are

$$\begin{aligned} Y_0 &= \frac{k_y}{\omega\mu_0} \\ Y_2 &= \frac{\Gamma_2}{j\omega\mu_0} \\ Y_3 &= \frac{\Gamma_3}{j\omega\mu_0} \end{aligned} \tag{4.35}$$

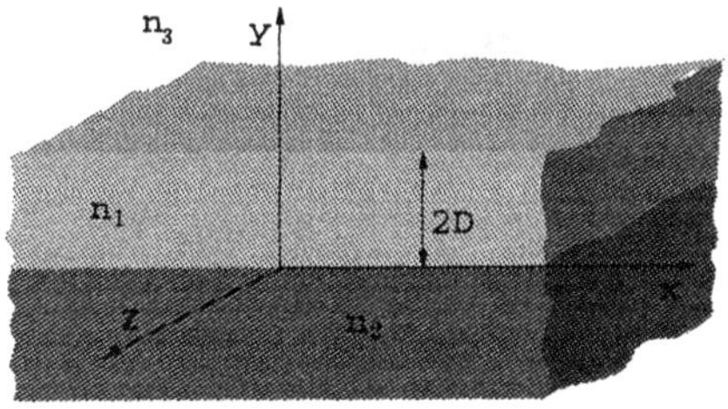

Figure 4.9 Application of EDC to the rib region (central region). The structure becomes homogeneous in the x–direction.

from which we get

$$\overrightarrow{Y}= \frac{\Gamma_3}{j\omega\mu_0} \tag{4.36}$$

and

$$\overleftarrow{Y}= Y_0 \frac{jY_0 \tan(2k_y D) + Y_2}{Y_0 + jY_2 \tan(2k_y D)} = \frac{k_y}{j\omega\mu_0} \frac{\Gamma_2 - k_y \tan(2k_y D)}{k_y + \Gamma_2 \tan(2k_y D)} \tag{4.37}$$

The transverse resonance condition is obtained when the total admittance is equal to zero, i.e. when

$$\overleftrightarrow{Y}=\overleftarrow{Y} + \overrightarrow{Y} \propto \Gamma_3 + k_y \frac{\Gamma_2 - k_y \tan(2k_y D)}{k_y + \Gamma_2 \tan(2k_y D)} = 0 \tag{4.38}$$

Equation (4.38) obtained by transverse resonance analysis is clearly the same as (4.32) already derived by the field matching technique.

Thus, by solving the system (4.34) we have accomplished the first step of the EDC process for what concerns the rib region. A similar procedure is now effected in the outer region of Figure 4.8. In this way, we compute a pair of effective dielectric constants. Since in the previous analysis we have considered a structure uniform along x ($k_x = 0$) we have for the central region

$$\varepsilon_{e1} = \varepsilon_1 - \left(\frac{k_y^2}{k_0^2}\right) = \varepsilon_3 + \left(\frac{\Gamma_3^2}{k_0^2}\right) = \varepsilon_2 + \left(\frac{\Gamma_2^2}{k_0^2}\right) \tag{4.39}$$

while in the outer region we have

$$\varepsilon_{e2} = \varepsilon_1 - \left(\frac{\overline{k}_y^2}{k_0^2}\right) = \varepsilon_3 + \left(\frac{\gamma_3^2}{k_0^2}\right) = \varepsilon_2 + \left(\frac{\gamma_2^2}{k_0^2}\right) \tag{4.40}$$

In the above expression we have used the propagation constants $\overline{k}_y, \gamma_2, \gamma_3$ found for the outer region in a manner similar to that detailed for the rib region. Once these dielectric constants have been computed, we analyze the structure varying in the x–direction, as shown in Figure 4.10. Note that the layering is now in the x–direction and the structure is homogeneous in the y–direction. Also note that we are now using the dielectric constant calculated from (4.39, 4.40) after solving the structure of Figure 4.9. Since in the structure of Figure 4.10 there is no variation along the y–direction, the field is TM in the x–direction, with only E_x, E_z, H_y present.

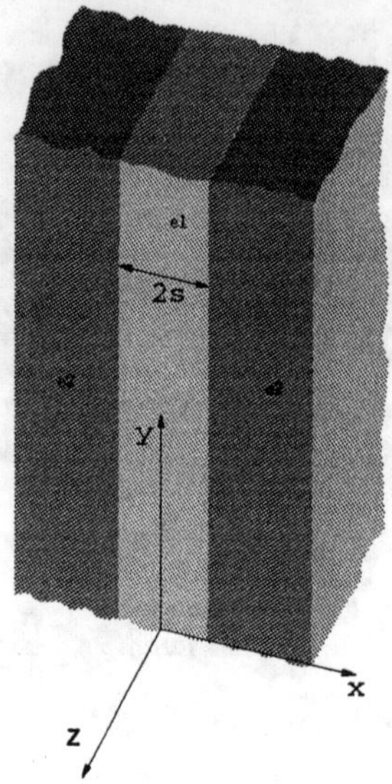

Figure 4.10 Structure analysed in the second step of the EDC method. The structure is homogeneous in the y–direction.

Therefore, recalling the dispersion equation for the symmetric slab for TM modes, we have

$$\begin{aligned} u \tan u &= \frac{\varepsilon_{e1}}{\varepsilon_{e2}} w \\ u^2 + w^2 &= v^2 \end{aligned} \tag{4.41}$$

with the usual positions

$$\begin{aligned} (\varepsilon_{e1} - \varepsilon_{e2})(sk_0)^2 &= v^2 \\ u &= sk_x \\ w &= s\delta \end{aligned} \tag{4.42}$$

where k_x is the wavenumber in the core region, δ the decay constant in the cladding. By solving (4.41) it is therefore possible to find values for k_x, d in the two regions of Figure 4.10.

Note that in this approach we may iterate the procedure, assuming at the beginning of each run the last value of k_x for each region, which was taken as the same in all regions in the original equation (4.26).

To give an idea of the accuracy of the EDC method, we have compared the propagation constant calculated with this approach with other calculations based on finite element [1] and finite difference [12, 11] methods as well as with the rigorous full–wave transverse resonance diffraction, which will be illustrated in Chapter 6. With reference to the structure of Figure 4.8 we have considered the following numerical values (see Table 4.1) for various different rib waveguides.

Table 4.1 Geometrical dimensions of rib waveguides.

Rib	n_1	n_2	n_3	D (mm)	d (mm)	2s (mm)	l (mm)
1	3.44	3.35	1.0	1.0	varying	3.0	1.15
2	3.44	3.40	1.0	1.0	varying	3.0	1.15
3	3.44	3.34	1.0	1.3	0.2	2.0	1.55
4	3.44	3.36	1.0	1.0	0.9	3.0	1.55
5	3.44	3.435	1.0	6.0	3.5	4.0	1.55
6	3.44	3.415	1.0	1.5	1.0	3.0	1.15

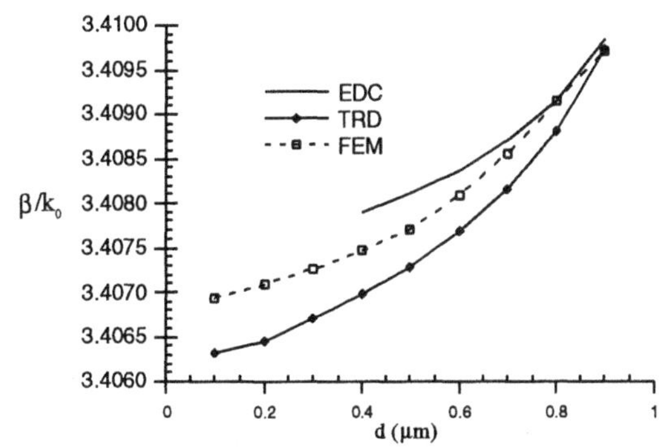

Figure 4.11 Propagation constant for structure 1 as derived by EDC method, transverse resonance diffraction and finite element method (FEM)[1].

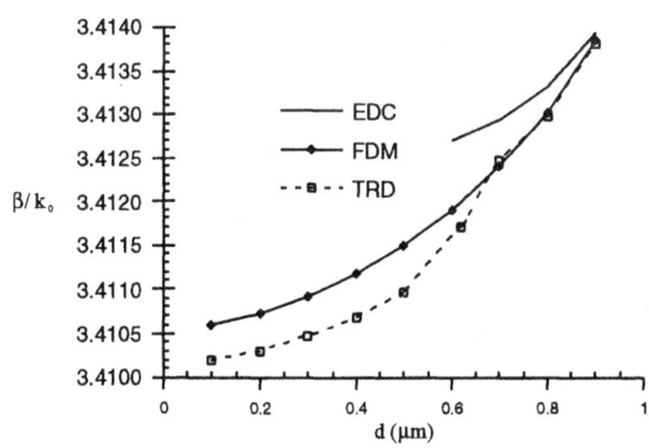

Figure 4.12 Propagation constant for structure 2 as derived by EDC method, transverse resonance diffraction and finite difference method (FDM)[12].

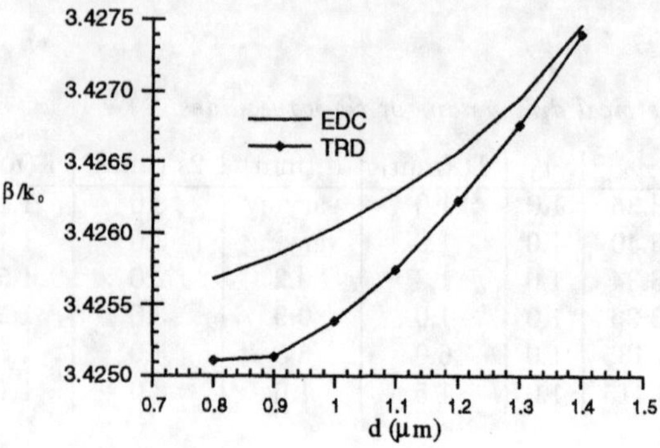

Figure 4.13 Propagation constant for structure 6 as derived by EDC method and transverse resonance diffraction.

Table 4.2 Values of β/k_0 as obtained from EDC, transverse resonance diffraction (TRD) and finite difference (FD) methods.

Rib guide	EDC	TRD	FD(1) [8]	FD(2) [8]	FD [12]
3	3.39032	3.38869	3.39062	3.39129	3.38693
4	3.39548	3.39533	3.39517	3.39543	3.39544
5	3.43733	3.43682	3.43684	3.43686	3.43681

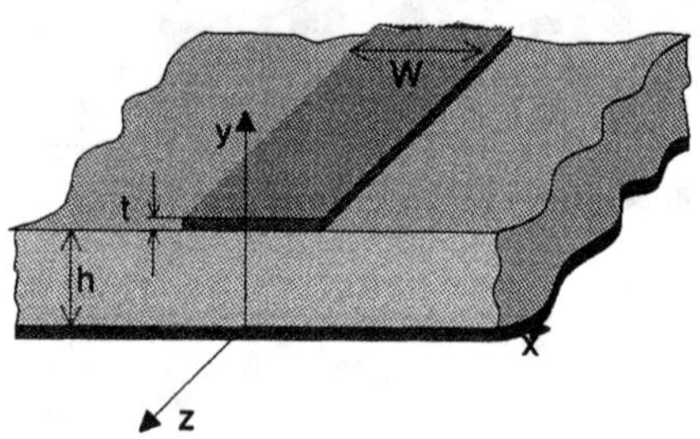

Figure 4.14 Microstrip geometry.

4.3 Microstrip lines by conformal mapping

The microstrip line is an open waveguide that, due to the presence of the dielectric–air interface and of the strip diffraction edges, does not support pure TEM, TE or TM modes. In fact, for this structure, the fundamental mode is quasi–TEM possessing all six components of the electric and magnetic fields. With reference to Figure 4.14 it is easy to see why the microstrip cannot support a pure TEM wave [2]. Continuity of the tangential component of the electric field E_x at the air–dielectric ($y = h$) interface gives

$$E_x(x, y = h^-) = E_x(x, y = h^+) \tag{4.43}$$

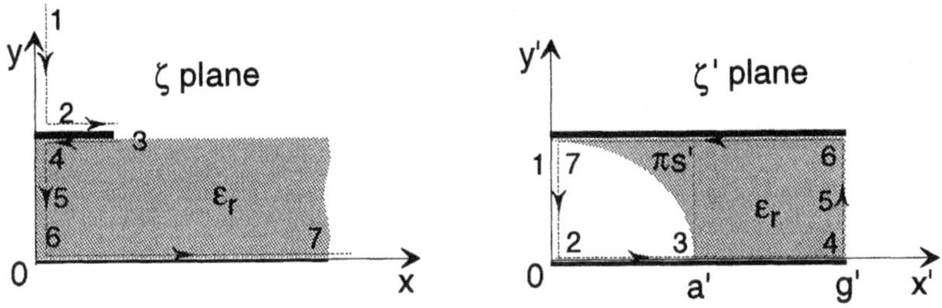

Figure 4.15 Conformal transformation for the evaluation of the capacitance per unit length.

By taking the x–component of the curl equation

$$\nabla \times \mathbf{H} = j\omega\varepsilon\mathbf{E} \tag{4.44}$$

one has

$$\frac{\partial H_z}{\partial y} - \frac{\partial H_y}{\partial z} = j\omega\varepsilon E_x \tag{4.45}$$

which, after considering the continuity of the normal component of magnetic field H_y at the dielectric interface gives, upon use of (4.43),

$$\varepsilon_r \frac{\partial H_z(x, y = h^+)}{\partial y} - \frac{\partial H_z(x, y = h^-)}{\partial y} = (\varepsilon_r - 1)\frac{\partial H_y}{\partial z} \tag{4.46}$$

As $\varepsilon_r \neq 1$ and also $\partial H_y/\partial z \neq 0$, (4.46) implies that also H_z is different from zero. Thus, it has been shown that when the right hand side of (4.44) does not vanish, i.e. for dynamical cases when $\omega \neq 0$, a longitudinal component of $\mathbf{H}$ is present. In a similar manner it is also possible to demonstrate the presence of a longitudinal component of $\mathbf{E}$. Consequently the microstrip field is a six–component (hybrid) field as shown in the figure in the first Chapter.

It is possible to get to the same conclusion also by considering what was said in the Section detailing the behaviour of fields near metallic edges in Chapter 2 and at dielectric interfaces. The latter implies that, with reference to Figure 4.14, only TE^y or TM^y fields, or their superpositions, may exist. On the other hand, due to the presence of the edge, an x–component of the electric or magnetic field also implies the presence of the y–component. Thus, if we consider present a TE^y field with an E_x component, the latter implies the E_y component, therefore contradicting the hypothesis.

However, in the static case, the presence of a TEM field becomes possible. It is quite obvious that, at lower frequencies, the microstrip field is not too different from the static one. In this case it is possible to calculate the microstrip characteristics from the electrostatic capacitance of the structure. This type of analysis has been found adequate for designing circuits at lower frequencies (generally below X–band) when the strip width and the substrate thickness are much smaller than the wavelength in the dielectric material [2]. The quasi–static field distribution of the microstrip line can be thought of as obtained from that of a capacitor, with the upper plane reduced. This field distribution is qualitatively shown in Chapter 1.

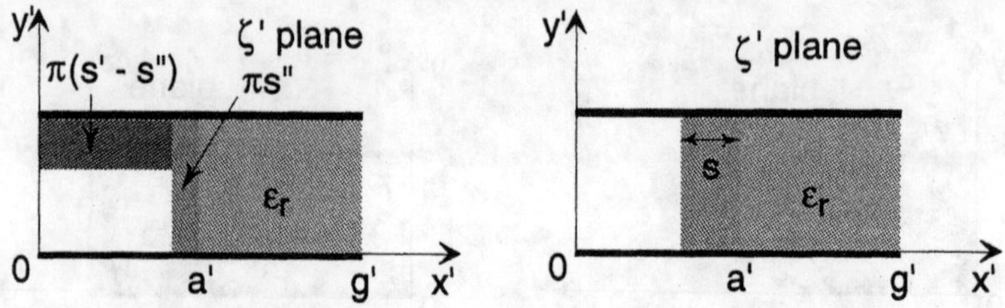

Figure 4.16 Evaluation of the effective dielectric constant.

4.3.1 Quasi–static analyses of microstrip

The quasi–static analysis of microstrip lines is based on considering first the microstrip without the dielectric layer. For this structure the propagation constant β_a and the characteristic impedance Z_a are simply given by

$$\beta_a = \frac{\omega}{c} \tag{4.47}$$

where c is the speed of electromagnetic waves in free–space, given by

$$c = \frac{1}{\sqrt{\mu_0 \varepsilon_0}} \tag{4.48}$$

while

$$Z_a = \frac{1}{cC_a} \tag{4.49}$$

In (4.49), C_a is the capacitance for unit length of the microstrip configuration with the dielectric substrate replaced by air. From the knowledge of the above quantities it is possible to compute the values of characteristic impedance Z_0 and phase constant β_0 of the actual microstrip line, in terms of the capacitance C of a unit length of microstrip line with the dielectric substrate present. In fact, we have

$$\begin{aligned} \beta_0 &= \beta_a \sqrt{\frac{C}{C_a}} \\ Z_0 &= Z_a \sqrt{\frac{C_a}{C}} \end{aligned} \tag{4.50}$$

It is apparent from the above discussion that, in order to compute the latter quantities, knowledge of C and C_a is required. Although there are several approaches for the calculation of these capacitances, the most widely used technique is that introduced by Wheeler [13, 14] in 1964–65. This technique makes use of a conformal transformation for the evaluation of C_a and also makes use of the concept of effective dielectric constant for evaluation of C. The conformal transformation used for the wide strip ($W/h > 2$) is such as to transform the microstrip configuration into a parallel plate capacitor and is given by

$$\zeta = j\pi + d \tanh(\zeta'/2) - \zeta' \tag{4.51}$$

Note that the variable ζ refers to the plane of the microstrip cross–section while ζ' refers to the parallel plate plane. The transformation is shown in Figure 4.15.

When the microstrip is filled with air, it is straightforward to evaluate the capacitance C_a from that of the parallel plate. Unfortunately, when dielectric is present, the dielectric–air interface is mapped into an elliptically–shaped curve as shown in Figure 4.15. This fact makes necessary the introduction of the effective dielectric constant ε_{reff} in order to evaluate C. The problem is now: how to compute such an EDC? To this end one approximates the elliptical contour of Figure 4.15 with an equivalent rectangular one as shown in Figure 4.16. The area $\pi s'$ over the curve is considered as the sum of a parallel area $\pi s''$ and a series area $\pi(s' - s'')$. These series and parallel areas are finally written in terms of an equivalent parallel area s given by (see Figure 4.16)

$$s = s'' + (s' - s'')/\varepsilon_r \tag{4.52}$$

The effective filling fraction is thus obtained as

$$q = (g' - a' + s)/g' \tag{4.53}$$

and the EDC is related to the effective filling fraction by

$$\varepsilon_{reff} = (1 - q) + q\varepsilon_r \tag{4.54}$$

The approximations derived for ε_{reff} depend on the strip width. For narrow strips we have

$$\varepsilon_{reff} = \frac{\varepsilon_r + 1}{2} + \frac{\varepsilon_r - 1}{2} \frac{\ln\frac{\pi}{2} + \frac{1}{\varepsilon_r}\ln\frac{4}{\pi}}{\ln\left(\frac{8h}{W}\right)} \tag{4.55}$$

while for wide strips ($W/h > 2$) one obtains [10],

$$\begin{aligned} q = 1 - \tfrac{1}{\delta}\ln\tfrac{\delta+\gamma}{\delta-\gamma} + \tfrac{0.732}{\delta\varepsilon_r}\left[\ln\tfrac{\delta+\gamma}{\delta-\gamma} - \cosh^{-1}(0.358\delta + 0.595)\right] + \\ \tfrac{\varepsilon_r - 1}{\varepsilon_r\delta}\left[0.386 - \tfrac{1}{2(\delta-1)}\right] \end{aligned} \tag{4.56}$$

with

$$\delta = 1 + \sqrt{1 + \gamma^2} \tag{4.57}$$

and γ is found by solving

$$\frac{\pi}{2}\frac{W}{h} = \gamma - \sinh^{-1}\gamma \tag{4.58}$$

From the above formulae it is possible to derive explicitly the impedance, as a function of the microstrip geometry, which is useful for application purposes. Moreover, it is also possible to give formulae which provide the W/h ratio as a function of the relative dielectric constant and of the impedance level. The latter relationships are useful for design purposes, and are summarised together with the previous one by means of the following expressions:

$$\textit{Wide Strip} : W/h > 2$$

$$Z_0 = \frac{377}{\sqrt{\varepsilon_r}}\left\{\frac{W}{h} + 0.883 + \frac{1+\varepsilon_r}{\pi\varepsilon_r}\left[\ln\left(\frac{W}{2h} + 0.94\right) + 1.451\right] + 0.165\frac{\varepsilon_r - 1}{\varepsilon_r^2}\right\}^{-1}$$

$$\begin{aligned} \frac{W}{2h}\pi = \frac{377\pi}{2Z_0\sqrt{\varepsilon_r}} - 1 - \ln\left[\frac{377\pi}{Z_0\sqrt{\varepsilon_r}} - 1\right] + \\ \frac{\varepsilon_r - 1}{2\varepsilon_r}\left\{\ln\left[\frac{377\pi}{2Z_0\sqrt{\varepsilon_r}} - 1\right] + 0.293 - \frac{0.517}{\varepsilon_r}\right\} \end{aligned}$$

$$Narrow\ Strip : W/h < 2$$

$$Z_0 = \frac{377}{2\pi\sqrt{\frac{\varepsilon_r+1}{2}}}\left\{\ln\left(\frac{8h}{W}\right) + \frac{1}{8}\left(\frac{W}{2h}\right)^2 - \frac{1}{2}\frac{\varepsilon_r-1}{\varepsilon_r+1}\left[\ln\frac{\pi}{2} + \frac{1}{\varepsilon_r}\ln\frac{4}{\pi}\right]\right\}$$

$$\frac{2h}{W} = \frac{1}{4}e^{h'} - \frac{1}{2}e^{-h'}$$

$$h' = \sqrt{\frac{\varepsilon_r+1}{2}\frac{Z_0}{60}} + \frac{\varepsilon_r-1}{\varepsilon_r+1}\left(0.226 + \frac{0.120}{\varepsilon_r}\right)$$

It is noted that, since Wheeler's analysis does not provide a closed form expression for the EDC when $W/h > 2$, the formulas appearing above have been derived empirically by curve fitting of the numerical data.

Finally, some simple observations are in order. First, note that the impedance decreases as W/h increases; in fact a larger W, or a smaller h, increases the line capacitance. Practical values of the level of impedance attainable have been given in the relative table in the first Chapter. Second, note that the thickness of the microstrip line is ignored, although the latter can be taken into account by more sophisticated conformal mapping transformations. A significant feature of the quasi–static model described above, is that it does not account for the non–TEM nature of the microstrip mode. This non–TEM behaviour makes the effective dielectric constant and the impedance frequency dependent. In particular, the EDC variations are quite significant. In order to take into account the full–wave behaviour of the microstrip line, one can make use of the full–wave analysis or employ an approximate, often semi–empirical technique, leading to a closed form solution for the dependence of the EDC and impedance on frequency. Several of these approximate techniques are described in [2] and are not repeated here.

In a similar way, it is also possible to deal with other waveguides in use in the microwave and millimetre–wave frequency range. Approximate formulae for slotline bound modes, conformal mapping analysis of CPW lines and conformal transformations for CPWs with finite thickness, have been used extensively; the interested reader is referred to [2].

Bibliography

[1] T. M. Benson and J. Buus. Optical guiding in III-V semiconductor rib structures. *IEE Conf. Publ. 227, Cardiff*, pages 17–20, April 1983.

[2] K. C. Gupta, R. Garg, and I. J. Bahl. *Microstrip Lines and Slotlines*. Artech House, Washington, 1979.

[3] R. G. Hunsperger. *Integrated Optics: Theory and Technology*. Springer-Verlag, Berlin, 1991.

[4] T. Itoh. In J. C. Wiltse, editor, *Infrared and Millimeter Waves*, volume 4. Academic Press, Orlando, Florida, 1981.

[5] M. R. Knox and P. P. Toulios. Integrated circuits for the millimetre through optical frequency range. *Proceedings of the symposium on submillimetre waves. Polytechnic, N.Y.*, pages 497–516, 1970.

[6] D. L. Lee. *Electromagnetic Principles of Integrated Optics*. John Wiley & Sons, Singapore, 1986.

[7] E. A. J. Marcatili. Dielectric rectangular waveguide. *Bell Syst. Tech. J.*, 48(9):2071–2102, September 1969.

[8] M. J. Robertson, S. Ritchie, and P. Dayan. Semiconductor waveguides: analysis of optical propagation in single rib structures and directional couplers. *IEE Proc., Pt. J*, 132(6), December 1985.

[9] T. Rozzi, T. Itoh, and L. Grun. Two-dimensional analysis of the Ga As d.h. stripe geometry laser. *Radio Science*, 2(4):543–549, July 1977.

[10] H. Sobol. Application of integrated circuit technology to microwave frequencies. *Proc. IEEE*, 59:1200–1211, 1971.

[11] M. S. Stern. Semivectorial polarized finite difference method for optical waveguides with arbitrary index profile. *IEE Proc., Pt. J*, 135(1):56–63, February 1988.

[12] M. S. Stern. Semivectorial polarized H field solutions for dielectric waveguides with arbitrary index profiles. *IEE Proc., Pt. J*, 135(5):333–338, October 1988.

[13] H. A. Wheeler. Transmission line properties of parallel wide strips by conformal mapping approximation. *IEEE Trans. on Microwave Theory Tech.*, 12:280–289, 1964.

[14] H. A. Wheeler. Transmission line properties of parallel strips separated by a dielectric sheet. *IEEE Trans. on Microwave Theory Tech.*, 13:172–185, 1965.

Chapter 5

Discontinuities in planar dielectric waveguides

In this Chapter we begin to examine some simple discontinuity problems by using the example of the slab waveguide. The known modal structure of this waveguide and its one–dimensional cross-section allows us to concentrate on the discontinuity problem itself. From a practical viewpoint the problems dealt with in this Chapter are relevant in integrated optics and millimetrics as the slab waveguide may be considered a good approximation of the guides used in common practice.

The methodology applied to deal with discontinuity problems finds a wide application in the following Chapters of this book. In fact, in order to recover the spectra of waveguides with two–dimensional non separable cross–sections, a discontinuity problem must be solved. This holds true not only for the bound modes, but also for the continuous part of the spectrum.

Moreover, once the spectrum is found, the junction of two different waveguides of 2D cross–sections is again a discontinuity problem, which can be considered and solved with the techniques introduced in this Chapter. As an example, the microstrip step discontinuity may be considered as a junction of two microstrips with different metal widths. The scattering parameters of this discontinuity may be obtained by matching their relative spectra at the junction.

The problem of a single step discontinuity is described in Section 5.1. We start from the simplest approach, i.e. we consider the junction between two slabs of different heights as two lines with different impedances. As a refinement of this model we then introduce the variational equivalent circuit and, finally, we consider the rigorous solution for both the TE and TM cases. Different integral equations, with different convergence properties, can be established when solving this problem. The relative merits and drawbacks of each formulation are discussed in the Section on the alternative integral equation formulation. Also the problem of the choice of the set of basis functions is considered with some detail and numerical results are provided in order to illustrate the different convergence properties.

A particular case of step discontinuities is that considered in Section 5.2 where the slab waveguide is terminated in air. In this case several analytical developments are feasible and are reported for reference purposes.

Section 5.3 considers interacting double step discontinuities. In this case a very effective approach is developed by describing the region between the two discontinuities in terms of either impedance or admittance operators. It is shown that it is possible to model the interacting double step discontinuity problem by means of an equivalent network, where the various elements are operators. This type of approach is also used in the following Chapters when modelling the rib waveguide, etc.

It is then natural to extend the analysis to cascaded step discontinuities as detailed in Section 5.4 and also to infinite periodic structures.

5.1 Single step discontinuity

Step discontinuities in planar dielectric waveguides are commonly used in integrated circuits ranging from sub–millimetre to optical frequencies. The step discontinuity in dielectric slab waveguides is, in fact, a basic one for several components such as distributed feedback lasers, gratings, transformers, antenna feeds, and others. It is thus important to have accurate and reliable theoretical predictions of the behaviour of this discontinuity. The need for theoretical models is made more acute by the difficulties of tuning integrated components once they are built. In addition, theoretical understanding of the scattering properties of a single step discontinuity provides considerable insight into the role played by the continuous spectrum in discontinuity problems.

The diffraction problem at an abrupt discontinuity can be solved by using a field matching technique, which requires field description on either side of the discontinuity in terms of modes. As we have seen in Chapter 3, the complete field propagating in an open slab waveguide can be resolved into a finite set of surface wave modes and a continuum of radiative and reactive modes. The modal expansion transforms the two–dimensional field problem into a one–dimensional problem of conventional transmission line form. With this transformation, one obtains a circuit representation of the field problem by introducing equivalent voltages and currents as measures of modal amplitudes for the electric and magnetic fields, respectively. This situation is thus similar to discontinuities in closed waveguides, where the fields are expressed by a sum over the discrete modes. The main difference is that, for open waveguides, the discrete sum is replaced by an integral. This integral term is bounded, although its rigorous analytical evaluation is often difficult. Direct numerical integration of the integral is fraught with convergence problems since the continuous spectrum describes not only the radiated field from the guide but also the infinitely fine field structure (possibly weakly singular) near the discontinuity. It is therefore essential to formulate the discontinuity problem in an effective way, so as to avoid convergence problems. Such rigorous solutions will be dealt with later in this Section, after a brief discussion of how to describe the step problem in terms of a simple equivalent circuit and after the introduction of the more refined variational equivalent circuit.

5.1.1 *A simple equivalent circuit*

A simple and effective equivalent circuit for the small step between two monomode slabs has been given in [30]. In fact, with reference to the step geometry of Figure 5.1, when $0.4 < d/D < 1$, the main effect of the step is a change of impedance. In

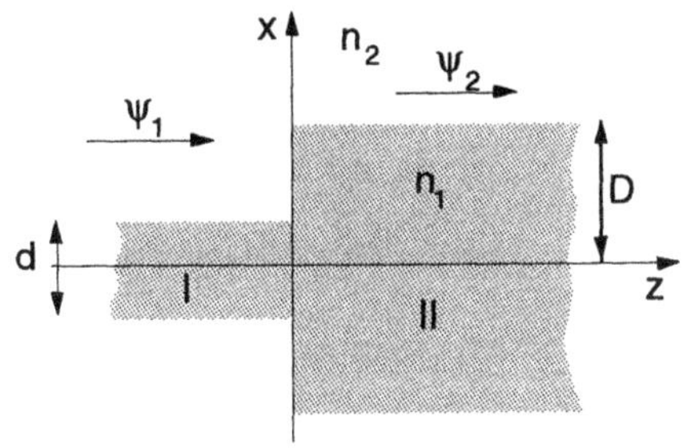

Figure 5.1 Single step discontinuity in planar dielectric waveguide.

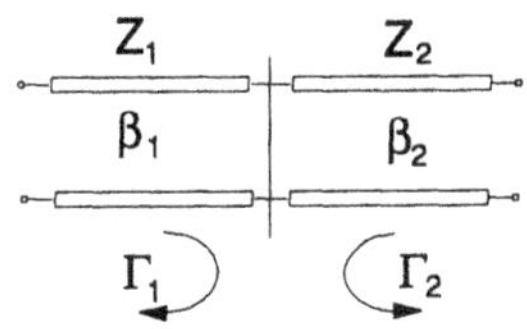

Figure 5.2 Simple equivalent circuit for small step discontinuity.

this case, the moduli of the reflection coefficients for incidence from the left (Γ_1) and from the right (Γ_2) are almost identical (see Figure 5.2). As an example, for the TE case, by indicating with β_1, β_2 the propagation constants in regions I and II respectively, since the impedances are given by

$$\begin{aligned} Z_1 &= \frac{\omega\mu}{\beta_1} \\ Z_2 &= \frac{\omega\mu}{\beta_2} \end{aligned}$$

we have the reflection coefficients

$$\Gamma_2 = -\Gamma_1 = \frac{Z_1 - Z_2}{Z_1 + Z_2} = \frac{\beta_2 - \beta_1}{\beta_2 + \beta_1} \tag{5.1}$$

For more pronounced steps, however, considerable mode mixing takes place at the junction and should be accounted for.

5.1.2 A variational equivalent circuit: electric field formulation

An effective way to consider such mode mixing is to derive an equivalent circuit using a variational approach, analogously to what is done for closed waveguides [45], [12, p.581-585]. To this end, with reference to Figure 5.1, let us assume that only the surface wave ψ_1 exists in region 1 and ψ_2 is present in region 2. Let us also indicate by ϕ_1, ϕ_2 the mode functions pertaining to the continuous spectrum in regions 1 and 2, respectively. An incident field ψ_1 impinging from the left with unit amplitude excites the following field at the left of the discontinuity

$$E_{left} = (1+R)\,\psi_1(x) + \int_0^\infty A(k_x)\phi_1(x;k_x)dk_x \tag{5.2}$$

where in the above we have indicated by R the reflection coefficient of the surface wave while $A(k_x)$ are the amplitudes of each modal component. Similarly, the following field is excited at the right side of the step

$$E_{right} = T\psi_2(x) + \int_0^\infty B(k_x)\phi_2(x;k_x)dk_x \tag{5.3}$$

where in the above T is the transmission coefficient of the surface wave in region 2 and $B(k_x)$ are the amplitudes of each modal component in this region. From the latter two equations it is apparent that A, B, R and T can be computed once the electric field E at the discontinuity is known. In fact, the orthonormality of modal sets yields

$$\begin{aligned} 1+R &= \int_{-\infty}^{\infty} E(x)\psi_1(x;k_x)dx \\ A(k_x) &= \int_{-\infty}^{\infty} E(x)\phi_1(x;k_x)dx \\ T &= \int_{-\infty}^{\infty} E(x)\psi_2(x;k_x)dx \\ B(k_x) &= \int_{-\infty}^{\infty} E(x)\phi_2(x;k_x)dx \end{aligned} \tag{5.4}$$

However, the transverse magnetic field at each side of the discontinuity must be continuous, thus requiring

$$\begin{aligned} &(1-R)\,Y_1\psi_1(x) - \int_0^\infty Y_1(k_x)A(k_x)\phi_1(x;k_x)dk_x = \\ &TY_2\psi_2(x) + \int_0^\infty Y_2(k_x)B(k_x)\phi_2(x;k_x)dk_x \end{aligned} \tag{5.5}$$

The main point in this variational approach is to avoid direct solution of the above integral equation and to get, instead, a stationary expression. Such an expression can be obtained by multiplying (5.5) by E, by integrating over x, and by dividing by $(1+R)^2$ as given by (5.4), thus yielding

$$\begin{aligned} \overline{Y}_{in} = \frac{1-R}{1+R} &= \frac{Y_2}{Y_1}\left[\frac{\int_{-\infty}^{\infty} E(x)\psi_2(x;k_x)dx}{\int_{-\infty}^{\infty} E(x)\psi_1(x;k_x)dx}\right]^2 + \\ &\frac{1}{Y_1}\frac{\int_0^\infty Y_1(k_x)\left[\int_{-\infty}^{\infty} E(x)\phi_1(x;k_x)dx\right]^2 dk_x}{\left[\int_{-\infty}^{\infty} E(x)\psi_1(x;k_x)dx\right]^2} + \\ &\frac{\int_0^\infty Y_2(k_x)\left[\int_{-\infty}^{\infty} E(x)\phi_2(x;k_x)dx\right]^2 dk_x}{\left[\int_{-\infty}^{\infty} E(x)\psi_1(x;k_x)dx\right]^2} \end{aligned} \tag{5.6}$$

By inserting into (5.6) a suitably chosen *trial field* $E(x)$ we obtain, thanks to the stationary properties of (5.6), the characterization of the step discontinuity accurate to the second order. It is expedient to rewrite the normalised (with respect to Y_1) input admittance (5.6) in the following form

$$\overline{Y}_{in} = n^2 \frac{Y_2}{Y_1} + \frac{Y_0}{Y_1} \tag{5.7}$$

with

$$Y_0 = G_0 + jB_0 \tag{5.8}$$

suggesting the equivalent circuit of Figure 5.3 with the various quantities given as follows:

$$\begin{aligned}
n^2 &= \left[\frac{\int_{-\infty}^{\infty} E(x)\psi_2(x;k_x)dx}{\int_{-\infty}^{\infty} E(x)\psi_1(x;k_x)dx}\right]^2 \\
Y_0 &= \frac{\int_0^{\infty} Y_1(k_x)\left[\int_{-\infty}^{\infty} E(x)\phi_1(x;k_x)dx\right]^2 dk_x}{\left[\int_{-\infty}^{\infty} E(x)\psi_1(x;k_x)dx\right]^2} + \\
&\quad \frac{\int_0^{\infty} Y_2(k_x)\left[\int_{-\infty}^{\infty} E(x)\phi_2(x;k_x)dx\right]^2 dk_x}{\left[\int_{-\infty}^{\infty} E(x)\psi_1(x;k_x)dx\right]^2}
\end{aligned} \tag{5.9}$$

It is noted that in this circuit radiation effects are also properly modelled and the radiated conductance is given by

$$\begin{aligned}
G_0 &= \frac{\int_0^{n_2 k_0} Y_1(k_x)\left[\int_{-\infty}^{\infty} E(x)\phi_1(x;k_x)dx\right]^2 dk_x}{\left[\int_{-\infty}^{\infty} E(x)\psi_1(x;k_x)dx\right]^2} + \\
&\quad \frac{\int_0^{n_2 k_0} Y_2(k_x)\left[\int_{-\infty}^{\infty} E(x)\phi_2(x;k_x)dx\right]^2 dk_x}{\left[\int_{-\infty}^{\infty} E(x)\psi_1(x;k_x)dx\right]^2}
\end{aligned} \tag{5.10}$$

while reactive effects are accounted for by modes which cannot propagate, i.e.

$$\begin{aligned}
jB_0 &= \frac{\int_{n_2 k_0}^{\infty} Y_1(k_x)\left[\int_{-\infty}^{\infty} E(x)\phi_1(x;k_x)dx\right]^2 dk_x}{\left[\int_{-\infty}^{\infty} E(x)\psi_1(x;k_x)dx\right]^2} + \\
&\quad \frac{\int_{n_2 k_0}^{\infty} Y_2(k_x)\left[\int_{-\infty}^{\infty} E(x)\phi_2(x;k_x)dx\right]^2 dk_x}{\left[\int_{-\infty}^{\infty} E(x)\psi_1(x;k_x)dx\right]^2}
\end{aligned} \tag{5.11}$$

It is also noted that a crucial point in the above variational method is a prudent choice of the field E on the discontinuity. As an example, by choosing $E = \psi_1$, i.e. equal to the incident surface wave field, due to the orthogonality of modal functions, we completely neglect radiation in region 1. Similarly, by choosing as trial field $E = \psi_2$, radiation in region 2 is neglected. From these considerations it transpires that a "good" trial field could be a suitable combination of the surface waves in regions 1 and 2. Indeed, we will later see that excellent results are obtained by selecting such a combination.

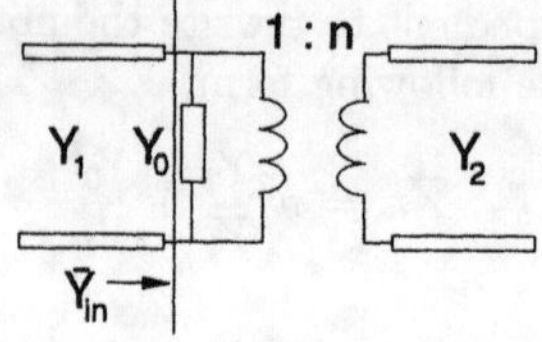

Figure 5.3 Variational equivalent circuit for a step discontinuity.

5.1.3 Rigorous solution

When more accurate results are desired, it becomes necessary to actually solve the integral equation (5.5) arising from field matching. In the past, the problem of the abrupt step [11, 17, 28, 2, 48] has been generally treated by "discretizing" the continuum. Following this approach, in [11] a discrete mode–matching has been applied just to the surface waves at either side of a small step. However, this approximation is useful only for small steps where no appreciable coupling between the surface waves and the radiation spectrum is present. Another suggested alternative has been the introduction of a metal enclosure as proposed in [2]; this approach, although adequate in some respects, alters essentially the boundary condition for radiation. In the following, we will solve the integral equation arising from field matching in a rigorous manner [40], thus considering the whole spectrum of the slab guides.

Particular attention will also be placed on the choice of the expanding functions, which are designed to explicitly account for each surface wave mode, for radiation modes, and for edge singularity of fields, when present. To this end, a set of expanding functions which are not orthogonal, but complete, is also investigated. This set offers better convergence particularly for the TM field incidence case where the field is known to be weakly singular at the dielectric corner of an abrupt step, as shown in Chapter 2.

Scattering matrix formulation (TE case)

For a TE mode excitation with no variation along the y–direction, only three field components exist and they are related as follows :

$$H_x(x,z) = \frac{1}{j\omega\mu}\frac{\partial}{\partial z}E_y(x,z) = \frac{\beta}{\omega\mu}E_y(x,z) = \frac{1}{z_0}E_y(x,z) \tag{5.12}$$

$$H_z(x,z) = -\frac{1}{j\omega\mu}\frac{\partial}{\partial x}E_y(x,z) \tag{5.13}$$

where $E_y(x,z)$ may be expressed as a modal expansion

$$E_y(x,z) = \left\{\sum_k a_k\psi_k(x) + \int_0^\infty b(k_x)\phi(x,k_x)dk_x\right\} e^{j(\omega t-\beta z)} \tag{5.14}$$

a_k and $b(k_x)$ are the unknown amplitudes of the surface and continuum modes respectively. In the following we consider the problem for the even TE case, whereas the corresponding problem for odd modes can be easily tackled along the same lines.

Surface wave modes

The surface mode function $\psi_k(x)$ has been derived in Chapter 3 for a symmetrical slab waveguide and is here repeated for ease of reference:

$$\psi(x) = \begin{cases} A\cos\kappa x & : \quad x \le d \\ A\cos\kappa d \; e^{-\gamma_x(x-d)} & : \quad x \ge d \end{cases} \tag{5.15}$$

where A is the normalisation constant such that

$$\int_0^\infty \psi^2(x)dx = 1 \tag{5.16}$$

$$A = \sqrt{\frac{2}{d + \frac{1}{\gamma_x}}} \tag{5.17}$$

Furthermore, the transverse (x–direction) propagation constants γ_x and κ satisfy the eigenvalue equation

$$\kappa \tan\kappa d = \gamma_x \tag{5.18}$$

as well as the conservation of wavenumbers

$$-\gamma_x^2 + \beta^2 = n_2^2 k_0^2 \qquad \text{in air} \tag{5.19}$$

$$\kappa^2 + \beta^2 = n_1^2 k_0^2 \qquad \text{in the slab waveguide} \tag{5.20}$$

Continuous modes

The normalised mode functions pertaining to the continuum are given as

$$\varphi(x) = \left\{ \begin{array}{lll} \sqrt{\frac{2}{\pi}}\frac{1}{C}\cos qx & \text{in the slab} & (d \ge x \ge 0) \\ \sqrt{\frac{2}{\pi}}\cos\left[k_x(x-d)+\alpha\right] & \text{in air} & (x \ge d) \end{array} \right\} \tag{5.21}$$

where

$$C = \sqrt{1 + \left(\frac{v}{k_x}\right)^2 \sin^2(qd)} \tag{5.22}$$

$$\tan\alpha = \frac{q}{k_x}\tan qd \tag{5.23}$$

The transverse wavenumbers q and k_x satisfy the relations

$$\beta^2 = n_1^2 k_0^2 - q^2 = n_2^2 k_0^2 - k_x^2 \tag{5.24}$$

Clearly, for $0 < k_x < n_2 k_0$, β is real and positive and the continuum modes in this range propagate to give the radiation modes. For $k_x > n_2 k_0$, β is negative and imaginary, and the modes are evanescent, thus representing the reactive part of the continuum. The normalization of the continuous spectrum is such that

$$\int_0^\infty \varphi(x;k_x)\varphi(x;k'_x)dx = \delta(k_x - k'_x) \tag{5.25}$$

Scattering formulation for the single step discontinuity: magnetic field formulation

Let us consider a steady–state and source free problem, with two different semi–infinite slab waveguides forming a step discontinuity at $z = 0$ (see Figure 5.1). The incident field considered here will be composed of surface waves only. The case of incident radiation fields is dealt with separately in a later Section where the problem of coupling of incident radiation fields to the guided–wave fields of a slab–waveguide is discussed. For now, let us assume that there are n_ℓ surface modes which are capable of propagating in guide I (left), and n_r surface modes which can propagate in guide II (right), with the total number of propagating surface waves given by n_i, with

$$n_i = n_\ell + n_r \tag{5.26}$$

Continuity of the electric field $E_y(x,z)$ and the magnetic field $H_x(x,z)$ at $z=0$ is expressed as

$$\begin{aligned} E_y(x,0) = \sum_{k=1}^{n_\ell} \left(V_k^i + V_k^r\right)\psi_k(x) + \int_0^\infty V^I(k_x)\phi^I(x;k_x)dk_x = \\ \sum_{k=n_\ell+1}^{n_i} \left(V_k^i + V_k^r\right)\psi_k(x) + \int_0^\infty V^{II}(k_x)\phi^{II}(x;k_x)dk_x \end{aligned} \tag{5.27}$$

$$\begin{aligned} H_x(x,0) = \sum_{k=1}^{n_\ell} -\tfrac{1}{z_{0k}}\left(V_k^i - V_k^r\right)\psi_k(x) + \int_0^\infty \tfrac{V^I(k_x)\phi^I(x;k_x)}{Z_0^I(k_x)}dk_x = \\ \sum_{k=n_\ell+1}^{n_i} -\tfrac{1}{z_{0k}}\left(-V_k^i + V_k^r\right)\psi_k(x) - \int_0^\infty \tfrac{V^{II}(k_x)\phi^{II}(x;k_x)}{Z_0^{II}(k_x)}dk_x \end{aligned} \tag{5.28}$$

Since a scattering formulation is sought, the incident V_k^i and the reflected V_k^r surface wave amplitudes are made explicit whereas $V^I(k_x)$ and $V^{II}(k_x)$ represent the amplitudes of the (scattered) continuum fields in guide I and guide II, respectively. It is interesting to note that matching of magnetic fields $H_x(x,z)$ at $z=0$ is equivalent to matching the gradient of electric fields $E_y(x,z)$ at $z=0$. By using orthogonality of the modal functions in (5.28), one may express the unknown modal amplitudes in terms of the unknown magnetic field $H_x(x,0)$ as

$$V_k^r = V_k^i + s_k z_{0k}\int_0^\infty \psi_k(x)H_x(x,0)dx \tag{5.29}$$

$$\left\{ \begin{array}{c} V^I(k_x) \\ V^{II}(k_x) \end{array} \right\} = \int_0^\infty \left\{ \begin{array}{c} \phi^I(x;k_x) \\ \phi^{II}(x;k_x) \end{array} \right\} z_0(k_x)H_x(x,0)dx \tag{5.30}$$

where

$$s_k = \left\{ \begin{array}{rl} 1 & \text{for} \quad k \le n_\ell \\ -1 & \text{for} \quad k > n_\ell \end{array} \right\} \tag{5.31}$$

Upon substituting the above expressions into (5.27) and re–arranging, one obtains

$$\sum_{k=1}^{n_\ell} s_k V_k^i \psi_k(x) = \int_0^\infty Z(x,x')H_x(x')dx' \tag{5.32}$$

where

$$\begin{aligned} Z(x,x') = \tfrac{1}{2}\sum_{k=1}^{n_i} z_{0k}\psi_k(x)\psi_k(x') + \\ \tfrac{1}{2}\int_0^\infty \left[Z_0^I(k_x)\phi^I(x;k_x)\phi^I(x';k_x) + Z_0^{II}(k_x)\phi^{II}(x;k_x)\phi^{II}(x';k_x)\right]dk_x \end{aligned} \tag{5.33}$$

The summation on the left hand side of (5.32) represents the total incident electric field impinging on either side of the discontinuity. The term $Z(x, x')$ is a Green's function which may be viewed as the "impedance" of the step discontinuity. This "impedance" function is constituted of known modal functions and modal impedances and completely characterizes the step discontinuity. From this and from the known incident fields, the unknown modal amplitudes can be found through the use of (5.29) and (5.32). The linear relationship between scattered (reflected and transmitted) and incident modal amplitudes in (5.29) suggests that we may obtain a scattering matrix formulation of the form

$$\mathbf{V}^r = \mathbf{S}\mathbf{V}^i \tag{5.34}$$

To this end, let us first consider the case of a single surface wave incident on the discontinuity. Since the amplitude of the incident surface mode is arbitrary, it is possible to set

$$V_\ell^i = 1 \qquad V_{k\neq\ell}^i = 0 \tag{5.35}$$

and (5.34) reduces to

$$V_k^r = S_{k\ell} \tag{5.36}$$

Let the corresponding scattered magnetic field be $h_\ell(x)$; then according to (5.32) we have

$$s_\ell V_\ell^i \psi_\ell(x) = \int_0^\infty Z(x, x') h_\ell(x') dx' \tag{5.37}$$

In the above equation, $h_\ell(x)$ is the unknown function to be determined by discretization of equation (5.37) by means of the Ritz–Galerkin procedure. To this end, an orthonormal basis functions set, $f_n(x)$, is introduced in the interval $0 < x < \infty$ and the magnetic field is represented as

$$h_\ell(x) = \sum_{n=0}^{\infty} \lambda_{n\ell} f_n(x) \tag{5.38}$$

By using the above equation in (5.37) and by testing the latter equation with the weight functions $f_k(x)$, we obtain

$$s_\ell V_\ell^i \int_0^\infty f_k(x)\psi_\ell(x)dx = \sum_{n=0}^{\infty} \lambda_{n\ell} \int_0^\infty \int_0^\infty f_k(x) Z(x, x') f_n(x) dx' \tag{5.39}$$

which, after introduction of the following quantities

$$Z_{kn} = \int_0^\infty \int_0^\infty f_k(x) Z(x, x') f_n(x) dx' \tag{5.40}$$

$$Q_{k\ell} = \int_0^\infty f_k(x)\psi_\ell(x)dx \tag{5.41}$$

and of the vectors

$$\mathbf{Q}_\ell = \begin{bmatrix} Q_{1\ell} \\ Q_{2\ell} \\ \\ Q_{n\ell} \end{bmatrix} \qquad \lambda_\ell = \begin{bmatrix} \lambda_{1\ell} \\ \lambda_{2\ell} \\ \\ \lambda_{n\ell} \end{bmatrix} \tag{5.42}$$

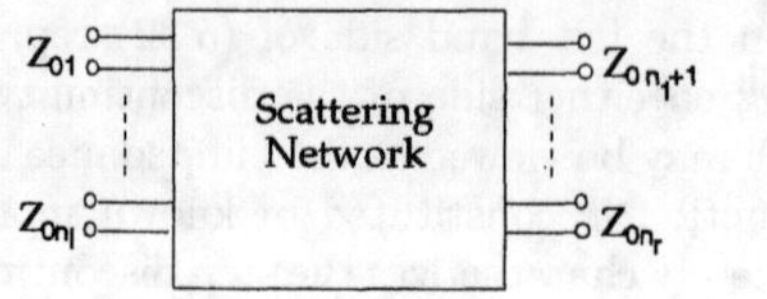

Figure 5.4 Generalized 2n–port scattering network.

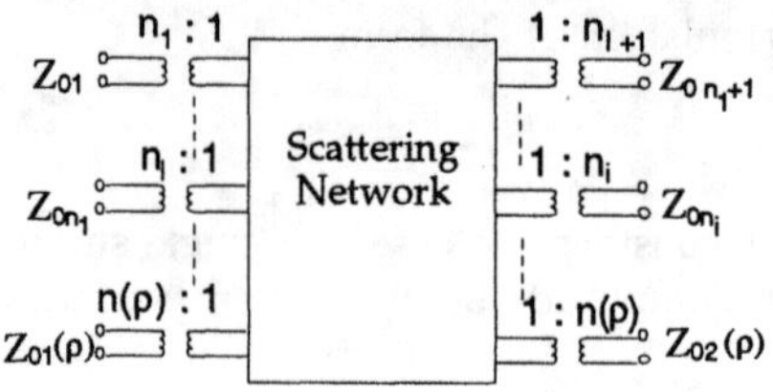

Figure 5.5 Normalised scattering network with embedded ideal transformers.

can be written in matrix form as

$$\lambda_\ell = s_\ell \mathbf{Z}^{-1}\mathbf{Q}_\ell \tag{5.43}$$

The scattering matrix is obtained from (5.29), (5.36) and (5.43) as

$$S_{k\ell} = \delta_{k\ell} + s_k z_{0k} \int_0^\infty \psi_k(x) \sum_{n=0}^{\infty} \lambda_{n\ell} f_n(x) dx = \delta_{k\ell} + s_k s_\ell z_{0k} \mathbf{Q}_k{}^T \mathbf{Z}^{-1}\mathbf{Q}_\ell \tag{5.44}$$

where $\delta_{k\ell}$ is the Kronecker symbol and the superscript T denotes transposition. Equation (5.44) specifies the scattering coefficient of the incident ℓ-th surface mode to the k-th surface mode. In the Ritz–Galerkin approach, the infinite column matrices $\mathbf{Q}_k$ and $\mathbf{Q}_\ell$ and square matrix $\mathbf{Z}$ are replaced by their finite truncations $(0 < n < N)$ and (5.44) becomes a finite matrix equation. In principle the solution can thus be approached as accurately as desired by increasing N. However, since matrix sizes increase with N the numerical calculations will become increasingly time–consuming and possibly ill–conditioned for large N; it is thus advantageous if the solution converges with just a few terms. As a consequence of the lossy nature (the presence of radiation) of the problem, the Green's impedance matrix Z is complex and the exact solution will represent neither a maximum nor minimum. As a result, the solution will not converge in a monotonic fashion with increasing order N. On the other hand, by a careful choice of the series of expanding functions, the oscillations in the solutions will decrease rapidly with increasing order and convergence is quickly achieved. The step discontinuity is therefore represented by a generalised $(n_\ell + n_r)$ port scattering network as shown in Figure 5.4.

However, the step discontinuity is a reciprocal junction, and its scattering matrix ought to be symmetrical: this will only be so if all ports are terminated by the same matching impedances and these are all normalised to unity. Hence, the normalised scattering matrix is obtained by introducing ideal transformers connected at each port as shown in Figure 5.5 such that

$$\overline{z} = 1 = n^2 z_{0k} \tag{5.45}$$

giving the transformation ratio

$$n = \frac{1}{\sqrt{z_{0k}}} \tag{5.46}$$

where z_{0k} is the characteristic impedance of the k-th surface mode. This means that the new incident ($\overline{\mathbf{V}}^i$) and reflected ($\overline{\mathbf{V}}^r$) wave amplitudes are related to the old values by the linear transformation

$$\begin{aligned} \mathbf{V}^i &= diag\mathbf{Z}_0^{1/2}\overline{\mathbf{V}}^i \\ \mathbf{V}^r &= diag\mathbf{Z}_0^{1/2}\overline{\mathbf{V}}^r \end{aligned} \tag{5.47}$$

which, by using (5.34), implies

$$\overline{\mathbf{V}}^r = \mathbf{Z}_0^{-1/2}\mathbf{S}\mathbf{Z}_0^{1/2}\overline{\mathbf{V}}^i \tag{5.48}$$

The normalised scattering matrix is obtained by inserting (5.47) into (5.34) thus giving

$$\overline{\mathbf{S}} = \mathbf{Z}_0^{-1/2}\mathbf{S}\mathbf{Z}_0^{1/2} \tag{5.49}$$

or in the form of (5.44) as

$$\overline{S}_{k\ell} = \delta_{k\ell} + s_k s_\ell \sqrt{z_{0k} z_{0\ell}}\, \mathbf{Q}_k{}^T \mathbf{Z}^{-1} \mathbf{Q}_\ell \tag{5.50}$$

which now displays the symmetry required by the reciprocity of the junction.

Owing to the orthogonality of modes, the scattering formulation between continuous modes can be easily derived,

$$S_{k_x k_x'} = \delta(k_x - k_x') + s_k s_k' \sqrt{z_0(k_x) z_0(k_x')}\, \mathbf{Q}^T(k_x)\mathbf{Z}^{-1}\mathbf{Q}(k'_x) \tag{5.51}$$

and between continuous modes and surface modes

$$S_{\ell, k_x'} = s_k s_k' \sqrt{z_{0\ell} z_0(k_x')}\, \mathbf{Q}_\ell{}^T \mathbf{Z}^{-1}\mathbf{Q}(k'_x) \tag{5.52}$$

The extension to the case where the incident field consists of surface modes and radiation modes is obvious by the theory of superposition.

Radiation loss at the step

The radiation loss at the step is a consequence of excitation of the radiative part of the continuum modes from the incident guided field, i.e. from the surface mode. However, unlike in closed waveguides, radiation modes are also propagating causing interaction with neighbouring circuit elements. When this is the case, radiation modes must also be considered accessible. The difference in power between the incident surface mode(s) and the scattered surface mode(s) provides the amount of the radiation loss. As a consequence, the radiation loss can be computed directly from knowledge of the scattering matrix since the power loss to radiation is given by

$$P_{rad} = 1 - \frac{1}{n}\sum_i \sum_j S_{ij} S_{ij}^* \tag{5.53}$$

In (5.53) n is the total number of incident surface waves. However, we can also compute the radiated power as the integral over the radiative part of the continuous spectrum, thus as

$$P_{rad} = \sum_{k=1}^{n} \int_{0}^{n_2 k_0} S_{kk_x} S_{kk_x}^{*} dk_x \tag{5.54}$$

From a numerical point of view, the use of (5.53) is preferable since there is no need to calculate the integral appearing in (5.54). On the other hand, comparison between (5.53) and (5.54) provides a valuable check of the results.

5.1.4 Choice of basis functions for the TE case

The accuracy and efficiency of the above procedure depends on a prudent choice of the set of basis functions. It is desirable to find a set which provides convergence with just a few terms out of a potentially infinite set. In the following, several different choices of basis functions are considered, each with increasing sophistication, and each providing further insight into the problem of field representation at the step discontinuity.

Cosine–Laguerre hybrid set

Most of the energy of the field in the slab waveguides is carried by surface waves. Therefore it seems natural to select basis functions which model well this particular part of the spectrum. Modal functions, as given by (5.15), show that the surface waves have a cosine shape in the guide and an exponential decay outside. This suggests a possible appropriate choice of the expanding functions for symmetric fields as :

$$C_m(x) = \sqrt{\frac{\varepsilon_m}{d}} \cos \frac{m\pi x}{d} \qquad \varepsilon_m = \left\{ \begin{matrix} 1 & m = 0 \\ 2 & m \neq 0 \end{matrix} \right\} \tag{5.55}$$

for the region $0 < x < d$ (inside the slab) and

$$L_n(x-d) = \frac{1}{\sqrt{x_0}} L_n\left(\frac{x-d}{x_0}\right) e^{-\frac{x-d}{2x_0}} \tag{5.56}$$

for the region $x > d$ (outside the slab). $L_n\left(\frac{x-d}{x_0}\right)$ are the general Laguerre polynomials with an arbitrary scale variable x_0. The above functions are orthogonal in their intervals of definition so that

$$\int_0^d C_m(x) C_n(x) dx = \delta_{mn}$$

$$\int_d^{\infty} L_m(x-d) L_n(x-d) dx = \delta_{mn} \tag{5.57}$$

However, in order to compute (5.40), overlapping integrals between the basis functions and the modal fields must be evaluated. For the sets of selected basis functions these evaluations can be carried out analytically giving the following overlapping integrals. The overlapping integral with surface modes inside the slab is given by

$$Q_m = \int_0^d \psi(x) C_m(x) dx = \int_0^d A\cos(\kappa x)\, C_m(x) dx =$$
$$A\sqrt{\varepsilon_m d}(-1)^m \left[\frac{\kappa d \sin(\kappa d)}{(\kappa d)^2 - (m\pi)^2}\right]; \qquad \kappa d \neq m\pi$$
$$Q_m = \int_0^d \psi(x) C_m(x) dx = A\sqrt{\varepsilon_m \frac{d}{4}}; \qquad \kappa d = m\pi \tag{5.58}$$

while the overlapping integral in air is given by

$$\overline{Q}_m = \int_0^d \psi(x) L_m(x-d) dx = A\sqrt{x_0}\cos(\kappa d)\left(qx_0 - \frac{1}{2}\right)^m \left(qx_0 + \frac{1}{2}\right)^{-m-1} \tag{5.59}$$

The overlapping integrals with continuous modes inside the slab are given by

$$P_m(k_x) = \int_0^d \phi(x;k_x) C_m(x) dx = \sqrt{\frac{2\varepsilon_m}{\pi d}}\frac{(-1)^m}{C}\left[\frac{qd\sin(qd)}{(qd)^2 - (m\pi)^2}\right]; \qquad qd \neq m\pi$$
$$P_m(k_x) = \int_0^d \phi(x;k_x) C_m(x) dx = \sqrt{\frac{\varepsilon_m}{\pi d}}\frac{1}{C}; \qquad qd = m\pi \tag{5.60}$$

while the overlapping integral in air is given by

$$\overline{P}_m(k_x) = \int_d^\infty \phi(x;k_x) L_m(x-d) dx = \sqrt{\frac{1}{2\pi x_0}}(-1)^m \left(\frac{1}{4} + k_x^2 x_0^2\right)^m$$
$$\left[e^{j\alpha}\left(\frac{1}{2} - jk_x x_0\right)^{-2m-1} + e^{-j\alpha}\left(\frac{1}{2} + jk_x x_0\right)^{-2m-1}\right] \tag{5.61}$$

There are two useful criteria for minimising the error in the representation of the field. The first one is the straightforward minimum square law approximation of the modal function for a finite truncation. The second criterion requires the continuity of the expanded fields at the dielectric interface ($x = d$). The above condition is met by optimising the free parameter x_0.

A combination of surface modes

As we will see later, with reference to the numerical behaviour of the solution, the cosine series of the previous Section provides a good fit of the field at the discontinuity with a few terms. Nevertheless, it is still desirable to obtain such a good fit with only one or two terms. To achieve this result, one should select a "trial field" which already has the same shape as the surface modes on either side of the step. This "trial field" could be another surface mode function. However, the surface modal functions on either side of the step are not generally suitable as

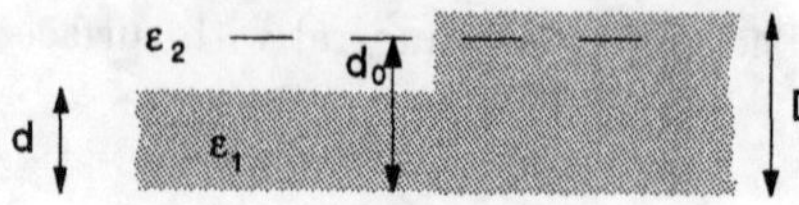

Figure 5.6 Intermediate guide as a transition between two waveguides.

they are orthogonal to their respective continuum modes. It is only for small steps, excited by a given surface wave, that it is reasonable to take the latter as a trial field and neglect radiation on the input side. In fact, in this case, radiation takes place mainly in the forward direction with negligible mode mixing. In the general case, however, it is expedient to use the modal functions of a guide of intermediate thickness, i.e. with its thickness d_0 such that $d < d_0 < D$ (see Figure 5.6). This is similar to making the intermediate guide an *EDC* (effective dielectric constant) approximation of the two actual waveguides which constitute the step discontinuity, where we require $\varepsilon_{eff} = \sqrt{\varepsilon_1\varepsilon_2}$. In a multimode situation the extension of this reasoning leads one to consider n_i surface waves as the first n_i functions of our basis. Moreover, in order to make use of a complete set, the modal "trial fields" are complemented by the addition of a set of orthonormal exponentially weighted Laguerre polynomials up to the chosen order of the variational solution $\overline{N}$, i.e.

$$\{G_n : 1 \leq n \leq \overline{N}\} = \left\{\psi_1 ... \psi_i, L_0 ... L_{\overline{N}-i-1}\right\} \tag{5.62}$$

Unlike the set of basis functions used in the previous section, each of the functions ψ and L is now defined over the entire interval $0 \leq x < \infty$, so that the above set is not orthogonal. On the other hand, the Ritz–Galerkin approach only requires the completeness of the set. The scale parameter x_0 still occurs in the above equation. The optimising criteria for the Laguerre series in this case is that they should be made as orthogonal to the intermediate modes as possible. This is done by choosing a scaling factor x_0 which minimises the functional

$$\sum_{m=0}^{\overline{N}-i-1} \sum_{k=0}^{i} \mid \int_0^\infty \psi_k(x) L_m(x) dx \mid^2 \tag{5.63}$$

The values of the overlapping integrals between this basis function set and the modal functions is given in the following.

The labour of computing the overlapping integrals between the surface modes and the slab spectrum can be greatly reduced if the following properties of the solutions of the wave equations are employed. Let us consider two independent solutions of the transverse wave equation ϕ_1, pertaining to the symmetric distribution of relative permittivity ε_{1r}, and ϕ_2 pertaining to ε_{2r}, respectively. Consequently we have

$$\begin{aligned} \phi_1'' + (\varepsilon_{1r}k_0^2 - \beta_1^2)\phi_1 = 0 \\ \phi_2'' + (\varepsilon_{2r}k_0^2 - \beta_2^2)\phi_2 = 0 \end{aligned} \tag{5.64}$$

By multiplying the first by ϕ_2, the second by ϕ_1, integrating over all the cross-section, $0 \leq x < \infty$, and subtracting the second equation from the first one, we obtain

$$< \phi_1, \phi_2 >= \frac{1}{\beta_1^2 - \beta_2^2} \left[\int_0^\infty (\phi_1''\phi_2 - \phi_2''\phi_1) dx + k_0^2 \int_0^\infty \Delta\varepsilon_r(x)\phi_1\phi_2 dx \right] \tag{5.65}$$

where $\Delta\varepsilon_r(x) = \varepsilon_{1r}(x) - \varepsilon_{2r}(x)$. If the permittivity distribution is piecewise constant, the above term can be taken out of the second integral and the integral itself is nonvanishing only over the domain D where $\Delta\varepsilon_r \neq 0$. Moreover, we can integrate by parts the first integral on the right hand side of the above equation obtaining

$$<\phi_1, \phi_2>= \frac{1}{\beta_1^2 - \beta_2^2}\left\{[(\phi_1'\phi_2 - \phi_2'\phi_1)]_0^\infty + k_0^2\Delta\varepsilon_r \int_D \phi_1\phi_2 dx\right\} \qquad (5.66)$$

If at least ϕ_1 or ϕ_2 is a bound mode, the integrated term vanishes, since it decays exponentially at infinity and either $\phi_{1,2}$ or $\phi_{1,2}'$ vanishes at $x = 0$. We recall, moreover, that the definition of EDC implies

$$\begin{aligned} \beta_1^2 &= n_{e1}^2 k_0^2 \\ \beta_2^2 &= n_{e2}^2 k_0^2 \end{aligned} \qquad (5.67)$$

Finally, we may rewrite (5.66) as

$$<\phi_1, \phi_2>= \frac{\varepsilon_{1r} - \varepsilon_{2r}}{n_{e1}^2 - n_{e2}^2}\int_D \phi_1\phi_2 dx \qquad (5.68)$$

which shows that the integral over the whole cross–section $0 \leq x < \infty$ may be reduced to one just over D. This result is easily extended to $2D$ cross–sections.

A "trial field" with edge effects

As remarked in Chapter 2 the electromagnetic field on the plane of the step exhibits a dependence like ρ^α (ρ being the distance from the wedge and α being the degree of the field singularity) near the dielectric edges at the step discontinuity where 90° and 270° corners are present. The actual value taken by α can be determined by the tables provided in Chapter 2. From a numerical point of view, it is very advantageous to include such behaviour when choosing the trial field. For TE solutions, as considered in this Section, the field remains finite near the edge so that the introduction of the correct type of radial dependence, although helpful toward improving convergence of the solution, is not essential. On the other hand, as will be shown later, the TM electric field becomes singular at the dielectric step. In this case, in order to avoid the Gibbs' phenomenon, arising when using a regular set of basis functions, the use of a singular trial field becomes mandatory.

In order to include the effect of this singularity in the solution, a trial field may be developed, based on the surface mode function of the intermediate guide of the previous Section, but with the appropriate singularities added in at $x = d, D$. This results in a trial function of the form:

$$\overline{\psi}(x) = \begin{cases} \frac{1}{\varepsilon_2} e^{-\gamma_0(x-d_0)}(x-D)^\alpha & : \quad x > D \\ \frac{1}{\varepsilon_2} e^{-\gamma_0(x-d_0)}(D-x)^\alpha & : \quad D > x \geq d_0 \\ \frac{1}{\varepsilon_1}\frac{\cos(\kappa_0 x)}{\cos(\kappa_0 d)}(x-d)^\alpha\left(\frac{D-d_0}{d_0-d}\right)^\alpha & : \quad d_0 \geq x > d \\ \frac{1}{\varepsilon_1}\frac{\cos(\kappa_0 x)}{\cos(\kappa_0 d)}(d-x)^\alpha\left(\frac{D-d_0}{d_0-d}\right)^\alpha & : \quad d > x \geq 0 \end{cases} \qquad (5.69)$$

where $\overline{\psi}(x)$ represents E_y or dE_y/dx (see below) for the TE case, E_x for the TM case. The trial function is applied in exactly the same manner as the surface mode functions of the previous Section.

Table 5.1 S_{11} *for a step with* $d/D = 0.5$ *at 10 GHz for a perspex slab with permittivity* $\varepsilon_1 = 2.57$ *(TE polarization).*

	Cosine–Laguerre		*Intermediate Surface mode*		*Laguerre Polynomials*	
m	$\lvert S_{11}\rvert$	$\angle S_{11}$	$\lvert S_{11}\rvert$	$\angle S_{11}$	$\lvert S_{11}\rvert$	$\angle S_{11}$
1	6.9392e-2	9.9758e-1	6.8197e-2	1.0132	6.6442e-2	9.8332e-1
2	6.8015e-2	1.0101	6.8015e-2	1.0124	6.7027e-2	9.8407e-1
3	6.8015e-2	1.0124	6.8015e-2	1.0125	6.7793e-2	9.8656e-1
4	6.8014e-2	1.0125	6.8014e-2	1.0126	6.8419e-2	9.8873e-1
5	6.8014e-2	1.0127	6.8014e-2	1.0125	6.8759e-2	9.8967e-1

Table 5.2 Scattering matrix as obtained by using a cosine–Laguerre set; $d/D = 0.2$, $n_2k_0D = 1$, $vD = 2$.

N	$\lvert S_{11}\rvert$ %	$\angle S_{11}$	$\lvert S_{21}\rvert$ %	$\angle S_{21}$	$\lvert S_{22}\rvert$ %	$\angle S_{22}$	Rad loss %
0	20.88	-.0865	94.16	-.0149	26.55	.0413	6.87
1	20.77	-.0920	93.77	-.0135	25.28	.0540	7.77
2	20.72	-.1038	93.70	-.0099	24.73	.0532	7.92
3	20.49	-.0802	94.10	-.0096	24.82	.0648	7.25
4	20.36	-.0832	94.14	-.0082	24.83	.0737	7.13
5	20.41	-.0859	94.15	-.0076	24.79	.0759	7.19
6	20.40	-.0858	94.15	-.0076	24.78	.0759	7.19
7	20.40	-.0858	94.16	-.0075	24.79	.0763	7.18
8	20.40	-.0860	94.16	-.0075	24.79	.0764	7.18
20	20.40	-.0860	94.16	-.0075	24.79	.0764	7.18

5.1.5 *Numerical and experimental results for the TE case*

It is interesting to compare the convergence of the solutions obtained when using the various sets of basis functions introduced in the previous Section. From Table 5.1 of [5], it is apparent that rapid convergence is achieved when using the cosine–Laguerre set or the intermediate surface mode set. In fact, in both cases, 4 significant digits are obtained by using just two terms. On the other hand, if a different basis function set is selected, such as the Laguerre polynomials alone, several terms have to be used before achieving convergence. This is confirmed by looking at the last two columns of Table 5.1

A further example of convergence behaviour is given in Table 5.2 from [40] where the scattering coefficients obtained in a monomode slab by using the cosine–Laguerre set are reported. In Table 5.3 [40] are reported the results relative to the same structure as Table 5.2 but now using the complete, but non orthogonal, basis introduced previously. In particular, in Table 5.3 $N = 0$ means that the discontinuity field is assumed to be that of the incoming surface wave. This approximation, although useful for small steps, is inadequate for small d/D ratios. A noticeably better approximation is obtained by considering the discontinuity field as a linear combination of the incident and transmitted surface waves. Note that for N large enough ($N > 5$) the same results as in Table 5.2 are recovered.

Table 5.3 Scattering matrix as obtained by using a combination of surface modes complemented by Laguerre functions; d/D = 0.2, n_2k_0D = 1, vD=2.

N	$\|S_{11}\|$ %	$\angle S_{11}$	$\|S_{21}\|$ %	$\angle S_{21}$	$\|S_{22}\|$ %	$\angle S_{22}$	Rad loss %
0	19.68	-.1282	91.33	-.0210	30.27	.0485	12.7
1	20.54	-.0779	93.92	-.0097	24.43	.0650	7.57
2	20.61	-.0840	93.87	-.0084	24.39	.0697	7.64
3	20.42	-.0825	94.13	-.0086	24.76	.0703	7.22
4	20.39	-.0835	94.17	-.0083	24.81	.0722	7.16
5	20.39	-.0862	94.16	-.0074	24.79	.0765	7.18
6	20.40	-.0861	94.16	-.0075	24.78	.0764	7.18
7	20.40	-.0862	94.16	-.0074	24.78	.0766	7.19
8	20.40	-.0861	94.16	-.0075	24.79	.0765	7.18

After having checked convergence properties of the solution, it is now interesting to observe the scattering behaviour of the step discontinuity. In Figure 5.7 is illustrated the variation of scattering parameters for varying width d_1 of the smaller guide, as obtained when using $N=5$ basis functions. Since the example refers to a slab with fairly high refractive index, radiation losses are low for small steps. As noted in the introduction, for small values of step height (with reference to Figure 5.7 for $d_1/d_2 > 0.3$), the reflection coefficient is the same for waves incident from either side of the step and can be obtained by (5.1). However, as d_1 goes to zero, $|S_{22}|$ approaches the limiting value relative to a truncated slab (this case is treated in greater detail in a later Section). The transmission coefficient also decreases for increasing step size, as energy leaks out in the form of radiation. It is also apparent that more power is radiated for waves incident from the left, i.e. from the thinner guide. The physical interpretation is that, the guided mode being closer to dielectric cutoff, less reflected power is captured by the reflected guided wave and lost through backward radiation. In addition it is interesting to note that S_{11} also remains fairly small since for small values of d_1 the surface wave is no longer guided.

In the next example the behaviour of the scattering parameters versus normalised frequency is investigated. In Figures 5.8 and 5.9 are compared a 5th order variational solution obtained by using a cosine–Laguerre basis [40] with the results published in [2] which are referred to by an asterisk. The results are relative to the structure in the inset of Figure 5.9. This structure allows multimodal propagation, as two surface waves are present in the wider guide. Consequently, a wave impinging on port 1 also excites the mode relative to port 3 as shown in the inset of Figure 5.9. Even before cut–off is reached, the imminent presence of the second mode is felt, as shown by the inflection of $|S_{11}|$ and $|S_{12}|$. This is accompanied by larger radiation losses.

5.1.6 *Alternative integral equation formulation*

A similar analysis of the step discontinuity may be carried out starting again from (5.27) but taking the electric field E_y as unknown instead of the magnetic field H_x,

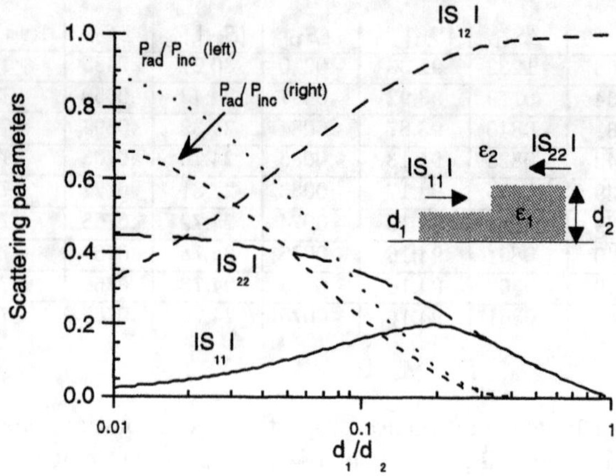

Figure 5.7 Scattering parameters for varying step ratio d_1/d_2; the geometrical dimensions are: $k_0 d_2$ = 1.0, ε_1 = 5, ε_2 = 1.

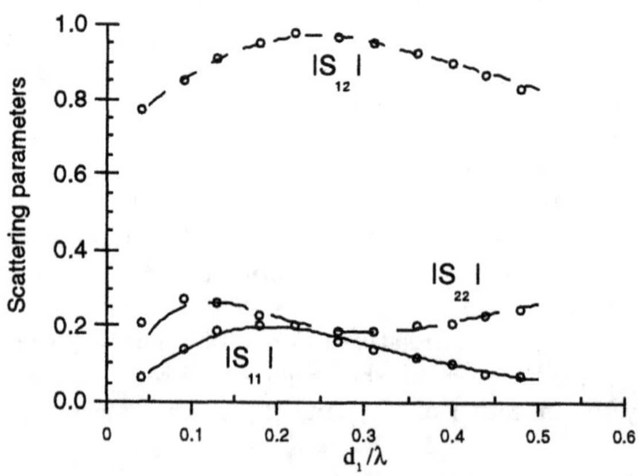

Figure 5.8 Scattering parameters vs. normalised frequency; the geometrical dimensions are: d_1/d_2 = 0.2, $k_0 d_2$ = 1.0, ε_1 = 5, ε_2 = 1.

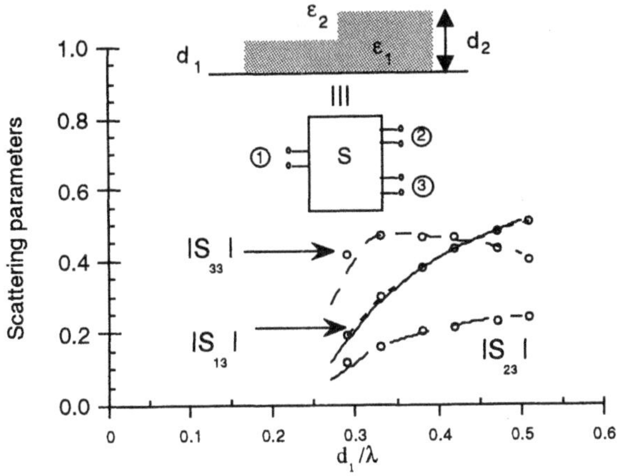

Figure 5.9 Scattering parameters versus normalised frequency; the geometrical dimensions are: $d_1/d_2 = 0.2$, $k_0 d_2 = 1.0$, $\varepsilon_1 = 5$, $\varepsilon_2 = 1$.

thus obtaining the following equation instead of (5.32)

$$\sum_{k=1}^{n_\ell} s_k I_k^i \psi_k(x) = \int_0^\infty Y(x,x')E_y(x')dx' \tag{5.70}$$

where the admittance kernel has the form

$$Y(x,x') = \tfrac{1}{2} \sum_{k=1}^{n_i} y_{0k}\psi_k(x)\psi_k(x') +$$
$$\tfrac{1}{2}\int_0^\infty \left[Y_0^I(k_x)\phi^I(x;k_x)\phi^I(x';k_x) + Y_0^{II}(k_x)\phi^{II}(x;k_x)\phi^{II}(x';k_x)\right] dk_x \tag{5.71}$$

In the above, the admittances are

$$Y_0^I(k_x) = Y_0^{II}(k_x) = \left\{ \begin{array}{ll} \frac{\sqrt{n_2^2k_0^2 - k_x^2}}{\omega\mu} & n_2k_0 > k_x \\ \frac{\sqrt{k_x^2 - n_2^2k_0^2}}{j\omega\mu} & n_2k_0 < k_x \end{array} \right\} \tag{5.72}$$

Hence, when imposing the continuity of the tangential fields at the step two different formulations are possible. In the previous Section we have taken the magnetic field as unknown, and equated the tangential components of the electric fields, thus obtaining the so called electric field integral equation (EFIE) with a kernel of impedance type. The other possibility is to consider as unknown the electric field and, by equating the tangential components of the magnetic fields, one obtains a magnetic field integral equation (MFIE) with a kernel of admittance type.

The two possibilities are briefly summarised in Table 5.4, where E, H stand for the (unknown) electric and magnetic field at the step interface, while E_{inc}, H_{inc} are the incident (known) fields.

However, the use of one or the other formulation leads to different convergence properties. In particular, the presence of k_x in the numerator in (5.72) poses convergence problems when evaluating the spectral integral, i.e. the integral in dk_x in

Table 5.4 Types of integral equation arising from the analysis of the step discontinuity.

MFIE (Magnetic Field Integral Equation) Kernel: Admittance type	$H_{inc} = \int_0^\infty Y(x,x')E(x')dx'$
EFIE (Electric Field Integral Equation) Kernel: Impedance type	$E_{inc} = \int_0^\infty Z(x,x')H(x')dx'$

Table 5.5 Asymptotic form of immitances for large k_x.

Immittance	TE	TM
$Z = 1/Y$	$\frac{\omega\mu}{k_x}$	$\frac{-k_x}{\omega\varepsilon}$
Y	$\frac{k_x}{\omega\mu}$	$-\frac{\omega\varepsilon}{k_x}$

(5.71). In fact, as it stands (5.70) is not suitable for numerical computation. This situation is similar to the case of a thin inductive iris in a rectangular waveguide (see [12, p.579]).

In order to improve convergence, it is convenient to take the indefinite integral of our integral equation as, for example for the MFIE case,

$$\int H_{inc}dx = \int \int_0^\infty Y(x,x')E(x')dx'dx \tag{5.73}$$

and perform an integration by parts of

$$\begin{gathered}\int_0^\infty Y(x,x')E(x')dx' = \\ -\int_0^\infty \left[\int Y(x,x')dx'\right]\tfrac{dE}{dx'}dx' + \left[\int Y(x,x')dx'E(x')\right]_0^\infty\end{gathered} \tag{5.74}$$

As E vanishes at infinity and $\int Y(x,x')dx'$ vanishes at zero, the last term in the above expression disappears yielding, by substitution in (5.73), the following equation

$$\int H_{inc}dx = -\int_0^\infty \left[\int\int Y(x,x')dxdx'\right]\frac{dE}{dx'}dx' \tag{5.75}$$

Now, by recalling the expression for the modal functions ϕ (for example, see (5.21)), it is noted that the double indefinite integration of the kernel provides a factor k_x^2 in the denominator, thus enhancing convergence of the spectral integrals appearing in (5.71) as k_x goes to infinity.

Therefore, when establishing the integral equation, we are free to choose between the EFIE or MFIE approaches for this type of discontinuity. However, in the TE case, when a MFIE is chosen one has to adopt the above procedure in order to achieve proper convergence. It seems therefore more convenient to use, in the TE case, the EFIE approach. Naturally, the reverse is true for the TM case, which will be treated in the next Section. In fact, for TM fields, the approach leading directly to having k_x in the denominator is the use of a kernel of admittance type, thus the

MFIE. The situation is summarised for reference purposes in Table 5.5, where the more convenient formulation is the one having k_x in the denominator.

5.1.7 TM case: scattering matrix formulation

The analysis of the TM case closely follows that for the TE case and as such the description will be kept brief. The main difference is that for TM modes, in order to avoid the convergence difficulties outlined in the previous Section, it is convenient to set up an integral equation with a Green's function of the admittance type (MFIE).

With reference to Figure 5.1 on page 191, assuming that no variations are present in the y–direction, the three field components of the TM mode can be written as shown in Chapter 3 and here repeated for convenience

$$\begin{aligned} E_x(x,z) &= \frac{\beta}{\omega\varepsilon_0\varepsilon(x)}H_y(x,z) = \frac{Z_0}{\varepsilon(x)}H_y(x,z) \\ E_z(x,z) &= \frac{1}{j\omega\varepsilon_0\varepsilon(x)}\frac{\partial}{\partial x}H_y(x,z) \end{aligned} \tag{5.76}$$

with

$$\varepsilon(x) = \left\{ \begin{array}{ccc} n_1^2 & ; & x < d \\ n_2^2 & ; & x > d \end{array} \right\} \tag{5.77}$$

The $E_x(x,z)$ component of the field is discontinuous at the interfaces between the slabs and air, that is for $x = d, D$. This is due to the discontinuous nature of $\varepsilon(x)$. Hence, the modal expansion of $E_x(x,z)$ may best be expressed as

$$E_x(x,z) = \frac{1}{\varepsilon(x)}\left\{\sum_k a_k\psi_k(x) + \int_0^\infty b(k_x)\phi(x,k_x)dk_x\right\}e^{-j\beta z} \tag{5.78}$$

with the $\varepsilon(x)$ term in evidence. The explicit expression for even and odd TM modes has been given in Chapter 3 and is repeated here for the reader's convenience.

Surface modes

The normalised surface mode function $\psi_k(x)$ for the even case (even H_y, which means an electric wall at $x = 0$) is given by

$$\psi(x) = \left\{ \begin{array}{ccc} A\cos\kappa x & : & x \le d \\ A\cos\kappa d e^{-\gamma_x(x-d)} & : & x \ge d \end{array} \right\} \tag{5.79}$$

where the normalisation constant A is now fixed by the requirement

$$\int_0^\infty \frac{1}{\varepsilon(x)}\psi^2(x)dx = 1 \tag{5.80}$$

which gives

$$A = \sqrt{\frac{2}{d + \frac{n_1^2 n_2^2}{\gamma_x}\frac{\kappa^2+\gamma_x^2}{n_2^4\kappa^2+n_1^4\gamma_x^2}}} \tag{5.81}$$

The transverse propagation constants γ_x and κ satisfy the eigenvalue equation

$$\kappa\tan\kappa d = \frac{n_1^2}{n_2^2}\gamma_x \tag{5.82}$$

while the conservation of wavenumbers still requires

$$\begin{aligned}\kappa^2 + \gamma_x^2 &= \left(n_1^2 - n_2^2\right) k_0^2 = v^2 \\ \beta^2 &= n_1^2 k_0^2 - \kappa^2 = n_2^2 k_0^2 + \gamma_x^2\end{aligned} \tag{5.83}$$

Continuous modes

The normalised modal functions of the continuum are given by

$$\varphi(x; k_x) = \begin{cases} \sqrt{\frac{2}{\pi}} \frac{n_2}{C} \cos qx & \text{in the slab} \quad (d \geq x \geq 0) \\ \sqrt{\frac{2}{\pi}} n_2 \cos\left[k_x(x-d) + \alpha\right] & \text{in air} \quad (x \geq d) \end{cases} \tag{5.84}$$

with

$$C = \sqrt{\cos^2(qd) + \left(\frac{n_2}{n_1}\right)^4 \left(\frac{q}{k_x}\right)^2 \sin^2(qd)} \tag{5.85}$$

$$\tan\alpha = \frac{q}{k_x}\left(\frac{n_2}{n_1}\right)^2 \tan qd \tag{5.86}$$

The quantities q and k_x are the transverse wavenumber in the slab and in air, respectively. They are linked to the propagation constant in the usual way as

$$\begin{aligned}\left(n_1^2 - n_2^2\right) k_0^2 = q^2 - k_x^2 = v^2 \\ \beta^2 = n_1^2 k_0^2 - q^2 = n_2^2 k_0^2 - k_x^2\end{aligned} \tag{5.87}$$

The normalising condition for this part of the spectrum is given by

$$\int_0^\infty \frac{1}{\varepsilon(x)} \varphi(x; k_x) \varphi(x; k'_x) dx = \delta(k_x - k'_x) \tag{5.88}$$

5.1.8 Integral equation for TM case

In order to obtain the relationship relating the incident surface wave(s) with fields at the step discontinuity, we proceed in a manner analogous to that followed for the TE case. By taking E_x as unknown and by equating H_y at the step discontinuity, we derive the following integral equation

$$\sum_{k=1}^{n_i} V_k^i \psi_k(x) = \int_0^\infty Y(x, x') E_x(x') dx' \tag{5.89}$$

with the kernel $Y(x, x')$ representing a Green's function of admittance type given by

$$\begin{aligned}Y(x, x') = \tfrac{1}{2} \sum_{k=1}^{n_i} Y_{0k} \psi_k(x) \psi_k(x') + \\ \tfrac{1}{2} \int_0^\infty \left[Y_0^I(k_x) \phi^I(x; k_x) \phi^I(x'; k_x) + Y_0^{II}(k_x) \phi^{II}(x; k_x) \phi^{II}(x'; k_x)\right] dk_x\end{aligned} \tag{5.90}$$

The solution procedure is analogous to that for the TE case. We introduce a set of expanding functions to represent E_x at the interface, and we use the same set of functions to discretise the integral equation (5.89) into a matrix equation, thus following the Ritz-Galerkin variational approach. The main difference with respect to the previous TE case stems from the discontinuous nature of E_x at the dielectric interface ($x = d$) since

$$n_1^2 E_x(d^-) = n_2^2 E_x(d^+) \tag{5.91}$$

To take into account this behaviour of the field, it is convenient, for the TM case, to introduce a piecewise continuous expansion set. That is, we introduce an expanding functions set of the type

$$\begin{array}{llll} C_m(x) & : & 0 \le x \le d & m = 0, 1...M \\ L_n(x-d) & : & x > d & n = 0, 1...N \end{array} \tag{5.92}$$

which is complete as M, N approach infinity. It is also convenient to introduce the following functional scalar product:

$$\left\langle \frac{\psi(x)}{\varepsilon(x)}, f(x) \right\rangle = \int_0^\infty \frac{\psi(x)}{\varepsilon(x)} f(x) dx =$$

$$\frac{1}{n_1^2} \int_0^d \psi_1(x) f_1(x) dx + \frac{1}{n_2^2} \int_d^\infty \psi_2(x) f_2(x) dx \tag{5.93}$$

where it is assumed that the function $f(x)$ is possibly discontinuous at $x = d$, but it is continuous over each subinterval $(0, d), (d, \infty)$ where it takes the form $f_1(x)$ and $f_2(x)$, respectively. The unknown electric field is thus represented as

$$E_x(x) = f_1(x) + f_2(x) \tag{5.94}$$

with $f_1(x)$ and $f_2(x)$ given by

$$f_1(x) = \sum_{m=0}^{M} F_{1m} C_m(x) \qquad 0 \le x \le d$$

$$f_2(x) = \sum_{n=0}^{N} F_{2n} L_n(x-d) \qquad x \ge d \tag{5.95}$$

It is also expedient to introduce a vector $\mathbf{F}$ composed of the vectors $\mathbf{F}_1$ and $\mathbf{F}_2$ as

$$\begin{pmatrix} \frac{1}{n_1^2}\mathbf{F}_1 \\ \frac{1}{n_2^2}\mathbf{F}_2 \end{pmatrix} = \mathbf{F} \tag{5.96}$$

Obviously, the knowledge of the vector $\mathbf{F}$ allows reconstruction of the electric field E_x at the interface. In a manner completely similar to the TE case, coupling coefficients between the surface waves and the basis function are defined as

$$\begin{aligned} Q_{km}^{(1)} &= \int_0^d C_m(x) \psi_{1k}(x) dx \\ Q_{kn}^{(2)} &= \int_d^\infty L_n(x-d) \psi_{2k}(x) dx \end{aligned} \tag{5.97}$$

which also allows us to build the vector $\mathbf{Q}_k$ as

$$\begin{pmatrix} \frac{1}{n_1^2}\mathbf{Q}_k^{(1)} \\ \frac{1}{n_2^2}\mathbf{Q}_k^{(2)} \end{pmatrix} = \mathbf{Q}_k \tag{5.98}$$

Analogous coefficients and vectors are defined for the coupling between modes of the continuous spectrum and basis function in region I ($z < 0$)

$$\begin{aligned} P_m^{(1)}(k_x) &= \int_0^d C_m(x)\phi_1(x;k_x)dx \\ P_n^{(2)}(k_x) &= \int_d^\infty L_n(x-d)\phi_2(x;k_x)dx \\ \begin{pmatrix} \frac{1}{n_1^2}\mathbf{P}_1 \\ \frac{1}{n_2^2}\mathbf{P}_2 \end{pmatrix} &= \mathbf{P} \end{aligned} \tag{5.99}$$

while coefficients and vectors describing the coupling between modes of the continuous spectrum and basis function in region II ($z > 0$) are denoted by $\mathbf{P}'$. By using the above definitions the discretised version of (5.89) is written as

$$\sum_{k=1}^{n_i} V_k^i \mathbf{Q}_k = \mathbf{YF} \tag{5.100}$$

with the matrix $\mathbf{Y}$ given by the following expression

$$\mathbf{Y} = \frac{1}{2}\left\{\sum_{k=1}^{n_i} Y_{0k}\mathbf{Q}_k\mathbf{Q}_k^T + \int_0^\infty \left[Y_0^I(k_x)\mathbf{P}\mathbf{P}^T + Y_0^{II}(k_x)\mathbf{P}'\mathbf{P}'^T\right] dk_x\right\} \tag{5.101}$$

By following the same procedure as for the TE case, the normalised scattering matrix is obtained in matrix form

$$\overline{\mathbf{S}} = -\mathbf{I} + \sqrt{\mathbf{Y}_0}\mathbf{Q}_k^T\mathbf{Y}^{-1}\mathbf{Q}_\ell\sqrt{\mathbf{Y}_0} \tag{5.102}$$

with $\mathbf{I}$ denoting the identity matrix and

$$\sqrt{\mathbf{Y}_0} = \begin{pmatrix} \sqrt{Y_{01}} & & 0 \\ & \ddots & \\ 0 & & \sqrt{Y_{0n_i}} \end{pmatrix} \tag{5.103}$$

representing the diagonal matrix of the square roots of surface wave admittances. It is now only necessary to choose an appropriate basis function set and compute the coupling integrals in (5.97), (5.99) to evaluate (5.102).

5.1.9 *Some results for TM case*

Analogously to what has been done for the TE case, it is interesting to compare the convergence of the solutions as obtained when using the various sets of basis functions previously introduced. From Table 5.6 it is apparent that rapid convergence is achieved when using the intermediate surface mode (see equation 5.69) or the cosine-Laguerre set. Note that the analysis with just one intermediate surface mode corrected for the presence of edge singularity provides $|S_{11}| = 1.779e-2$, $\angle S_{11} = 0.033$. On the other hand, if a different basis function set is selected as the Laguerre polynomials alone, several terms have to be used before achieving convergence. In particular, it is noted that a single singular surface mode provides the same accuracy as using ten Laguerre polynomials.

Table 5.6 S_{11} *for a step with* $d/D = 0.5$ *at 10 GHz for a perspex slab with permittivity* $\varepsilon_1 = 2.57$ *(TM polarization).*

	Cosine-Laguerre		*Intermediate Surface mode*		*Laguerre Polynomials*	
m	$\|S_{11}\|$	$\angle S_{11}$	$\|S_{11}\|$	$\angle S_{11}$	$\|S_{11}\|$	$\angle S_{11}$
1	1.8225e-2	0.0308	1.8116e-2	0.0312	1.9242e-2	0.0332
2	1.7916e-2	0.0294	1.8001e-2	0.0303	1.7027e-2	0.0247
3	1.7733e-2	0.0142	1.7715e-2	0.0164	1.7493e-2	0.0565
4	1.7694e-2	0.0125	1.7684e-2	0.0122	1.7119e-2	0.0921
5	1.7683e-2	0.0107	1.7688e-2	0.0103	1.7335e-2	0.0765
10	1.7683e-2	0.0092	1.7684e-2	0.0101	1.7735e-2	0.0533

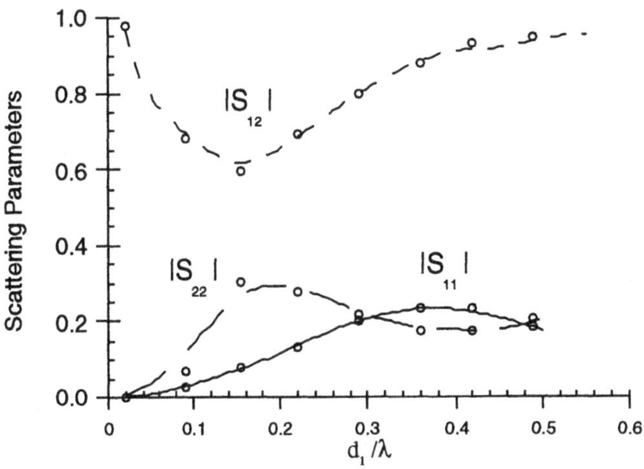

Figure 5.10 TM scattering parameters of the single step discontinuity.

It is now interesting to consider the scattering behaviour of the step discontinuity under TM incidence. Figures 5.10 and 5.11 show the results obtained for the TM case by using a 5th order cosine Laguerre basis compared with the results of [2] denoted by circles. The agreement between the two approaches is generally excellent; only a minor disagreement is present for small values of normalised frequency. In this case, in fact, the field tails of surface modes extend at considerable distance away from the slab guide. Since in [2] the solution has been obtained by considering a structure shielded by metallic walls, the effect of these walls becomes noticeable for lower frequency values when the field spreads out.

5.2 Abruptly terminated slab

A classical problem in open waveguide theory is that of a dielectric waveguide terminated by an abrupt transverse discontinuity, i.e. the facet problem. This type

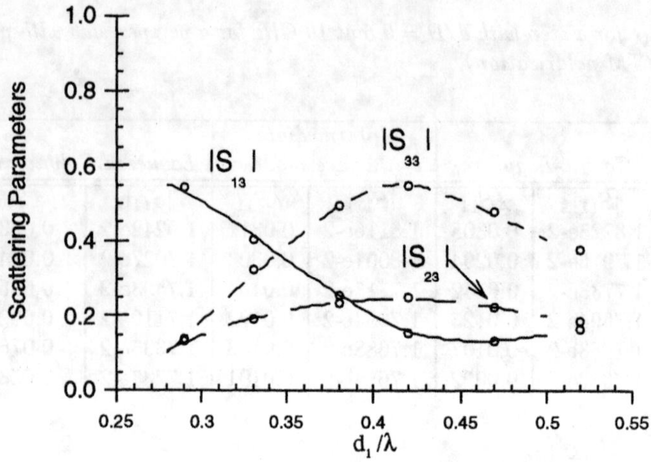

Figure 5.11 TM scattering parameters of the single step discontinuity.

of configuration finds practical application in a double heterostructure (d.h.) laser, where the mirror reflectivity is a crucial parameter of the laser, since it determines its oscillation condition; other important applications are in endfire antennas for millimetric waves. This type of discontinuity may be considered as a special case of the step discontinuity. Since on one side a homogeneous dielectric is present, considerable analytical progress is feasible, as illustrated in the next pages.

A variational formulation of the diffraction of TM waves by a semi–infinite slab was given in [1] as far back as 1957. In this treatment, the incident surface wave is used as a trial field. An analogous approach was used in [19, 17] to discuss the diffraction of TE and TM waves, perpendicular to the p–n junction, at the mirror of a d.h. laser.

A different approach based on a Fourier transform of the integral equation was introduced in [38]. An elegant approximate analysis neglecting mode conversion at the facet leads to an explicit expression of the far field, as elaborated in [3], whereas in [26] a theoretical comparison of various approximations is presented. A comprehensive review of the subject with numerous references can be found in [4, 25].

5.2.1 TE case

With respect to the step discontinuity analysed in the previous Section, in the facet problem we have a homogeneous right half–space for $z > 0$. Moreover, in the TE case, when the effective frequency $vd = \sqrt{n_1^2 - n_2^2}k_0 d$ goes to zero, the slab Green's function of the MFIE reduces to that of a homogeneous half–space, i.e. to the Hankel function. The integral of the latter with the basis function of the previous Section can be analytically evaluated. It is thus possible to compute the impedance matrix elements by closed expressions plus a small correction term to be determined numerically.

The geometry of the problem is described in Figure 5.12, where the two complex

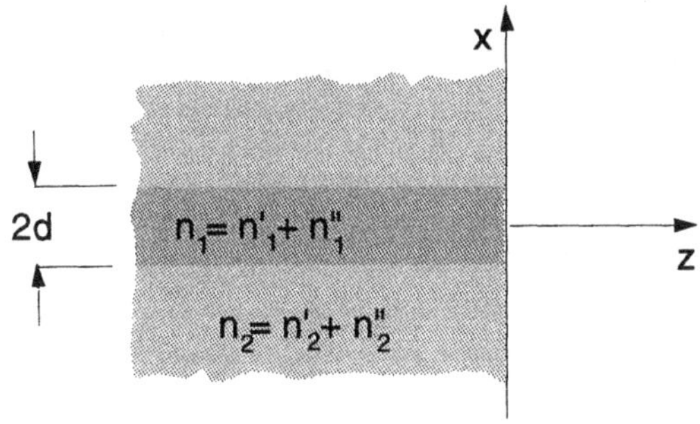

Figure 5.12 Longitudinal cross–section of laser, showing facet with mirror.

refractive indices $n_1 = n_1' + jn_1''$, for the active slab, and $n_2 = n_2' + jn_2''$, with $n_1' > n_2'$, for the semi–infinite layers, are used. The structure is terminated abruptly at $z = 0$, since for $z > 0$ we have an air half–space. However, owing to symmetry, we need only consider the region for $x \geq 0$. We can set up the EFIE exactly as done for the TE step discontinuity; the only difference being that for $z > 0$ we have the free–space eigenfunctions instead of the eigenfunctions of the slab waveguide of thickness $2D$. Since we are considering the even case, the eigenfunctions are cosine functions and the integral equation is written as

$$V_\ell \psi_\ell(x) = \int_0^\infty Z(x, x') h_\ell(x') dx' \tag{5.104}$$

with the Green's impedance $2Z(x, x')$ given by (see (5.33) on page 196)

$$\sum_{k=1}^{n} z_{0k} \psi_k(x) \psi_k(x') + \int_0^\infty Z_0^I(k_x) \phi^I(x; k_x) \phi^I(x'; k_x) dk_x + \\ \frac{2}{\pi} \int_0^\infty \frac{\omega\mu}{\sqrt{k_0^2 - k_x^2}} \cos(k_x x) \cos(k_x x') dk_x \tag{5.105}$$

The last integral in (5.105), i.e. the contribution from free–space, may be written in terms of the Hankel function of the second type $H_0^{(2)} = J_0 - jN_0$ by recalling the plane wave expansion of the Hankel function, i.e.

$$\frac{1}{2} \sum_{\pm} H_0^{(2)}(k_0 |x \pm x'|) = \frac{2}{\pi} \int_0^\infty \frac{1}{\sqrt{k_0^2 - k_x^2}} \cos(k_x x) \cos(k_x x') dk_x \tag{5.106}$$

For the region $z < 0$ no close form expression is available to express compactly the other integral appearing in (5.105). Nonetheless, by introducing the "difference" kernel

$$t(x, x'; k_x) = \phi^I(x; k_x) \phi^I(x'; k_x) - \frac{2}{\pi} \cos(k_x x) \cos(k_x x') \tag{5.107}$$

we can write the integral referring to the $z < 0$ region as

$$\int_0^\infty \frac{\phi^I(x; k_x) \phi^I(x'; k_x)}{\sqrt{n_2^2 k_0^2 - k_x^2}} dk_x =$$

$$\frac{1}{2}\sum_{\pm} H_0^{(2)}(n_2k_0|x \pm x'|) + \int_0^\infty \frac{t(x,x';k_x)}{\sqrt{n_2^2k_0^2 - k_x^2}}dk_x \tag{5.108}$$

where we have extracted the limit of lightly trapped waves, that is the first term on the right hand side. By inserting (5.106) and (5.108) into (5.105), and by considering the expression for the TE impedance, we have the total impedance Green's function for the facet discontinuity written as

$$\begin{aligned} Z(x,x') = \frac{\omega\mu}{2}\sum_{k=1}^{n}\frac{1}{\beta_k}\psi_k(x)\psi_k(x') + \\ \frac{\omega\mu}{4}\sum_{\pm}\left[H_0^{(2)}(k_0|x \pm x'|) + H_0^{(2)}(n_2k_0|x \pm x'|)\right] + \\ \frac{\omega\mu}{2}\int_0^\infty \frac{t(x,x';k_x)}{\beta(k_x)}dk_x \end{aligned} \tag{5.109}$$

It is to be observed that an analogous result holds for the odd case, where the free–space eigenfunctions are sine functions. With respect to the step discontinuity the main difference is represented by the fact that now, when applying the Ritz–Galerkin method with the Laguerre functions used to represent the unknown magnetic field, we can evaluate analytically the matrix elements, i.e. the elements

$$\begin{aligned} H_{mn}(u) = \\ \tfrac{1}{2}\int_0^\infty\int_0^\infty L_m(x)\left[H_0^{(2)}(u|x+x'|) + H_0^{(2)}(u|x-x'|)\right]L_n(x')dxdx' \end{aligned} \tag{5.110}$$

The evaluation of (5.110) is detailed, for reference purposes, at the end of this paragraph. By proceeding in a manner similar to that followed for the step discontinuity we obtain

$$\lambda_\ell = \mathbf{Z}^{-1}\mathbf{Q}_\ell \tag{5.111}$$

where λ_ℓ is the vector containing the amplitude of the expansion functions and $\mathbf{Q}$ has the same meaning as in the previous Section. The matrix $\mathbf{Z}$ is now given by

$$\begin{aligned} \mathbf{Z} = \tfrac{\omega\mu}{2}\textstyle\sum_{k=1}^{n}\tfrac{1}{\beta_k}\mathbf{Q}_k\mathbf{Q}_k^T + \\ \tfrac{\omega\mu}{4}\left[H_{mn}(k_0) + H_{mn}(n_2k_0)\right] + \\ \tfrac{\omega\mu}{2}\int_0^\infty \tfrac{t_{mn}(k_x)}{\beta(k_x)}dk_x \end{aligned} \tag{5.112}$$

The elements of the normalised scattering matrix between surface waves are thus given by

$$\overline{S}_{k\ell} = \delta_{k\ell} - \sqrt{Z_{0k}Z_{0\ell}}\mathbf{Q}_k{}^T\mathbf{Z}^{-1}\mathbf{Q}_\ell \tag{5.113}$$

Evaluation of the overlapping integral between Hankel function and Laguerre polynomials

It is required to calculate (5.110), i.e. the following integral

$$\int_0^\infty e^{-x/2x_0}L_p\left(\frac{x}{x_0}\right)\left[H_0^{(2)}(k_0|x \pm x'|)\right]e^{-x'/2x_0}L_q\left(\frac{x'}{x_0}\right)\frac{dx}{\sqrt{x_0}}\frac{dx'}{\sqrt{x_0}} \tag{5.114}$$

To this end it is advantageous to introduce the expansion

$$H_0^{(2)}(k_0|x - x'|) = \frac{e^{-|x-x'|/2x_0}}{x_0}\sum_{k=0}^{\infty}a_kL_k\left(\frac{|x-x'|}{x_0}\right) \tag{5.115}$$

The coefficients a_k appearing in (5.115) will be determined later. Using (5.115) in the previous equation we obtain

$$H_{pq}^{-}(k_0) = \sum_{k=0}^{\infty} a_k I_k \tag{5.116}$$

with

$$\begin{gathered} I_k = \\ \int_0^\infty \int_0^\infty e^{-x/2x_0} L_p\left(\frac{x}{x_0}\right) e^{-|x-x'|/2x_0} L_k\left(\frac{|x-x'|}{x_0}\right) e^{-x'/2x_0} L_q\left(\frac{x'}{x_0}\right) \frac{dx}{x_0}\frac{dx'}{x_0} \\ = \int_0^\infty e^{-x/x_0} L_p\left(\frac{x}{x_0}\right) \frac{dx}{x_0} \int_0^x L_q\left(\frac{x'}{x_0}\right) L_k\left(\frac{x-x'}{x_0}\right) \frac{dx'}{x_0} + \\ + \int_0^\infty L_p\left(\frac{x}{x_0}\right) \frac{dx}{x_0} \int_x^\infty e^{-x'/x_0} L_q\left(\frac{x'}{x_0}\right) L_k\left(\frac{x'-x}{x_0}\right) \frac{dx'}{x_0} \equiv I_k' + I_k'' \end{gathered} \tag{5.117}$$

Hence, by setting $\frac{x-x'}{x_0} = u$ we have for the first integral in (5.117)

$$\begin{gathered} I_k' = \int_0^\infty e^{-x/x_0} L_p\left(\frac{x}{x_0}\right) \frac{dx}{x_0} \int_0^{x/x_0} L_q\left(\frac{x}{x_0} - u\right) L_k(u)\, du = \\ \int_0^\infty e^{-x/x_0} L_p\left(\frac{x}{x_0}\right) \frac{dx}{x_0} \left[L_{k+q}\left(\frac{x}{x_0}\right) - L_{k+q+1}\left(\frac{x}{x_0}\right)\right] = \delta_{p,k+q} - \delta_{p,k+q+1} \end{gathered} \tag{5.118}$$

Similarly, setting $\frac{x'-x}{x_0} = u$ in the second integral in (5.117) we get

$$I_k'' = \int_0^\infty e^{-x/x_0} L_p\left(\frac{x}{x_0}\right) \frac{dx}{x_0} \int_0^\infty e^{-u} L_q\left(\frac{x}{x_0} + u\right) L_k(u)\, du \tag{5.119}$$

Using integration by parts, the above equation is written as

$$\begin{gathered} \int_0^\infty e^{-u} L_q\left(\frac{x}{x_0} + u\right) L_k(u)\, du = \\ \int_0^\infty L_q\left(\frac{x}{x_0} + u\right) d\left\{e^{-u}\left[L_{k-1}(u) - L_k(u)\right]\right\} = \\ \int_0^\infty L_{q-1}'\left(\frac{x}{x_0} + u\right) e^{-u}\left[L_{k-1}(u) - L_k(u)\right] du \end{gathered} \tag{5.120}$$

Taking advantage of the following result [16, p.1038],

$$L_{q-1}'\left(\frac{x}{x_0} + u\right) = \sum_{m=0}^{q-1} L_m(u)\, L_{q-1-m}\left(\frac{x}{x_0}\right) \tag{5.121}$$

the previous equation can be expressed as

$$\begin{gathered} \sum_{m=0}^{q-1} L_{q-1-m}\left(\frac{x}{x_0}\right) \int_0^\infty L_m(u)\, e^{-u}\left[L_{k-1}(u) - L_k(u)\right] du = \\ = \sum_{m=0}^{q-1} L_{q-1-m}\left(\frac{x}{x_0}\right)\left[\delta_{m,k-1} - \delta_{m,k1}\right] = L_{q-k}\left(\frac{x}{x_0}\right) - L_{q-k-1}\left(\frac{x}{x_0}\right) \end{gathered} \tag{5.122}$$

Hence,

$$I_k'' = \delta_{p,q-k} - \delta_{p,q-k-1} \tag{5.123}$$

which, together with (5.118) provides

$$H_{pq}^{-} = a_{p-q} - a_{p-q-1} + a_{q-p} - a_{q-p-1} \tag{5.124}$$

In the above $a_k = 0$ for $k < 0$. The same type of calculation, only changing the minus sign into plus, also yields

$$H^+_{pq} = a_{p+q} - a_{p+q+1} \tag{5.125}$$

Hence the result

$$\begin{gathered} H_{pq}(k_0) \equiv \tfrac{1}{2}\left(H^+_{pq} + H^-_{pq}\right) = \\ \tfrac{1}{2}\left(a_{p+q} - a_{p+q+1}\right) + \tfrac{1}{2}\left(a_{p-q} - a_{p-q-1} + a_{q-p} - a_{q-p-1}\right) \end{gathered} \tag{5.126}$$

It still remains to evaluate the coefficients a_k, which have the form

$$a_k = \int_0^\infty H_0^{(2)}\left(k_0 x_0 u\right) e^{-u/2} L_k\left(u\right) du \tag{5.127}$$

By using the power expansion of the Laguerre polynomials [16, p.1037] we obtain

$$\begin{aligned} a_k &= x_0 \sum_{m=0}^{k} (-1)^m \binom{k}{k-m} \tfrac{1}{m!} \int_0^\infty H_0^{(2)}\left(k_0 x_0 u\right) e^{-u/2} u^m du \\ &= x_0 \sum_{m=0}^{k} (-1)^m \binom{k}{k-m} \left(\tfrac{1}{4} + k_0^2 x_0^2\right)^{-1/2(m+1)} \\ &\left\{P_m\left[\left(1+4k_0^2x_0^2\right)^{-1/2}\right] + j\tfrac{2}{\pi} Q_m\left[\left(1+4k_0^2x_0^2\right)^{-1/2}\right]\right\} \end{aligned} \tag{5.128}$$

In the above equation the definition of the Hankel function and [16, pp.7-11] have been used. P and Q are the Legendre polynomials of order m. Let

$$\begin{aligned} \phi &= \arctan 2k_0 x_0 \\ \cos\phi &= \left(1+4k_0^2x_0^2\right)^{-1/2} \end{aligned} \tag{5.129}$$

Hence, $a_k = 0$ for $k < 0$, while for $k \geq 0$ we have

$$a_k = x_0 \sum_{m=0}^{k} (-1)^m \binom{k}{k-m} \left(2\cos\phi\right)^{(m+1)} \left\{P_m\left[\cos\phi\right] + j\tfrac{2}{\pi} Q_m\left[\cos\phi\right]\right\} \tag{5.130}$$

It is now interesting to observe the numerical convergence of the solution for increasing order of the variational solution. This is shown in Table 5.7 relative to the following structure: d = 0.3 μm; $\lambda_0 = 0.9$ μm; $n_1 = n'_1 = 3.61$; $n_2 = n'_2 = 3.40$ (from [43]). The modulus of the reflection coefficient agrees quite well with that given in [19, 14], where the incident TE_0 mode is assumed as a trial field. In this manner in [19, 14] the coupling of the incident mode with the continuous spectrum of the slab guide is completely neglected. Thus, the backward radiation and the reactive storage at the junction in the $z < 0$ region, are not properly modelled. On the other hand, owing to the variational nature of the reflection coefficient, a fairly good agreement is found.

5.3 The double step discontinuity

We will consider now the case of cascaded discontinuities in a planar dielectric waveguide. The geometry of this problem is shown in Figure 5.13. A cascade of steps finds application in passive and active components for integrated optics and optical communication, such as the grating coupler, the transformer/echelon, and

Table 5.7 Convergence of the reflection coefficient for an incident TE_0 mode.

N	$\lvert\Gamma\rvert$	$\angle\Gamma$ (degrees)
0	0.598	4.41
1	0.624	3.04
2	0.620	2.52
3	0.624	2.75
4	0.623	2.98
5	0.622	2.92

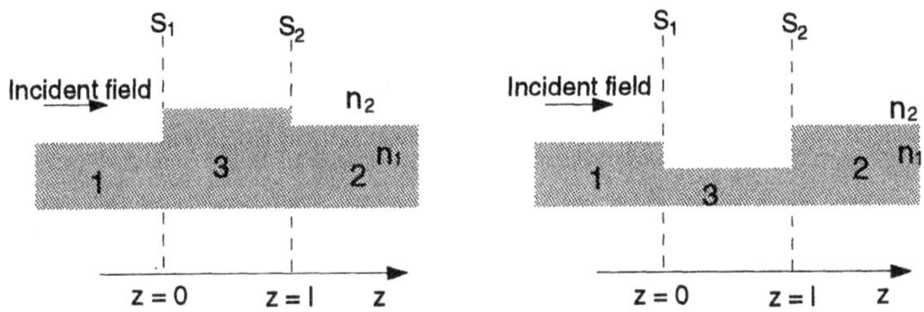

Figure 5.13 Geometry of the double step discontinuity: (a) rib, (b) groove

the distributed feedback laser. Corrugated dielectric waveguides are also used for millimetre waves and as microwave antenna feeds. The extension from the case of a single step discontinuity to a double step poses an interesting problem. Let us consider a surface wave incident from the left in Figure 5.13; it will be scattered by the step S_1, creating all surface and continuous modes allowable on either side of S_1. A finite range of the continuous modes will be propagating, radiating away from either side of the step. The rest are reactive and, as such, non–propagating. They represent localised energy storage in the neighbourhood of the step. The propagating surface and continuous modes that reach the second step S_2 will be further scattered into all allowable modes on either side of S_2. If the distance between the two steps S_1 and S_2 is short compared with the wavelength of the incident field there will be considerable interaction between the scattered waves of the two steps.

For the simple case when the field is well guided and step sizes are small, we may disregard the interaction due to the continuous modes. The problem reduces to that of two simple steps in cascade. This analysis can be accomplished by a model consisting of two discrete multiports connected by transmission lines with each transmission line corresponding to a single surface mode. This is analogous to the representation of discontinuities on closed waveguides.

Unfortunately, for many open structures of interest such as grating couplers and leaky wave antennas, it is precisely the strong interaction of the step discontinuities with the continuum modes that is required for these devices to be effective. Hence, the simple model just mentioned cannot be used to describe them.

The double step discontinuities in slab waveguides have been investigated by several authors (see e.g. [42, 18, 13, 24, 47, 49, 53] to cite just a few) and several

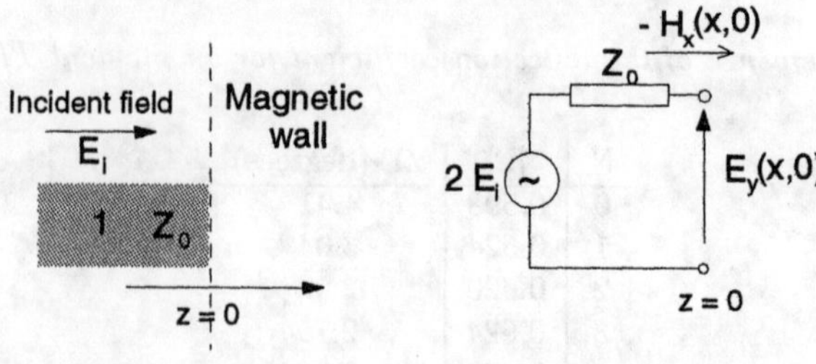

Figure 5.14 Thevenin's equivalent circuit of a semi–infinite slab guide.

variational techniques have also been proposed [42, 13]. In [13] the coupled integral equations are solved by direct numerical integration in conjunction with repeated iterations to achieve convergence. This approach couples an analytically complex variational method with a brute force numerical integration and iteration process, which is quite time consuming.

In [18], the entire mode spectrum has been discretised by replacing the problem of an open waveguide with that of a periodic multilayer optical waveguide. This was claimed as a generalisation of a similar method of mode discretization employed by Brooke and Kharadly [2] where a variable bound approach was first proposed. Essentially the method of [18] reduces the problem of the interacting double steps to that of two simple steps in cascade. However, in this formulation, an infinite, discrete, number of waveguide modes is present in each region and an approximate solution will be obtained by using a finite number of such modes. Typically the method requires a large number of modes before convergence is achieved. In this case, in the convergence test it took 20 modes to obtain a solution with a relative error of 4.5%, and more than 70 modes to achieve a relative error of 0.05%.

It will be shown here that, by using the variational approach and the basis functions presented in the previous paragraph, a worst case convergence with relative error better than 1.0% can be achieved with only a few terms of the expansion set. Hence the numerical evaluation of the solutions is both efficient and quick.

5.3.1 Interacting double steps

Consider the simple case of a single mode semi–infinite slab. It can be modelled by an equivalent semi–infinite transmission line whose characteristic impedance Z_0 is that of the single mode field travelling in the slab waveguide. By network theory, the Thevenin's equivalent circuit for the transmission line is then as shown in Figure 5.14. This equivalent circuit may be used to represent regions (1) and (2) of Figure 5.13 for each mode travelling in these regions.

The total electric and magnetic fields in region (1) can be written in modal form as

$$\begin{aligned} E_y(x,0) &= \sum_{k=1}^{n} (a_k + b_k)\,\psi_k(x) + \int_0^{\infty} b(k_x)\phi(x;k_x)dk_x \\ H_x(x,0) &= \sum_{k=1}^{n} \frac{(a_k - b_k)}{z_{0k}}\psi_k(x) - \int_0^{\infty} \frac{b(k_x)}{z_0(k_x)}\phi(x;k_x)dk_x \end{aligned} \qquad (5.131)$$

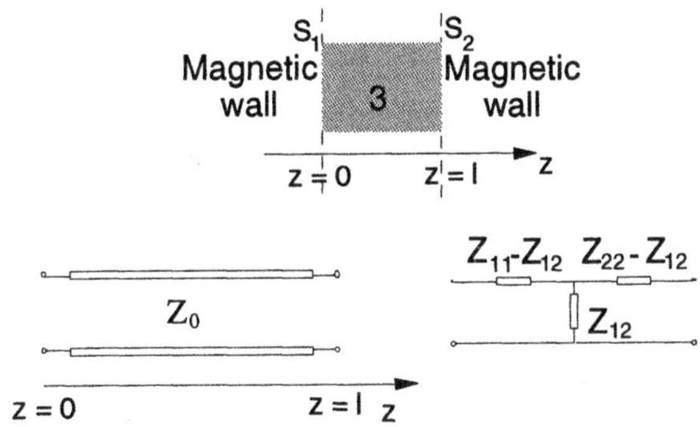

Figure 5.15 Equivalent network representation of a finite length of slab waveguide.

where a_k is the amplitude of the incident modal fields and $b_k, b(k_x)$ are the amplitudes of the reflected modal fields for the surface and continuous modes, respectively. The characteristic impedances of the surface modes are z_{0k} and $z_0(k_x)$ the corresponding impedance of a component of the continuum. By using orthogonality of modes $\psi_k(x)$ and $\phi(x;k_x)$ in the above equation, we obtain

$$\begin{aligned} b_k &= a_k - z_{0k}\int_0^\infty \psi_k(x')H_x(x')dx' \\ b(k_x) &= -z_0(k_x)\int_0^\infty \phi(x';k_x)H_x(x')dx' \end{aligned} \tag{5.132}$$

and, upon substituting $b_k, b(k_x)$ into the first of (5.131), the scattered field representation at $z = 0$ is given as :

$$\begin{aligned} E_y(x,0) = \sum_{k=1}^{n} 2a_k\psi_k(x) + \textstyle\int_0^\infty Z_1(x,x')H_x(x',0)dx' = \\ 2E_1 - \hat{Z}_1 \cdot [-H_x(x',0)] \end{aligned} \tag{5.133}$$

with

$$Z_1 = \sum_{k=1}^{n} z_{0k}\psi_k(x)\psi_k(x') + \int_0^\infty z_0(k_x)\phi(x;k_x)\phi(x';k_x)dk_x \tag{5.134}$$

whereas $\hat{Z}_1$ is the Green's impedance integral operator (which may also be viewed as the driving–point impedance) of a semi–infinite slab, E_1 is the incident field. The dot in the above equation is understood as an integration over x'. By a similar argument, for a field E_2 incident from the right in region (2), we get

$$E_y(x,\ell) = -2E_2 + \hat{Z}_2 \cdot [-H_x(x',\ell)] \tag{5.135}$$

where the positive sign convention is from left to right.

Region (3), which is delimited by magnetic walls on both sides, may be similarly modelled by finite lengths of transmission lines (open circuited at both ends) with each pair of transmission lines corresponding to a mode in the region. Developing this analogy further, each pair of transmission lines may in turn be represented by an equivalent T–network as shown in Figure 5.15. The equivalent values for Z_{11}, Z_{12} and Z_{22} are

$$
\begin{aligned}
Z_{11} = Z_{22} &= Z_0 \coth \gamma\ell \\
Z_{12} &= \frac{Z_0}{\sinh \gamma\ell}
\end{aligned}
\tag{5.136}
$$

where ℓ is the length of region (3) and γ, Z_0 are respectively the propagation constant and the characteristic impedance of a mode in the region. By applying this analogy to every field mode in region (3) we may infer from Figure 5.15 the relationship of the fields at $z = 0$ and $z = \ell$ of the open circuited section of a slab as

$$
\begin{bmatrix} E_y(x,0) \\ E_y(x,\ell) \end{bmatrix} = \begin{bmatrix} \hat{Z}_{11} & -\hat{Z}_{12} \\ \hat{Z}_{12} & -\hat{Z}_{11} \end{bmatrix} \cdot \begin{bmatrix} -H_x(x,0) \\ -H_x(x,\ell) \end{bmatrix}
\tag{5.137}
$$

where

$$
\begin{aligned}
Z_{11} = \sum_{k=1}^{n_3} z_{0k} \coth(\gamma_k\ell)\, \psi_k(x)\psi_k(x') + \\
\int_0^\infty z_0(k_x) \coth(\gamma\ell)\, \phi(x;k_x)\phi(x';k_x)dk_x
\end{aligned}
\tag{5.138}
$$

$$
\begin{aligned}
Z_{12} = \sum_{k=1}^{n_3} \frac{z_{0k}}{\sinh(\gamma_k\ell)} \psi_k(x)\psi_k(x') + \\
\int_0^\infty \frac{z_0(k_x)}{\sinh(\gamma\ell)} \phi(x;k_x)\phi(x';k_x)dk_x
\end{aligned}
\tag{5.139}
$$

The integrals over k_x in $\hat{Z}_{11}$ and $\hat{Z}_{12}$ do not contain branch cuts and are evaluated analytically by the method of residues. Equation (5.137) represents a "two–port" impedance operator for the open circuited section of region (3). Consider now that this open circuited section is "loaded" by the semi–infinite slab guides of regions (1) and (2) and that there are propagating fields E_1 incident from the left and E_2 from the right. The "loading" of the open circuited section of the guide in circuit terms is equivalent to the imposition of field continuity at S_1 and S_2. The corresponding network representation is shown in Figure 5.16. Hence, from (5.132), (5.133) and (5.135), the Green's impedance operator of the double step is obtained:

$$
\begin{bmatrix} E_y(x,0) \\ E_y(x,\ell) \end{bmatrix} = \frac{1}{2} \begin{bmatrix} \hat{Z}_1 + \hat{Z}_{11} & -\hat{Z}_{12} \\ -\hat{Z}_{12} & \hat{Z}_2 + \hat{Z}_{11} \end{bmatrix} \cdot \begin{bmatrix} -H_x(x,0) \\ -H_x(x,\ell) \end{bmatrix}
\tag{5.140}
$$

The coupling of the modes between the interacting steps is essentially represented by the $\hat{Z}_{12}$ term in the above equation. In this form it is noted that the modes in the intervening length of guide are no longer accessible.

Discretization of the integral equation

The integral equations (5.140) can be reduced to a finite matrix equation by the usual discretization procedure. The application of the moment method in its Ritz–Galerkin version has been described previously and need not be repeated here. There is, however, one point that is worthy of mention. That is, two different sets of expansion functions may be used, one for each step S_1 and S_2. This is possible because the modal field terms in regions (1) and (2) never mix in the scattering formulation of (5.140). This is a considerable advantage especially if region (1) is

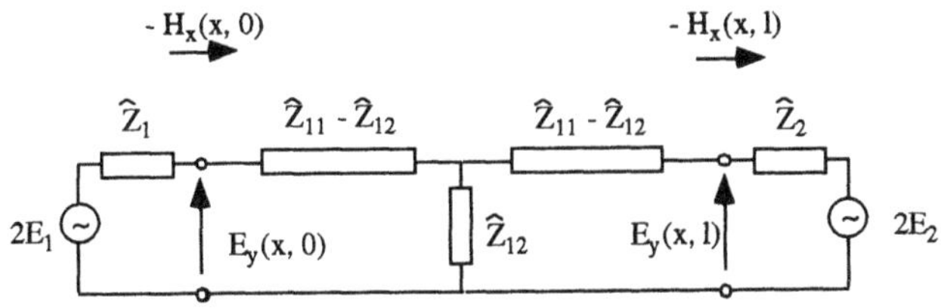

Figure 5.16 Equivalent network model of an interacting double step discontinuity.

very different from region (2), requiring a very different "fit" for the two sets of basis functions. Moreover we may write the discretised form of (5.140) as

$$\begin{bmatrix} \mathbf{E}_1 \\ \mathbf{E}_2 \end{bmatrix} = \frac{1}{2} \begin{bmatrix} \mathbf{Z}_1 + \mathbf{Z}_{11} & -\mathbf{Z}_{12} \\ -\mathbf{Z}_{12} & \mathbf{Z}_2 + \mathbf{Z}_{11} \end{bmatrix} \begin{bmatrix} -\mathbf{H}_1 \\ -\mathbf{H}_2 \end{bmatrix} = \mathbf{Z}_S \begin{bmatrix} -\mathbf{H}_1 \\ -\mathbf{H}_2 \end{bmatrix} \tag{5.141}$$

The derivation of the normalised scattering matrix elements $S_{k\ell}$ is formally the same as for the single step, providing

$$S_{k\ell} = \delta_{k\ell} - s_k s_\ell \sqrt{z_{0k} z_{0\ell}} \mathbf{U}_k^T \mathbf{Z}_S^{-1} \mathbf{U}_\ell \tag{5.142}$$

where

$$\delta_{k\ell} = \left\{ \begin{matrix} 1 & : & k = \ell \\ 0 & : & k \neq \ell \end{matrix} \right\}$$

$$1 \leq k, \ell \leq n_1 + n_2$$

with

$$s_k, s_\ell = \left\{ \begin{matrix} 1 & : & k \leq n_1 & \text{region 1} \\ -1 & : & k > n_1 & \text{region 2} \end{matrix} \right\}$$

and

$$\mathbf{U}_k = \left\{ \begin{matrix} \mathbf{P}_k & : & k \leq n_1 & \text{region 1} \\ \mathbf{R}_k & : & k > n_1 & \text{region 2} \end{matrix} \right\} \tag{5.143}$$

In the above $\mathbf{P}_k$ and $\mathbf{R}_k$ are the column vectors of the modal expansion coefficients in regions (1) and (2), respectively.

Numerical results for interacting steps

The results presented are relative to a symmetric slab excited by an even TE mode. As a demonstration of the rapid convergence of the variational solution using an intermediate mode basis function set complemented with Laguerre polynomials as discussed in the previous Section, the magnitude of S_{11} and S_{12} of a double step were plotted against frequency for varying order of solution N in Figure 5.17. The slab guides on either side of the double step begin to support a second mode at 21 GHz. It is evident from the figures that the onset of the second mode has a strong effect on the first order variational solution ($N = 1$). This effect is particularly evident on the phase of S_{11} and starts at around 16 GHz. This is because the thickness of the "intermediate" guide wherefrom the trial field is derived is given by $\sqrt{d_1 d_2}$, so that it is wider than the actual slab guide d_1 and it overmodes sooner. However, as soon as the second mode is included in the expansion set, the solution

Table 5.8 Convergence test of intermediate surface mode basis functions complemented with Laguerre polynomials at 21 GHz.

N	$\lvert S_{11}\rvert$	$\angle S_{11}$ (rad)	$\lvert S_{12}\rvert$	$\angle S_{12}$ (rad)	$1-\lvert S_{11}\rvert^2-\lvert S_{12}\rvert^2$
1	0.06621	2.1103	0.96742	1.0213	4.38376e-3
2	0.07047	1.3916	0.99513	1.0502	4.75026e-3
3	0.02295	1.2982	0.99878	1.0272	1.91181e-3
4	0.00330	1.4626	0.98172	1.0303	3.56991e-2
5	0.00338	1.5739	0.98340	1.0321	3.29130e-2
10	0.00335	1.5812	0.98331	1.0321	3.31099e-2

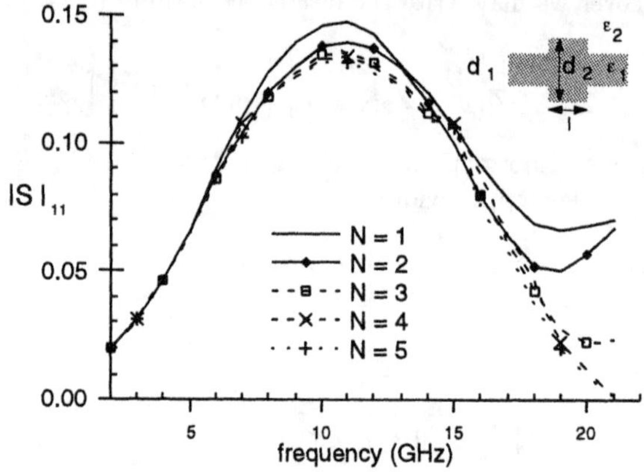

Figure 5.17 Convergence of $|S_{11}|$ for different order of solution N. Parameter values are $\ell/d_1 = 1$, $d_2/d_1 = 2$, $\varepsilon_1 = 2.57$, $\varepsilon_2 = 1$.

quickly converges. The numerical values of the scattering parameters at 21 GHz are given in Table 5.8, where the error of truncating at $N = 5$, relative to the 10th order solution, is 0.594%.

The degenerate case when the double step completely disappears leaving an air gap between two semi–infinite slabs was analysed in [42, 24]. In this case the surface modes in the mid–region no longer exist and coupling of the modes is through degenerate continuous modes. The results given by the present method in Figure 5.18 compare well with those of Koshiba [24]. While the transmission coefficient S_{12} decreases slowly with increasing gap length, the reflection coefficient S_{11} is seen to approach that of a semi–infinite slab guide radiating into air half space.

The variation of S_{11} and S_{12} with increasing step height is given in Figure 5.19, where the ratio of the radiated power to incident power is also shown. It is interesting to note that S_{11}, S_{12} and the radiated power all tend to a limit as the step height is increased. The reason is that for very large step height, the middle region effectively becomes a dielectric layer between two semi infinite slabs and the incident wave no longer feels the effect of the corners of the step. The phase of S_{11} also shows an interesting 180° flip as the middle region changes from a groove to a ridge. Another comparison with the results of [24] and [18] is given in Figure

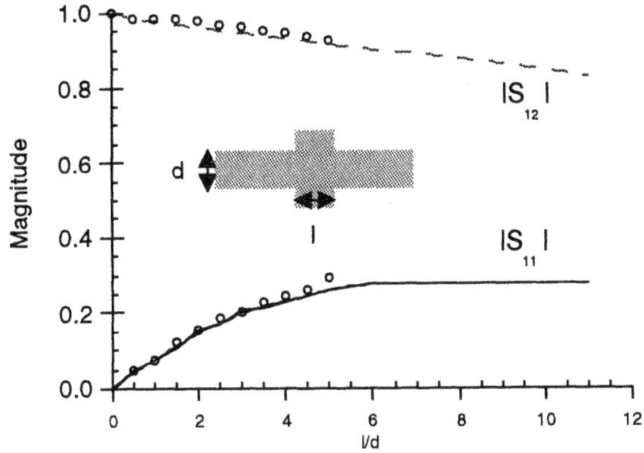

Figure 5.18 Scattering magnitude versus different gap lengths.

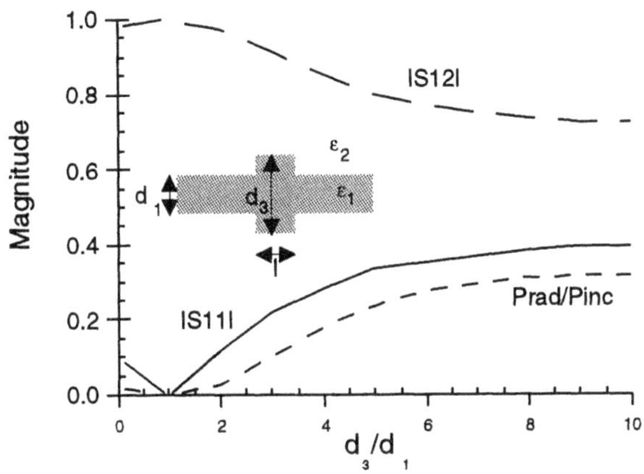

Figure 5.19 Scattering magnitude versus different rib heights.

5.20. The figure shows the results obtained by the present method, displaying a stronger interaction of the steps, with the peaks of S_{11} being higher than those obtained by Hosono [18] and Koshiba [24]. However Koshiba stated in his paper that he may have taken an insufficient number of elements in the computation for $\ell/d_1 > 3.5$. In Hosono's case, he had used only 30 modes for his results which by his own calculations gives 3-10% error. In any case, proper account of the continuous modes was not taken. On the other hand, a 5th order intermediate mode set with the present method is found to give a relative error of less than 0.8%. The stronger reflection peaks are likely to be due to the variational solution accounting for more of the interaction caused by the continuous modes. The peaks in the scattering values are the result of resonances in the middle region as its length is varied.

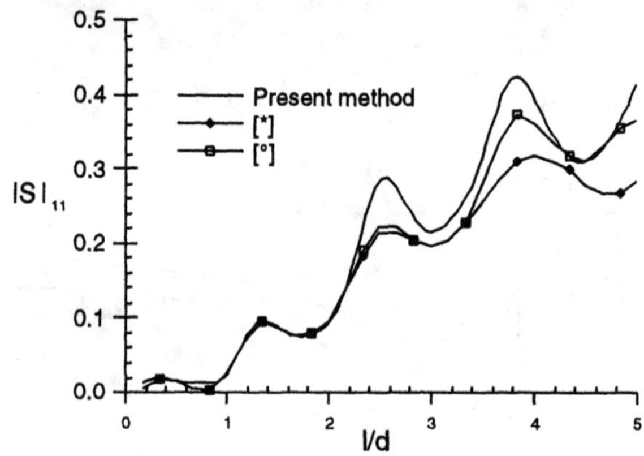

Figure 5.20 Scattering magnitude versus different values of ℓ/d (defined in the inset of the previous figure).

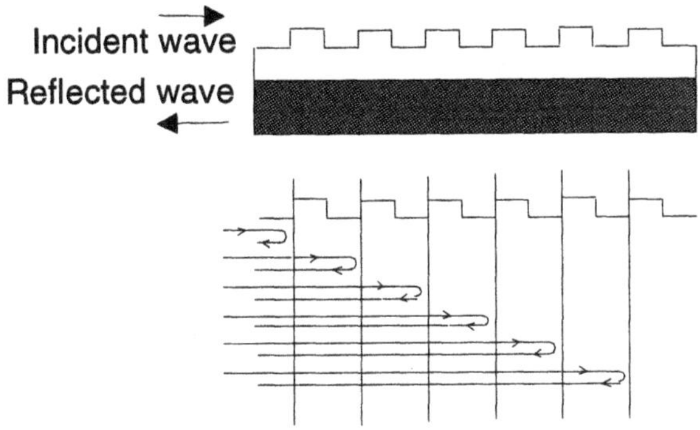

Figure 5.21 Distributed grating reflector where several small reflections arising from step discontinuities are coherently added.

5.4 Cascaded step discontinuities

Cascaded discontinuities, arranged to form periodic structures, have received considerable attention in the past. In the early 1970s, the new developments in fabrication of thin film optical waveguides found applications for periodic gratings in distributed feedback lasers (DFB) and distributed Bragg reflector (DBR) lasers [23, 10, 54, 8, 52], grating couplers [36], bandstop filters [20, 35, 34] and modulators [37, 15]. One advantage of these devices is their integrability with other components on a planar circuit. Figure 5.21 shows an example of a distributed grating reflector where the small reflections from a large number of grooves add up in phase to provide the desired reflection. Since the operation depends on the phase–synchronised cumulative effects of each individual discontinuity, these types of devices are also tolerant to random defects in the manufacturing process. Moreover, in the case of DFB lasers, they can also offer longitudinal mode selection and higher power handling capabilities.

More recently, it was recognised that the same type of design used for optical devices, could also be adopted in millimetre–wave applications. This led to the realization of bandstop filters on image guides [20, 35] and leaky–wave antennas [44, 22] which are very similar to grating couplers. The same approach has also suggested the use of DBR oscillators employing several diodes to achieve a high power output [21].

In fact, most of the techniques used in the analysis of such devices are common to integrated optics and millimetre waves. These techniques are based either on:

- approximate coupled modes;
- space harmonic expansion.

The latter approach, although more rigorous, involves the solution of a wave equation of the Hill's type.

Naturally, both approaches are effective only for solving periodic structures of infinite extent. To consider the structure as of infinite extent represents a good approximation for optical devices, where very fine periodicities occur (the number of discontinuities is of the order of 1000 or more). This approximation does not

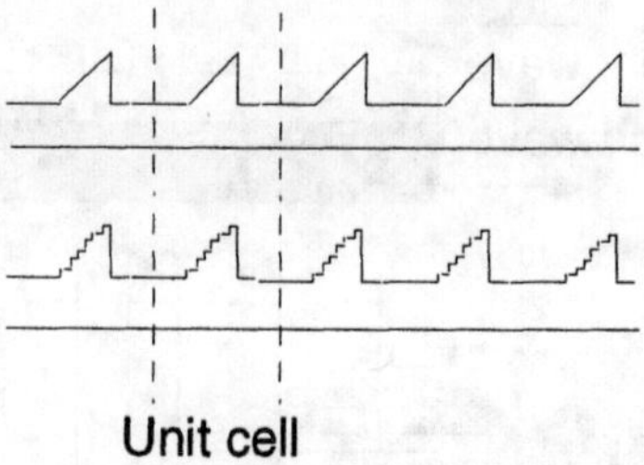

Figure 5.22 Blazed periodic grating, where each unit cell is approximated by steps.

normally apply to microwave components where the number of corrugations used is limited by the inherent losses and the requirement for small size. As a consequence, it is important to include a description of the discontinuities at either end, since these terminations will cause noticeable effects.

In this Section we present a method to deal with scattering problems posed by finite periodic structures, considered as a cascade of interacting step discontinuities. In this case, several finite lengths of waveguide are cascaded together to form what is called a unit cell. The field problem of a finite length of guide (i.e. of the interacting step discontinuities), which has been solved in the previous Section, is now used to develop a network approach to the problem of cascaded discontinuities.

In the previous Section the field problem of interacting steps was formulated by using impedance matrices. Now, since the periodic structure is described by an equivalent cascade of multi–port networks, it is expedient to use a transmission matrix representation. In this way the overall transmission matrix is easily obtained by the cascade multiplication of the individual transmission matrices. Finally, the impedance matrix is recovered by a reverse transformation. The scattering parameters, if desired, may be obtained in a manner similar to that described for the interactive double steps.

It is worthwhile to note that the present analysis also applies to periodic structures with different unit cells. In addition, each unit cell may itself be composed of cascades of other cells to build structures of arbitrary complexity. As an example, the blazed grating [33, 7] can be approximated by cascading unit cells of increasing height (see Figure 5.22).

Another virtue of this approach is to include the radiation coupling between cascaded cells. The discrete network matrices represent, in fact, the entire field. The resultant composite cell can then be used to build rigorously the periodic structure. Similarly, cell elements with variable periodicity can be cascaded together to give a chirped grating [51], which has the property of being able to focus its radiated power to a point outside the waveguide. Another common structure is the log–periodic grating. In this case the period and grating height of each element is increased in a log–periodic manner. The behaviour of this component is similar to that of the blazed grating, but showing a much wider bandwidth.

Finally, it may be pointed out that, although the present analysis is effected in two dimensions, the results obtained may still be valid for three dimensions if the guiding structure is more than a wavelength wide in the third dimension. This has been shown by [44] in which the effects of the width on the corrugated image guide were studied using the EDC (effective dielectric constant) method.

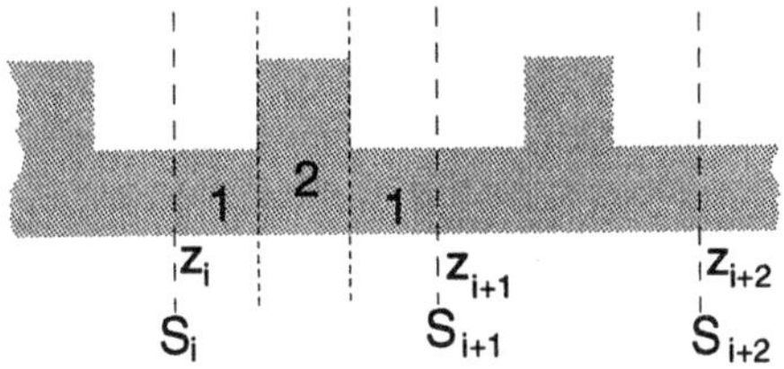

Figure 5.23 Cascaded steps; the regions 1,2,1 form a symmetrical cell.

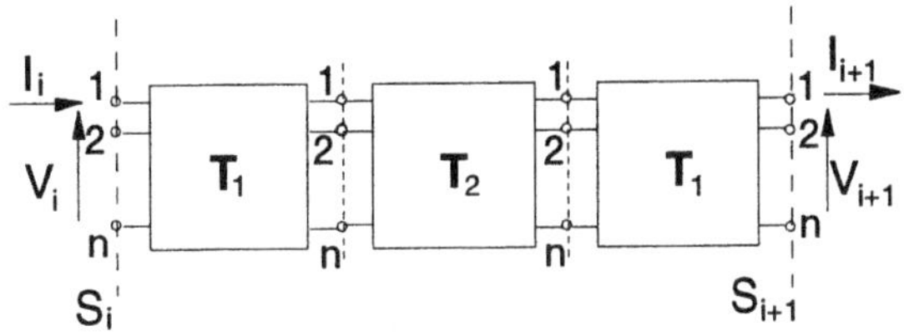

Figure 5.24 Equivalent network for a symmetrical cell.

5.4.1 *Impedance matrix of the unit cell*

A symmetrical unit cell is defined as the finite length of guide between the planes S_i and S_{i+1} in Figure 5.23. By placing magnetic walls at S_i and S_{i+1}, we can give an impedance type description of the unit cell, where the electric and magnetic fields are related by the expression

$$\begin{pmatrix} E(x, z_i) \\ E(x, z_{i+1}) \end{pmatrix} = \begin{pmatrix} \hat{\mathbf{Z}}_{11}^{(i)} & \hat{\mathbf{Z}}_{12}^{(i)} \\ \hat{\mathbf{Z}}_{12}^{(i)} & \hat{\mathbf{Z}}_{11}^{(i)} \end{pmatrix} \cdot \begin{pmatrix} -H(x, z_i) \\ -H(x, z_{i+1}) \end{pmatrix} \tag{5.144}$$

where $\hat{\mathbf{Z}}_{11}^{(i)}$ and $\hat{\mathbf{Z}}_{12}^{(i)}$ are the two port impedance operators of the i–th section. By employing an appropriate set of expanding functions the 2 x 2 operator form of (5.144) is transformed to a discrete matrix representation,

$$\begin{pmatrix} \mathbf{E}_1 \\ \mathbf{E}_2 \end{pmatrix} = \begin{pmatrix} \mathbf{Z}_{11} & \mathbf{Z}_{12} \\ \mathbf{Z}_{12}^t & \mathbf{Z}_{11} \end{pmatrix} \begin{pmatrix} \mathbf{H}_1 \\ \mathbf{H}_2 \end{pmatrix} \tag{5.145}$$

which can be represented by the equivalent network of Figure 5.24. Continuity of the transverse fields at z_i,

$$\begin{aligned} E(x, z_i^-) &= E(x, z_i^+) \\ H(x, z_i^-) &= H(x, z_i^+) \end{aligned} \tag{5.146}$$

then becomes the continuity of voltages and currents at the corresponding planes. To reflect this, it is convenient to write (5.145) in terms of voltages and currents as,

$$\begin{pmatrix} \mathbf{V}_1 \\ \mathbf{V}_2 \end{pmatrix} = \begin{pmatrix} \mathbf{Z}_{11} & \mathbf{Z}_{12} \\ \mathbf{Z}_{12}^t & \mathbf{Z}_{11} \end{pmatrix} \begin{pmatrix} \mathbf{I}_1 \\ -\mathbf{I}_2 \end{pmatrix} \tag{5.147}$$

Note that (5.144) relates the total field at the planes S_i and S_{i+1} and any interactions by the continuum within the finite section of guide are built into the model.

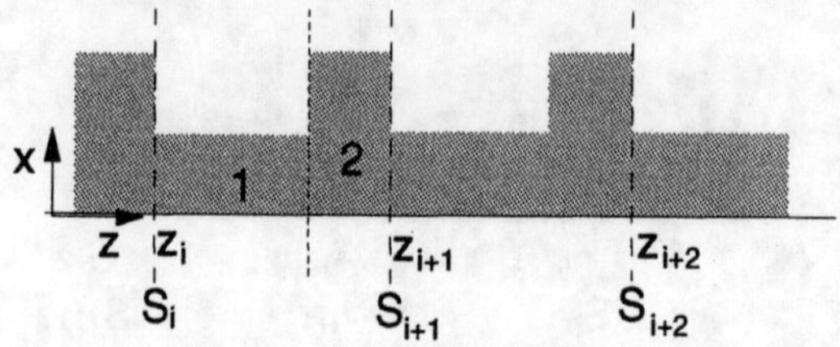

Figure 5.25 Cascaded steps arranged in asymmetrical unit cells. The unit cell is the region of space between the planes S_i and S_{i+1}.

The problem of the cascaded structure, now becomes one of cascaded "2N–ports". This is most conveniently solved by cascade multiplication of the transmission matrix of each individual block.

5.4.2 Transmission matrix

With reference to Figure 5.24, the transmission or $ABCD$ matrix is written as,

$$\begin{pmatrix} \mathbf{V}_1 \\ \mathbf{I}_1 \end{pmatrix} = \begin{pmatrix} \mathbf{A} & \mathbf{B} \\ \mathbf{C} & \mathbf{D} \end{pmatrix} \begin{pmatrix} \mathbf{V}_2 \\ \mathbf{I}_2 \end{pmatrix} \tag{5.148}$$

However, the above representation can be easily related to the impedance description. In fact, use of (5.147) yields

$$\begin{pmatrix} \mathbf{V}_1 \\ \mathbf{I}_1 \end{pmatrix} = \begin{pmatrix} \mathbf{Z}_{11}(\mathbf{Z}_{12}^t)^{-1} & \mathbf{Z}_{11}(\mathbf{Z}_{12}^t)^{-1}\mathbf{Z}_{22} - \mathbf{Z}_{12} \\ (\mathbf{Z}_{12}^t)^{-1} & (\mathbf{Z}_{12}^t)^{-1}\mathbf{Z}_{22} \end{pmatrix} \begin{pmatrix} \mathbf{V}_2 \\ \mathbf{I}_2 \end{pmatrix} \tag{5.149}$$

Vice versa, the impedance matrix formulation in terms of the $ABCD$ parameters can be similarly derived and is given by:

$$\begin{pmatrix} \mathbf{V}_1 \\ \mathbf{V}_2 \end{pmatrix} = \begin{pmatrix} \mathbf{A}\mathbf{C}^{-1} & \mathbf{A}\mathbf{C}^{-1}\mathbf{D} - \mathbf{B} \\ \mathbf{C}^{-1} & \mathbf{C}^{-1}\mathbf{D} \end{pmatrix} \begin{pmatrix} \mathbf{I}_1 \\ -\mathbf{I}_2 \end{pmatrix} \tag{5.150}$$

In the following, it will be convenient to start from the impedance representation of the individual unit cells and transform them to $ABCD$ representation, compute then the cascaded transmission matrix by simple matrix multiplication and, finally, transform back to impedances.

Note that for a symmetrical two–port,

$$\begin{gathered} Z_{11} = Z_{11}^t = Z_{22} \\ Z_{12} = Z_{12}^t \end{gathered} \tag{5.151}$$

From equations (5.149) and (5.151), one can further deduce that,

$$\begin{gathered} \mathbf{D} = \mathbf{A}^t \\ \mathbf{I} = \mathbf{A}\mathbf{C}^{-1}\mathbf{D}\mathbf{C} - \mathbf{B}\mathbf{C} \end{gathered} \tag{5.152}$$

Depending on the choice of the reference planes the unit cell may be symmetrical (as in Figure 5.23) or asymmetrical (Figure 5.25).

For uniform periodic structures it is possible to choose the reference planes so that the unit cell is always symmetrical. In the general case of non–uniform periodic structures the cell will be asymmetrical. The latter case requires a quite different approach and will be illustrated later. For the symmetrical case, as evident from the equivalent network of Figure 5.24, we have the following relationship between currents and voltages at the reference planes

$$\begin{pmatrix} \mathbf{V}_i \\ \mathbf{I}_i \end{pmatrix} = \mathbf{T}_1\mathbf{T}_2\mathbf{T}_3 \begin{pmatrix} \mathbf{V}_{i+3} \\ \mathbf{I}_{i+3} \end{pmatrix} = \mathbf{T} \begin{pmatrix} \mathbf{V}_{i+3} \\ \mathbf{I}_{i+3} \end{pmatrix} \tag{5.153}$$

where $\mathbf{T}_1$ and $\mathbf{T}_2$ are the transmission matrices of regions 1 and 2 in Figure 5.23, and $\mathbf{T}$ is the overall transmission matrix of the unit cell.

5.4.3 Canonical form of the transmission matrix

The transmission matrix of the cascade of n equal unit cells is obtained by raising $\mathbf{T}$ to the power n. It is noted that the evaluation of $\mathbf{T}^n$ may pose numerical problems, particularly for large n, since $\mathbf{T}$ is a $2N \times 2N$ matrix for an Nth order solution. To overcome this problem it is convenient to make use of the following canonical decomposition,

$$\mathbf{T} = \mathbf{M}\,\mathbf{S}\,\mathbf{M}^{-1} \tag{5.154}$$

where $\mathbf{M}$ is the modal matrix whose columns are the eigenvectors of $\mathbf{T}$ and $\mathbf{S}$ is the diagonal spectral matrix whose elements are the eigenvalues of $\mathbf{T}$. Then,

$$\mathbf{T}^n = \mathbf{M}\,\mathbf{S}^n\,\mathbf{M}^{-1} \tag{5.155}$$

Since $\mathbf{S}^n$ only involves raising the individual eigenvalue to the power n, the problem of error propagation in computing the series product of $\mathbf{T}$ is avoided.

5.4.4 Transmission matrix of the first order

It is instructive to consider the simple case when $\mathbf{T}$ is just a 2×2 matrix, i.e. when the latter is obtained from a Ritz–Galerkin solution of the first order. The extension to the general Nth order solution is given next and, although formally more involved, it does not add further physical insight.

For a cascade of n identical two–ports, we have,

$$\begin{pmatrix} V_1 \\ I_1 \end{pmatrix} = \mathbf{T}^n \begin{pmatrix} V_n \\ I_n \end{pmatrix} \tag{5.156}$$

or in expanded form,

$$\begin{pmatrix} V_1 \\ I_1 \end{pmatrix} = \begin{pmatrix} A & B \\ C & D \end{pmatrix}^n \begin{pmatrix} V_n \\ I_n \end{pmatrix} \tag{5.157}$$

In order to express $\mathbf{T}$ in the canonical form (5.155), it is necessary to determine eigenvalues and eigenvectors. The eigenvalues are simply given by the roots of the characteristic equation,

$$\det(\mathbf{T} - \lambda\mathbf{I}) = \lambda^2 - (A+D)\lambda + (AD - BC) = 0 \tag{5.158}$$

which, since $AD - BC = \det\mathbf{T} = 1$, reduces to

$$\lambda^2 - (A+D)\lambda + 1 = 0 \tag{5.159}$$

The solution of the above quadratic equation yields,

$$\begin{aligned}\lambda_1 &= t + \sqrt{t^2 - 1} \\ \lambda_2 &= t - \sqrt{t^2 - 1}\end{aligned} \tag{5.160}$$

where the parameter t has been defined as

$$t = \frac{A + D}{2}$$

Naturally, the eigenvalues have the following properties

$$\begin{aligned}\lambda_1 \lambda_2 &= \det \mathbf{T} = 1 \\ \lambda_1 + \lambda_2 &= \mathrm{trace}\mathbf{T} = A + D\end{aligned} \tag{5.161}$$

In particular, the first condition above means that the eigenvalues can be written in the form,

$$\begin{aligned}\lambda_1 &= e^{\gamma \ell} \\ \lambda_2 &= e^{-\gamma \ell}\end{aligned} \tag{5.162}$$

where γ can be identified with the propagation constant of the periodic structure. Equation (5.162) is a statement of the reciprocity of the transmission matrix. The corresponding eigenvectors are then obtained from the requirement,

$$\begin{pmatrix} A & B \\ C & D \end{pmatrix} \begin{pmatrix} x_1 \\ x_2 \end{pmatrix} = \lambda \begin{pmatrix} x_1 \\ x_2 \end{pmatrix} \tag{5.163}$$

It is clear from the above equation that only the ratio x_1/x_2 of the eigenvectors can be determined, which is given as

$$Z_\nu = \left(\frac{x_1}{x_2} \right) = \frac{\lambda_\nu - D}{C} = \frac{A - D}{2C} \pm \frac{\sinh \gamma \ell}{C} \qquad \nu = 1, 2 \tag{5.164}$$

If the four–terminal network is symmetrical (or reversible), then by (5.152),

$$Z_1 = -Z_2 = \frac{\sinh \gamma \ell}{C} = \sqrt{\frac{B}{C}} = Z_0 \tag{5.165}$$

From the equations (5.148) and (5.149) it is apparent that Z_1 and Z_2 have dimensions of impedance. Z_0 is commonly known as the "image" impedance of $\mathbf{T}$. Since only the ratios of the eigenvectors are specified it is possible to recover $\mathbf{M}$ as

$$\mathbf{M} = \begin{pmatrix} Z_1 & Z_2 \\ 1 & 1 \end{pmatrix} \tag{5.166}$$

and its inverse is

$$\mathbf{M}^{-1} = \frac{1}{Z_1 - Z_2} \begin{pmatrix} 1 & -Z_2 \\ -1 & Z_1 \end{pmatrix} \tag{5.167}$$

It is straightforward to apply (5.155) in order to calculate the complete transmission matrix of the n cascaded unit cells as

$$\begin{aligned}\mathbf{T}^n &= \tfrac{1}{Z_1 - Z_2} \begin{pmatrix} Z_1 & Z_2 \\ 1 & 1 \end{pmatrix} \begin{pmatrix} e^{n\gamma\ell} & 0 \\ 0 & e^{-n\gamma\ell} \end{pmatrix} \begin{pmatrix} 1 & -Z_2 \\ -1 & Z_1 \end{pmatrix} \\ &= \tfrac{1}{Z_1 - Z_2} \begin{pmatrix} Z_1 e^{n\gamma\ell} - Z_2 e^{-n\gamma\ell} & -2 Z_1 Z_2 \sinh n\gamma\ell \\ 2 \sinh n\gamma\ell & Z_1 e^{-n\gamma\ell} - Z_2 e^{n\gamma\ell} \end{pmatrix}\end{aligned} \tag{5.168}$$

When the network is symmetrical, application of (5.165) in the above yields the important special case,

$$\mathbf{T}^n = \begin{pmatrix} \cosh n\gamma\ell & Z_0 \sinh n\gamma\ell \\ \frac{1}{Z_0}\sinh n\gamma\ell & \cosh n\gamma\ell \end{pmatrix} \tag{5.169}$$

and from (5.157),

$$\begin{pmatrix} V_n \\ I_n \end{pmatrix} = \mathbf{T}^{-n}\begin{pmatrix} V_1 \\ I_1 \end{pmatrix} = \begin{pmatrix} \cosh n\gamma\ell & -Z_0 \sinh n\gamma\ell \\ -Z_0^{-1}\sinh n\gamma\ell & \cosh n\gamma\ell \end{pmatrix}\begin{pmatrix} V_1 \\ I_1 \end{pmatrix} \tag{5.170}$$

Formula (5.169) shows that a periodic structure composed of n equal, symmetrical, unit cells, behaves as a transmission line of propagation constant γ, length $n\ell$ and characteristic impedance Z_0.

When an infinite structure is considered by using (5.170) and (5.150), and by taking the limit as n goes to infinity, the input impedance of the periodic structure is seen to be the characteristic impedance Z_0; so that,

$$V_n = Z_0 I_n$$

Using this in (5.170) gives,

$$\begin{aligned} V_n &= [\cosh n\gamma\ell - \sinh n\gamma\ell]\, V_1 = e^{-n\gamma\ell}V_1 \\ I_n &= [\cosh n\gamma\ell - \sinh n\gamma\ell]\, \tfrac{V_1}{Z_0} = e^{-n\gamma\ell}\tfrac{V_1}{Z_0} \end{aligned} \tag{5.171}$$

Thus, a cascade of symmetrical networks behaves like a finite length of transmission line with characteristic impedance Z_0 and a propagation constant γ. If the networks are asymmetrical however, the impedances will be different for the two different directions of propagation and their values are given by (5.164).

5.4.5 The general Nth order solution

Symmetrical case

As stipulated by the reciprocity condition (5.161), the $2N \times 2N$ spectral matrix $\mathbf{S}$ must have the form,

$$\mathbf{S} = \begin{pmatrix} e^{\gamma_1\ell} & & & & & \\ & \ddots & & & & \\ & & e^{\gamma_N\ell} & & & \\ & & & e^{-\gamma_1\ell} & & \\ & & & & \ddots & \\ & & & & & e^{-\gamma_N\ell} \end{pmatrix} \tag{5.172}$$

The expansion of the modal fields by means of N basis terms has introduced N propagation constants for the periodic structure. However, these propagation constants do not correspond to the basis terms but to an alternative system of eigenvectors. The eigenvectors may be viewed as representing a different coordinate system. Thus, (5.154) may be interpreted as a process of successive transformations whereby the field is first resolved into the eigen–directions by the transformation $\mathbf{M}^{-1}$. Then, multiplication by $\mathbf{S}$ effectively scales each resultant eigen–component by the corresponding eigenvalue. Hence their interpretation as the propagation constants of each corresponding eigen-component. Finally, the scaled values are retransformed back by $\mathbf{M}$ to the original representation in terms of the basis functions. Furthermore, with the modal matrix understood in the sense of (5.156), it is convenient to express the Nth order case in partitioned form as,

$$\mathbf{M} = \begin{pmatrix} \mathbf{V}_{m1} & \mathbf{V}_{m2} \\ \mathbf{I}_{m1} & \mathbf{I}_{m2} \end{pmatrix} \tag{5.173}$$

where $\mathbf{V}_{m1}, \mathbf{V}_{m2}, \mathbf{I}_{m1}$ and $\mathbf{I}_{m2}$ are each $N \times N$ matrices, which in the symmetrical case, reduces to

$$\mathbf{M} = \begin{pmatrix} \mathbf{V}_{m1} & \mathbf{V}_{m1} \\ \mathbf{I}_{m1} & -\mathbf{I}_{m1} \end{pmatrix} \tag{5.174}$$

The minus sign of $\mathbf{I}_{m1}$ is a reflection of the current convention that was adopted in (5.148) and (5.150). The choice of the above notation is now self–evident, for the canonical form of $\mathbf{T}$ becomes

$$\mathbf{T} = \begin{pmatrix} \mathbf{V}_{m1} & \mathbf{V}_{m1} \\ \mathbf{I}_{m1} & -\mathbf{I}_{m1} \end{pmatrix} \begin{pmatrix} \text{diag}\left[e^{\gamma\ell}\right] & \\ & \text{diag}\left[e^{-\gamma\ell}\right] \end{pmatrix} \begin{pmatrix} \mathbf{V}_{m1} & \mathbf{V}_{m1} \\ \mathbf{I}_{m1} & -\mathbf{I}_{m1} \end{pmatrix}^{-1} \tag{5.175}$$

Using the condition in (5.152), $\mathbf{T}$ may also be given as

$$\mathbf{T} = \begin{pmatrix} \mathbf{A} & \mathbf{B} \\ \mathbf{C} & \mathbf{A}^t \end{pmatrix} \tag{5.176}$$

Hence, it may be proved that the inverse of $\mathbf{T}$ is

$$\mathbf{T}^{-1} = \begin{pmatrix} \mathbf{A} & -\mathbf{B} \\ -\mathbf{C} & \mathbf{A}^t \end{pmatrix} \tag{5.177}$$

The two last equations may now be used to obtain the two diagonal forms,

$$\begin{aligned} \tfrac{1}{2}\left(\mathbf{T} + \mathbf{T}^{-1}\right) &= \begin{pmatrix} \mathbf{A} & \mathbf{0} \\ \mathbf{0} & \mathbf{A}^t \end{pmatrix} \\ \tfrac{1}{2}\left(\mathbf{T} - \mathbf{T}^{-1}\right) &= \begin{pmatrix} \mathbf{0} & \mathbf{B} \\ \mathbf{C} & \mathbf{0} \end{pmatrix} \end{aligned} \tag{5.178}$$

Performing the same operations on (5.175) and then comparing with the above equations, gives the equivalence,

$$\begin{aligned} \mathbf{A} &= \mathbf{V}_{m1}\text{diag}\left[\cosh\gamma\ell\right]\mathbf{V}_{m1}^{-1} \\ \mathbf{B} &= \mathbf{V}_{m1}\text{diag}\left[\sinh\gamma\ell\right]\mathbf{I}_{m1}^{-1} \\ \mathbf{C} &= \mathbf{I}_{m1}\text{diag}\left[\sinh\gamma\ell\right]\mathbf{V}_{m1}^{-1} \\ \mathbf{D} &= \mathbf{I}_{m1}\text{diag}\left[\sinh\gamma\ell\right]\mathbf{I}_{m1}^{-1} \end{aligned} \tag{5.179}$$

In order to obtain $\mathbf{T}^n$, γ is simply replaced by $n\gamma$ in the above equation. The form of $\mathbf{T}^n$ is in fact very similar to the 2×2 case of (5.169). There remains one problem in the evaluation of $\mathbf{T}^n$. Since γ is nearly always complex and may be written as,

$$\gamma = \alpha + j\beta \tag{5.180}$$

we see that for $|\alpha| > 1$, the elements of $\mathbf{T}^n$ may become extremely large for large n since each element grows as $e^{n|\alpha|}$. This condition arises when the periodic structure is radiating strongly or when the structure is inherently very lossy. Moreover, for a high order solution (i.e. large N), α for the higher order terms will tend to be large and negative. This is because the higher order terms are rapidly attenuated in the direction of propagation; numerically, however, they grow as $e^{n|\alpha|}$ and will completely mask the values of the lower terms, which represent most of the field. Fortunately, it is not required to evaluate $\mathbf{T}^n$ numerically since the transmission matrix formulation is only a means to obtain the overall T–network representation of the periodic structure. The impedances are recovered from (5.150) and (5.179), to give,

$$\begin{aligned}\mathbf{Z}_{11} &= \mathbf{AC}^{-1} = \mathbf{V}_{m1}\mathrm{diag}\left[\coth\gamma\ell\right]\mathbf{I}_{m1}^{-1}\\ \mathbf{Z}_{12} &= \mathbf{C}^{-1} = \mathbf{V}_{m1}\mathrm{diag}\left[\mathrm{cosech}\gamma\ell\right]\mathbf{I}_{m1}^{-1}\end{aligned} \tag{5.181}$$

Since the structure is symmetrical, only $\mathbf{Z}_{11}$ and $\mathbf{Z}_{12}$ are required in order to completely specify the characteristics of the T–junction. Now, for large $|\alpha|$, $\mathbf{Z}_{11}$ is seen to tend to a constant value while $\mathbf{Z}_{12}$ tends to zero.

Asymmetrical case

The modal matrix for the asymmetrical unit cell is given by (5.173). In this case no simplification is possible so that the inverse of the modal matrix will have to be evaluated numerically. Again it is useful to write the modal matrix and its inverse in partitioned form; this time using a different notation as follows

$$\begin{aligned}\mathbf{M} &= \begin{pmatrix}\mathbf{m}_{11} & \mathbf{m}_{12}\\ \mathbf{m}_{21} & \mathbf{m}_{22}\end{pmatrix}\\ \mathbf{M}^{-1} &= \begin{pmatrix}\tilde{\mathbf{m}}_{11} & \tilde{\mathbf{m}}_{12}\\ \tilde{\mathbf{m}}_{21} & \tilde{\mathbf{m}}_{22}\end{pmatrix}\end{aligned} \tag{5.182}$$

Then the partitioned matrix multiplication of the canonical form of (5.175) gives,

$$\begin{aligned}\mathbf{A} &= \mathbf{m}_{11}\mathrm{diag}\left[e^{\gamma\ell}\right]\tilde{\mathbf{m}}_{11} + \mathbf{m}_{12}\mathrm{diag}\left[e^{-\gamma\ell}\right]\tilde{\mathbf{m}}_{21}\\ \mathbf{C} &= \mathbf{m}_{21}\mathrm{diag}\left[e^{\gamma\ell}\right]\tilde{\mathbf{m}}_{11} + \mathbf{m}_{22}\mathrm{diag}\left[e^{-\gamma\ell}\right]\tilde{\mathbf{m}}_{21}\\ \mathbf{D} &= \mathbf{m}_{21}\mathrm{diag}\left[e^{\gamma\ell}\right]\tilde{\mathbf{m}}_{12} + \mathbf{m}_{22}\mathrm{diag}\left[e^{-\gamma\ell}\right]\tilde{\mathbf{m}}_{22}\end{aligned} \tag{5.183}$$

Note that only three blocks of the transmission matrix are uniquely defined as the fourth one (in this case $\mathbf{B}$), is determined from the requirement of reciprocity:

$$\det \mathbf{T} = 1 \tag{5.184}$$

The impedances are then given by the expressions

$$\begin{aligned}\mathbf{Z}_{11} = \mathbf{AC}^{-1} &= \left(\mathbf{m}_{11}\mathbf{m}_{21}^{-1} + \mathbf{m}_{12}\mathrm{diag}\left[e^{-\gamma\ell}\right]\tilde{\mathbf{m}}_{21}\mathbf{m}_{11}^{-1}\mathrm{diag}\left[e^{-\gamma\ell}\right]\mathbf{m}_{21}^{-1}\right)\\ &\quad\left(\mathbf{I} + \mathbf{m}_{22}\mathrm{diag}\left[e^{-\gamma\ell}\right]\tilde{\mathbf{m}}_{21}\tilde{\mathbf{m}}_{11}^{-1}\mathrm{diag}\left[e^{-\gamma\ell}\right]\tilde{\mathbf{m}}_{21}^{-1}\right)^{-1}\\ \mathbf{Z}_{12} = \mathbf{C}^{-1} &= \left(\tilde{\mathbf{m}}_{11}^{-1}\mathrm{diag}\left[e^{-\gamma\ell}\right]\mathbf{m}_{21}^{-1}\right)\\ &\quad\left(\mathbf{I} + \mathbf{m}_{22}\mathrm{diag}\left[e^{-\gamma\ell}\right]\tilde{\mathbf{m}}_{21}\tilde{\mathbf{m}}_{11}^{-1}\mathit{diag}\left[e^{-\gamma\ell}\right]\mathbf{m}_{21}^{-1}\right)\\ \mathbf{Z}_{22} = \mathbf{C}^{-1}\mathbf{D} &= \left(\mathbf{I} + \tilde{\mathbf{m}}_{11}^{-1}\mathrm{diag}\left[e^{-\gamma\ell}\right]\mathbf{m}_{21}^{-1}\mathbf{m}_{22}\mathrm{diag}\left[e^{-\gamma\ell}\right]\tilde{\mathbf{m}}_{21}\right)^{-1}\\ &\quad\left(\tilde{\mathbf{m}}_{11}^{-1}\tilde{\mathbf{m}}_{12} + \tilde{\mathbf{m}}_{11}^{-1}\mathrm{diag}\left[e^{-\gamma\ell}\right]\mathbf{m}_{21}^{-1}\mathbf{m}_{22}\mathrm{diag}\left[e^{-\gamma\ell}\right]\tilde{\mathbf{m}}_{22}\right)\end{aligned} \tag{5.185}$$

Again for the case of $\mathbf{T}^n$, then γ is just replaced by $n\gamma$. Note that all the impedances only contain terms with $e^{-\gamma\ell}$ which will tend to zero for large $|\alpha|$ and n.

5.4.6 Infinite periodic structure, the Nth order solution for symmetrical cells

For an infinite periodic structure, the Nth order solution is obtained from equation (5.179), which gives

$$\mathbf{T}^{-1} = \begin{pmatrix} \mathbf{V}_{m1}\text{diag}\left[\cosh n\gamma\ell\right]\mathbf{V}_{m1}^{-1} & -\mathbf{V}_{m1}\text{diag}\left[\sinh n\gamma\ell\right]\mathbf{I}_{m1}^{-1} \\ -\mathbf{I}_{m1}\text{diag}\left[\sinh n\gamma\ell\right]\mathbf{V}_{m1}^{-1} & \mathbf{I}_{m1}\text{diag}\left[\cosh n\gamma\ell\right]\mathbf{I}_{m1}^{-1} \end{pmatrix} \tag{5.186}$$

Using the above in (5.170) yields,

$$\begin{aligned} \mathbf{V}_n &= \mathbf{V}_{m1}\text{diag}\left[\cosh n\gamma\ell\right]\mathbf{V}_{m1}^{-1}\mathbf{V}_1 - \mathbf{V}_{m1}\text{diag}\left[\sinh n\gamma\ell\right]\mathbf{I}_{m1}^{-1}\mathbf{I}_1 \\ \mathbf{I}_n &= -\mathbf{I}_{m1}\text{diag}\left[\sinh n\gamma\ell\right]\mathbf{V}_{m1}^{-1}\mathbf{V}_1 + \mathbf{I}_{m1}\text{diag}\left[\cosh n\gamma\ell\right]\mathbf{I}_{m1}^{-1}\mathbf{I}_1 \end{aligned} \tag{5.187}$$

As for the first order case, the characteristic impedance matrix of the infinite periodic structure is given by,

$$\mathbf{Z}_0 = \mathbf{V}_{m1}\mathbf{I}_{m1}^{-1} \tag{5.188}$$

Applying the above expression to the previous equation, gives

$$\begin{aligned} \mathbf{V}_n &= \mathbf{V}_{m1}\left\{\text{diag}\left[\cosh n\gamma\ell\right] - \text{diag}\left[\sinh n\gamma\ell\right]\right\}\mathbf{V}_{m1}^{-1}\mathbf{V}_1 \\ &= \mathbf{V}_{m1}\text{diag}\left[e^{-n\gamma\ell}\right]\mathbf{V}_{m1}^{-1}\mathbf{V}_1 \\ \mathbf{I}_n &= \mathbf{I}_{m1}\left\{\text{diag}\left[\cosh n\gamma\ell\right] - \text{diag}\left[\sinh n\gamma\ell\right]\right\}\mathbf{I}_{m1}^{-1}\mathbf{I}_1 \\ &= \mathbf{I}_{m1}\text{diag}\left[e^{-n\gamma\ell}\right]\mathbf{I}_{m1}^{-1}\mathbf{I}_1 \end{aligned} \tag{5.189}$$

Except for the similarity transformation by the partial eigenvectors, the above relations are the same as for the simpler 2×2 case of (5.171). As before, the multiplication by $\mathbf{V}_{m1}^{-1}$ and $\mathbf{I}_{m1}^{-1}$ can be viewed as a coordinate transformation. It is clear that the results of the last equation are exactly the same as for a finite periodic structure of length ℓ terminated by its own characteristic impedance Z_0.

On the basis of the above theoretical approach, a fairly large number of different structures were analyzed and a wealth of numerical results can be found in [5, chapter 5] and [6].

5.5 Bends in planar dielectric waveguides

As observed in the pioneering work of [39] the curvature of an open waveguide introduces radiation. In this work also formulas for the radiative attenuation of bent dielectric slabs were provided. Later analyses were concerned with similar problems, such as radio wave propagation around the Earth, while the problem of curved optical guides has received considerable attention from 1969 onwards after the work of [29, 31]. An extensive survey of the related literature can be found in existing textbooks such as [32] and especially in the fundamental book on electromagnetic waves and curved structures, [27]. In the latter book, as well as in several other papers, the radiation from curved open structures is mainly considered by using a perturbation approach, that is by treating the curvature as a small perturbation of the straight configuration.

However, the perturbational approach is not entirely suited for the analysis of relatively sharp bends, such as those required in integrated optics and especially short millimetre waves. Moreover, when the junction between a straight and a curved dielectric slab is considered, mode–matching needs to be adopted [27, chapter 11].

The latter obviously requires knowledge of the modal spectra of both waveguides. This type of problem can be solved in an elegant way by using the concept of local modes, similarly to what has been done by [46] for the case of propagation on a finitely conducting surface.

5.5.1 The curved dielectric slab

Let us consider the simple waveguide bend of a slab dielectric waveguide independent of the y coordinate as shown in Figure 5.28 on page 243. The field in the slab waveguide may be described by a suitable potential $\phi(x, z)$ which satisfies the Helmholtz equation, i.e.

$$\left(\partial_x^2 + \partial_z^2\right)\phi(x, z) + \varepsilon k_0^2 \phi(x, z) = 0 \tag{5.190}$$

In particular, for a *straight* waveguide section the potential may be expressed in the following way

$$\phi(x, z) = e^{-jh_n z}\psi_n(x) \tag{5.191}$$

by considering a harmonic variation along the z coordinate. The above is the potential relative to a *straight* waveguide section and it also satisfies the following equation

$$\partial_x^2 \psi_n(x) + \left(\varepsilon k_0^2 - h_n^2\right)\psi_n(x) = 0 \tag{5.192}$$

where we consider incident and reflected waves, corresponding to positive and negative values of n, respectively, and where

$$\begin{aligned} h_n &= -h_{-n} \\ \psi_n &= \psi_{-n} \end{aligned} \tag{5.193}$$

It is now useful to express the field in the bend region too as a superposition of solutions of the straight waveguide modes, the so–called *local modes*. Expanding the potential in terms of local modes we have

$$\phi = \sum_{\nu=-N}^{N} c_\nu \psi_\nu + \sum_{\mu=\pm} \int_0^\infty c_\mu(\chi)\psi_\mu(\chi)d\chi \tag{5.194}$$

and for its derivative,

$$\frac{d\phi}{dz} = -j\sum_{\nu=-N}^{N} h_\nu c_\nu \psi_\nu - j\sum_{\mu=\pm}\int_0^\infty h_\mu(\chi)c_\mu(\chi)\psi_\mu(\chi)d\chi \tag{5.195}$$

Equation (5.194) follows from the completeness of the modes of the straight waveguide, while (5.195) follows from the property of the expansion to be local. It is now expedient to modify the above expressions considering the specific geometry of a circular bend. In this case we have:

$$\begin{aligned} \partial_\rho &= \partial_x \\ \rho &= R + x \\ \frac{1}{\rho}\partial_\varphi &= \partial_z \\ \partial_x^2 + \partial_z^2 &= -\frac{1}{\rho}\partial_\rho(\rho\partial_\rho\psi) + \frac{1}{\rho^2}\partial_\varphi^2 \end{aligned} \tag{5.196}$$

Hence, in circular cylinder coordinates, the wave equation (5.190) becomes

$$\left(\partial_\rho^2+\frac{1}{\rho}\partial_\rho+\frac{1}{\rho^2}\partial_\varphi^2\right)\phi+\varepsilon k_0^2\phi=0 \tag{5.197}$$

Direct differentiation with respect to z of (5.194), yields

$$\begin{aligned}\tfrac{1}{\rho}\partial_\varphi\phi=\tfrac{1}{\rho}\sum_{\nu=-N}^{N}(c'_\nu\psi_\nu+c_\nu\psi'_\nu)+\\ \tfrac{1}{\rho}\sum_{\mu=\pm}\int_0^\infty\left[c'_\mu(\chi)\psi_\mu(\chi)+c_\mu(\chi)\psi'_\mu(\chi)\right]d\chi\end{aligned} \tag{5.198}$$

where the prime denotes differentiation with respect to φ. Equating (5.198) and (5.195) and recalling that $\rho=R+x$, yields

$$\begin{aligned}&\sum_{\nu=-N}^{N}(c'_\nu+jRh_\nu c_\nu)\psi_\nu+\sum_{\mu=\pm}\int_0^\infty\left[c'_\mu(\chi)+jRh_\mu(\chi)c_\mu(\chi)\right]\psi_\mu(\chi)d\chi\\ &=-j\sum_{\nu=-N}^{N}h_\nu c_\nu x\psi_\nu-j\sum_{\mu=\pm}\int_0^\infty h_\mu(\chi)c_\mu(\chi)x\psi_\mu(\chi)d\chi\\ &\quad-\sum_{\nu=-N}^{N}c_\nu\psi'_\nu-\sum_{\mu=\pm}\int_0^\infty c_\mu(\chi)\psi'_\mu(\chi)d\chi\end{aligned} \tag{5.199}$$

Substituting (5.194) and (5.195) together in the wave equation (5.197), yields

$$\begin{aligned}&\sum_{\nu=-N}^{N}\left(\partial_\rho^2+\tfrac{1}{\rho}\partial_\rho\right)c_\nu\psi_\nu+\sum_{\mu=\pm}\int_0^\infty\left(\partial_\rho^2+\tfrac{1}{\rho}\partial_\rho\right)c_\mu(\chi)\psi_\mu(\chi)d\chi\\ &\tfrac{-j}{\rho}\left\{\sum_{\nu=-N}^{N}\partial_\varphi\left(h_\nu c_\nu\psi_\nu\right)+\sum_{\mu=\pm}\int_0^\infty\partial_\varphi\left[h_\mu(\chi)c_\mu(\chi)\psi_\mu(\chi)\right]d\chi\right\}\\ &+\sum_{\nu=-N}^{N}\varepsilon k_0^2c_\nu\psi_\nu+\sum_{\mu=\pm}\varepsilon k_0^2\int_0^\infty c_\mu(\chi)\psi_\mu(\chi)d\chi=0\end{aligned} \tag{5.200}$$

By multiplying by $\rho=R+x$, carrying out the φ differentiation explicitly and upon use of (5.196) and (5.192) the last equation becomes

$$\begin{aligned}&\sum_{\nu=-N}^{N}h^2Rc_\nu\psi_\nu+\sum_{\mu=\pm}\int_0^\infty h^2(\chi)Rc_\mu(\chi)\psi_\mu(\chi)d\chi\\ &-j\sum_{\nu=-N}^{N}h'_\nu c_\nu\psi_\nu+h_\nu c'_\nu\psi_\nu+h_\nu c_\nu\psi'_\nu\\ &-j\sum_{\mu=\pm}\int_0^\infty\left[h'_\mu(\chi)c_\mu(\chi)\psi_\mu(\chi)+h_\mu(\chi)c'_\mu(\chi)\psi_\mu(\chi)\right]d\chi\\ &+\int_0^\infty\left[h_\mu(\chi)c_\mu(\chi)\psi'_\mu(\chi)\right]d\chi\\ &+\sum_{\nu=-N}^{N}xh_\nu^2c_\nu\psi_\nu+\sum_{\mu=\pm}\int_0^\infty xh_\mu(\chi)c_\mu(\chi)\psi_\mu(\chi)d\chi\\ &+\sum_{\nu=-N}^{N}c_\nu\partial_x\psi_\nu+\sum_{\mu=\pm}\int_0^\infty c_\mu(\chi)\partial_x\psi_\mu(\chi)d\chi=0\end{aligned} \tag{5.201}$$

Multiplying (5.199) by ψ_n and integrating, we get

$$\begin{aligned}\sum_{\nu=-n} (c'_\nu + jRh_\nu c_\nu) = -j \sum_{\nu=-N}^{N} h_\nu c_\nu < \psi_n x \psi_\nu > \\ -j \sum_{\mu=\pm} \int_0^\infty h_\mu(\chi) c_\mu(\chi) < \psi_n x \psi_\mu(\chi) > d\chi \\ - \sum_{\nu=-N}^{N} c_\nu < \psi_n \psi'_\nu > - \sum_{\mu=\pm} \int_0^\infty c_\mu(\chi) < \psi_n \psi'_\mu(\chi) > d\chi\end{aligned} \tag{5.202}$$

where $<>$ denotes integration over the cross–section; similarly, from (5.201)

$$\begin{aligned}\sum_{\nu=-n} \pm (c'_\nu + jRh_\nu c_\nu) \\ - \sum_{\nu=-N}^{N} \tfrac{1}{h_{|n|}} [h'_\nu \delta_{|n|\nu} + h_\nu < \psi_n \psi'_\nu > + j h_\nu^2 < \psi_n x \psi_\nu > \\ + j < \psi_n \partial_x \psi_\nu >] c_\nu \\ - \sum_{\mu=\pm} \tfrac{1}{h_{|n|}} \int_0^\infty [h_\mu(\chi) < \psi_n \psi'_\mu(\chi) > + j h^2(\chi) < \psi_n x \psi_\mu(\chi) > \\ + j < \psi_n \partial_x \psi_\mu(\chi) >] c_\mu(\chi) d\chi = 0\end{aligned} \tag{5.203}$$

Recalling that $h_{-\nu} = -h_\nu$, upon adding and subtracting (5.202 and 5.203) we obtain the integro-differential equations for the amplitudes of the single waves

$$c'_n + jRh_n c_n = \sum_{\nu=-n} S_{n\nu} c_\nu + \sum_{\mu=\pm} \int_0^\infty S_{n\mu}(\chi) c_\mu(\chi) d\chi \tag{5.204}$$

where $n = N, \cdots, 1, -1, \cdots, -N$ and

$$\begin{aligned}S_{n\nu} = -\tfrac{1}{2} \left[j h_\nu < \psi_n x \psi_\nu > + < \psi_n \psi'_\nu > + \tfrac{h'_\nu}{h_{|n|}} \delta_{\nu|n|} \right] \\ -\tfrac{1}{2} \left[j \tfrac{h_\nu}{h_{|n|}} < \psi_n \psi'_\nu > + j \tfrac{h_\nu^2}{h_{|n|}} < \psi_n x \psi_\nu > + j \tfrac{1}{h_{|n|}} < \psi_n \partial_x \psi_\nu > \right] \\ S_{n\mu}(\chi) = -\tfrac{1}{2} \left[j h_\mu(\chi) < \psi_n x \psi_\mu(\chi) > + \tfrac{h_\mu(\chi)}{h_{|n|}} < \psi_n \psi'_\mu(\chi) > \right] \\ -\tfrac{1}{2} \left[< \psi_n \psi'_\mu(\chi) > + j \tfrac{h^2(\chi)}{h_{|n|}} < \psi_n x \psi_\mu(\chi) > + j \tfrac{1}{h_{|n|}} < \psi_n \partial_x \psi_\mu(\chi) > \right]\end{aligned} \tag{5.205}$$

An equation similar to (5.202), but for the amplitudes of the continuous spectrum, is obtained by multiplying (5.199) by $\psi_\mu(\sigma)$ and integrating:

$$\begin{aligned}\sum_{m=\pm} (c'_m(\sigma) + jRh_m(\sigma) c_m(\sigma)) = -j \sum_{\nu=-N}^{N} h_\nu c_\nu < \psi(\sigma) x \psi_\nu > \\ -j \sum_{\mu=\pm} \int_0^\infty h_\mu(\chi) c_\mu(\chi) < \psi(\sigma) x \psi_\mu(\chi) > d\chi \\ - \sum_{\nu=-N}^{N} c_\nu < \psi(\sigma) \psi'_\nu > - \sum_{\mu=\pm} \int_0^\infty c_\mu(\chi) < \psi(\sigma) \psi'_\mu(\chi) > d\chi\end{aligned} \tag{5.206}$$

Similarly, from (5.201)

$$\begin{aligned} &\sum_{m=\pm} \pm(c'_m(\sigma) + jRh_m(\sigma)c_m(\sigma)) \\ &- \sum_{\nu=-N}^{N} \tfrac{1}{h(\sigma)} \left[h_\nu < \psi(\sigma)\psi'_\nu > + jh_\nu^2 < \psi(\sigma)x\psi_\nu > + j < \psi(\sigma)\partial_x\psi_\nu >\right] c_\nu \\ &-j \sum_{\mu=\pm} \tfrac{1}{h(\sigma)} \int_0^\infty \left[h'_\mu(\chi)\delta(\sigma-\chi) + h_\mu(\chi) < \psi(\sigma)\psi'_\mu(\chi) >\right] c_\mu(\chi)d\chi \\ &+ \sum_{\mu=\pm} \tfrac{1}{h_{|n|}} \int_0^\infty \left[jh_\mu^2(\chi) < \psi(\sigma)x\psi_\mu(\chi) > + j < \psi(\sigma)\partial_x\psi_\mu(\chi) >\right] c_\mu(\chi)d\chi = 0 \end{aligned} \tag{5.207}$$

Hence, by adding and subtracting (5.206) and (5.207) we obtain:

$$c'_m(\sigma) + jRh_m(\sigma)c_m(\sigma) = \sum_{\nu=-N}^{N} S'_{\nu\mu}(\sigma)c_\nu + \sum_{\mu=\pm} \int_0^\infty S_\mu(\sigma,\chi)c_\mu(\chi)d\chi \tag{5.208}$$

where $m = \pm$ and

$$\begin{aligned} S'_{\nu\mu} &= -\tfrac{1}{2}\left[jh_\nu < \psi(\sigma)_n x\psi_\nu > + < \psi(\sigma)\psi'_\nu >\right] \\ &-\tfrac{1}{2}\left[\tfrac{h_\nu}{h(\sigma)} < \psi(\sigma)\psi'_\nu > + j\tfrac{h_\nu^2}{h(\sigma)} < \psi(\sigma)x\psi_\nu > + j\tfrac{1}{h(\sigma)} < \psi(\sigma)\partial_x\psi_\nu >\right] \end{aligned} \tag{5.209}$$

$$\begin{aligned} &S_\mu(\sigma,\chi) = -\tfrac{1}{2} \\ &\left[\left(jh_\mu(\chi) < \psi(\sigma)x\psi_\mu(\chi) > + < \psi(\sigma)\psi'_\mu(\chi) > + \tfrac{h_\mu(\chi)}{h(\sigma)} < \psi(\sigma)\psi'_\mu(\chi) >\right)\right. \\ &\left. - \left(\tfrac{h'_\mu(\chi)}{h(\sigma)}\delta(\sigma-\chi) + j\tfrac{h^2(\chi)}{h(\sigma)} < \psi(\sigma)x\psi_\mu(\chi) > + j\tfrac{1}{h(\sigma)} < \psi(\sigma)\partial_x\psi_\mu(\chi) >\right)\right] \end{aligned}$$

5.5.2 Orthonormal properties

By definition, the normalization of the local modes is φ-independent:

$$\begin{aligned} < \psi_n\psi_\nu > &= \delta_{n|\nu|} \\ < \psi_n\psi(\chi) > &= 0 \\ < \psi(\sigma)\psi(\chi) > &= \delta(\sigma-\chi) \end{aligned} \tag{5.210}$$

Hence by differentiation of the above equations, we have

$$\begin{aligned} < \psi_n\psi'_\nu > &= - < \psi'_n\psi_\nu > \\ < \psi_n\psi'_{\pm n} > &= 0 \\ < \psi_n\psi'(\chi) > &= - < \psi'_n\psi(\chi) > \\ < \psi(\sigma)\psi'(\chi) > &= - < \psi'(\sigma)\psi(\chi) > \\ < \psi(\sigma)\psi'(\sigma) > &= 0 \end{aligned} \tag{5.211}$$

whereas for a waveguide uniform with respect to φ,

$$h'_n = h'(\chi) = \psi'_\nu = \psi'(\chi) = 0 \tag{5.212}$$

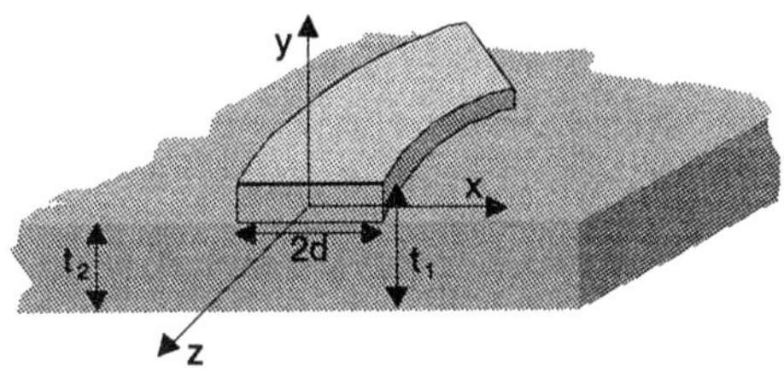

Figure 5.26 Curved rib geometry.

5.5.3 Curved rib waveguide: coupling of LSE and LSM modes

Bends in dielectric waveguides occur frequently in integrated optics since they are commonly used in order to achieve lateral guidance. The bending losses are generated from the following, nearly independent, mechanisms:

- coupling between the operating LSE (or LSM) mode and the parasite LSM (or LSE); this type of loss can be referred to as cross–polarisation coupling
- coupling between the operating mode and the local radiative spectrum of the same polarization
- effects of corrugations, due to wall imperfections, which also generate radiation.

In this Section, in order to provide an example of the application of the local modes approach, we investigate the cross–polarisation coupling by the method of local modes [9]. The coupling between the operating mode and the local radiative spectrum and the effects of corrugations have been dealt with in [41] by the method of local modes.

As observed previously, the field in a straight dielectric rib waveguide although hybrid, can be approximated for practical aspect ratios as pure LSE or LSM. On the contrary, a curved dielectric waveguide of two–dimensional cross–section causes the two cross-polarisations to couple with each other through the newly imposed boundary conditions at the bend. The experimental verification of this effect has been reported in the literature for several dielectric waveguides, such as the nonradiative waveguide [55].

It is noted that cross–polarisation effects cannot be described in terms of three–component TE and TM modes, since they have no common component. To characterise this effect we need at least a five–component field such as the LSE, LSM; in this case the LSE and LSM polarisations have four common components (E_x, E_z), (H_x, H_z). For the straight guide the LSE (LSM) five–component mode can be expressed in terms of the scalar, y–directed, Hertzian potential Π_h (Π_e) which is proportional to $H_y(E_y)$. This potential satisfies the wave equations as

$$\partial_x^2\Pi_{h,e}(x,y) + \partial_y^2\Pi_{h,e}(x,y) + \left(\varepsilon_r k_0^2 - \beta^2\right)\Pi_{h,e}(x,y) = 0 \qquad (5.213)$$

For small aspect ratios, defined with reference to Figure 5.26 as $(t_1 - t_2)/2d$, we can neglect the ∂_x^2 term while considering the y–variation of $\Pi_{e,h}(x,y)$. This is a reasonable approximation for most practical cases. We can now proceed to the

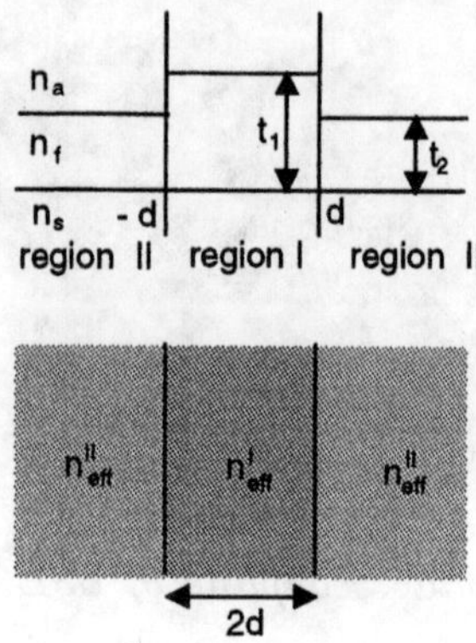

Figure 5.27 Cross–section of the rib guide (top) and its equivalent slab model (bottom).

application of EDC to the rib waveguide, thus reducing it to an equivalent dielectric slab waveguide as illustrated in Figure 5.27. We are now in a position to describe the curved rib waveguide of Figure 5.26 by means of its equivalent slab model, i.e. to replace the structure of Figure 5.26 with the simpler structure of Figure 5.28, showing the curved guide as seen from the top. We can now investigate the effect of the radius of curvature on the cross–polarisation by using the local mode approach. The local modes, introduced in the previous Section, are defined [46] as the eigenfunctions of a locally straight guide, the cross–section of which is the same as that of the curved guide at any point along its curved axis. Therefore, local modes are solutions of the transverse wave equation

$$\partial_x^2\psi_n + \partial_y^2\psi_n + \left(\varepsilon_r k_0^2 - \beta^2\right)\psi_n = 0 \tag{5.214}$$

By neglecting the radiation modes of the local rib waveguide, the global transverse field of the rib bend at any point $z = z_0$ can be expressed as a superposition of the discrete components in the two polarisations [50]

$$\begin{aligned}\mathbf{E}_t(x,y,z) &= \sum_n V_n(z)\mathbf{E}_n(x,y)\\ \mathbf{H}_t(x,y,z) &= \sum_n I_n(z)\mathbf{H}_n(x,y)\end{aligned} \tag{5.215}$$

while Maxwell's equations are expressed in the curvilinear coordinate system of Figure 5.28 as

$$\begin{aligned}\partial_z\mathbf{E}_t &= \frac{1}{j\omega\varepsilon}\nabla_t e_3\nabla_t\cdot(\mathbf{H}_t\times\hat{\mathbf{z}}) - j\omega\mu e_3\mathbf{H}_t\times\hat{\mathbf{z}}\\ \partial_z\mathbf{H}_t &= \frac{1}{j\omega\mu}\nabla_t e_3\nabla_t\cdot(\hat{\mathbf{z}}\times\mathbf{E}_t) - j\omega\varepsilon e_3\hat{\mathbf{z}}\times\mathbf{E}_t\end{aligned} \tag{5.216}$$

where

$$e_3 = 1 + \frac{x}{R} \tag{5.217}$$

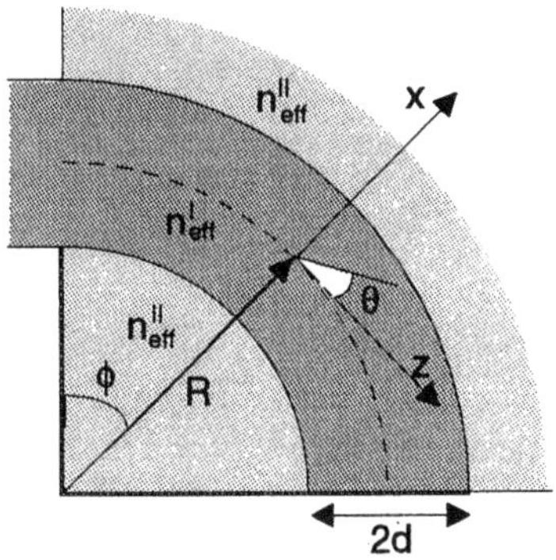

Figure 5.28 Equivalent two–dimensional structure after application of EDC.

In (5.215) the local eigenmodes $\mathbf{E}_n(x, y)$, $\mathbf{H}_n(x, y)$ are normalised to carry unit power on the guide cross–section S, i.e.

$$\int_S \mathbf{E}_m \times \mathbf{H}_n \cdot \hat{\mathbf{z}} ds = \begin{Bmatrix} 1 & m = n \\ 0 & m \neq n \end{Bmatrix} \tag{5.218}$$

Inserting (5.215) into (5.216), scalar multiplication of the first of (5.216) by $\mathrm{H_m} \times \hat{\mathbf{z}}$ and the second of (5.216) by $\hat{z} \times \mathbf{E}_{\mathrm{m}}$, and integrating over the whole cross–section yields the following form of the Telegrapher's equations:

$$\begin{aligned} \tfrac{dV_m(z)}{dz} &= -j\sum_n X_{mn} I_n(z) \\ \tfrac{dI_m(z)}{dz} &= -j\sum_n B_{mn} V_n(z) \end{aligned} \tag{5.219}$$

with the transmission line parameters defined as:

$$\begin{aligned} X_{mn} &= \omega \int_S \left(1 + \tfrac{x}{R}\right) \left[\varepsilon E_{zm} E_{zn} + \mu \mathbf{H}_m \cdot \mathbf{H}_n\right] ds \\ B_{mn} &= \omega \int_S \left(1 + \tfrac{x}{R}\right) \left[\mu H_{zm} H_{zn} + \varepsilon \mathbf{E}_m \cdot \mathbf{E}_n\right] ds \end{aligned} \tag{5.220}$$

It is clear that as R goes to infinity the modes are decoupled, since the term x/R, which represents the lateral displacement over the radius of curvature, gives the coupling between the lines. In order to describe cross–polarisation in terms of the forward (a_n) and backward (b_n) travelling waves, we take "1Ω" normalised wave impedances thus setting:

$$\begin{aligned} a_n &= \tfrac{(V_n + I_n)}{2} \\ b_n &= \tfrac{(V_n - I_n)}{2} \end{aligned} \tag{5.221}$$

When obtaining (5.219) we have, in this context, neglected coupling to the continuum of radiation. Rigorous solution of (5.219) provides the coupling between all the discrete modes (both LSE and LSM) of the structure. Nonetheless, most of the cross–polarisation is due to the coupling between the assumed fundamental LSE (operating mode) and the fundamental LSM (parasite mode). Moreover, also the coupling between forward and backward waves can be neglected, so that we reduce our problem to just two coupled transmission lines. Using subscript 1 to designate the fundamental LSE and associating subscript 2 with the fundamental LSM, after substitution of (5.221) into (5.219), we obtain

$$\begin{aligned} \frac{da_1}{dz} &= -j\beta_1 a_1 - jca_2 \\ \frac{da_2}{dz} &= -jca_1 - j\beta_2 a_2 \end{aligned} \tag{5.222}$$

where

$$\begin{aligned} \beta_1 &= \frac{(X_{11}+B_{11})}{2} \\ \beta_2 &= \frac{(X_{22}+B_{22})}{2} \\ c &= \frac{(X_{12}+B_{12})}{2} \end{aligned} \tag{5.223}$$

and obviously $X_{12} = X_{21}, B_{12} = B_{21}$. By assuming just fundamental LSE mode incidence we have the initial conditions $a_1(0) = 1$ and $a_2(0) = 0$ which produces the following solution of (5.222)

$$\begin{aligned} a_1(z) &= \left[\cos\left(\frac{\Gamma z}{2}\right) - j\frac{\beta_d}{\Gamma}\sin\left(\frac{\Gamma z}{2}\right)\right] e^{-j\beta_s z} \\ a_2(z) &= -j\frac{cz}{\Gamma}\sin\left(\frac{\Gamma z}{2}\right) e^{-j\beta_s z} \end{aligned} \tag{5.224}$$

where we have introduced

$$\begin{aligned} \beta_s &= \frac{\beta_1 + \beta_2}{2} \\ \beta_d &= \frac{\beta_1 - \beta_2}{2} \\ \Gamma &= \sqrt{\beta_d^2 + 4c^2} \end{aligned} \tag{5.225}$$

Since the operating mode has amplitude

$$|a_1(z)| = \sqrt{1 - \frac{4c^2}{\Gamma^2}\sin^2\left(\frac{\Gamma z}{2}\right)} \tag{5.226}$$

the cross-polarisation coupling of the rib bend, is given by

$$L(dB) = -10\log|a_1|^2 \tag{5.227}$$

and the maximum coupling loss can be calculated as

$$L_{max} = 20\log\frac{\Gamma}{|\beta_d|} \tag{5.228}$$

The above theory was applied to calculating the cross–polarisation coupling of the rib bend giving interesting numerical results as discussed in [9].

Bibliography

[1] C. Angulo. Diffraction of surface waves by a semi-infinite dielectric slab. *IRE Trans. Antennas Propagat.*, 5:100–109, January 1957.

[2] C. Brooke and M. Kharadhly. Step discontinuities on dielectric waveguides. *Electron. Lett.*, 12:471–473, September 1976.

[3] J. K. Butler and J. Zoroofchi. Radiation fields of GaAs injection lasers. *J. Quantum Electron.*, 10:809–815, October 1974.

[4] H. C. Casey and M. B. Panish. *Heterostructure lasers.* Academic, New York, 1978.

[5] C. S. Chang. *A Rigorous analysis of cascaded step discontinuities in open waveguides.* PhD thesis, University of Bath, Bath, U.K., 1993.

[6] C. S. Chang and T. Rozzi. Rigorous analysis of finite periodic structures in dielectric waveguides. *16th Microwave Conference, Dublin*, September 1986.

[7] K. C. Chang and T. Tamir. Simplified approach to surface-wave scattering by blazed dielectric gratings. *Applied Opt.*, 19:282–288, 1980.

[8] W. S. C. Chang. Periodic structures and their applications in integrated optics. *IEEE Trans. on Microwave Theory Tech.*, 21:775–785, 1974.

[9] C. Cheung, T. Rozzi, and G. Cerri. Cross–polarisation coupling in curved dielectric rib waveguides. *IEE Proc. Pt. J*, 135:281–284, June 1988.

[10] S. R. Chinn. Effects of mirror reflectivity in a distributed feedback laser. *IEEE J. Quantum Electron.*, 9:574–580, 1973.

[11] P. Clarricoats and A. Sharpe. Modal matching applied to a discontinuity in a planar surface waveguide. *Electron. Lett.*, 8:28–29, January 1972.

[12] R. E. Collin. *Field Theory of Guided Waves.* IEEE Press, New York, 1991.

[13] P. G. Cottis and N. K. Uzunoglu. Analysis of longitudinal discontinuities in dielectric slab waveguides. *J.Opt. Soc. Am. A*, 1:206–215, February 1984.

[14] P. J. de Waard. Calculation of the far field half power width and mirror reflection coefficients of d.h. lasers. *Electron. Lett.*, 11:11–12, January 1975.

[15] D. P. Gianrusso and J. H. Harris. Electro-optic modulation in a thin-film waveguide. *Appl. Opt.*, 10:2786–2788, 1971.

[16] I. S. Gradshteyn and I. M. Ryzhik. *Table of integrals, series and products.* Academic Press, New York, 1980.

[17] G. A. Hockham and A. B. Sharpe. Dielectric waveguide discontinuities. *Electron. Lett.*, 8:230–231, May 1972.

[18] T. Hosono, T. Hinata, and A. Inoue. Numerical analysis of the discontinuities in slab waveguides. *Radio Science*, 17:75–83, January 1982.

[19] T. Ikegami. Reflectivity of mode at facet and oscillation mode in double heterostructure injection lasers. *IEEE J. Quantum Electronics*, 6:470–476, June 1972.

[20] T. Itoh. Application of gratings in a dielectric waveguide for leaky-wave antennas and band-rejection filters. *IEEE Trans. on Microwave Theory Tech.*, 25:1134–1138, December 1977.

[21] T. Itoh and J. Hsu. Distributed Bragg reflector Gunn oscillators for dielectric millimeter-wave integrated circuits. *IEEE Trans. on Microwave Theory Tech.*, 27:514–518, May 1979.

[22] K. L. Klohn, R. E. Horn, H. Jacobs, and E. Freibergs. Silicon waveguide frequency scanning linear array antennas. *IEEE Trans. on Microwave Theory Tech.*, 26:764–773, October 1978.

[23] H. Kogelnik and C. V. Shank. Couple-wave theory of distributed feedback lasers. *J. Appl Phys*, 43:2327–2335, 1972.

[24] M. Koshiba, T. Miki, K. Ooishi, and M. Suzuki. On finite-element solutions of the discontinuity problems in a bounded dielectric slab waveguide. *I IECE Trans. vol. E*, 66:250–251, April 1983.

[25] H. Kressel and J. K. Butler. *Semiconductor Lasers and Heterojunction LEDs.* Academic, New York, 1977.

[26] L. Lewin. A method for the calculation of the radiation pattern and mode conversion properties of a solid-state heterojunction laser. *IEEE Trans. on Microwave Theory Tech.*, 23:576–585, July 1975.

[27] L. Lewin, D. C. Chang, and E. F. Kuester. *Electromagnetic Waves and Curved Structures.* Peter Peregrinus LTD., London, 1977.

[28] S. Mahmoud and J. Beal. Scattering of surface waves at a dielectric discontinuity on a planar waveguide. *IEEE Trans. on Microwave Theory Tech.*, 23:193–198, February 1975.

[29] E. A. J. Marcatili. Bends in optical dielectric guides. *Bell Syst. Tech. J.*, 48(9):2103–2133, September 1969.

[30] D. Marcuse. Radiation losses of tapered dielectric slab waveguides. *Bell Syst. Tech. J.*, 49:273–287, February 1970.

[31] D. Marcuse. Bending losses of the asymmetric slab waveguide. *Bell Syst. Tech. J.*, 50:2551–2563, 1971.

[32] D. Marcuse. *Light Transmission Optics.* Van Nostrand Reinhold, New York, 1972.

[33] D. Marcuse. Exact theory of TE-wave scattering from blazed dielectric gratings. *Bell Syst. Tech. J.*, 55:1295–1317, 1976.

[34] G. L. Matthaei. A note concerning modes in dielectric waveguide gratings for filter applications. *IEEE Trans. on Microwave Theory Tech.*, 31:309–312, March 1983.

[35] G. L. Matthaei, C. E. Harris, D. C. Park, and Y. Kim. Dielectric waveguide filters using parallel-coupled grating resonators. *Electronics Lett.*, 18:509–510, 1982.

[36] L. Ogawa, W. S. C. Chang, B. L. Sopori, and F. J. Rosenbaum. A theoretical analysis of etched grating couplers for integrated optics. *IEEE J. Quantum Electron.*, 9:29–42, January 1973.

[37] J. N. Polky and J. H. Harris. Electro-optic thin film modulators. *Appl. Phys. Lett.*, 21:307–309, 1972.

[38] F. K. Reinhart, I. Hayashi, and M. Panish. Mode reflectivity and waveguide properties of double heterostructure injection lasers. *J. Appl. Phys.*, 42:4466–4479, October 1971.

[39] R. D. Richtmyer. Dielectric resonators. *J. Appl. Phys.*, 10:391–398, June 1939.

[40] T. Rozzi. Rigorous analysis of the step discontinuity in a planar dielectric waveguide. *IEEE Trans. on Microwave Theory Tech.*, 26:738–746, October 1978.

[41] T. Rozzi, G. Cerri, F. Chiaraluce, R. De Leo, and R. F. Ormondroyd. Finite curvature and corrugations in dielectric ridge waveguides. *IEEE Trans. Microwave Theory Tech*, 36:68–79, January 1988.

[42] T. Rozzi and G. H. in't Veld. Field and network analysis of interacting step discontinuities in planar dielectric waveguides. *IEEE Trans. on Microwave Theory Tech.*, 27:303–309, April 1979.

[43] T. Rozzi and G. H. in't Veld. Variational treatment of the diffraction at the facet of d.h. lasers and of dielectric millimeter wave antennas. *IEEE Trans. on Microwave Theory Tech.*, 28:61–73, February 1980.

[44] F. K. Schwering and S. T. Peng. Design of dielectric grating antennas for millimeter-wave applications. *IEEE Trans. on Microwave Theory Tech.*, 31:199–208, February 1983.

[45] J. Schwinger and D. S. Saxon. *Discontinuities in Waveguides.* Gordon and Breach, New York, 1968.

[46] V. V. Shevchenko. *Continuous Transition in Open Waveguides.* Golem Press, Boulder, Colorado, 1971.

[47] H. Shigesawa and M. Tsuji. A completely theoretical design method of dielectric image guide gratings in the Bragg reflection region. *IEEE Trans. on Microwave Theory Tech.*, 34:420–426, April 1986.

[48] H. Shigesawa, M. Tsuji, and K. Takiyama. Mode propagation through a step discontinuity in dielectric planar waveguide. *IEEE MTT-S Digest*, pages 121–123, 1984.

[49] H. Shigesawa, M. Tsuji, and K. Takiyama. Microwave network representation of discontinuity in open dielectric waveguides and its applications to periodic structures. *IEEE MTT-S INT. Microwave Symposium*, June 1985.

[50] E. W. Snyder and J. D. Love. *Optical Waveguide Theory*. Chapmann and Hall, 1983.

[51] W. Streifer, R. D. Burnham, and D. R. Scifres. Analysis of grating-coupled radiation in GaAs: GaAlAs lasers and waveguides: Blazing effects. *IEEE J. Quantum Electron.*, 12:494–499, August 1976.

[52] W. Streifer, D. R. Scifres, and R. D. Burnham. Coupled wave analysis of DFB and DBR lasers. *IEEE J. Quantum Electron.*, 13:134–141, 1977.

[53] M. Tsuji, S. Matsumoto, H. Shigesawa, and K. Takiyama. Guided-wave experiments with dielectric waveguides having finite periodic corrugations. *IEEE Trans. on Microwave Theory Tech.*, 31:337–344, April 1983.

[54] S. Wang. Principles of distributed feedback and distributed Bragg reflector lasers. *IEEE J. Quantum Electron.*, 10:413–426, 1974.

[55] T. Yoneyama, H. Tamaky, and S. Nishida. Analysis and measurements of nonradiative waveguide bends. *IEEE Trans. Microwave Theory Tech.*, 34:876–882, 1986.

Chapter 6

Bound modes by transverse resonance diffraction

The purpose of computation is insight, not numbers.
Richard Hamming

In this Chapter we study, in a rigorous manner, the bound modes of several transmission lines which find applications in both the microwave and millimetre–wave frequency ranges, as well as in integrated optics. The bound modes of these transmission lines play a fundamental role in circuit design. As noted in Chapter 4, in several instances it is possible to derive their properties by using suitable approximations, e.g. the effective dielectric constant method. However, when propagation constants must be known with greater accuracy, or when the approximations break down, or when the field shape is of interest as in discontinuity problems, it becomes necessary to employ rigorous procedures such as those described hereafter. Note also that, as new structures are considered, it is sometimes unclear what approximations are feasible; in these cases a prudent approach suggests using rigorous full–wave analyses.

As opposed to modes pertaining to the continuum, which extend up to infinity, bound modes are necessarily confined near the guiding structure. Hence, it is possible to introduce a shielding enclosure, placed at a sufficient distance from the guide, without significantly perturbing the field of the bound modes. Thanks to this device, several numerical methods can be employed to determine the propagation characteristics, each having its own specific merits and disadvantages. In this respect, the reader is advised to look at the excellent book of Itoh, [16], as well as at papers [41, 15].

The approach followed throughout this Chapter is based on the transverse resonance diffraction (TRD), as often used in the case of closed, nonseparable, waveguides. This is, in the authors' opinion, the most analytical approach available so far and, as such, it has the merit of providing a greater insight into the guide operation. It also allows us to derive various levels of approximation both in analytical terms (we are referring to the possibility of considering just an LSE or LSM analysis) and from the point of view of equivalent circuits. In fact, the approach also provides equivalent networks which make for easier interpretation of the phenomena taking place in the guide.

Table 6.1 Waveguides analysed in this Chapter.

Integrated optics waveguides	Microwave/millimetre–wave waveguides
Image Guide	Slotline
Dielectric Rib Waveguide	Inset Waveguide
Coupled Rib Waveguides	

In the following, TRD is applied to different structures of increasing analytical complexity. Although not explicitly considered, transmission lines useful in the high millimetre regions, such as those proposed in [28], can also be examined by TRD.

A synopsis of the waveguides considered in this Chapter is provided in Table 6.1. The first structure considered is the image line, an open waveguide with a finite, discrete, spectrum of bound modes complemented by a continuum. The latter part of the spectrum plays an important role in discontinuity and radiation problems arising in circuit components and antenna applications and is dealt with in the next Chapter. Here, we focus on the discrete part of the spectrum, the bound modes, of this two–dimensional nonseparable waveguide. In particular, although the problem is essentially hybrid in nature, we derive the guided modes under LSE/LSM assumption. This hypothesis greatly simplifies the mathematical details, thus making it easier to understand the method employed. The hybrid content, in fact, decreases very rapidly with the aspect ratio (h/a, see figure on page 252), so that the much simpler pure LSE/LSM approximation does apply very well down to the dielectric cut–off for all aspect ratios smaller than unity.

The same type of approximation is thereafter used for the analysis of the dielectric rib waveguide, which constitutes a key transmission medium in millimetrics and in integrated optics. As such, it has been the object of extensive investigations and several texts are now available on this subject [35, 31]. Although various approximate analyses of the EDC type exist, these break down for many practical configurations. More complete TRD formulations involve mode matching with a partially discrete, partially continuous spectrum. In addition, the bound modes of the rib guide can also be obtained by finite differences, finite elements, or other similar numerical approaches. Unfortunately, the latter are numerically expensive while providing only a modest amount of understanding of the field behaviour. It seems fair to say that the above numerical methods, while useful in order to get numerical values for the propagation constants and to provide some field visualisation, are of modest use with a view to developing physical understanding of complex electromagnetic phenomena. Nonetheless, their importance for the numerical investigation of fairly irregular structures is beyond discussion.

In Section 6.2 we focus on the pure LSE/LSM cases, this time deriving a highly accurate TRD variational solution of the problem, of first order (thus a scalar dispersion equation). To this end, we assume at the transverse step discontinuity a single function "trial field" which incorporates the physical properties of the solution. This is, in fact, the surface wave mode of a slab waveguide of height intermediate between that of the rib and that of the cladding slab, also including dielectric edge singularities in the LSM case. The height of the "intermediate guide" is obtained by optimising the overlapping integral with the slab mode in the rib and in the cladding, as already done in the previous Chapter. This criterion turns out to be equivalent to choosing an intermediate guide whose EDC is the geometric mean of those of the rib and cladding. In the case of the rib waveguide numerical results are also shown and compared with those obtainable by other methods.

The general problem of multiple coupled rib waveguides, where energy may be leaked from one guide to the other via the substrate and radiation mode, is of great practical and theoretical importance. This problem is dealt with in Section 6.3, where rigorous results, including substrate and air mode coupling, are provided for the general case of coupling of two or more different guides. The analysis is developed in terms of a cascade of transverse steps, utilising a variational solution with a single trial function and making explicit use of edge singularities at the dielectric corners in order to produce an effective and rigorous solution. Multiple coupled rib guides are then reduced to a cascade of interacting step discontinuities in the transverse direction. The numerical results are compared with those obtained by FEM/FDM showing excellent agreement.

We consider next, in Section 6.4, the hybrid propagation in slotline at microwave and millimetre–wave frequencies. In this case a full–wave solution is described for infinitely thin conductors. The analysis is carried out rigorously in the space domain involving the variational solution of an integral equation for an E field tangential to the slot. Moreover, it highlights the onset of leakage into a sufficiently thick substrate due to the excitation of a TM surface wave. This sets a high–frequency limit for lossless operation.

A similar analysis may be accomplished for the coplanar waveguide (CPW) and for the microstrip line. However, since the method remains the same and since these structures have been thoroughly analyzed in several texts (e.g. in [16]), it seems unnecessary to derive in detail the bound modes of these transmission lines.

The case of thick conductors is considered for the inset dielectric guide (IDG), which is an easy to fabricate alternative to image line with reduced radiation losses. This structure is investigated in Section 6.5, where a rigorous variational analysis in the space domain, based on TRD, is presented. Also in this case the theory follows the one presented for slotline, although in the discretization process, the presence of $r^{-1/3}$ singularities at the rectangular metallic corners, is accounted for.

Another interesting transmission line, not considered here for reasons of brevity, is the microstrip loaded inset dielectric waveguide, that was studied in [39]. In this case one has to account for the simultaneous presence of an $r^{-1/2}$ singularity near the microstrip edge (assumed as infinitely thin) and of an $r^{-1/3}$ singularity at the 90° metallic edge.

6.1 Bound modes of image line

In spite of practical difficulties in fabrication, image line is an open waveguide of considerable theoretical interest, not only on its own merits, but also as a basic example of an open waveguide of two–dimensional, nonseparable, cross–section [23]. Not surprisingly, the bound modes of image line received considerable attention, and were the object of numerical [12] and analytical [27] treatment, or by enclosing the guide in walls "far removed" [30], or by retaining the open waveguide configuration [25]. It is fair to say that accurate results for the bound modes are

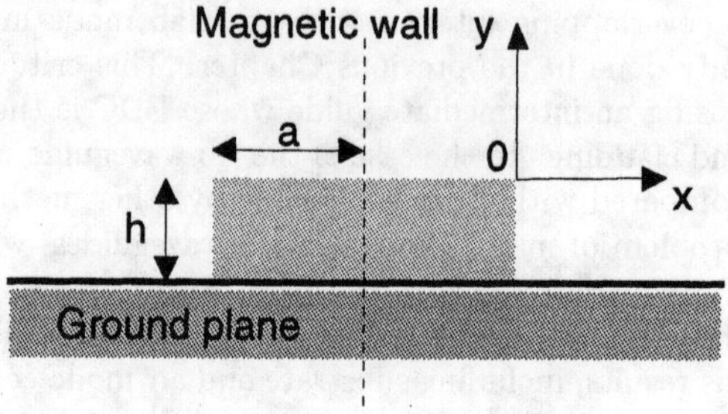

Figure 6.1 Image line cross-section (even E_y modes).

obtained by all of the above methods.

The bound modes of image line have already been found, by using the EDC method, in Chapter 4. However, on that occasion, we made some approximations that limit the accuracy as well as the range of applicability of the analysis. In the following, we present a more accurate way of finding the bound modes of image line. Note that the problem of finding the modes is essentially complicated by the nonseparable nature of the field in this structure, caused by the presence of the 90° diffraction edges (the dielectric corners in air), near which the transverse field becomes singular. It is the very presence of these corners that causes the discrete modes to be essentially hybrid. The hybrid content, however, decreases very rapidly with aspect ratio (h/a of Figure 6.1) so that the much simpler pure LSM/LSE approximation does apply very well down to and close to the dielectric cut–off, for all aspect ratios smaller than unity, provided that edge conditions are still accounted for in the field distribution.

In this Section, the analysis is developed for the even LSM (TM^y) polarisation, having $H_y = 0$ and E_y as the main electric field component, whereas the LSE analysis proceeds along dual lines. Use is made of the symmetry about the plane $x = -a$, by considering just half of the structure ($-a \leq x$) with magnetic wall boundary conditions on the plane of symmetry. As readily noted, the limiting case of the image guide for small aspect ratios is a dielectric slab over a ground plane. Apart from a change of coordinates, the potential is the same already seen in Chapter 3 and repeated in Chapter 5. For ease of reference, this potential is rewritten in the following as

$$\varphi_s(y) = \left\{ \begin{array}{lcl} A_s \frac{\cos k_s(y+h)}{\cos k_s h} & : & y \leq 0 \\ A_s e^{-\gamma_s y} & : & y \geq 0 \end{array} \right\} \tag{6.1}$$

with

$$k_s \tan k_s h = \varepsilon_r \gamma_s \tag{6.2}$$

and normalization constant

$$A_s = \sqrt{\frac{2\varepsilon_r}{h + \frac{\varepsilon_r}{\gamma_s}\left[1 + \frac{\gamma_s^2}{k_s^2}(1 + \varepsilon_r \gamma_s h)\right]}} \tag{6.3}$$

The continuum is described by means of the following potential

$$\varphi(y;k_y) = \begin{Bmatrix} \sqrt{\frac{2}{\pi}}\frac{\cos\alpha}{\cos kh}\cos k(y+h) & y \le 0 \\ \sqrt{\frac{2}{\pi}}\cos\left[k_y y + \alpha\right] & y \ge 0 \end{Bmatrix} \tag{6.4}$$

where the phase shift α is expressed as

$$\tan\alpha = \frac{k}{k_y \varepsilon_r}\tan kh \tag{6.5}$$

The above expressions are obtained from those reported in Chapter 5 by setting $n_2 = 1$, $\varepsilon_1 = \varepsilon_r$ and

$$\psi \to \varphi_s, \kappa \to k_s, \gamma_x \to \gamma_s, q \to k, k_x \to k_y, A \to A_s/\cos k_s h \tag{6.6}$$

This change in notation is expedient since we are now looking for the spectrum of image line, which is no longer uniform in the x–direction (see Figure 6.1). As a result, the field obtained from a y–directed electric Hertzian potential $\psi_e = \psi(x,y)e^{-j\beta z}$, given by

$$\begin{aligned} E_x &= \partial_x \partial_y \psi_e \\ E_y &= \left(\varepsilon_r k_0^2 + \partial_y^2\right)\psi_e \\ E_z &= -j\beta\partial_y\psi_e \\ H_x &= -\omega\varepsilon\psi_e \\ H_y &= 0 \\ H_z &= j\omega\varepsilon\partial_x\psi_e \end{aligned} \tag{6.7}$$

is a five–component field and not a three–component one as in the 2D case. The plane $x = 0$ (see Figure 6.1) is the discontinuous interface. At the right of this interface ($x > 0$) we have an empty half–space, while for $-a < x < 0$ we see a slab waveguide closed on an open circuit. In fact, in view of the symmetry of the image line, we have placed a magnetic wall in the middle ($x = -a$) of the guide. As discussed in the previous Chapter, it is advantageous to keep the transverse electric field singularity at the corner as the unknown and, to this end, we proceed with an admittance formulation in terms of the pair E_y, H_z. The field E_y of a bound mode in the region $x \ge 0$ is describable by using a spectral expansion in terms of the y–dependent free-space (air) eigenfunctions as

$$E_y(x,y) = \int_0^\infty V_a(k_y)\sqrt{\frac{2}{\pi}}\cos k_y(y+h)e^{-|k_x|x}dk_y \equiv \psi_a(x,y) \tag{6.8}$$

with

$$k_x^2 = k_0^2 - \beta_s^2 - k_y^2 \equiv k_t^2 - k_y^2 < 0 \tag{6.9}$$

as for bound modes $k_t^2 < 0$. In the above equation, β_s is the propagation constant, as yet unknown, of the bound mode. Due to modal orthogonality the amplitudes V_a are obtained from (6.8) as

$$V_a(k_y) = \int_{-h}^\infty E_y(x,y')\sqrt{\frac{2}{\pi}}\cos k_y(y'+h)dy' \tag{6.10}$$

It is now straightforward, by using (6.8) and the above, to calculate H_z as a function of E_y in the air half space $x > 0$; we have

$$H_z(0,y) = -\frac{2}{\pi}\int_0^\infty \frac{j\omega\varepsilon k_x}{k_0^2 - k_y^2}\int_{-h}^\infty E_y(y')\cos k_y(y'+h)dy'\cos k_y(y+h)dk_y \quad (6.11)$$

which, by introducing the following Green's function of the air space $x > 0$:

$$G_a(y,y') = \frac{2}{\pi}\int_0^\infty \frac{j\omega\varepsilon k_x}{k_0^2 - k_y^2}\cos k_y(y'+h)\cos k_y(y+h)dk_y \quad (6.12)$$

is written as

$$H_z(x=0,y) = -\int_{-h}^\infty G_a(y,y')E_y(y')dy' \quad (6.13)$$

A similar procedure may also be followed to express the magnetic field H_z as a function of the electric field E_y in the region $x < 0$. However, this region is limited by a magnetic wall placed at $x = -a$, so that the field E_y is described as

$$E_y(x,y) = V_s\frac{\varphi_s(y)}{\varepsilon_r(y)}\chi_s(x) + \int_0^\infty V'(\rho)\frac{\varphi(y,\rho)}{\varepsilon_r(y)}\chi(x)d\rho \equiv \psi_s(x,y) \quad (6.14)$$

where $\varphi_s(y)$ is the normalised distribution of a TM surface wave of the slab, considered monomode, and $\varphi(y,\rho)$ is that of its continuum. In the above equation the presence of the magnetic wall at $x = -a$ is accounted for by the terms

$$\begin{aligned}\chi_s(x) &= \frac{\cos q_s(x+a)}{\cos q_s a}\\ \chi(x) &= \frac{\cos q(x+a)}{\cos qa}\end{aligned} \quad (6.15)$$

for the guided mode (surface wave) and for the continuum, respectively. As usual the propagation constants satisfy the following equations

$$\begin{aligned}q_s^2 = \varepsilon_r k_0^2 - \beta_s^2 - k_s^2 &= k_0^2 - \beta_s^2 + \gamma_s^2\\ q^2 &= k_0^2 - \beta_s^2 - \rho^2\end{aligned} \quad (6.16)$$

For ease of reference the above quantities are also illustrated in Figure 6.2, where the propagation constant for the slab guide, i.e. the guide present at $x < 0$, and the free-space guide (the one at $x > 0$) are represented. By proceeding as in the case of (6.13), it is possible to express the magnetic field H_z for $x < 0$ as

$$H_z(x=0,y) = \int_{-h}^\infty \left[G_s(y,y') + G'_s(y,y')\right]E_y(y')dy' \quad (6.17)$$

where we have introduced the following Green's functions

$$\begin{aligned}G_s(y,y') &= j\omega\varepsilon\frac{q_s\tan q_s a}{q_s^2+\beta_s^2}\varphi_s(y)\varphi_s(y')\\ G'_s(y,y') &= j\omega\varepsilon\int_0^\infty \frac{q\tan qa}{k_0^2-\rho^2}\varphi(y,\rho)\varphi(y',\rho)d\rho\end{aligned} \quad (6.18)$$

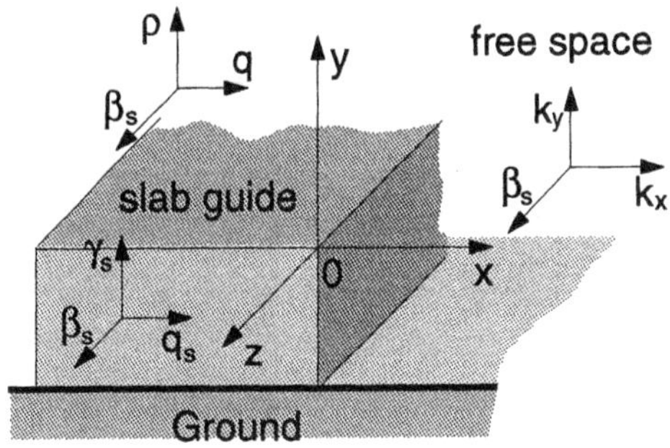

Figure 6.2 Wavenumbers in the air ($x > 0$) and slab ($x < 0$) waveguides.

The continuity of H_z at $x = 0$ thus provides the following equation

$$\int_{-h}^{\infty} [G_a + G_s + G'_s] E_y dy = 0 \tag{6.19}$$

The latter is solved by multiplying by E_y, integrating over y, and by dividing by $\langle E_y, \varphi_s \rangle^2$, where the bracket notation denotes integration over y. In this manner the following scalar dispersion equation is obtained,

$$Y_a + Y_s + Y'_s = 0 \tag{6.20}$$

which is illustrated by the equivalent circuit of Figure 6.3 and where

$$\begin{aligned} Y_a &= \frac{\langle E_y, G_a, E_y \rangle}{\langle E_y, \varphi_s \rangle^2} \\ Y_s &= j\omega\varepsilon \frac{q_s \tan q_s a}{q_s^2 + \beta_s^2} \\ Y'_s &= \frac{\langle E_y, G'_s, E_y \rangle}{\langle E_y, \varphi_s \rangle^2} \end{aligned} \tag{6.21}$$

Y_s is the driving point admittance seen by the slab surface wave, Y_a that of the air region ($x > 0$) and Y'_s that of the continuous spectrum of the slab. It is interesting to note that:

1. Y_a and Y'_s are purely imaginary, since even when q is purely imaginary we still have

$$q \tan qa = -|q| \tanh |q| a \tag{6.22}$$

2. The Green's functions G_a and G'_s contain principal value integrals. The poles of the tan function, arising from χ_s, have to be excluded. In fact, for a bound mode $\beta_s > k_0$, q is purely imaginary but $q \tan qa$ remains real. Simple poles seem also to occur at $k_y = k_0$ in (6.12) and at $\rho = k_0$ in (6.18). At these points, however, owing to (6.8), we have $V_a(k_0) = V'(k_0) = 0$, that is, their corresponding residues vanish.

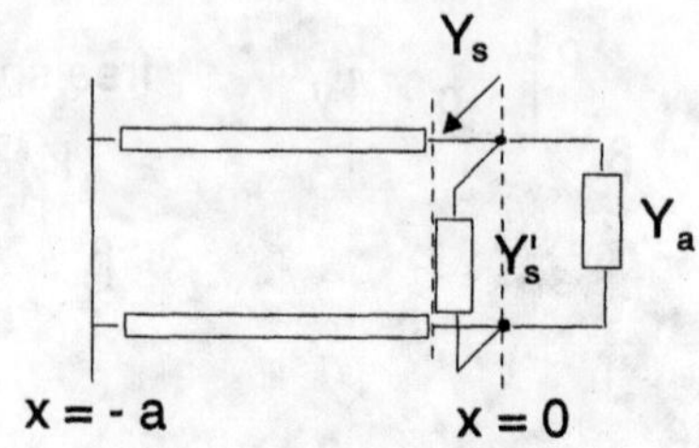

Figure 6.3 Equivalent circuit for transverse resonance equation. A magnetic wall is present at $x = -a$.

3. Y_a and Y_s, the admittances in (6.20), are functions of the unknown β_s. The latter is determined by solving the dispersion equation (6.20) for β_s.

Once β_s is found, the field components are determined from the potential according to (6.8). By following the above procedure fairly accurate numerical results are obtained as shown, for instance, in [25].

6.2 Bound modes of the rib waveguide

Rib waveguide is the most used guide in integrated optics, with applications also in the higher millimetric region. As such, it has been the object of several investigations in recent years. Accurate results for the bound modes of the single rib guide have been obtained by using purely numerical methods (finite differences, finite elements [32], etc.) which have the drawback of being very expensive in computer time. As a result, the effective dielectric constant (EDC), used in Chapter 4, has been the accepted approximation method to characterise rib waveguides in millimetrics and integrated optics [22]. The simplicity, ease of implementation and, unfortunately, the limits of EDC are now well recognised. To overcome these limits, transverse resonance and network methods were combined in studying dielectric waveguides [30, 29]. In particular, transverse resonance has been formulated for the rib guide in approximate form in [4], while the more rigorous hybrid form has been used in [24] where mode matching has been applied at the transverse step discontinuity ($x = 0$ in Figure 6.4). In this way the approach is similar to that used in the previous case for the image line. In both cases the problem is seen as one of diffraction by a transverse step discontinuity, where TRD is employed to derive the propagation constant.

As in longitudinal step discontinuity problems, direct mode matching should be applied with due care, in order to take into account the true diffraction field at the step and the presence of a continuum. The LSE (TE to y, with reference to Figure 6.4) case is somewhat easier since the field is regular at the transverse step discontinuity. Correspondingly, all rigorous methods tend to yield similar accurate values of the propagation constant. On the other hand, in the LSM (TM to y) case, the 90° and 270° dielectric corners play an important role in the field distribution at the plane of the transverse step. The knowledge of the field behaviour on the latter plane is the key to an efficient numerical solution.

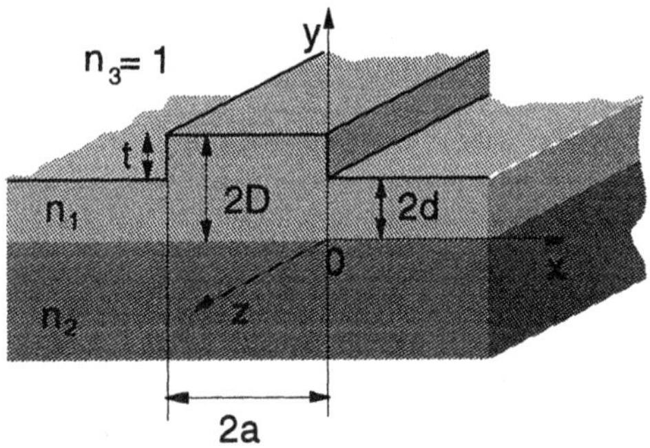

Figure 6.4 Geometry of the rib waveguide.

Actually, the above dielectric corners also cause the field to be essentially hybrid. Nonetheless, for typical aspect ratios and low refractive index differences, the hybrid content is minor, while the effect of the singularity of the field, i.e. its nonseparability, is a stronger feature. Moreover, restricting attention to the case of pure polarization introduces a considerable simplification in the analysis. In the next Section we will investigate the hybrid case, thus removing this simplification.

When solving the TRD problem, we will adopt a variational formulation of the Rayleigh–Ritz type; in particular, as stated previously in this Chapter, we will use a single trial field at the transverse step discontinuity, including all known physical features of the field. It is obvious that the success of this approach entirely relies on the prudent choice of the single function trial field. To this end, the experience acquired when studying the single step discontinuity in Chapter 5 becomes precious. For instance, as already observed in Chapter 5, using just the surface mode to the right or left of the step is inaccurate because of the orthogonality to its continuum.

It was observed that a very expedient choice is provided by the field of a slab of height intermediate between that of the rib and of the cladding layer, a "transition" function. Moreover, in the LSM case, it is also necessary to include explicitly the proper singularity of the electric field at the corners. As a consequence of this choice, it is possible to recover a scalar dispersion equation yielding results for the propagation constants and discontinuity fields which are at least as accurate as those obtained by purely numerical methods, but at a fraction of the required computer time. Moreover, the simple two–port transverse equivalent network, as derived from a single function trial field, provides considerable insight into the physics of the problem.

We will now investigate first the LSE (TE to y) and then the LSM (TM to y) cases. In both analyses we assume that we are looking for the bound modes with even parity (with respect to x), so that we can insert a magnetic wall in the middle of our structure, in a manner completely similar to what has already been done for the image guide.

6.2.1 LSE analysis

The analysis starts from a knowledge of the complete spectrum of an asymmetric slab waveguide in the y–direction. The latter was studied in Section 3.6 and includes the following contributions:

1. Surface waves, of orthonormalised distribution $\phi_s(y)$.
2. Air modes, of distribution $\phi(y, \rho)$, where ρ is a continuous wavenumber $0 \leq \rho < \infty$. Both the so-called even and odd waves need to be included in view of the lack of symmetry of the problem in the y-direction. Where necessary, these are distinguished by the notation ϕ_μ : μ= even, odd.
3. Substrate waves, $\phi_{sb}(y, \sigma), 0 \leq \sigma \leq \sqrt{\varepsilon_2 - 1}k_0$

The spectral expression of the y-directed Hertzian potential $\pi_h(x, y)$ at either side of the step comprises all the above contributions. In this case, the fields E_z and H_y constitute a pair of transverse fields at the discontinuous interface at $x = 0$. At each side of the step, these fields are derived from the potential as

$$\begin{aligned} E_z(x,y) &= j\omega\mu\ \partial_x \pi_h(x,y) \\ H_y(x,y) &= \left(\varepsilon_r k_0^2 + \partial_y^2\right) \pi_h(x,y) \end{aligned} \tag{6.23}$$

By introducing the spectral expression of the potential in the above expressions, it is possible to obtain, in a manner completely similar to that followed in the previous Section for image line, an integral relationship, relating the transverse fields to the left of $x = 0$, of the type

$$E_z(0,y) = \hat{Z}_L \cdot H_y(0,y) \tag{6.24}$$

where the kernel $\hat{Z}_L(y, y')$ of the integral operator is given by the impedance Green's function:

$$Z_L(y,y') = \frac{1}{j\omega\varepsilon_0}\left[\begin{array}{c} -\frac{q_s \tan q_s a}{\varepsilon_e}\phi_s(y)\phi_s(y')+ \\ \sum_\mu \int_0^\infty \frac{\gamma \tanh \gamma a}{1-\left(\frac{\rho}{k_0}\right)^2}\phi_\mu(y,\rho)\phi_\mu(y',\rho)d\rho+ \\ \int_0^\nu \frac{\eta \tanh \eta a}{\varepsilon_2-\left(\frac{\sigma}{k_0}\right)^2}\phi_{sb}(y,\sigma)\phi_{sb}(y',\sigma)d\sigma \end{array}\right] \tag{6.25}$$

In (6.25), the principal value of the last integral is taken. The wavenumber for the bound mode in x is given by

$$\begin{aligned} k_0^2 - \beta^2 - \rho^2 &= -\gamma^2 \\ \varepsilon_2 k_0^2 - \beta^2 - \sigma^2 &= -\eta^2 \\ q_s^2 + \beta^2 &= \varepsilon_e k_0^2 \end{aligned} \tag{6.26}$$

In (6.25) we have also assumed that just one surface wave (in y), $\phi_s(y)$, may be supported by the slab waveguide left of the step. A relationship analogous to (6.24) exists at the right-hand side of the step, namely

$$E_z(0,y) = -\hat{Z}_R \cdot H_y(0,y) \tag{6.27}$$

where the Green's impedance $\hat{Z}_R(y, y')$ has a form analogous to (6.25), after due modification for the semi-infinite region, i.e.

$$Z_R(y,y') = \frac{1}{j\omega\varepsilon_0}\left[\begin{array}{c} \frac{\gamma_s'}{\varepsilon_e}\psi_s(y)\psi_s(y')+ \\ \sum_\mu \int_0^\infty \frac{\gamma}{1-\left(\frac{\rho}{k_0}\right)^2}\psi_\mu(y,\rho)\psi_\mu(y',\rho)d\rho+ \\ \int_0^\nu \frac{\eta}{\varepsilon_2-\left(\frac{\sigma}{k_0}\right)^2}\psi_{sb}(y,\sigma)\psi_{sb}(y',\sigma)d\sigma \end{array}\right] \tag{6.28}$$

where $q'_s = -j\gamma'_s$ corresponds to q_s, ε'_e to ε_e, and ψ to ϕ in the region $x > 0$. With reference to Figure 6.4, it is noted that $\psi(y)$ has the same form as $\phi(y)$, where the appropriate wavenumbers and geometrical dimensions (d instead of D) are used.

Finally, by adding (6.24) and (6.27), we obtain the sought dispersion equation of the rib waveguide in the form of an integral equation:

$$\hat{Z} \cdot H_y(0, y) = 0 \tag{6.29}$$

where

$$\hat{Z} = \hat{Z}_L + \hat{Z}_R \tag{6.30}$$

Variational solution

In the course of a variational development, when choosing expanding functions for the unknown field $H(y) = H_y(0, y)$, it is very advantageous to be able to capture most of the solution with just one term, i.e. a first–order solution. The above discussion seems to suggest that a prudent choice for H may consist in taking the surface wave, u_s, of a slab waveguide with core height $\overline{d}$ intermediate between d and D ("transition function").

For small steps there is little difference between the spectra of the three waveguides; for larger steps it seems intuitively reasonable that an intermediate guide with EDC may provide a suitable "transition" between the two waveguides. More precisely, the transition function u_s overlaps the spectra of both slabs. We seek to optimise the overlap with ϕ_s and ψ_s, where the latter exists, as the surface waves carry most of the power. Consequently, we choose $\overline{d}$ so as to minimise the error function:

$$\varepsilon(\overline{d}) = 2 - (P_s^2 + {P'}_s^2) > 0 \tag{6.31}$$

As it turns out, $\overline{d}$ resulting from this condition is but little different from that corresponding to $\overline{\varepsilon}_e \approx \sqrt{\varepsilon_e {\varepsilon'}_e}$.

It is noted that for d/D sufficiently small, no guidance is present in the cladding slab and the above criterion breaks down. From this point onward we extrapolate for $\overline{d}$ as a function of d. The evaluation of the overlapping integrals P and P' between u_s and the spectra to either side of the step is facilitated by the following result (see page 202), for calculating the overlapping of two solutions of the wave equation:

$$\begin{aligned} P_s = \int_{-\infty}^{\infty} u_s \phi_s dy &= \frac{\varepsilon_r - 1}{\overline{\varepsilon}_e - \varepsilon_e} \int_{-\overline{d}}^{0} u_s \phi_s dy \\ {P'}_s = \int_{-\infty}^{\infty} u_s \psi_s dy &= \frac{\varepsilon_r - 1}{{\varepsilon'}_e - \overline{\varepsilon}_e} \int_{-d}^{\overline{d}} u_s \psi_s dy \end{aligned} \tag{6.32}$$

and similarly for the remaining overlapping integrals $P(\rho)$ and $P_{sb}(\sigma)$ with air and substrate modes respectively.

Let us assume the solution $H(y)$ of (6.29) to be known and multiply this equation on the left by $H(y)$, integrate over y, and divide across by

$$\langle \phi_s, H \rangle^2 = \left[\int_{-\infty}^{\infty} \phi_s(y) H(y) dy \right]^2 \tag{6.33}$$

Isolate then from Z_L, as given in (6.25), and from Z_R, given in (6.28), the contributions of the surface waves at each side. The following scalar dispersion equation results:

$$\frac{q_s}{\varepsilon_e}\tan q_s a = z_a + z_{sb} + n^2\frac{\gamma'_s}{\varepsilon'_e} \tag{6.34}$$

where

$$\begin{aligned} z_a &= \sum_{\mu=e,o}\int_0^\infty \frac{\gamma\tanh\gamma a P_\mu^2(\rho) + \gamma P'^2_\mu(\rho)}{1-\left(\frac{\rho}{k_0}\right)^2}d\rho \\ z_{sb} &= \frac{1}{\varepsilon_2}\int_0^\nu \frac{\eta\tanh\eta a P_\mu^2(\sigma) + \gamma P'^2_\mu(\sigma)}{1-\left(\frac{\sigma}{\varepsilon_2 k_0}\right)^2}d\sigma \end{aligned} \tag{6.35}$$

Its transverse equivalent network interpretation is illustrated in Figure 6.5. In (6.35) and in this figure one recognises four contributions:

1. The impedance of the surface wave under the rib, represented by an open–circuited transmission line of electrical length $q_s a$ and characteristic impedance $q_s/(j\omega\varepsilon_0\varepsilon_e)$.

2. The impedance of the air modes at both sides of the junction as seen by the above surface wave, given by z_a.

3. The contribution of the substrate modes at both sides of the junction as seen by the surface wave under the rib, given by z_{sb}.

4. The contribution of the surface wave mode (if any), to the right of the junction, i.e. a load of impedance $(1/j\omega\varepsilon_0)\gamma'_s/\varepsilon'_e$ with real γ'_s for a bound mode in the x–direction. This load is seen by the surface wave mode under the rib through an ideal transformer

$$n = P'_s/P_s \tag{6.36}$$

The representation given by (6.34) is exact in principle and enjoys variational properties. Although its accuracy depends on an actual knowledge of H, it is noted that the latter, being transverse to the dielectric wedge along the z axis, is nonsingular. Hence, any reasonable assumption will lead to relatively good results.

If, at this point, the small–step approximation is made:

$$\phi_s \approx \psi_s \approx H \tag{6.37}$$

then, by orthogonality, the contributions of air and substrate modes vanish; i.e. $z_a = z_{sb} = 0$ in (6.34) and we are reduced to the EDC calculation. It is worth remarking that even when the step is not small, so that ϕ_s and ψ_s may differ from each other and from H, nonetheless P_s, and P'_s of the ideal transformer are both less than unity and their ratio is still fairly close to unity. This fact explains the remarkable accuracy of the EDC calculation in the LSE case not too close to cut–off, where the continuous spectra can no longer be overlooked. The above observation is at the basis of a very accurate first–order variational solution of the step problem.

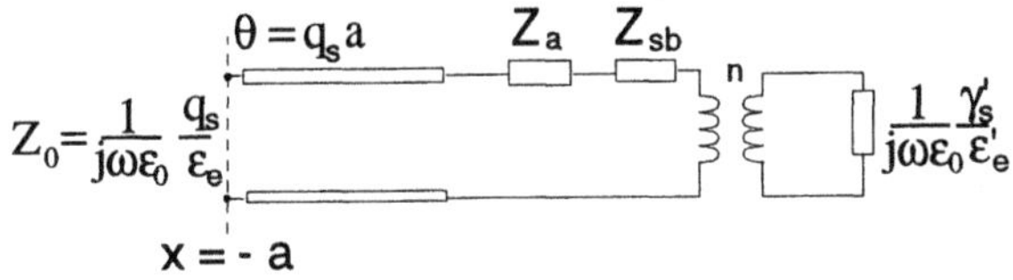

Figure 6.5 Equivalent network for the LSE case.

Table 6.2 Parameters of the semiconductor rib waveguide structures which were analysed.

Rib guide	n_1	n_2	n_3	$D(\mu m)$	$d(\mu m)$	$2s(\mu m)$	$\lambda(\mu m)$
1	3.44	3.35	1.0	1.0	varying	3.0	1.15
2	3.44	3.40	1.0	1.0	varying	3.0	1.15
3	3.44	3.34	1.0	1.3	0.2	2.0	1.55
4	3.44	3.36	1.0	1.0	0.9	3.0	1.55
5	3.44	3.435	1.0	6.0	3.5	4.0	1.55
6	3.44	3.415	1.0	1.5	1.0	3.0	1.15

Numerical results for the LSE case

The theory developed in the previous Section was applied to the investigation of the propagation constants of a rib waveguide whose parameters are illustrated in Table 6.2. We report in Table 6.3 the results of modal indices for structure 1 where the thickness of the rib (D) is kept constant while that of the cladding layer d is varied from 0.1 to 0.9 μm. Results obtained by the standard EDC are given in column 2 and compared with those obtained by the above analysis, column 3, but neglecting the effect of the continuum at either side. It is noted that the EDC approximation, as well as that of neglecting the continuum, breaks down for $d < 0.3\,\mu$m. Column 4 reports present results valid for any value of d. These are compared with column 6, the values obtained in [1] by finite element analysis on an electrical field basis, for varying values of d.

Results on three additional guiding structures (structures 3, 4, 5), given in Table 6.4, are found to be in excellent agreement with those of [42] and [33], obtained using either the finite difference or the finite element method.

6.2.2 LSM analysis

The even LSM modes (TM to y) of the rib waveguide are obtained by placing a magnetic wall on the plane of symmetry $x = -a$, by means of the Hertzian electric potential $\pi_e(x, y)$. The pair of fields, transverse to x, to be used in the analysis are E_y and H_z, which are derived from the potential as

$$\begin{aligned} E_y(x,y) &= \left(\varepsilon_r k_0^2 + \partial_y^2\right)\pi_e(x,y) \\ H_z(x,y) &= j\omega\varepsilon\partial_x\pi_e(x,y) \end{aligned} \tag{6.38}$$

Table 6.3 Propagation constant, β/k_0, for structure 1 as derived by EDC method, transverse resonance diffraction (TRD) with and without continuum, and finite element method (FEM).

$d(\mu m)$	EDC	TRD without continuum	TRD with continuum	Finite Element [1] (electric field)
0.1			3.40633	3.40693
0.2			3.40645	3.40708
0.3			3.40671	3.40725
0.4	3.4078903	3.40773	3.40697	3.40746
0.5	3.4080994	3.40807	3.40727	3.40770
0.6	3.4083595	3.40835	3.40768	3.40808
0.7	3.4086952	3.40869	3.40816	3.40856
0.8	3.4091444	3.40914	3.40881	3.40914
0.9	3.4098396	3.40984	3.40972	3.40971

Table 6.4 Values of β/k_0 as obtained from EDC, transverse resonance diffraction (TRD) and finite difference (FD) methods.

Rib guide	EDC	TRD	FD(1) [33]	FD(2) [33]	FD [42] (electric field)
3	3.39032237	3.3886900	3.3906177	3.3912917	3.3869266
4	3.39547825	3.3953275	3.3951666	3.3954298	3.3954405
5	3.43732738	3.4368203	3.4368425	3.4368635	3.4368112

By a development analogous to that of the LSE case, we arrive at an integral equation for the unknown E_y field at the interface $x = 0$:

$$\hat{Y} \cdot E_y(0, y) \quad = \quad 0 \tag{6.39}$$

where

$$\hat{Y} \quad = \quad \hat{Y}_L + \hat{Y}_R \tag{6.40}$$

and

$$\begin{aligned}
Y_L(y, y') &= \frac{1}{j\omega\mu_0}\left[\begin{array}{c} -\frac{q_s \tan q_s a}{\varepsilon_e}\phi_s(y)\phi_s(y')+ \\ \sum_\mu \int_0^\infty \frac{\gamma \tanh \gamma a}{1-\left(\frac{\rho}{k_0}\right)^2}\phi_\mu(y, \rho)\phi_\mu(y', \rho)d\rho+ \\ \int_0^\nu \frac{\eta \tanh \eta a}{\varepsilon_2-\left(\frac{\sigma}{k_0}\right)^2}\phi_{sb}(y, \sigma)\phi_{sb}(y', \sigma)d\sigma \end{array}\right] \\
Y_R(y, y') &= \frac{1}{j\omega\mu_0}\left[\begin{array}{c} \frac{\gamma_s}{\varepsilon_e}\psi_s(y)\psi_s(y')+ \\ \sum_\mu \int_0^\infty \frac{\gamma}{1-\left(\frac{\rho}{k_0}\right)^2}\psi_\mu(y, \rho)\psi_\mu(y', \rho)d\rho+ \\ \int_0^\nu \frac{\eta}{\varepsilon_2-\left(\frac{\sigma}{k_0}\right)^2}\psi_{sb}(y, \sigma)\psi_{sb}(y', \sigma)d\sigma \end{array}\right]
\end{aligned} \tag{6.41}$$

Again, by a development similar to that followed for the LSE case in the previous paragraph, one derives the scalar dispersion equation

$$\frac{q_s}{\varepsilon_e}\tan q_s a \quad = \quad y_a + y_{sb} + n^2\frac{\gamma'_s}{\varepsilon'_e} \tag{6.42}$$

corresponding to the transverse equivalent circuits of Figure 6.6, which is analogous to Figure 6.5. In the above equation, y_a, y_{sb}, and n_2 are defined analogously to (6.35) and (6.36), respectively, in terms now of the unknown transverse electric field at the interface $E_y(0,y) = E(y)$ and of the wavenumbers appropriate to the TM case. It is noted that, in contrast to the LSE case, the electric field transverse to a dielectric wedge is singular. Hence, $y = d$ and $y = D$ are singular points and an appropriate choice of E has more bearing on the accuracy of the solution.

The second feature of (6.42) is, in fact, its close resemblance to (6.34), which is a dispersion equation for a TM line in x. According to the standard EDC procedure, however, the transmission line representing propagation in the x–direction for an LSM mode ought to be of the pure TE type. This apparent incongruity is explained by noting that in the EDC approximation one uses a separable potential in each region. Then, one has

$$Y^{TE} = \frac{H_z}{E_y} \quad = \quad \frac{\omega\varepsilon_0\varepsilon_r k_x}{\varepsilon_r k_0^2 - k_y^2} = \frac{\varepsilon_r k_x}{\omega\mu_0\varepsilon_e} \tag{6.43}$$

The characteristic admittance of a TE wave is recovered from the above equation only if $\varepsilon_r = \varepsilon_e$. It is, therefore, fair to say that while in the LSE case the EDC is the most basic level of approximation to the rigorous dispersion equation as obtained from transverse resonance diffraction, in the LSM case a stronger approximation is effected. It is consequently to be expected that the EDC approximation in the LSM case will be further removed from the accurate solution than in the LSE case.

The field's singularity at the dielectric corners is accounted for by the following choice of trial field

$$E_y \quad = \quad \begin{cases} \overline{A}e^{-\gamma_0(y-\overline{d})}(y-D)^\alpha & : \quad y > D \\ \frac{\overline{A}}{\varepsilon_1}e^{-\gamma_0(y-\overline{d})}(D-y)^\alpha & : \quad D > y \geq \overline{d} \\ \frac{\overline{A}}{\varepsilon_1}\frac{\cos(\overline{k}_y y-\overline{\psi})}{\cos(\overline{k}_y\overline{d}-\overline{\psi})}(y-d)^\alpha\left(\frac{D-\overline{d}}{\overline{d}-d}\right)^\alpha & : \quad \overline{d} \geq y > d \\ \frac{\overline{A}}{\varepsilon_1}\frac{\cos(\overline{k}_y y-\overline{\psi})}{\cos(\overline{k}_y\overline{d}-\overline{\psi})}(d-y)^\alpha\left(\frac{D-\overline{d}}{\overline{d}-d}\right)^\alpha & : \quad d > y \geq 0 \\ \frac{\overline{A}}{\varepsilon_2}\frac{\cos\overline{\psi}}{\cos(\overline{k}_y\overline{d}-\overline{\psi})}e^{\overline{\gamma}_2 y}d^\alpha\left(\frac{D-\overline{d}}{\overline{d}-d}\right)^\alpha & : \quad y < 0 \end{cases} \tag{6.44}$$

where

$$\begin{aligned} \tan\left(\overline{k}_y\overline{d}-\overline{\psi}\right) &= \varepsilon_1\frac{\overline{\gamma}_0}{\overline{k}_y} \\ \tan\overline{\psi} &= \frac{\varepsilon_1}{\varepsilon_2}\frac{\overline{\gamma}_2}{\overline{k}_y} \end{aligned} \tag{6.45}$$

We note, in fact, that the above form (6.44) takes into account each of the two singularities at $y = d, D$ and includes the dielectric constant step at $y = 0$. Moreover, it is physically clear that no other discontinuity occurs between the two

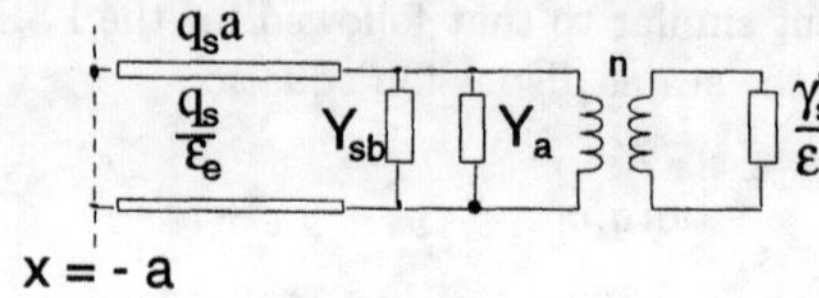

Figure 6.6 Equivalent network for the LSM case.

Table 6.5 Comparison of β/k_0 for the LSM polarisation for structure 2.

$d(\mu m)$	TRD with singularity	TRD without singularity	Finite Difference [42]	EDC
0.1	3.4102127	3.410083	3.41060	
0.2	3.4102402	3.410091	3.41073	
0.3	3.4102597	3.410119	3.41092	
0.4	3.4107675	3.410268	3.41117	
0.5	3.4112158	3.410925	3.41150	
0.6	3.4118191	3.411380	3.41190	3.412707
0.7	3.4121408	3.411476	3.41241	3.412928
0.8	3.4128545	3.412496	3.41303	3.413308
0.9	3.4137921	3.413784	3.41385	3.413937

singularities at $y = d, D$. Hence, we assume a continuous function in this interval. For $y > D$, we include the effect of the singularity at $y = D$, whereas the dielectric constant is that of the "air region" above the rib.

Moreover, the above trial field also allows evaluation of the scalar products such as P_s analytically in terms of gamma functions and the scalar products are numerically calculated in rapidly converging series form. For large values of the argument, in fact, the series reduces to a single term. This analyticity of the scalar products results in high numerical efficiency.

Numerical results for the LSM case

As previously noted, the LSM case is complicated by the presence of the field singularity. The accuracy of the above approach, based on the transition function, is ascertained by comparing the results with those of [42] in Table 6.5 and with EDC results in Table 6.6. It is interesting to observe that the modal indices, as computed by taking into account the field singularities, are always greater than those computed without considering singularities. Moreover, when singularities are accounted for, the results are closer to the finite difference analysis of [42]. This is a confirmation of the importance of explicitly considering the field singularities. Further numerical results on the approach outlined in this Section can be found in [36].

Table 6.6 Values of β/k_0 for the LSM polarisation for structure 6.

$d(\mu m)$	TRD with singularity	TRD without singularity	EDC
0.1	3.423084	3.422931	
0.2	3.423113	3.422933	
0.3	3.423332	3.422936	
0.4	3.423447	3.422960	
0.5	3.423517	3.423047	
0.6	3.423759	3.423123	
0.7	3.423955	3.423163	
0.8	3.424381	3.423188	3.425677
0.9	3.424406	3.423212	3.425839
1.0	3.424751	3.423406	3.426036
1.1	3.425256	3.423901	3.426278
1.2	3.425725	3.425062	3.426573
1.3	3.426515	3.426468	3.426946
1.4	3.427225	3.427199	3.427455

6.3 Multiple coupled rib waveguides

Coupled structures play a very important role in the field of millimetre waves and integrated optics. Arrays of coupled dielectric antennas are interesting for millimetric applications, while laser arrays may be realised using coupled rib guides. The feasibility of fabricating such devices, using coupled guides, has been demonstrated and the near field and propagation characteristics of such arrays are already of great practical importance in integrated optics. As a consequence, accurate theoretical analyses of coupled structures are essential to the design of devices that meet specific requirements and criteria.

Coupled rib guides in a parallel plate configuration were studied in [3] by transverse mode matching. Until recently there were no truly exact analytical methods available for the analysis of multiple coupled rib waveguides in the open configuration. Approximate methods mostly relied on either the EDC approximation [22, 26, 43], or coupled mode theory which is appropriate for weakly coupled systems [45]. The approach in [45] assumes the individual guided modes to be orthogonal in the coupled structure. An improved coupled mode theory [9, 11] which removes the assumption is also available. However, the EDC and the coupled mode theory both fail to model the effects arising from leakage of energy via the substrate and the air regions, particularly in the millimetric case. This effect may also be significant at optical frequencies, especially if the rib height is large, where the interaction effects via the continuum cannot be neglected. For more accurate results, sophisticated numerical techniques such as the finite element method or finite difference method in the frequency and in the time–domain are used [33]. As in previous cases, the most important drawback of this approach is the high computational effort needed for accurate results, as the area and hence the number of mesh points is doubled

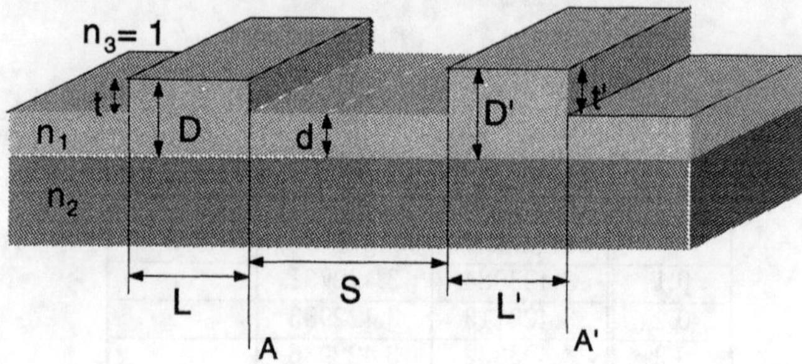

Figure 6.7 Geometry of the coupled rib waveguides.

with respect to that needed to analyse a single guide. Moreover, FEM/FDM becomes extremely cumbersome for situations more complex than that of two identical closely coupled guides, as shown for instance in Figure 6.7: this is the typical problem we are addressing in this Section. We consider pure LSE and LSM polarizations but extension to the hybrid case is straightforward, though more cumbersome. The approach here adopted is consistent with the one used in the previous paragraph for the analysis of the single rib guide. Application follows to the case of the directional coupler, three guide coupler, closely spaced unequal guide and waveguide arrays. Moreover, results of the present method are compared to other published results to get a feeling for its accuracy. It is shown that this approach yields accurate results with a computational effort only slightly greater than that needed to analyse a single guide as done in the previous Section.

6.3.1 Theoretical approach

The cross sectional geometry of two coupled rib waveguides is shown in Figure 6.7. In the framework of TRD, this configuration is seen as that of four cascaded step discontinuities in the transverse direction. We make use of a variational solution of the step problem that employs a single test function, the "transition function" developed in the previous Chapter in order to expand the unknown interface field. Using a single test function at the interface yields a simple two port equivalent network representation of the coupled structure such as shown in Figure 6.8. Consequently, the analysis of the transverse cascade reduces to solving the two port network, which is easily effected by standard network technique. In order to derive the variational expression for the parameters of the two port network shown in Figure 6.8, let us consider the cross–sectional geometry of Figure 6.9 which represents the basic building component of the coupled structure in Figure 6.7. In this case, the analysis is complicated by the presence of the two step discontinuities at $x = 0$ and $x = L$ (see Figure 6.9). Considering the transverse propagation of LSE and LSM modes in such a structure, at each step discontinuity, modes of the continuous spectrum are excited and multiple reflections of these modes occur between the two steps. If, when applying standard transverse resonance to this situation, interaction via the continuous modes is ignored, i.e. a small step approximation is assumed, then the uniform region and the step discontinuity may be represented by a transmission line and ideal transformer, respectively. Interaction via the continuum between the steps at $x = 0$ and $x = L$ of Figure 6.9, however, is not representable by means of

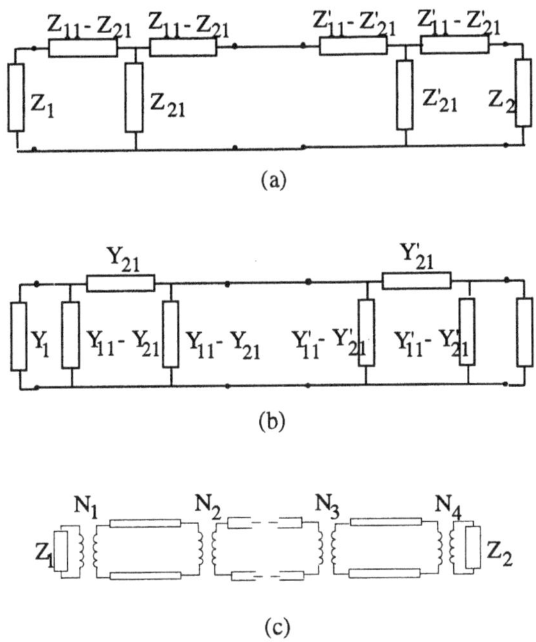

Figure 6.8 Equivalent network representation of the coupled structure for the LSE case.

a finite number of transmission lines coupled at the discontinuities as in a closed waveguide.

Hence, instead of the transmission line analogy we adopt a two–port, "T" representation of each region between two successive steps. Its open–circuit impedance, short–circuit admittance parameters are obtained by placing a magnetic/electric wall at $x = 0, L$ in turn. Under "open circuit" and "short circuit" conditions at $x = 0, L$ integral operators are found relating the total $\mathbf{E}$ and $\mathbf{H}$ fields at various ports and then these are used to relate the total fields at each port to one another. This operator method has already been used in Chapter 5 for analyzing cascaded step discontinuities in slab waveguides. As already done in that Chapter, as well as in the previous analysis of the single rib waveguide, the interface field is approximated by a single "transition function", also inclusive of the edge conditions in the LSM case. As a consequence, the numerical complexity of the problem is substantially reduced, avoiding the use of matrices altogether. Furthermore, this formulation leads to an equivalent network model of the structure represented by a cascade of "T" or "π" two port networks as shown in Figure 6.8a, b.

6.3.2 LSE analysis

Consider just the situation where slabs in regions 1 and 3 of Figure 6.9 are semi–infinite. The relationship between the transverse electric field E (z–directed) and

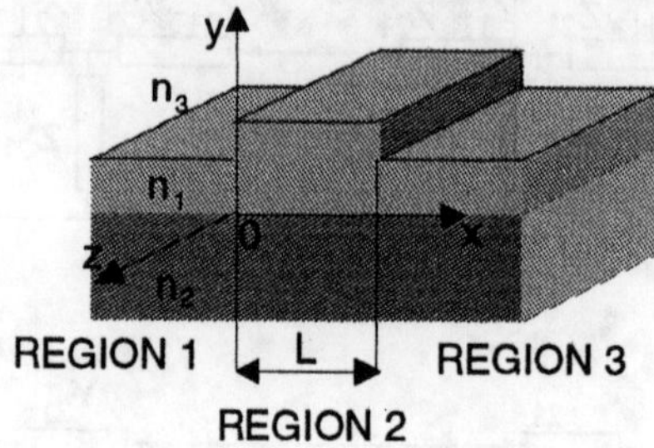

Figure 6.9 Unit cell of the coupled structure showing double step discontinuity in the transverse x–direction.

the transverse magnetic field H (y–directed) at $x = 0^-$ is given by

$$E(0,y) = -\int_{-\infty}^{\infty} Z_1(y,y')\,[-H(0,y')]\,dy' \equiv -\hat{Z}_1 \cdot [-H(0,y')] \qquad (6.46)$$

where

$$\begin{aligned} Z_1(y,y') &= -z_{01}\psi_1(y)\psi_1(y') - \sum_{\mu=e,o}\int_0^{\infty} z_{01}(\rho)\psi_{\mu 1}(y,\rho)\psi_{\mu 1}(y',\rho)d\rho \\ &\quad - \int_0^{\nu} z_{01}(\sigma)\psi_1(y,\sigma)\psi_1(y',\sigma)d\sigma \end{aligned} \qquad (6.47)$$

$Z_1(y,y')$ is the Green's impedance function of the semi infinite slab 1. $\psi_1(y)$ is the modal field of the surface wave with characteristic impedance z_{01}; $\psi_{e1}(y,\rho)$ and $\psi_{o1}(y,\rho)$ represent the even and odd air modes with characteristic impedance $z_{01}(\rho)$, and $\psi_1(y,\sigma)$ the substrate mode with characteristic impedance $z_{01}(\sigma)$, $0 \le \sigma \le \nu$.

In the above equations, we have explicitly assumed that just one surface wave (in y) may be supported by the slab waveguides, and the various characteristic impedances are given by

$$\begin{aligned} z_{0i} &= \frac{j\omega\mu_0\gamma_{si}}{\beta^2-\gamma_{si}^2} \quad i=1,2,3 \\ z_0(\rho) &= \frac{j\omega\mu_0\gamma}{k_0^2-\rho^2} \\ z_0(\sigma) &= \frac{j\omega\mu_0\eta}{\varepsilon_2 k_0^2-\sigma^2} \end{aligned} \qquad (6.48)$$

with

$$\begin{aligned} \varepsilon_{ei}k_0^2 - \beta^2 &= -\gamma_{si}^2 \quad i=1,2,3 \\ k_0^2 - \beta^2 - \rho^2 &= -\gamma^2 \\ \varepsilon_2 k_0^2 - \beta^2 - \sigma^2 &= -\eta^2 \qquad (6.49) \\ \nu &= k_0\sqrt{\varepsilon_2 - \varepsilon_3} \qquad (6.50) \end{aligned}$$

where $\varepsilon_{e1}, \varepsilon_{e2}, \varepsilon_{e3}$ are the effective dielectric constants (TE polarization) of the slabs 1, 2, and 3, respectively.

Similarly at $x = L^+$, we have

$$E(L,y) = -\int_{-\infty}^{\infty} Z_3(y,y')\,[-H(L,y')]\,dy' \equiv \hat{Z}_3 \cdot [-H(L,y')] \tag{6.51}$$

where

$$\begin{aligned} Z_3(y,y') = z_{03}\psi_3(y)\psi_3(y') + \sum_{\mu=e,o} \int_0^{\infty} z_{03}(\rho)\psi_{\mu 3}(y,\rho)\psi_{\mu 3}(y',\rho)d\rho \\ + \int_0^{\nu} z_{03}(\sigma)\psi_3(y,\sigma)\psi_3(y',\sigma)d\sigma \end{aligned} \tag{6.52}$$

$Z_3(y,y')$ is the Green's impedance function of the semi–infinite slab 3. $\psi_h(y)$ is the modal field of the surface wave with characteristic impedance z_{01}; $\psi_{e3}(y,\rho)$ and $\psi_{o3}(y,\rho)$ represent the even and odd air modes with characteristic impedance $z_{03}(\rho)$, and $\psi_3(y,\sigma)$ the substrate mode with characteristic impedance $z_{03}(\sigma)$. When the heights are the same, also the characteristic impedances for air and substrate modes of slabs 1 and 3 are the same and they are given by

$$\begin{aligned} z_{01}(\rho) = z_{03}(\rho) = z_0(\rho) \\ z_{01}(\sigma) = z_{03}(\sigma) = z_0(\sigma) \end{aligned} \tag{6.53}$$

$\hat{Z}_1$ and $\hat{Z}_3$ can be viewed as the driving-point impedance operators of the semi–infinite slabs 1 and 3. In order to relate the total fields at $x = 0$ and $x = L$ via an open circuit impedance operator, a magnetic wall is placed at $x = 0$ and $x = L$ in turn and the relationship between the transverse E and the transverse H field for each condition is derived. With a magnetic wall at $x = L$, the transverse electric field $E(0,y)$ excited by magnetic field $H(0,y)$ is given by

$$E(0,y) = -\int_{-\infty}^{\infty} Z_{11}(y,y')\,[-H(0,y')]\,dy' \equiv \hat{Z}_{11} \cdot [-H(0,y')] \tag{6.54}$$

where

$$\begin{aligned} Z_{11}(y,y') = z_{02}\coth(\gamma_{s2}L)\,\phi_2(y)\phi_2(y') + \\ \sum_{\mu=e,o} \int_0^{\infty} z_0(\rho)\coth(\gamma L)\,\phi_{\mu 2}(y,\rho)\phi_{\mu 2}(y',\rho)d\rho + \\ \int_0^{\nu} z_0(\sigma)\coth(\eta L)\,\phi_2(y,\sigma)\phi_2(y',\sigma)d\sigma \end{aligned} \tag{6.55}$$

Moreover, under the same boundary conditions, we have

$$E(L,y) = -\int_{-\infty}^{\infty} Z_{21}(y,y')\,[-H(0,y')]\,dy' \equiv \hat{Z}_{21} \cdot [-H(0,y')] \tag{6.56}$$

where

$$\begin{aligned} Z_{21}(y,y') = z_{02}\text{cosech}(\gamma_{s2}L)\,\phi_2(y)\phi_2(y') + \\ \sum_{\mu=e,o} \int_0^{\infty} z_0(\rho)\text{cosech}(\gamma L)\,\phi_{\mu 2}(y,\rho)\phi_{\mu 2}(y',\rho)d\rho + \\ \int_0^{\nu} z_0(\sigma)\text{cosech}(\eta L)\,\phi_2(y,\sigma)\phi_2(y',\sigma)d\sigma \end{aligned} \tag{6.57}$$

When a magnetic wall is placed at $x = 0$, we have by symmetry

$$E(L,y) \equiv \hat{Z}_{22} \cdot [-H(L,y')] \tag{6.58}$$

and

$$E(0,y) \equiv \hat{Z}_{12} \cdot [-H(L,y')] \tag{6.59}$$

where $\hat{Z}_{22} = -\hat{Z}_{11}$ and $\hat{Z}_{12} = -\hat{Z}_{21}$.

Combining (6.54), (6.56), (6.58), and the above we obtain a "two port" Green's open circuit impedance operator for the section $0 \leq x \leq L$ given by

$$\begin{bmatrix} E(0,y) \\ E(L,y) \end{bmatrix} = \begin{bmatrix} \hat{Z}_{11} & -\hat{Z}_{21} \\ \hat{Z}_{21} & -\hat{Z}_{11} \end{bmatrix} \cdot \begin{bmatrix} -H(0,y) \\ -H(L,y) \end{bmatrix} \tag{6.60}$$

Having obtained the two port impedance operators for each uniform section of the waveguide as given by (6.60), we are now in a position to consider the cascade of steps in Figure 6.7. From (6.60) the relationship between the transverse field at x_i and x_{i+1} can be written as

$$\begin{bmatrix} E(x_i,y) \\ E(x_{i+1},y) \end{bmatrix} = \begin{bmatrix} \hat{Z}_{11}^{(i)} & -\hat{Z}_{21}^{(i)} \\ \hat{Z}_{21}^{(i)} & -\hat{Z}_{11}^{(i)} \end{bmatrix} \cdot \begin{bmatrix} -H(x_i,y) \\ -H(x_{i+1},y) \end{bmatrix} \tag{6.61}$$

If the continuity of the transverse field is translated into the continuity of voltages and currents at the interface plane x_i and the interface field is approximated by a scalar function, then the cascade of steps in Figure 6.7 can be represented by an equivalent "T" network as shown in Figure 6.8a. In this formulation the interaction via the propagating air and substrate modes between non adjacent discontinuities is built into the model, as $E(x_i, y)$ in (6.61) represents the total transverse field at x_i. If we disregard interaction via propagating continuum modes, then the equivalent network reduces to that of Figure 6.8c. The application of the above theory to the analysis of coupled and multiple coupled guides is postponed until the development of the LSM case hereafter.

6.3.3 LSM analysis

By placing an electric wall at $x = 0$ and $x = L$ in Figure 6.9, and by a development analogous to that of the LSE case we obtain a "two–port" Green's short circuit admittance operator for the length of the waveguide $0 \leq x \leq L$ given by

$$\begin{bmatrix} -H(0,y) \\ -H(L,y) \end{bmatrix} = \begin{bmatrix} \hat{Y}_{11} & \hat{Y}_{12} \\ \hat{Y}_{21} & \hat{Y}_{22} \end{bmatrix} \cdot \begin{bmatrix} E(0,y) \\ E(L,y) \end{bmatrix} \tag{6.62}$$

where $\hat{Y}_{22} = -\hat{Y}_{11}$ and $\hat{Y}_{12} = -\hat{Y}_{21}$.

The various quantities occurring in (6.62) correspond to those occurring in (6.60) with

$$\begin{aligned} Y_{11}(y,y') = y_{02} \coth(\gamma_{s2}L)\, \phi_2(y)\phi_2(y') + \\ \sum_{\mu=e,o} \int_0^\infty y_0(\rho) \coth(\gamma L)\, \phi_{\mu 2}(y,\rho)\phi_{\mu 2}(y',\rho) d\rho + \\ \int_0^\nu y_0(\sigma) \coth(\eta L)\, \phi_2(y,\sigma)\phi_2(y',\sigma) d\sigma \end{aligned} \tag{6.63}$$

and

$$\begin{aligned} Y_{21}(y,y') = y_{02} \operatorname{cosech}(\gamma_{s2}L)\, \phi_2(y)\phi_2(y') + \\ \sum_{\mu=e,o} \int_0^\infty y_0(\rho) \operatorname{cosech}(\gamma L)\, \phi_{\mu 2}(y,\rho)\phi_{\mu 2}(y',\rho) d\rho + \\ \int_0^\nu y_0(\sigma) \operatorname{cosech}(\eta L)\, \phi_2(y,\sigma)\phi_2(y',\sigma) d\sigma \end{aligned} \tag{6.64}$$

The driving point admittance operators of slabs 1 and 3 are given by

$$\begin{aligned} Y_1(y,y') = y_{01}\psi_1(y)\psi_1(y') - \sum_{\mu=e,o} \int_0^\infty y_0(\rho)\psi_{\mu 1}(y,\rho)\psi_{\mu 1}(y',\rho) d\rho \\ - \int_0^\nu y_0(\sigma)\psi_1(y,\sigma)\psi_1(y',\sigma) d\sigma \end{aligned} \tag{6.65}$$

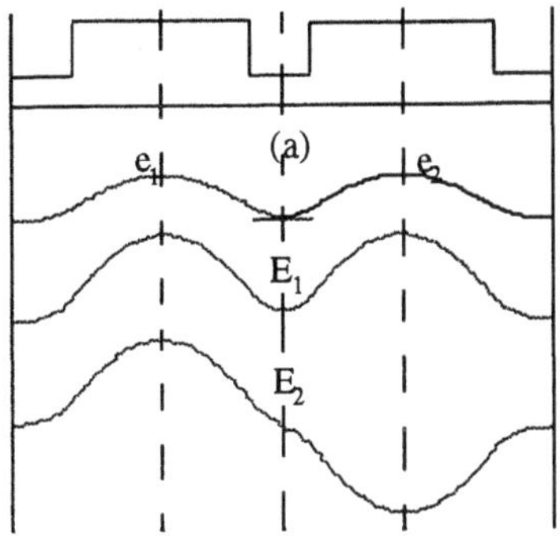

Figure 6.10 a) Cross–sectional geometry of straight rib waveguide directional coupler; b) the even (E_1) and odd (E_2) field distribution of the structure.

and

$$\begin{aligned} Y_3(y,y') = y_{03}\psi_3(y)\psi_3(y') + \sum_{\mu=e,o} \int_0^\infty y_0(\rho)\psi_{\mu 3}(y,\rho)\psi_{\mu 3}(y',\rho)d\rho \\ + \int_0^\nu y_0(\sigma)\psi_3(y,\sigma)\psi_3(y',\sigma)d\sigma \end{aligned} \tag{6.66}$$

It is emphasised that the wavenumbers and modal functions occurring in the above equations are now those appropriate to the TM case, where the admittances are given by

$$\begin{aligned} y_{0i} &= \frac{j\omega\varepsilon_0\gamma_{si}}{\beta^2-\gamma_{si}^2} \quad i=1,2,3 \\ y_0(\rho) &= \frac{j\omega\varepsilon_0\gamma}{k_0^2-\rho^2} \\ y_0(\sigma) &= \frac{j\omega\varepsilon_0\eta}{\varepsilon_2 k_0^2-\sigma^2} \end{aligned} \tag{6.67}$$

and the wavenumbers by

$$\begin{aligned} \varepsilon_{ei}k_0^2-\beta^2 &= -\gamma_{si}^2 \quad i=1,2,3 \\ k_0^2-\beta^2-\rho^2 &= -\gamma^2 \\ \varepsilon_2 k_0^2-\beta^2-\sigma^2 &= -\eta^2 \\ \nu &= k_0\sqrt{\varepsilon_2-\varepsilon_3} \end{aligned} \tag{6.68}$$

$\varepsilon_{e1}, \varepsilon_{e2}, \varepsilon_{e3}$ are the effective dielectric constants (TM polarization) of the slabs 1, 2 and 3, respectively.

Using (6.62) and translating the continuity of the transverse fields at each port into the continuity of voltages and currents and approximating the interface field by a scalar function we recover the equivalent network for the cascade of steps as shown in Figure 6.8b.

6.3.4 Numerical solution of coupled rib waveguides

Directional couplers

The cross-sectional geometry of a straight rib waveguide directional coupler is shown in Figure 6.10a, where the two guides are identical. The field distribution of this structure, shown in Figure 6.10b, comprises even and odd modes propagating along the coupler with propagation constants β_e and β_o, respectively. Including time and z-dependence the even and odd mode amplitudes may be expressed as:

$$\begin{array}{ll} \text{for the even mode} & E(z) \approx \cos(\omega t - \beta_e z) \\ \text{for the odd mode} & E(z) \approx \cos(\omega t - \beta_o z) \end{array} \tag{6.69}$$

By linearity, the response of the structure is given by the superposition of the even and odd mode responses. If the even and odd modes are excited with equal amplitudes we have, for guide 1,

$$\begin{aligned} E_1(z) &= A\cos(\omega t - \beta_e z) + A\cos(\omega t - \beta_o z) \\ &= 2A\cos(\omega t - \tfrac{\beta_e+\beta_o}{2} z)\cos(\tfrac{\beta_e-\beta_o}{2} z) \\ &= 2A\cos(\omega t - \beta z)\cos(cz) \end{aligned} \tag{6.70}$$

and for guide 2,

$$\begin{aligned} E_2(z) &= A\cos(\omega t - \beta_e z) - A\cos(\omega t - \beta_o z) \\ &= -2A\cos(\omega t - \tfrac{\beta_e+\beta_o}{2} z)\sin(\tfrac{\beta_e-\beta_o}{2} z) \\ &= -2A\cos(\omega t - \beta z)\sin(cz) \end{aligned} \tag{6.71}$$

At some distance L_t along the length of the coupler, the relative phases of the two modes will reverse and the power is transferred to and fro between the two guides. The transfer length L_t for maximum power transfer is

$$L_t = \frac{n\pi}{\beta_e - \beta_o} \tag{6.72}$$

If the even and odd modes are not excited with equal amplitudes, when the two modes are out of phase cancellation in one of the guides will not be complete. Hence, a small amount of power will be left which may contribute to crosstalk. In the following, emphasis is given to the determination of transfer length, a useful parameter required to characterise the coupler, and the problems of crosstalk are not pursued further. Most practical structures are considered to be strongly coupled: for example, short transfer lengths are desired for small and fast switches. In order to evaluate the transfer length of a coupler the calculation of β_e and β_o is required. Hence, an appropriate technique, accurate for the strongly coupled situation, is highly desirable to determine the propagation constants. This is done using the theory developed in the previous Section. In this case, the problem of solving the coupled structure reduces to that of solving the equivalent network. This is achieved more effectively by cascade multiplication of the transfer matrices of the individual networks. The transfer matrix relating voltages and currents at the left hand side ports to the right hand side is given by

$$T = \begin{bmatrix} Z_{11} \cdot Z_{21}^{-1} & -Z_{11} \cdot Z_{22} \cdot Z_{21} + Z_{12}^{-1} \\ Z_{21}^{-1} & Z_{21}^{-1} \cdot Z_{22} \end{bmatrix} \tag{6.73}$$

Table 6.7 Parameters of the directional coupler which have been analysed. Dimensions are in μm.

RIB	n_1	n_2	n_2	D	d	L	S	λ
1	3.44	3.34	1.0	1.3	0.2	2.0	2.0	1.55
2	3.44	3.36	1.0	1.0	0.9	3.0	2.0	1.55
3	3.44	3.435	1.0	6.0	3.5	4.0	2.0	1.55

The z–directed propagation constants β_e and β_o of the even and odd modes, respectively, are determined from the transverse resonance condition applied at a chosen reference point. By imposing transverse resonance and choosing the reference plane at $x = 0$, we obtain the scalar dispersion equation given by

$$Z_1 + Z_{in} = 0 \tag{6.74}$$

on an impedance basis, or

$$Y_1 + Y_{in} = 0 \tag{6.75}$$

on an admittance basis. In the above, Z_{in} and Y_{in} are the input impedance and admittance respectively looking to the right of $x = 0$. Solving the dispersion equation yields the propagation constants β_e and β_o.

In order to demonstrate the accuracy of the present technique we consider first the directional couplers analysed by [33] using the finite difference method. The parameters describing the directional couplers analysed in [33] are given in Table 6.7. The results of the present analysis together with those of FDM and the EDC method are summarised in Table 6.8. Simulations indicate that both present and FDM analysis results agree very well and prove the accuracy of the present method. The results of the EDC agree very well with the present and FDM analysis only for small rib height. As shown in Table 6.8, the present simulation does not produce the odd mode index for the third structure. This is because the odd mode for this structure is below cutoff and the mode index is complex. This argument is justified by considering that the index of the odd mode predicted by FDM analysis β_o/k_0 is less than the EDC of the outer slab $\sqrt{\varepsilon_e'}$. The computer program used in the present analysis is developed to predict a real mode index β_o/k_0 in the range $\sqrt{\varepsilon_e} < \beta_o/k_0 < \sqrt{\varepsilon_e'}$. However, this is not a limitation of the method and if complex algebra is used in the computer program it should be able to yield a complex index for this mode. This problem is not pursued further as emphasis is here focussed on the determination of the transfer length of coupled guides. In order to assess the effect of the propagating continuous modes on the transfer length of coupled guides at optical frequencies, we compare the results of the analysis using equivalent networks in Figure 6.8a and Figure 6.8c. The parameters of the structure examined in this exercise are $n_1 = 3.44$, $n_2 = 3.40$, $n_3 = 1.0$, D = 1 μm; L = 3 μm, l = 1.15 μm, S = 2 μm and varying d. The results of the simulations are summarised in Table 6.9. Simulations indicate that the effect of the propagating continuous modes on the transfer length is small for well confined modes. This observation shows that at optical frequencies with $n_1 >> n_3$, the interaction via the continuous modes can be neglected. Hence, it is possible to obtain accurate results by considering

Table 6.8 Modal refractive indices of even and odd modes in directional coupler structures.

rib	Mode indices	TRD Figure 6.8a	Using Figure 6.8c	FDM [33]	EDC
1	β_e/k_0	3.388841027	3.388838939	3.391472377	3.390322424
	β_o/k_0	3.388841027	3.388838350	3.391470106	3.390322421
	L_t (mm)		1314.57	341.0	300.0
2	β_e/k_0	3.395739365	3.395739353	3.396053426	3.395914969
	β_o/k_0	3.394802846	3.394802837	3.395097562	3.394954357
	L_t (mm)	0.827	0.827	0.811	0.807
3	β_e/k_0	3.436820329	3.436806957	3.437034755	3.437661171
	β_o/k_0			3.436459357	3.437072754
	L_t (mm)			1.347	1.317

Table 6.9 Comparison of modal refractive indices of even and odd modes and transfer lengths of directional coupler structures for different outer slab heights.

d μm	Mode indices	Model Figure 6.8a	Model Figure 6.8c	Difference %
0.3	β_e/k_0	3.412597677	3.412595607	
	β_o/k_0	3.412551746	3.412549877	
	L_t (mm)	12.5272	12.5739	0.37
0.6	β_e/k_0	3.413685884	3.413685555	
	β_o/k_0	3.413535726	3.413535438	
	L_t (mm)	3.831	3.830	0.025
0.9	β_e/k_0	3.415891467	3.415891456	
	β_o/k_0	3.415203392	3.415203383	
	L_t (mm)	0.835666	0.835666	0.

interaction via the surface modes only and using the simpler equivalent network Figure 6.8c.

Having established the accuracy of the technique, we are in a position to consider some numerical results that can be utilised as design data. The variation of the modal indices of the even and odd modes as a function of the guide separation S, as well as the parameters of the coupled structure considered, are shown in Figure 6.11. The even and odd mode indices are above and below the single guide index respectively. For small guide separation S, i.e. for a strongly coupled system, the differences of even and odd mode indices from the single guide index are not equal and hence $\beta \neq \frac{\beta_e+\beta_o}{2}$. As the guide separation S increases both β_e/k_0 and β_o/k_0 appear to approach the single guide index and for sufficiently large S they become almost equally spaced from β/k_0. This behaviour is consistent with the prediction of coupled mode theory and β can be accurately predicted by $\beta = \frac{\beta_e+\beta_o}{2}$. This occurs when the coupling is loose. This observation shows that, for a strongly coupled system, simple coupled mode theory is less than accurate and more rigorous analyses such as the present method ought to be used. The transfer lengths can be calculated

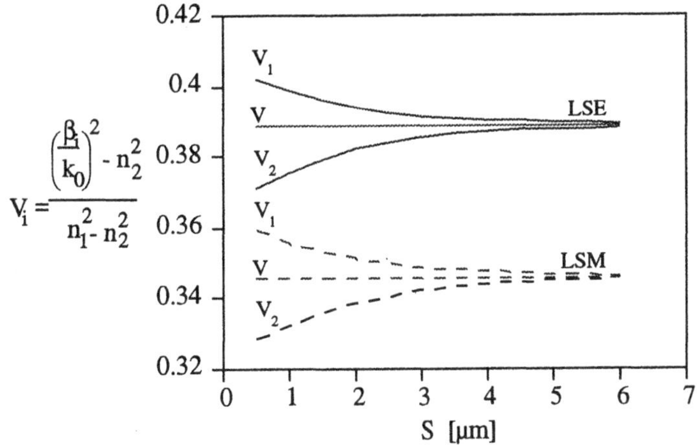

Figure 6.11 Variation of the normalised modal indices of the even and odd modes as a function of the guide separation S for directional coupler. V_1 - even mode, V_2 - odd mode, V - mode of single guide. The values of the parameters of the directional coupler ST4 are: $n_1 = 3.44$, $n_2 = 3.40$, $n_3 = 1.0$, $D = 1$ μm, $d = 0.9$ μm, $L = 3.0$ μm, $l = 1.15$ μm.

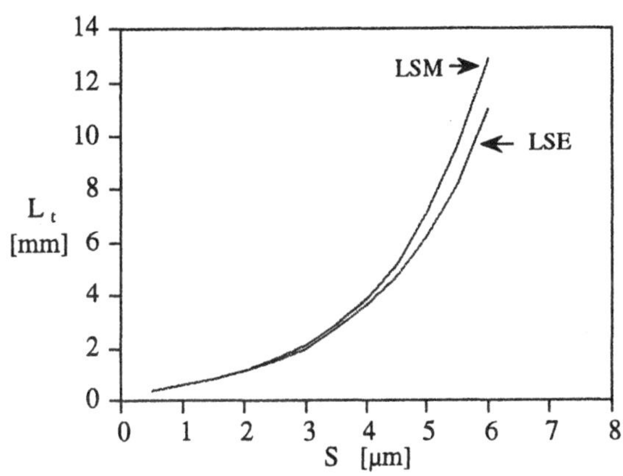

Figure 6.12 Variation of the transfer length L_t as a function of guide separation S for directional coupler ST4.

using the data in Figure 6.11 and the plot of L_t versus S is shown in Figure 6.12. The transfer length increases exponentially with increasing S, whereas coupling decreases exponentially as S is increased. This behaviour is a direct consequence of the fact that fields decay exponentially outside the guides. Their decay constants may be predicted from the gradient of the plot of $\log(L_t)$ versus S given in Figure 6.13.

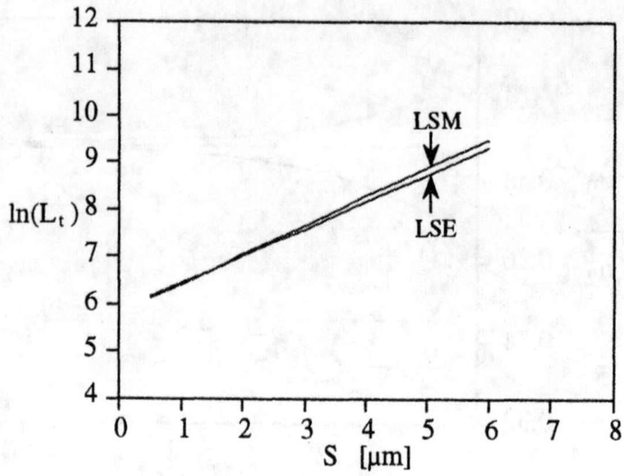

Figure 6.13 Variation of $ln(L_t)$ as a function of guide separation S for directional coupler ST4.

Three guide–coupler

A symmetrical three guide–coupler composed of three identical rib waveguides in close proximity is shown in Figure 6.14a. The individual waveguides are assumed to be single moded. Before applying the present method of analysis, it is instructive to consider the modal characteristics of this structure. The transverse field profiles of the overlapping guide modes and the modes of the coupled structure in Figure 6.14a are shown in Figure 6.14b. The coupled structure has three bound modes; two are even with propagation constants β_1 and β_2 and one is odd with propagation constant β_3. In [5], a similar configuration has been analysed using the EDC technique and it has been shown that the propagation constants of the three modes obey the following conditions,

$$\begin{array}{ll} \beta_1 > \beta & \\ \beta_2 < \beta & \\ \beta_3 \lesseqgtr \beta; & \text{depending on guide parameters} \\ (\beta_3 - \beta) \leq (\beta_1 - \beta) \text{ or } (\beta_2 - \beta) & \end{array} \tag{6.76}$$

where β is the modal propagation constant of a single isolated rib waveguide.

There are two possible ways of exciting the structure in Figure 6.14a. In the first case, the structure is excited symmetrically through the centre guide and this configuration may be used as a power divider. In this case, only modes 1 and 2 are excited, so that at $z = 0$ their fields are in phase i.e. they add in the centre guide and subtract in the external guides. At some distance z down the line, modes 1 and 2 get out of phase, their fields interfere destructively in the centre guide and constructively in the external guides, accounting for the transfer of power from the centre guide to the outside guides. This occurs at multiples of the coupling length given by [5],

$$L_{ct} = \frac{\pi}{\beta_1 - \beta_2} \tag{6.77}$$

In the second case, the three guide coupler is excited antisymmetrically through either one of the outer guides. Such a configuration may be used to transfer power

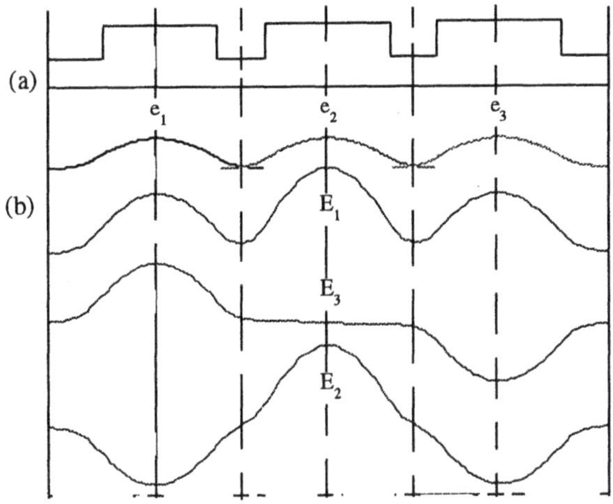

Figure 6.14 a) Cross–sectional geometry of three identical rib waveguide coupler; b) the transverse field profiles of the overlapping guide modes and the modes of the coupled structure.

from one of the outer guides to the other. The feasibility of using such a device as an improved sampler has been demonstrated in [10]. At $z = 0$, modes 1 and 2 are excited so as to be in antiphase in the centre guide and in phase in the outer guides. Mode 3 is excited so as to be in phase with modes 1 and 2 in the input outer guide and in antiphase in the other outer guide. In this case, the situation is more complex and beats periodic with distance along the length of the coupler can only be obtained with high power transfer efficiency between guides if the phase velocities of the modes are related to each other in a simple manner. Under the condition of loose coupling most of the power will be transferred from one outer guide to the other at a length given by [5],

$$L_{ot} = \frac{2\pi}{\beta_1 - \beta_2} \tag{6.78}$$

i.e. the coupling length is twice that required to transfer power symmetrically from the centre guide to the two outer guides.

Again the key factor in the analysis of the three guide coupler is the accurate evaluation of β_1, β_2, and β_3. In this numerical example, three guides of the geometry used in the previous example will be analyzed. The resulting equivalent circuit is similar to that relative to Figure 6.10 but it includes two additional network sections corresponding to the third guide. Solving the equivalent circuit yields all the β values. Equation (6.76) is used as a reference in the identification of β values for the two even modes and odd mode. The variation of the effective indices of the even modes and odd mode as a function of separation between the guides S is shown in Figure 6.15. The straight line is the effective index of the single isolated rib waveguide. Using Figure 6.15 the length required to transfer power from the centre guide to the outer guides, L_{ct}, can be calculated using (6.77) and that to transfer power from one of the outer guides to the other, L_{ot} , can be obtained

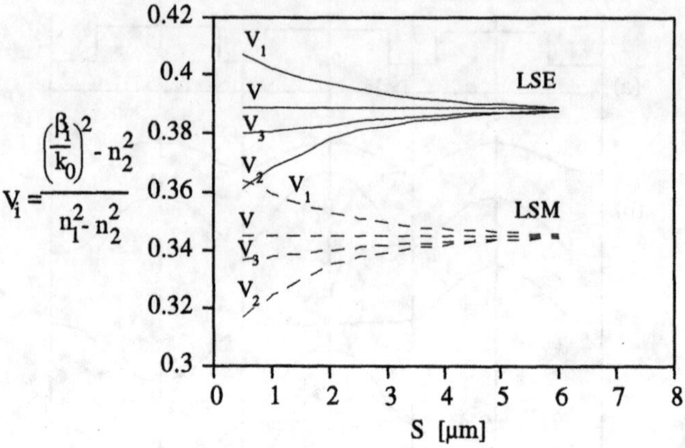

Figure 6.15 Variation of the normalised modal indices of the even and odd modes as functions of guide separation S for three guide couplers. V_1, V_2 *- even modes,* V_3 *- odd mode, V - mode of single guide. The values of the parameters of the three guide coupler ST5 are:* $n_1 = 3.44$, $n_2 = 3.40$, $n_3 = 1.0$, $D = 1\ \mu m$, $d = 0.9\ \mu m$, $L = 3.0\ \mu m$, $l = 1.15\ \mu m$.

from (6.78). The variation of $\ln(L_{ct})$ and $\ln(L_{ot})$ as a function of guide separation S is plotted in Figure 6.16. In Figure 6.17 the variation of the transfer length of the three guide coupler and of the directional coupler is shown as a function of guide separation S. Comparing the two curves, it can be seen that for strongly coupled structures the ratio of the two transfer lengths is between 1.45 and 1.5. Only for loosely coupled structures does the ratio become very close to $\sqrt{2}$, that is the prediction of coupled mode theory. There are two points to be noted from this observation. First, accurate analyses should be employed even for practical devices. Secondly, one must beware of using data derived from two guide couplers when designing devices involving three guide couplers. In fact, a simple $\sqrt{2}$ relationship between the transfer lengths predicted by coupled mode theory does not hold for strongly coupled situations.

Strongly coupled unequal guides

The theory developed in the previous Section applies not just to coupled identical rib waveguides but also to coupled rib guides of different dimensions. As an example, we investigate the propagation constants of strongly coupled unequal guides where the geometrical dimensions of guide 1 are fixed and the rib width or rib height of guide 2 varies. The other dimensions of guide 2 are fixed. The optical parameters of the guides are $n_1 = 3.44$, $n_2 = 3.40$, $n_3 = 1.0$, and $l = 1.15\ \mu m$. Their geometrical dimensions are shown in the figure. The results for the modal indices as functions of the guide separation are shown in Figure 6.18. From this figure, it can be seen that the fundamental modes of the two rib waveguides are not synchronous; in fact, each of the two modes favours one of the two rib waveguides. Thus, each mode is nearly equal to the mode of one or the other rib waveguide taken in isolation. For sufficiently large guide separation the two guides are decoupled. The

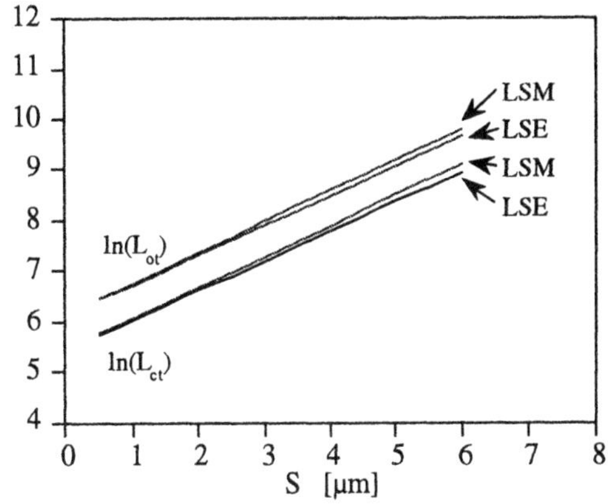

Figure 6.16 Variation of $\ln(L_{ct})$ *and* $\ln(L_{ot})$*, as a function of guide separation S for three guide coupler of previous figure.*

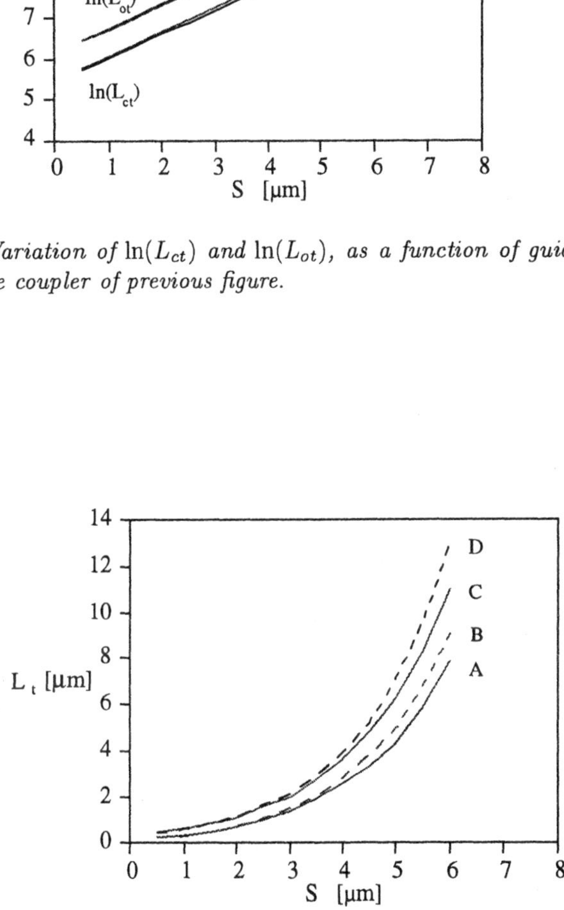

Figure 6.17 Variation of the transfer length of the three guide coupler ST5 L_{ct} *(curves A and B) and of the directional coupler ST4* L_t *(curves C and D) as a function of guide separation S. Continuous line refers to LSE analysis; dashed line refers to LSM analysis.*

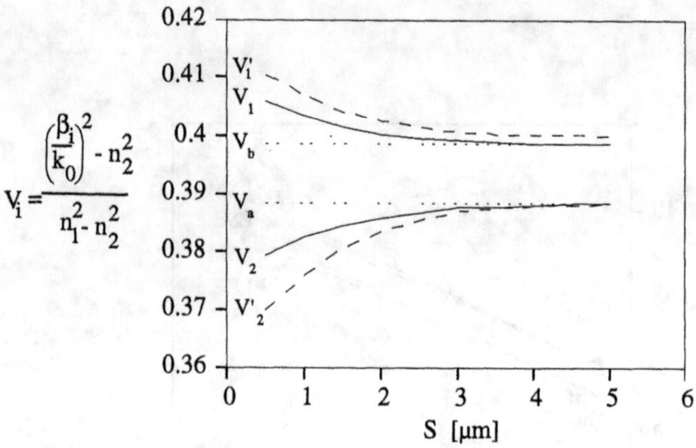

Figure 6.18 Variation of the normalised modal indices (LSE) of the modes as a function of guide spacing for closely coupled nonidentical guides. Curves V_a and V_b refer to uncoupled guides of L = 3.0 μm and 4.0 μm respectively and D = D' = 1.0 μm, d = 0.9 μm. Continuous line (V_1, V_2): L = 3.0 μm, L' = 4.0 μm, d = 0.9 μm, D = 1.0 μm, D' = 1.0 μm. Dashed line (V_1', V_2'): L= 3.0 μm, L' = 4.0 μm, d = 0.9 μm, D = 1.0 μm, D' = 1.2 μm.

fundamental modes of these two guides acquire even and odd symmetry when the two rib waveguides are identical. From this observation it can be seen that in order to realise a coupler using nonidentical guides, the two guides ought to be strongly coupled and in synchronism so that the two modes can be superimposed in such a way that at the input end of the coupler their fields nearly cancel in one guide while they reinforce each other in the other guide. At a distance $L_t = \frac{\pi}{\beta_2 - \beta_1}$ the mode fields reinforce each other in the opposite sense, accounting for the transfer of power between the two guides. In order to achieve synchronism of the two guides, the rib width or the rib height may be adjusted.

Waveguide arrays

An array of coupled rib waveguides can be used to design coupled laser arrays [44]. The operation of coupled waveguide arrays is more complicated than that of two or three–guide couplers because of the higher number of modes which can exist in these structures. If the individual guides are single moded, then the number of possible guided modes of coupled arrays is generally equal to the number of guides in the array [40]. The propagation constants of the structure can be obtained by using the theory previously developed. In the case of finite arrays, the numerical evaluation of the propagation constants is similar to that of a three guide coupler with little additional computation. The increase in computation time is small since we are dealing only with an additional multiplication of the transfer matrices and there is no need to calculate additional equivalent circuit parameters.

If the array is large and periodic we may approximate its propagation constant, in a manner similar to the determination of the propagation constant of a transmission line composed of cascaded sections. To solve a periodic structure let us consider the unit element of the periodic structure and its equivalent circuit. The relationship between voltages and currents at A and A' (see Figure 6.7) is given by

$$\begin{bmatrix} \mathbf{V} \\ \mathbf{I} \end{bmatrix} = T_1 \cdot T_2 \begin{bmatrix} \mathbf{V}' \\ \mathbf{I}' \end{bmatrix} \tag{6.79}$$

where T is the transfer matrix for each uniform section. The periodicity condition requires

$$\begin{bmatrix} \mathbf{V}' \\ \mathbf{I}' \end{bmatrix} = e^{-jk_x \ell} \begin{bmatrix} \mathbf{V} \\ \mathbf{I} \end{bmatrix} \tag{6.80}$$

where ℓ is the length of the period and k_x is the propagation constant of the periodic structure. Substituting the above into (6.79) gives the approximate eigenvalue equation for the periodic structure, that is

$$T_1^{-1} \cdot T_2^{-1} \begin{bmatrix} \mathbf{V} \\ \mathbf{I} \end{bmatrix} = e^{-jk_x \ell} \begin{bmatrix} \mathbf{V} \\ \mathbf{I} \end{bmatrix} \tag{6.81}$$

In conclusion, in this Section we have introduced a simple and effective rigorous solution for coupled rib waveguides. The problem is seen as that of cascaded step discontinuities in the transverse direction. The use of a "transition function" at the discontinuous interfaces not only simplifies the calculation, avoiding the use of matrices, but it also yields a simple cascaded two port network representation of the coupled structure.

The accurate determination of the propagation constants of the modes of the coupled structure is paramount in determining the coupling between waveguides. The method followed here is accurate and valid under all circumstances of both strong and weak coupling. The accuracy of the method has been validated by comparing the results with the results of FDM calculations. Numerical examples have been given and it has been shown that results compare very favourably with those produced by FDM. Numerical simulations also show that the results of coupled mode theory agree with the above accurate results only for weakly coupled structures. As most practical devices are tightly coupled, more accurate analyses such as those of the present method should be employed in their study. The additional computational effort required to analyse multiple guide couplers including arrays and nonidentical guides when using the present method is only marginally greater than that required for a single guide. This is a significant advantage of TRD over numerical methods where the computational effort increases with the cross–sectional area.

6.4 Slotline hybrid modes

We will now move to consider a fully hybrid problem arising in the microwave range. The ease of source integration in slotline and its convenient duality to microstrip, often sharing the same ground plane, are well appreciated [2, 6, 34], contributing to its popularity in the lower microwave range. Application in the microwave range required a better understanding of the fundamental mode [18, 21, 8, 20, 19]. Design information based on improved quasi–static formulas [8] as well as hybrid results [20] for a broad slot are also available. However, for applications where radiative

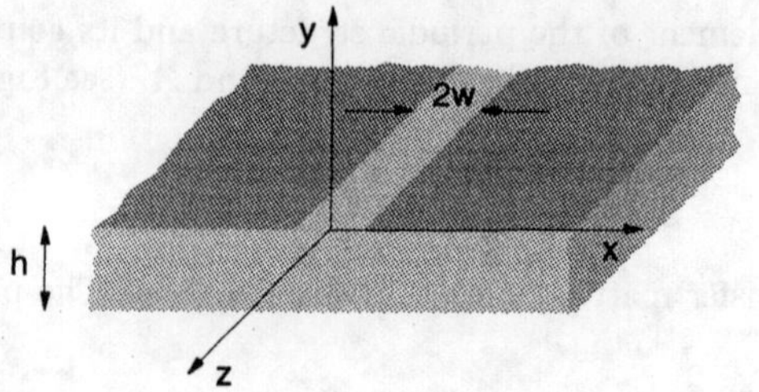

Figure 6.19 Slotline geometry and coordinate system.

and leakage effects play a role, such as discontinuities in the millimetric range, further characterization becomes necessary. In particular, the onset of leakage in the dielectric due to the excitation of a TM mode in the transverse direction (parallel to the substrate) sets an upper limit for lossless operation. In addition, radiation modes are excited by discontinuities such as diode mounts across the slot, impedance steps, and abrupt terminations arising in the circuitry. Radiation modes are examined in the next Chapter, while we are now interested in the application of transverse resonance in the space domain to the problem of the fundamental hybrid mode. To this end, we use expanding functions on the slot that implicitly take into account the $r^{-1/2}$ singularity at the metallic edges. The ensuing results fall between those obtained in [8] and [20]. Moreover, the analysis also shows the possible existence of TM dielectric surface wave poles on the path of integration in the wavenumber plane.

6.4.1 Analysis

The geometry of slotline is shown in Figure 6.19. With reference to this figure, from the continuity of the transverse to y electric fields, E_x and E_z, across the slot we obtain a pair of coupled integral equations for E_x and $E'_z = \partial E_z/\partial x$ on the slot aperture in the form

$$\begin{bmatrix} H_z \\ \frac{\pi}{w}\int H_x dx \end{bmatrix} = \begin{bmatrix} \hat{Y}_{11} & \hat{Y}_{12} \\ \hat{Y}_{21} & \hat{Y}_{22} \end{bmatrix} \cdot \begin{bmatrix} E_x(x,0) \\ \frac{w}{\pi}E'_z(x,0) \end{bmatrix} = 0 \tag{6.82}$$

Note that we use E'_z, which satisfies the same boundary and singular edge conditions as E_x in the gap, rather than E_z itself. A similar approach, i.e. the use of the derivative of the field instead of the field itself, was introduced in the previous Chapter in order to improve the convergence of the solution of the integral equation. The integral operators $\hat{Y}$, etc., denote the sum of the admittance operators of the two half spaces, i.e.

$$\hat{Y}_{11} = \hat{Y}_{11}^{a} + \hat{Y}_{11}^{s}$$

and their kernels $Y_{11}(x, x')$, etc. are obtained as follows.

First, we introduce the even and odd eigenfunctions in the x–direction, given respectively by

$$\phi(x; k_x) = \sqrt{\frac{2}{\pi}} \cos k_x x$$

$$\xi(x;k_x) = \sqrt{\frac{2}{\pi}}\sin k_x x \tag{6.83}$$

Next, we describe the field by the two Hertzian potentials

$$\begin{aligned}\mathbf{\Pi}_e &= \hat{\mathbf{y}}\psi_e(x,y)e^{-j\beta z}\\ \mathbf{\Pi}_h &= \hat{\mathbf{y}}\psi_h(x,y)e^{-j\beta z}\end{aligned} \tag{6.84}$$

which, given β and omitting the propagation factor along z, yields the following field components

$$\begin{aligned}E_x &= \partial_x\partial_y\psi_e + \omega\mu_0\beta\psi_h\\ E_y &= \partial_y^2\psi_e + \varepsilon_r k_0^2\psi_e\\ E_z &= -j\beta\partial_y\psi_e - j\omega\mu_0\partial_x\psi_h\\ H_x &= -\omega\varepsilon\beta\psi_e + \partial_x\partial_y\psi_h\\ H_y &= \partial_y^2\psi_h + \varepsilon_r k_0^2\psi_h\\ H_z &= j\omega\varepsilon\partial_x\psi_e - j\beta\partial_y\psi_h\end{aligned} \tag{6.85}$$

Due to the field symmetry with respect to x, the spectral expansion of the potentials in the homogeneous region $y > 0$ is written as

$$\begin{aligned}\psi_e(x,y) &= \frac{1}{j\omega\varepsilon_0}\int_0^\infty \frac{I_e(k_x)}{\sqrt{\beta^2+k_x^2}}e^{-jk_y y}\xi(x;k_x)dk_x\\ \psi_h(x,y) &= \frac{1}{j\omega\mu_0}\int_0^\infty \frac{V_h(k_x)}{\sqrt{\beta^2+k_x^2}}e^{-jk_y y}\phi(x;k_x)dk_x\end{aligned} \tag{6.86}$$

where I_e and V_h are the spectral equivalent currents and voltages, respectively. It is also convenient to introduce the Fourier (spectral) amplitudes of the fields, defined, for instance, by

$$\begin{aligned}\tilde{E}_x(k_x,y) &= \int_0^\infty E_x(x,y)\phi(x;k_x)dx\\ \tilde{E}_y(k_x,y) &= \int_0^\infty E_y(x,y)\xi(x;k_x)dx\end{aligned} \tag{6.87}$$

and the following quantities

$$\begin{aligned}\cos\theta &= \frac{k_x}{\sqrt{k_x^2+\beta^2}}\\ \sin\theta &= \frac{\beta}{\sqrt{k_x^2+\beta^2}}\end{aligned} \tag{6.88}$$

which have the clear geometrical meaning illustrated in Figure 6.20.

Finally, we introduce the characteristic admittance of TM^y waves in air, Y',and the characteristic admittance of TE^y waves in air, Y'', which are respectively given by

$$\begin{aligned}Y'(k_x) &= \frac{\omega\varepsilon_0}{k_y}\\ Y''(k_x) &= \frac{k_y}{\omega\mu_0}\end{aligned} \tag{6.89}$$

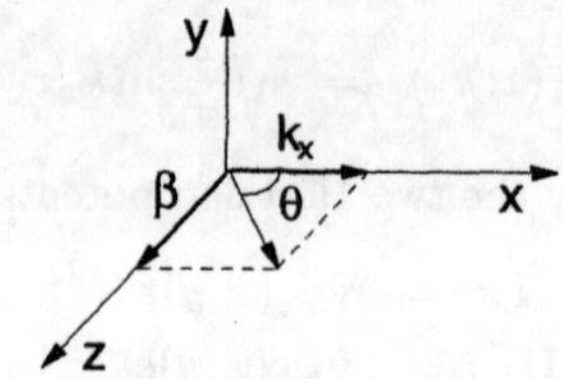

Figure 6.20 Definition of the angle θ.

As always, wavenumber conservation yields,

$$k_x^2 + k_y^2 + \beta^2 = k_0^2 \tag{6.90}$$

and, for a bound mode,

$$\begin{array}{c} k_0 < \beta < \sqrt{\varepsilon_r} k_0 \\ k_y^2 = -\gamma^2 < 0 \end{array} \tag{6.91}$$

With the above definitions, by considering the Fourier transform of (6.86), we obtain the following relationship for the transverse to y field components at $y = 0$

$$\begin{aligned} \tilde{E}_x &= \frac{\cos\theta}{Y'} I_e - j\sin\theta V_h \\ \tilde{E}'_z &= -\frac{j\sin\theta k_x}{Y'} I_e + k_x \cos\theta V_h \\ \int \tilde{H}_x dx &= -\frac{j\sin\theta}{k_x} I_e + Y'' \frac{\cos\theta}{k_x} V_h \\ \tilde{H}_z &= \cos\theta I_e - jY'' \sin\theta V_h \end{aligned} \tag{6.92}$$

It is noted that the indefinite integral of H_x is considered; the reason for this position will presently become apparent. By deriving I_e, V_h from the first two of (6.92) in terms of the electric field components as

$$\begin{aligned} I_e &= Y'\cos\theta \tilde{E}_x + jY' \frac{\sin\theta}{k_x} \tilde{E}'_z \\ V_h &= j\sin\theta \tilde{E}_x + \frac{\cos\theta}{k_x} \tilde{E}'_z \end{aligned} \tag{6.93}$$

and substituting back into the last two of (6.92), we obtain finally

$$\begin{pmatrix} \tilde{H}_z \\ \frac{\pi}{w} \int \tilde{H}_x dx \end{pmatrix} = \begin{pmatrix} Y'\cos^2\theta + Y''\sin^2\theta & j\frac{\pi}{wk_x} \sin\theta\cos\theta\,(Y' - Y'') \\ -j\frac{\pi}{wk_x} \sin\theta\cos\theta\,(Y' - Y'') & \left(\frac{\pi}{wk_x}\right)^2 [Y'\sin^2\theta + Y''\cos^2\theta] \end{pmatrix} \begin{pmatrix} \tilde{E}_x \\ \frac{w}{\pi} \tilde{E}'_z \end{pmatrix} \tag{6.94}$$

Obviously, in correspondence with the indefinite integral of H_x, the derivative of E_z appears as unknown in (6.94).

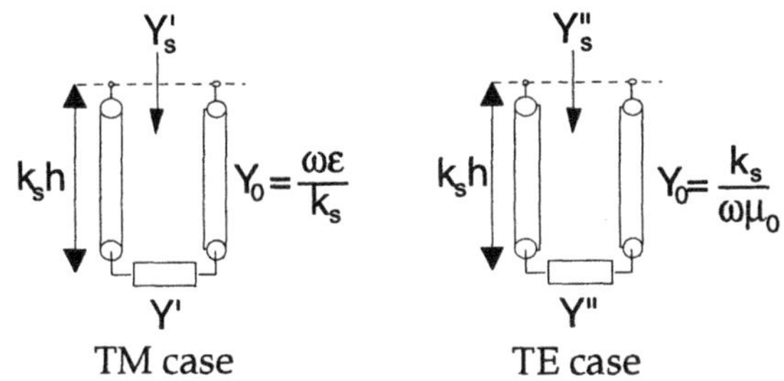

Figure 6.21 TM, TE transverse equivalent networks looking into the substrate ($y < 0$) for each Fourier component.

Air region

It is now straightforward to obtain the integral operators appearing in (6.82). From (6.94), the latter are given by

$$\begin{aligned} \hat{Y}^a_{11} &= \int_0^\infty \left(Y'\cos^2\theta + Y''\sin^2\theta\right)\phi(x,k_x)\phi(x',k_x)dk_x \\ \hat{Y}^a_{21} = -\hat{Y}^a_{12} &= j\int_0^\infty (Y' - Y'')\sin\theta\cos\theta\frac{\pi}{wk_x}\phi(x,k_x)\phi(x',k_x)dk_x \\ \hat{Y}^a_{22} &= \int_0^\infty \left(Y'\sin^2\theta + Y''\cos^2\theta\right)\left(\frac{\pi}{wk_x}\right)^2\phi(x,k_x)\phi(x',k_x)dk_x \end{aligned} \tag{6.95}$$

Substrate region

The main difference is constituted by the admittance seen from the slot. Each k_x component in the substrate region has an equivalent circuit as shown in Figure 6.21. Hence, the input admittance of each component, as seen by the slot, is given by

$$\begin{aligned} Y'_s &= \frac{\omega\varepsilon}{k_s}\frac{Y' + j\frac{\omega\varepsilon}{k_s}\tan k_s h}{\frac{\omega\varepsilon}{k_s} + jY'\tan k_s h} = j\frac{\omega\varepsilon}{k_s}\frac{k_s + \gamma\varepsilon_r\tan k_s h}{\gamma\varepsilon_r - k_s\tan k_s h} \\ Y''_s &= \frac{k_s}{\omega\mu_0}\frac{Y'' + j\frac{k_s}{\omega\mu_0}\tan k_s h}{\frac{k_s}{\omega\mu_0} + jY''\tan k_s h} = \frac{k_s}{j\omega\mu_0}\frac{\gamma - k_s\tan k_s h}{k_s + \gamma\tan k_s h} \end{aligned} \tag{6.96}$$

where we have set $k_y = -j\gamma$ ($\gamma > 0$) in accordance with the fact that a bound mode must decay without propagation away from the slot, i.e. $\exp^{jk_y y} = \exp^{\gamma y}$: ($y < 0$).

Hence, for the substrate region the kernels of the integral operators are given by,

$$\begin{aligned} Y^s_{11} &= \int_0^\infty \left(Y'_s\cos^2\theta + Y''_s\sin^2\theta\right)\phi(x,k_x)\phi(x',k_x)dk_x \\ Y^s_{21} = -Y^s_{12} &= j\int_0^\infty (Y'_s - Y''_s)\sin\theta\cos\theta\frac{\pi}{wk_x}\phi(x,k_x)\phi(x',k_x)dk_x \\ Y^s_{22} &= \int_0^\infty \left(Y'_s\sin^2\theta + Y''_s\cos^2\theta\right)\left(\frac{\pi}{wk_x}\right)^2\phi(x,k_x)\phi(x',k_x)dk_x \end{aligned} \tag{6.97}$$

6.4.2 Dispersion equation for the bound modes of slotline

Integral equation (6.82) is the dispersion equation for the discrete modes of slotline, which we solve by discretization in the space domain, building the appropriate edge condition into the solution. It was noted that E'_z satisfies the same boundary and edge conditions as E_x. In fact, both fields are singular in the same way at the sharp metallic edges of the slot, namely as $r^{-1/2}$, r being the distance from the edge. Hence, we adopt a discrete expansion of the unknown fields in terms of functions orthogonal with respect to a weight function that describes the edge condition above. This is

$$\begin{aligned} E_x(x,0) &= \frac{1}{\sqrt{1-\left(\frac{x}{w}\right)^2}} \sum_{n=0,2,\cdots} X_n \sqrt{\frac{2\varepsilon_n}{\pi w}} T_n\left(\frac{x}{w}\right) \\ \frac{w}{\pi} E'_z(x,0) &= \frac{1}{\sqrt{1-\left(\frac{x}{w}\right)^2}} \sum_{n=2,4,\cdots} Z_n \sqrt{\frac{2\varepsilon_n}{\pi w}} T_n\left(\frac{x}{w}\right) \end{aligned} \tag{6.98}$$

The even parity in the above expressions is dictated by that of the E_x field, here assumed even. For fields of odd parity we take $n = 1, 3, 5, \ldots$. It is also noted that in the second of the above the sum begins at $n = 2$. This is so because if the term $n = 0$ is considered, we have

$$\frac{\partial E_z}{\partial x} = \frac{1}{\sqrt{1-\left(\frac{x}{w}\right)^2}} \tag{6.99}$$

and

$$E_z = -\cos^{-1}\left(\frac{x}{w}\right) \tag{6.100}$$

which is not allowed on physical grounds, since this would involve an E_z field sloping from one edge of the slot to the other. The Chebyshev polynomials

$$\sqrt{\tfrac{2\varepsilon_n}{\pi w}} T_n\left(\tfrac{x}{w}\right) \quad ; \quad \left\{ \begin{array}{ll} \varepsilon_n = 2 & : \quad n > 0 \\ \varepsilon_0 = 1 & \end{array} \right\} \tag{6.101}$$

are normalised so that

$$\int_0^w \sqrt{\frac{2\varepsilon_m}{\pi w}} T_m\left(\frac{x}{w}\right) \sqrt{\frac{2\varepsilon_n}{\pi w}} T_n\left(\frac{x}{w}\right) \frac{dx}{\sqrt{1-\left(\frac{x}{w}\right)^2}} = \delta_{mn} \tag{6.102}$$

Moreover, the above choice of expanding functions gives simple overlapping integrals with the ϕ's. In fact we have

$$\int_0^w \sqrt{\frac{2}{\pi}} \cos(k_x x) \sqrt{\frac{2\varepsilon_n}{\pi w}} T_n\left(\frac{x}{w}\right) \frac{dx}{\sqrt{1-\left(\frac{x}{w}\right)^2}} = \sqrt{\varepsilon_n w}(-1)^{n/2} J_n(k_x w) \tag{6.103}$$

We can now take the matrix element (m, n) of the admittance operators, using the above expanding functions. For instance, considering the first of (6.97) we have

$$\begin{aligned} &[Y^a_{11}]_{mn} = \\ &\int_0^w \int_0^w Y^a_{11}(x,x') \sqrt{\tfrac{2\varepsilon_m}{\pi w}} T_m\left(\tfrac{x}{w}\right) \frac{dx}{\sqrt{1-\left(\frac{x}{w}\right)^2}} \sqrt{\tfrac{2\varepsilon_n}{\pi w}} T_n\left(\tfrac{x'}{w}\right) \frac{dx'}{\sqrt{1-\left(\frac{x'}{w}\right)^2}} \\ &= j\omega\varepsilon_0(-1)^{(m+n)/2} \sqrt{\varepsilon_m \varepsilon_n} w \left[1 - \left(\frac{\beta}{k_0}\right)^2\right] \int_0^\infty \frac{J_m(k_x w) J_n(k_x w)}{\sqrt{\beta^2 - k_0^2 - k_x^2}} dk_x \end{aligned} \tag{6.104}$$

with the indices running as

$$\left\{ \begin{array}{c} m = 0, 2, \cdots \\ n = 0, 2, \cdots \end{array} \right\}$$

and also,

$$
\begin{aligned}
[Y_{12}^a]_{mn} &= -\omega\varepsilon_0(-1)^{(m+n)/2}\sqrt{\varepsilon_m\varepsilon_n}\pi\frac{\beta}{k_0^2}\int_0^\infty \frac{J_m(k_xw)J_n(k_xw)}{\sqrt{\beta^2-k_0^2-k_x^2}}dk_x \\
&\left\{ \begin{array}{c} m = 0, 2, \cdots \\ n = 2, 4, \cdots \end{array} \right\} \\
[Y_{22}^a]_{mn} &= j2\omega\varepsilon_0(-1)^{(m+n)/2}\sqrt{\varepsilon_m\varepsilon_n}\frac{\pi^2}{w}\int_0^\infty \left[\frac{1}{k_x^2} - \frac{1}{k_0^2}\right]\frac{J_m(k_xw)J_n(k_xw)}{\sqrt{\beta^2-k_0^2-k_x^2}}dk_x \\
&\left\{ \begin{array}{c} m = 2, 4, \cdots \\ n = 2, 4, \cdots \end{array} \right\}
\end{aligned}
\tag{6.105}
$$

Similar expressions are also obtained for the substrate region. It is also noted that

$$\lim_{k_x \to 0} J_n(k_xw) = \left[\frac{k_xw}{2}\right]^n \tag{6.106}$$

Hence, no pole occurs at $k_x = 0$ where $n > 0$. It is also noted that the unknown β appears in the integrand. This difficulty is overcome by tabulating the matrix elements as functions of β given over the range $1 < \beta/k_0 < \sqrt{\varepsilon_r}$ and interpolating numerically over the range.

Moreover, the numerical computation of the integrals in (6.105) can be significantly improved by using the following asymptotic formulas. In fact, as $k_x \to \infty$ the integral in the first of (6.104) assumes the following asymptotic form

$$
\begin{aligned}
I_{mn} &= \int_0^\infty \frac{J_m(k_xw)J_n(k_xw)}{k_x}dk_x = \\
&\frac{1}{\pi}\frac{\left(\frac{m+n}{2}-1\right)!}{(m+n)!}\frac{\sin\left(\pi\frac{m-n}{2}\right)}{(m-n)}, \; m+n > 0
\end{aligned}
\tag{6.107}
$$

[7, pp.692 and 937]. When $m = n = 0$, we make the rational approximation, valid for large k_x

$$\frac{1}{\sqrt{\beta^2 - k_0^2 - k_x^2}} \approx \frac{k_x}{\left(\frac{\beta^2-k_0^2}{2}\right) + k_x^2} \tag{6.108}$$

and we find from [7, p.678]

$$I_{00} = \int_0^\infty \frac{k_xJ_0^2(k_xw)}{\left(\frac{\beta^2-k_0^2}{2}\right) + k_x^2}dk_x = I_0\left(\sqrt{\frac{\beta^2 - k_0^2}{2}}w\right)K_0\left(\sqrt{\frac{\beta^2 - k_0^2}{2}}w\right) \tag{6.109}$$

Adding and subtracting I_{00} in (6.104), we get

$$
\begin{aligned}
[Y_{11}^a]_{00} &= j\omega\varepsilon_0w\left[1 - \left(\frac{\beta}{k_0}\right)^2\right] \\
&\left\{I_{00} + \int_0^\infty \left[\frac{1}{\sqrt{\beta^2-k_0^2-k_x^2}} - \frac{k_x}{\left(\frac{\beta^2-k_0^2}{2}\right)+k_x^2}\right]J_0^2(k_xw)\,dk_x\right\}
\end{aligned}
\tag{6.110}
$$

The integrand now decreases as $1/k_x^4$ for large k_x. Similar results are obtained by adding and subtracting I_{mn} to all matrix elements Y^a and Y^s. Thus, for $m+n > 0$, we have

$$
\begin{aligned}
&[Y_{11}^{a}]_{mn} = j\omega\varepsilon_0(-1)^{(m+n)/2}\sqrt{\varepsilon_m\varepsilon_n}w\left[1-\left(\frac{\beta}{k_0}\right)^2\right] \\
&\left\{I_{mn} + \int_0^\infty \left[\frac{1}{\sqrt{\beta^2-k_0^2-k_x^2}} - \frac{1}{k_x}\right] J_m\left(k_x w\right) J_n\left(k_x w\right) dk_x\right\}
\end{aligned}
\tag{6.111}
$$

and so forth.

Finally, having carried out the discretization process described above, the integral equation (6.82) is reduced to the following matrix equation:

$$
\begin{bmatrix} \mathbf{Y}_{11}(\beta) & \mathbf{Y}_{12}(\beta) \\ \mathbf{Y}_{21}(\beta) & \mathbf{Y}_{22}(\beta) \end{bmatrix} \begin{bmatrix} \mathbf{X} \\ \mathbf{Z} \end{bmatrix} = 0 \tag{6.112}
$$

where typically $\mathbf{X}$ is a two–component vector and $\mathbf{Z}$ a scalar. The vanishing of the determinant of (6.112) gives the propagation constant β of the guide, whereas $\mathbf{X}$ and $\mathbf{Z}$ yield the transverse field distributions within a normalization constant which is fixed by the transverse normalization of the fields.

6.5 Inset dielectric guide

Image line is a recognised low–loss transmission media, with the main disadvantage, besides manufacturing difficulties, posed by considerable radiation loss from all practical components. In order to better confine the field to the structure, trapped image guide has been proposed [17], but this is even harder to make, especially for small guide dimensions. Such manufacturing difficulties are avoided by the inset dielectric guide (IDG), shown in cross–section in Figure 6.22, which was proposed as a low–cost alternative. IDG, which is just a rectangular groove filled with dielectric, has many of the advantages of the trapped image guide without its fabrication problems.

The IDG structure has been analyzed by Zhou and Itoh [46], as an intermediate structure in the analysis of trapped image guide, by using the EDC method. In this manner useful approximate results for the fundamental mode have been obtained. However, a more accurate evaluation of the modal cutoffs and field distributions should consider the singularity imposed on the field by the 90° metal edges. In the following we consider a general solution to the IDG, which yields an accurate equivalent circuit for the fundamental mode.

We adopt a space domain approach, taking into account the singularity imposed on the guide field by the 90° metal edges. The problem is formulated so that only one set of basis functions is required, which results in a further increase in accuracy. As a result, the convergence is very rapid, and often only the first term of the expansion set is sufficient.

6.5.1 *Transverse resonance diffraction analysis of IDG*

The coordinate system used in this analysis is shown, along with the guide structure, in Figure 6.22. For each homogeneous region, we may write

$$
\varepsilon_r k_0^2 \quad = \quad k_x^2 + k_y^2 + \beta^2 \tag{6.113}
$$

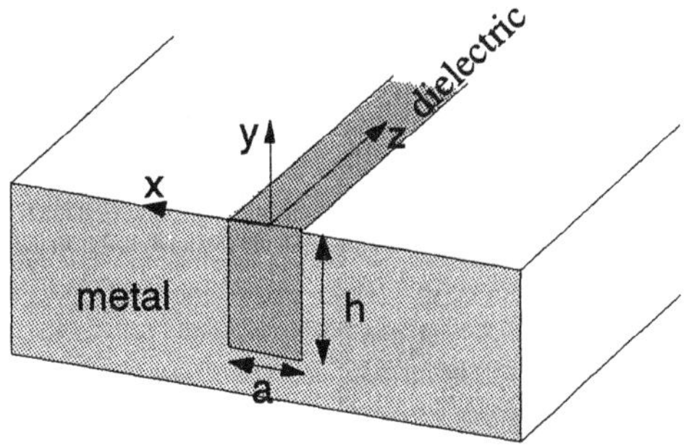

Figure 6.22 Geometry of inset dielectric waveguide and coordinate system.

where β is the z–directed propagation constant to be determined. For such a composite structure it is easier to solve for propagation in a direction transverse to the homogeneous boundaries, i.e. in the y-direction.

From boundary conditions and Maxwell's equations, integral equations for the transverse field components are set up. In order to solve these equations on a computer, they are transformed via Galerkin's technique into scalar equations. In this way, the integral equations are transformed into the space spanned by the set of functions used to discretise the unknown field components E_x and E_z. These functions are chosen to model as accurately as possible the field components, so that few terms are needed for adequate convergence. This involves taking into account the diffraction of the field due to the presence of the 90° metal edges at $x = \pm a/2, y = 0$.

The scalar equations so obtained describe propagation in the transverse direction that can be described by an equivalent network. The resonant frequencies of this equivalent network are those frequencies for which the total admittance at any point in the circuit vanishes. Thus, for a given value of k_0, the propagation constant is found as that value that causes the total transverse admittance to vanish [30, 25, 4, 24]. This approach is found to give fast convergence for the value of β. For most applications, a 2 by 2 or, at most, a 4 by 4 matrix is all that is required to be solved.

The transverse discontinuity at $y = 0$ in the guide cross–section (i.e., the diffraction due to metal edges and the change of dielectric) results in a hybrid mode structure which requires a full six–field analysis. The six field components are obtained from the superposition of the LSE and LSM modes. In view of the air–dielectric interface, it is expedient to derive the hybrid modes from y–directed electric and magnetic Hertzian vector potentials, similarly to what has been done for the slotline, i.e.

$$\begin{aligned} \mathbf{\Pi}_e &= \hat{\mathbf{y}}\psi_e(x,y)e^{-j\beta z} \\ \mathbf{\Pi}_h &= \hat{\mathbf{y}}\psi_h(x,y)e^{-j\beta z} \end{aligned} \tag{6.114}$$

From the potentials the fields are derived as

$$\begin{aligned} \mathbf{E} &= k^2\mathbf{\Pi}_e + \nabla\nabla\cdot\mathbf{\Pi}_e - j\omega\mu\nabla\times\mathbf{\Pi}_h \\ \mathbf{H} &= j\omega\varepsilon\nabla\times\mathbf{\Pi}_e + k^2\mathbf{\Pi}_h + \nabla\nabla\cdot\mathbf{\Pi}_h \end{aligned} \tag{6.115}$$

The scalars $\psi_e(x,y), \psi_h(x,y)$ must be chosen so as to satisfy the correct boundary conditions for the field components and have dimensional consistency. In the following analysis, the z–dependence $\exp^{-j\beta z}$ is suppressed.

The IDG structure can be considered to be a dielectric filled rectangular waveguide with one of the side walls removed. Hence, the field in the slot can be constructed from the superposition of discrete waveguide mode functions. On the contrary, in the air region, the superposition of a continuous spectrum of solutions is considered. The field in each region is thus derived from separate sets of potential functions, and continuity is imposed across the interface between the two regions.

The admittance operators are formulated from the transverse equivalent circuit. For propagation in the y–direction, the slot appears as a short–circuited length of dielectric–filled parallel-plate waveguide radiating into free space. Considering only even–mode solutions with respect to x, then the potential functions are chosen as follows.

In the slot region,

$$\begin{aligned}\psi_e(x,y) &= \frac{1}{j\omega\varepsilon}\sum_{n=0,2,\cdots}^{\infty}\frac{I_{en}}{\sqrt{\beta^2+(\frac{n\pi}{a})^2}}\frac{\cos k_n(y+h)}{\cos k_n y}\xi_n(x)\\ \psi_h(x,y) &= \frac{1}{j\omega\mu_0}\sum_{n=0,2,\cdots}^{\infty}\frac{V_{hn}}{\sqrt{\beta^2+(\frac{n\pi}{a})^2}}\frac{\sin k_n(y+h)}{\sin k_n y}\phi_n(x)\end{aligned} \tag{6.116}$$

where

$$\begin{aligned}\xi_n(x) &= \frac{2}{\sqrt{a}}\sin\left(\frac{n\pi}{a}x\right)\\ \phi_n(x) &= \sqrt{\frac{2\varepsilon_n}{a}}\cos\left(\frac{n\pi}{a}x\right)\end{aligned} \tag{6.117}$$

with

$$\begin{aligned}\varepsilon_0 &= 1\\ \varepsilon_n &= 2 \quad n \neq 0\end{aligned}$$

and the conservation of wavenumbers gives

$$\varepsilon_r k_0^2 = k_n^2 + \beta^2 + \left(\frac{n\pi}{a}\right)^2 \tag{6.118}$$

In the air region,

$$\begin{aligned}\psi_e(x,y) &= \frac{1}{j\omega\varepsilon_0}\int_0^{\infty}\frac{I_e(k_x)}{\sqrt{\beta^2+k_x^2}}e^{-jk_y y}\xi(x;k_x)dk_x\\ \psi_h(x,y) &= \frac{1}{j\omega\mu_0}\int_0^{\infty}\frac{V_h(k_x)}{\sqrt{\beta^2+k_x^2}}e^{-jk_y y}\phi(x;k_x)dk_x\end{aligned} \tag{6.119}$$

where

$$\begin{aligned}\phi(x;k_x) &= \sqrt{\tfrac{2}{\pi}}\cos k_x x\\ \xi(x;k_x) &= \sqrt{\tfrac{2}{\pi}}\sin k_x x\end{aligned} \tag{6.120}$$

and wavenumber conservation yields

$$k_0^2 = k_x^2 + k_y^2 + \beta^2 \tag{6.121}$$

The orthogonal sets $\phi_n(x)$, and $\phi(x;k_x)$, are normalised respectively in the intervals $[0,a/2]$ and $[0,\infty)$, as

$$\begin{aligned} \int_0^{a/2} \phi_m(x)\phi_n(x)dx &= \delta_{mn} \\ \int_0^{\infty} \phi(x;k_x)\phi(x;k'_x)dx &= \delta(k_x - k'_x) \end{aligned} \tag{6.122}$$

and similarly for $\xi_n(x)$ and $\xi(x;k_x)$. The amplitude functions are chosen as in the slotline case for the sake of convenience so as to provide the unknown amplitudes V_h and I_e with the dimensions of voltage and current, respectively. This proves useful particularly when circuit analogies are made. By placing the potential functions (6.116) into (6.115) the field components in the slot are found to be

$$\begin{aligned} E_x(x,y) &= \sum_{n=0,2,\cdots}^{N} E_{xn}\frac{\sin k_n(y+h)}{\sin k_n y}\phi_n(x) \\ E_y(x,y) &= \sum_{n=2,4,\cdots}^{N} E_{yn}\frac{\cos k_n(y+h)}{\cos k_n y}\xi_n(x) \\ E_z(x,y) &= \sum_{n=2,4,\cdots}^{N} E_{zn}\frac{\sin k_n(y+h)}{\sin k_n y}\xi_n(x) \\ H_x(x,y) &= \sum_{n=2,4,\cdots}^{N} H_{xn}\frac{\cos k_n(y+h)}{\cos k_n y}\xi_n(x) \\ H_y(x,y) &= \sum_{n=0,2,\cdots}^{N} H_{yn}\frac{\sin k_n(y+h)}{\sin k_n y}\phi_n(x) \\ H_z(x,y) &= \sum_{n=0,2,\cdots}^{N} H_{zn}\frac{\cos k_n(y+h)}{\cos k_n y}\phi_n(x) \end{aligned} \tag{6.123}$$

Analogous expressions are written in the air region where discrete sums are replaced by integrals. Note that the amplitudes E_{xn}, E_{yn}, etc. are as yet unknown.

It is desired that the E fields transverse to y in the above be retained as the unknown quantities, whereas the two remaining quantities transverse to y, i.e. H_z, H_x, be expressed in terms of the above two. By straightforward Fourier analysis of (6.123) over the slot, it is possible to express each set of amplitudes, e.g. H_{zn}, as a linear operation on the amplitudes E_{xn}, E_{zn}. This linear transformation is compactly written by using a linear integral operator acting on the fields $E_x(x,0)$, $E_z(x,0)$ to give the fields $H_z(x,0)$, namely

$$\begin{aligned} H_z(x,0) = \textstyle\int_0^{a/2} Y_{11}(x,x',y=y'=0)E_x(x',0)dx' + \\ \textstyle\int_0^{a/2} Y_{12}(x,x',y=y'=0)E_z(x',0)dx' \equiv \hat{Y}_{11}\cdot E_x + \hat{Y}_{12}\cdot E_z \end{aligned} \tag{6.124}$$

and similarly for $H_x(x,0)$. Analogous expressions hold in the air region. Thus we obtain for each region integral equations of the type

$$\pm\begin{bmatrix} H_z(x,0) \\ -H_x(x,0) \end{bmatrix} = \begin{bmatrix} \hat{Y'}^{s,a}_{11} & \hat{Y'}^{s,a}_{12} \\ \hat{Y'}^{s,a}_{21} & \hat{Y'}^{s,a}_{22} \end{bmatrix} \cdot \begin{bmatrix} E_x(x,0) \\ E_z(x,0) \end{bmatrix} \tag{6.125}$$

where the symbols s,a and the sign on the LHS refer to the slot and air region, respectively.

The signs are consistent with the power flow from the interface into each region. Continuity of the fields at the interface ($y = 0$) gives an integral equation in the unknowns $E_x(x,0)$, $E_z(x,0)$. It is convenient to expand both unknowns in terms of the same set of functions and this is possible when the two fields display the same x-dependence. This will be so if, instead of $E_z(x,0)$, the problem is formulated in terms of $dE_z(x,0)/dx$. As an added bonus, as already discussed for slotline, proper convergence of the admittance operators will also result from this transformation. Thus, integration by parts of (6.125) yields

$$\pm\begin{bmatrix} H_z(x,0) \\ -\frac{\pi}{a}\int H_x(x,0)dx \end{bmatrix} = \begin{bmatrix} \hat{Y}_{11}^{s,a} & \hat{Y}_{12}^{s,a} \\ \hat{Y}_{21}^{s,a} & \hat{Y}_{22}^{s,a} \end{bmatrix} \cdot \begin{bmatrix} E_x(x,0) \\ \frac{a}{\pi}\frac{dE_z(x,0)}{dx} \end{bmatrix} \tag{6.126}$$

As shown by Itoh in [14], a hybrid field can be resolved into pure TE or TM components in a vertical direction y by means of a coordinate rotation about the direction of propagation. Following this procedure, angles can be defined that give rise to pure TE or TM fields. In the slot region, these angles are discrete and defined as

$$\begin{aligned} \cos\theta &= \frac{\frac{n\pi}{a}}{\sqrt{\left(\frac{n\pi}{a}\right)^2+\beta^2}} \\ \sin\theta &= \frac{\beta}{\sqrt{\left(\frac{n\pi}{a}\right)^2+\beta^2}} \end{aligned} \tag{6.127}$$

while for the air region,

$$\begin{aligned} \cos\theta &= \frac{k_x}{\sqrt{k_x^2+\beta^2}} \\ \sin\theta &= \frac{\beta}{\sqrt{k_x^2+\beta^2}} \end{aligned} \tag{6.128}$$

It is noted that in this problem, the two half-spaces are different and the two sets of angles (6.127) and (6.128) are therefore separately defined, unlike in the standard spectral domain approach [14] where the cross-section does not vary along y. Using the above definitions, the Green admittances obtained from (6.126) are, in the slot region,

$$\begin{aligned} Y_{11}^s &= \sum_{n=0,2,}^{N} \left(Y'_n\cos^2\theta + Y''_n\sin^2\theta\right)\phi_n(x)\phi_n(x') \\ Y_{12}^s = -Y_{21}^s &= \sum_{n=2,4,}^{N} j\left(Y'_n - Y''_n\right)\frac{1}{n}\cos\theta\sin\theta\,\phi_n(x)\phi_n(x') \\ Y_{22}^s &= \sum_{n=2,4,}^{N} \left(Y'_n\sin^2\theta + Y''_n\cos^2\theta\right)\frac{1}{n^2}\phi_n(x)\phi_n(x') \end{aligned} \tag{6.129}$$

In the air region, the admittances are given by

$$Y_{11}^a = \int_0^\infty \left(Y'\cos^2\theta + Y''\sin^2\theta\right)\phi(x,k_x)\phi(x',k_x)dk_x$$

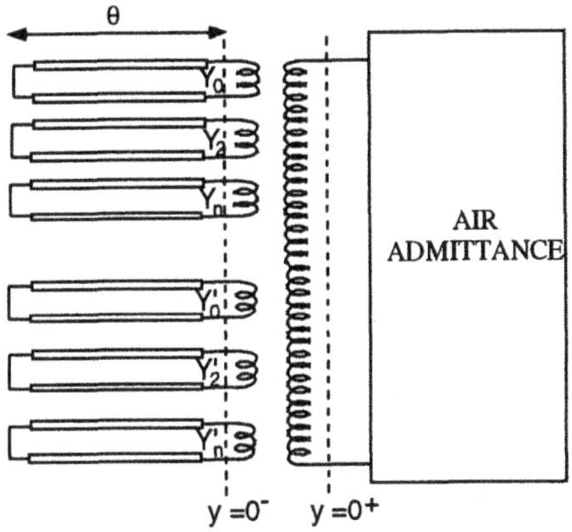

Figure 6.23 Transverse equivalent circuit representation for inset dielectric waveguide. The parallel plates are short–circuited at the electric distances $\theta_n = k_{yn}h$ left of the interface.

$$Y_{21}^a = -Y_{12}^a = j\int_0^\infty (Y' - Y'')\sin\theta\cos\theta\frac{\pi}{ak_x}\phi(x,k_x)\phi(x',k_x)dk_x \tag{6.130}$$
$$Y_{22}^a = \int_0^\infty \left(Y'\sin^2\theta + Y''\cos^2\theta\right)\left(\frac{\pi}{ak_x}\right)^2\phi(x,k_x)\phi(x',k_x)dk_x$$

where Y_n' and Y_n'' are the input admittances of the slot seen by the nth order TM and TE modes, respectively, i.e.,

$$Y'_n = -j\frac{\omega\varepsilon}{k_n}\cot(k_n h)$$
$$Y''_n = -j\frac{k_n}{\omega\mu_0}\cot(k_n h) \tag{6.131}$$

In the air region the admittance Y' of the TM waves is given by

$$Y'(k_x) = \frac{\omega\varepsilon_0}{k_y} \tag{6.132}$$

whereas that of TE^y waves in air, Y'' is,

$$Y''(k_x) = \frac{k_y}{\omega\mu_0} \tag{6.133}$$

This formulation produces the transverse equivalent circuit representation shown in Figure 6.23. The slot field is composed of an infinite number of TE and TM components which are transformed by the coordinate rotation into an admittance dyadic. It is noted that this spatial formulation, unlike the standard spectral domain approach, can easily account for changes in the transverse geometry of the structure. In this problem the following integral equation (6.134) has to be solved in the space domain, unlike in the previous example, slotline, where solution in either domain was equally feasible.

Discretization of the dispersion equation

Bearing in mind that the admittance operators have been defined for each region for positive power flow into each region from the interface, continuity of the fields at $(y = 0)$ gives the equation

$$\left[\hat{Y}\right] \cdot \begin{bmatrix} E_x(x,0) \\ \frac{a}{\pi}\frac{dE_z(x,0)}{dx} \end{bmatrix} = 0 \tag{6.134}$$

where

$$\hat{Y} = \hat{Y}^s + \hat{Y}^a \tag{6.135}$$

To solve the dispersion equation (6.134) it is necessary to discretise it; thanks to the set of functions used to expand the unknown $E_x(x,0)$ and $E_z'(x,0)$. The choice of a finite expanding set is crucial for achieving rapid convergence in the discretization of the dispersion equation. If the choice satisfies the edge condition, then the "scalar products" in (6.134) will converge rapidly and only a few terms are needed.

As seen in Chapter 2, the singularity from a 90° metal edge is of the order $r^{-1/3}$. Thus we seek an orthogonal set of functions that can be weighted by a term that takes into account the effect of the singularity. A function that shows the correct singularity behaviour at $x = \pm a/2$ and has a continuous derivative is given by

$$W(x) = \left[1 + \left(\frac{2x}{a}\right)\right]^{-1/3}\left[1 - \left(\frac{2x}{a}\right)\right]^{-1/3} = \left[1 - \left(\frac{2x}{a}\right)^2\right]^{-1/3} \tag{6.136}$$

With such a weight function, an obvious choice for the basis terms are the Gegenbauer polynomials since they are orthonormal over the interval $0 \leq \frac{2x}{a} \leq 1$ just with respect to that weight function. Thus we expand the field unknowns in terms of a weighted set of the normalised even Gegenbauer polynomials:

$$f_m(x) = \frac{1}{N_m} C_m^{1/6}\left(\frac{2x}{a}\right) \tag{6.137}$$

so that

$$\begin{aligned} E_x(x,0) &= W(x) \sum_{m=0,2,\cdots} X_m f_m(x) \\ \frac{a}{\pi}\frac{dE_z(x,0)}{dx} &= W(x) \sum_{m=2,4,\cdots} Z_m f_m(x) \end{aligned} \tag{6.138}$$

Note that, in order to satisfy the boundary conditions of the $E_z(x,0)$ field component, the summation in the second of the above contains no constant term and starts from $m = 2$. The coefficients X_m and Z_m are the as yet undetermined amplitudes. When introducing the last expansion into (6.134) the inner products P_{mn} and $P_m(k_x)$, defined as

$$\begin{aligned} P_{mn} &= \langle f_m(x), W(x)\phi_n(x)\rangle \equiv \int_0^{a/2} W(x) f_m(x)\phi_n(x)dx \\ P_m(k_x) &= \langle f_m(x), W(x)\phi(x,k_x)\rangle \equiv \int_0^{a/2} W(x) f_m(x)\phi(x,k_x)dx \end{aligned} \tag{6.139}$$

have to be evaluated across the half slot. This evaluation can be done analytically by using the result [7]

$$\int_0^1 \left(1-t^2\right)^{\nu-1/2} C_{2m}^{\nu}(t)\cos bt\,dt = \frac{(-1)^n \pi\Gamma(2n+2\nu)J_{\nu+2n}(b)}{(2n)!\Gamma(\nu)(2b)^{\nu}} \quad \left[\mathrm{Re}\nu > -\tfrac{1}{2}, b>0\right] \tag{6.140}$$

With the substitutions $\nu = 1/6$, $t = 2x/a$, $b = n\pi/2$ and $2n = m$ we can write

$$P_{mn} = \frac{1}{N_m}\frac{\delta_n\sqrt{a}(-1)^{m/2}\pi\Gamma\left(m+\frac{1}{3}\right)J_{m+1/6}\left(\frac{n\pi}{2}\right)}{(m)!\Gamma\left(\frac{1}{6}\right)(n\pi)^{1/6}} \tag{6.141}$$

where

$$N_m^2 = \frac{a2^{-4/3}\pi\Gamma\left(m+\frac{1}{3}\right)}{(m)!\left[\Gamma\left(\frac{1}{6}\right)\right]^2\left(m+\frac{1}{6}\right)} \tag{6.142}$$

These expressions are only valid for $n > 0$. Taking the limit as n goes to 0 gives

$$\begin{aligned} P_{00} &= \frac{1}{N_0}\frac{\sqrt{a}\pi\Gamma\left(\frac{1}{3}\right)}{\Gamma\left(\frac{1}{6}\right)2^{5/6}\Gamma\left(1+\frac{1}{6}\right)} \\ P_{m0} &= 0 \end{aligned} \tag{6.143}$$

Similar expressions hold for $P_m(k_x)$ with the substitution

$$\begin{aligned} P_m(k_x) &= \sqrt{\frac{a}{2\pi}}P_{m\overline{k}} \\ \overline{k} &= \frac{a}{2\pi}k_x \end{aligned} \tag{6.144}$$

Upon discretization we obtain from (6.134) the following matrix equation

$$[\mathbf{Y}]\begin{bmatrix}\mathbf{X}\\ \mathbf{Z}\end{bmatrix} = 0 \tag{6.145}$$

which has a nontrivial solution when

$$\det|\mathbf{Y}| = 0 \tag{6.146}$$

In order to recover a single line transverse equivalent circuit, it is convenient to consider the fundamental transverse propagating mode in the slot, seen as a parallel–plate waveguide terminated by a short–circuit at $y = -h$. This is incident upon a discontinuity (the transition between the two regions) and excites the radiation modes in the air region and the higher order nonpropagating transverse modes in the slot. In this manner, the fundamental slot mode can be isolated in (6.134) and all other contributions lumped together to give

$$\left[\hat{Y}\right]\cdot\begin{bmatrix}E_x(x,0)\\ \frac{a}{\pi}\frac{dE_x(x,0)}{dx}\end{bmatrix} = \begin{bmatrix}-H_{z0}(x,0)\phi_0(x)\\ 0\end{bmatrix} \tag{6.147}$$

where $\hat{Y}$ is now the slot admittance operator with the fundamental term removed. Also, this quantity will denote from now on the total admittance with the fundamental term removed. Upon discretization the latter equation becomes

$$[\mathbf{Y}]\begin{bmatrix}\mathbf{X}\\ \mathbf{Z}\end{bmatrix} = (-H_{z0})\begin{bmatrix}\mathbf{P}_0\\ \mathbf{0}\end{bmatrix} \tag{6.148}$$

Rearranging,

$$\begin{bmatrix} \mathbf{X} \\ \mathbf{Z} \end{bmatrix} = (-H_{z0})\,[\mathbf{Y}]^{-1} \begin{bmatrix} \mathbf{P}_0 \\ \mathbf{0} \end{bmatrix} \tag{6.149}$$

Multiplying both sides by $\mathbf{P}_0^T\ 0^T$ yields

$$[\mathbf{P}_0^T \quad 0^T] \begin{bmatrix} \mathbf{X} \\ \mathbf{Z} \end{bmatrix} = (-H_{z0})\,[\mathbf{P}_0^T \quad 0^T]\,[\mathbf{Y}]^{-1} \begin{bmatrix} \mathbf{P}_0 \\ \mathbf{0} \end{bmatrix} \tag{6.150}$$

However, from the first of (6.123) and (6.138), we have

$$\mathbf{P}_0^T \mathbf{X} = E_{x0} \tag{6.151}$$

Hence

$$\frac{-H_{z0}}{E_{x0}} = \left([\mathbf{P}_0^T \quad 0^T]\,[\mathbf{Y}]^{-1} \begin{bmatrix} \mathbf{P}_0 \\ \mathbf{0} \end{bmatrix} \right)^{-1} \tag{6.152}$$

The normalised admittance of the fundamental Fourier component in the slot (slot mode for short) when looking from $y = 0$ toward the short circuit is

$$\frac{H_{z0}}{E_{x0}} = -j \cot k_{y0} h \tag{6.153}$$

Equation (6.152) represents the admittance of all the higher order slot modes and air waves as seen from the interface. At resonance, the total admittance seen from both sides of the interface must be zero. Therefore the equation for resonance is

$$-j \cot k_{y0} h + \left([\mathbf{P}_0^T \quad 0^T] \begin{bmatrix} \mathbf{Y}_{11} & \mathbf{Y}_{12} \\ \mathbf{Y}_{21} & \mathbf{Y}_{22} \end{bmatrix}^{-1} \begin{bmatrix} \mathbf{P}_0 \\ \mathbf{0} \end{bmatrix} \right)^{-1} = 0 \tag{6.154}$$

When N basis terms are used, Y_{11} becomes an N by N matrix.

The elements of the admittance matrices of the slot region occurring in the latter equation can be found from (6.130) to be

$$(Y_{11}^s)_{km} = -ju \sum_{n=2,4,\cdots} \frac{\pi h}{a} \frac{\coth\sqrt{n^2 - u^2}}{\sqrt{n^2 - u^2}} P_{kn} P_{mn} \tag{6.155}$$

and so on, where

$$\begin{aligned} u &\equiv \tfrac{a}{\pi}\sqrt{\varepsilon_r k_0^2 - \beta^2} \\ k_n &= \tfrac{\pi}{a}\sqrt{u^2 - n^2} \end{aligned} \tag{6.156}$$

$$\begin{array}{llll} k = 0, 2, 4, & \ldots \quad m = 0, 2, 4, & \ldots & \text{for } Y_{11} \\ k = 0, 2, 4, & \ldots \quad m = 2, 4, 6, & \ldots & \text{for } Y_{12} \\ k = 2, 4, 6, & \ldots \quad m = 0, 2, 4, & \ldots & \text{for } Y_{21} \\ k = 2, 4, 6, & \ldots \quad m = 2, 4, 6, & \ldots & \text{for } Y_{22} \end{array} \tag{6.157}$$

Corresponding expressions are derived from (6.130) for the admittance matrices of the air region (see also (6.96)). The apparent pole singularity at $k_x = 0$ in (6.130) is in fact compensated by a zero of $P_m(k_x)$ there.

6.5.2 Analysis of the partially filled IDG

In order to minimise the field perturbation by the 90° edges, it may be helpful to lower the level of the dielectric filling in the slot. Such a change in transverse circuit is easily accommodated by the formulation. For this structure the resonance condition is still enforced by the plane $y = 0$, so that the air admittance operators are unchanged. The admittance operators for the slot region must be modified to take into account the new y–variation. The latter is given for V_{hn} and I_{en}, respectively, as the current and voltage variation along a cascaded section of transmission line. Hence, with suitable normalization, the potentials become

$$
\begin{aligned}
\psi_e(x,y) &= \frac{1}{j\omega\varepsilon}\sum_{n=0,2,\cdots}^{\infty}\frac{I_{en}}{\sqrt{\beta^2+\left(\frac{n\pi}{a}\right)^2}}f(y)\xi_n(x)\\
\psi_h(x,y) &= \frac{1}{j\omega\mu_0}\sum_{n=0,2,\cdots}^{\infty}\frac{V_{hn}}{\sqrt{\beta^2+\left(\frac{n\pi}{a}\right)^2}}g(y)\phi_n(x)
\end{aligned}
\tag{6.158}
$$

with

$$
\begin{aligned}
f(y) &= \frac{jY'_0\sin k_{n2}(y+h_2)+Y'_L\cos k_{n2}(y+h_2)}{jY''_0\sin k_{n2}h_2+Y'_L\cos k_{n2}h_2}\\
g(y) &= \frac{Y''_0\cos k_{n2}(y+h_2)+Y''_L\sin k_{n2}(y+h_2)}{jY''_0\cos k_{n2}h_2+Y''_L\sin k_{n2}h_2}
\end{aligned}
\tag{6.159}
$$

and where

$$
\begin{aligned}
Y'_L &= -jY'_0\cot k_{n1}h_1\\
Y''_L &= -jY''_0\cot k_{n1}h_1
\end{aligned}
\tag{6.160}
$$

The subscripts 1 and 2 refer to propagation in the dielectric– and air–filled regions of the slot, respectively. With these potential functions, the analysis follows as before. The modified admittance operators for the partially filled slot case can be found by inspection by replacing Y'_n and Y''_n with the TM and TE admittances for a cascaded section of transmission line, showing the flexibility of this approach.

Simplified transverse equivalent circuit.

By means of the rotation of the coordinate system established in [14], the hybrid field can be resolved into TE and TM components. This is useful in enabling us to develop an equivalent circuit model of the transverse network: it consists of short–circuited lengths of transmission line, one for each TM and TE slot component, that are mutually transformer–coupled via the coordinate rotation and coupled to a radiation admittance. The transverse equivalent circuit is shown in Figure 6.24.

When propagation in the transverse y–direction is considered, admittance terms like (6.155) can be characterised as follows: Y_{11} and Y_{22} correspond to the TE and TM contributions, respectively, whereas Y_{12} and Y_{21} express the coupling between the two polarizations. In general the dispersion equation (6.154) will contain both TE and TM contributions. However, when only the first basis term is considered in (6.154), the matrix equation reduces to the scalar equation

$$
-j\cot\left(\frac{\pi uh}{a}\right)+\frac{Y_{11}}{P_{00}^2} = 0 \tag{6.161}
$$

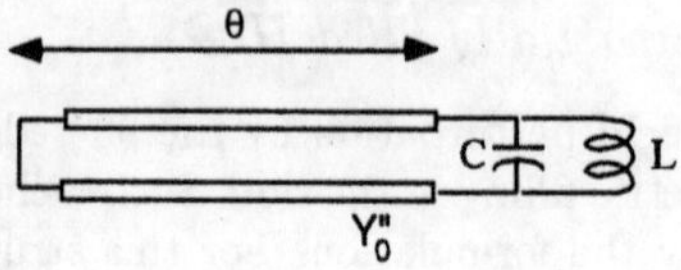

Figure 6.24 Simplified transverse equivalent circuit obtained by assuming that only the fundamental slot mode is propagating.

This equation contains only TE components, thus allowing a simplification to be made in the transverse equivalent circuit. Writing out the latter equation in full, we obtain

$$-j\cot\left(\frac{\pi u h}{a}\right)+ \\ ju\sum_{n=2}^{\infty}\frac{\coth\sqrt{n^2-u^2}\left(\frac{\pi h}{a}\right)}{\sqrt{n^2-u^2}}\left(\frac{P_{0n}}{P_{00}}\right)^2+\frac{w^2}{ju}\int_0^{\infty}\frac{1}{\sqrt{w^2-\overline{\rho}^2}}\left(\frac{P_0(\rho)}{P_{00}}\right)^2 d\rho=0 \qquad (6.162)$$

where

$$\begin{aligned} u &= \frac{a}{\pi}\sqrt{\varepsilon_r k_0^2-\beta^2} \\ w &= \frac{a}{\pi}\sqrt{\beta^2-k_0^2} \\ \overline{\rho} &= \frac{a}{\pi}\rho \end{aligned} \qquad (6.163)$$

This equation contains the "effective" frequency variable u, which is the y–directed wavenumber of the $n = 0$ Fourier component in the slot (fundamental slot mode). Thus, by considering the dependence of (6.162) with respect to u, the terms can be recognised as elements in a transverse equivalent circuit for the fundamental mode. In this way the terms in (6.162) can be identified as follows: the first term represents a short–circuited length of transmission line for the fundamental mode. The second term is capacitive and represents the energy storage in the slot due to the nonpropagating modes. The third term describes the energy storage of the nonpropagating TE modes in the air region. The admittance terms so identified give rise to the simple transverse equivalent circuit of Figure 6.24.

A wealth of numerical results demonstrating the accuracy of the proposed approach have been published in [13] and in [37].

Another interesting structure that can be analyzed in a quite similar manner is the microstrip loaded inset dielectric guide (see [38, 39]). In this case, the main difference, from a theoretical point of view, is the presence of two different types of field singularities of the field which leads to the use of Jacobi's polynomials in order to express the edge condition.

Bibliography

[1] T. M. Benson and J. Buus. Optical guiding in III-V semiconductor rib structures. *IEE Conf. Publ. 227, Cardiff, U.K*, pages 17–20, April 1983.

[2] S. B. Cohn. Slot line on a dielectric substrate. *IEEE Trans. on Microwave Theory Tech.*, 17(10):768–778, October 1969.

[3] U. Crombach. Analysis of single and coupled rectangular dielectric waveguides. *IEEE Trans. Microwave Theory Tech.*, 29(9):870–874, September 1981.

[4] N. Dagli and C. Fonstad. Analysis of rib dielectric waveguides. *IEEE J. Quantum Electron.*, 21:315–319, April 1985.

[5] J. P. Donnely, N. L. De Meo, and G. A. Ferrante. Three guide couplers in GaAs. *IEEE J. Lightwave Technol.*, 1(2):417–424, June 1983.

[6] E. A. Mariani, et al. Slotline characteristics. *IEEE Trans. Microwave Theory Tech.*, 17:1091–1096, 1969.

[7] I. S. Gradshteyn and I. M. Ryzhik. *Table of Integrals, Series and Products.* Academic Press, New York, 1980.

[8] K. C. Gupta, R. Garg, and I. J. Bahl. *Microstrip Lines and Slotlines.* Artech House, Washington, 1979.

[9] A. Hardy and W. Streifer. Coupled mode solutions of multiwaveguide systems. *IEEE J. Quantum Electron.*, 22:528–534, April 1986.

[10] H. A. Haus and C. G. Fonstad. Three waveguide couplers for improved sampling and filtering. *IEEE J. Quantum Electron.*, 17:2321–2325, December 1981.

[11] H. A. Haus, W. P. Huang, S. Kawakami, and N. A. Whitaker. Coupled mode theory of optical waveguides. *IEEE J. Lightwave Technol.*, 5:16–23, January 1987.

[12] K. Hayata, M. Koshiba, M. Eguchi, and M. Suzuki. Vectorial finite-element method without any spurious solutions for dielectric waveguiding problems using transverse magnetic-field component. *IEEE Trans. Microwave Theory Tech.*, 34:1120–1124, November 1986.

[13] S. J. Hedges. *The analysis of inset dielectric guide by transverse resonance diffraction.* PhD thesis, University of Bath, Bath, U.K., 1987.

[14] T. Itoh. Spectral domain immittance approach for dispersion characteristics of generalized printed transmission lines. *IEEE Trans. Microwave Theory Tech.*, 28:733–736, 1980.

[15] T. Itoh. *Planar Transmission Lines Structures.* IEEE Press, New York, 1987.

[16] T. Itoh. *Numerical Techniques for Microwave and Millimeter-Wave Passive Structures.* John Wiley & Sons, Singapore, 1989.

[17] T. Itoh and B. Adelseck. Trapped image guide for millimeter wave circuits. *IEEE Trans. Microwave Theory Tech.*, 28:1433–1436, December 1980.

[18] T. Itoh and R. Mittra. Dispersion characteristics of slotlines. *Electronics Lett.*, 7:364–365, 1971.

[19] J. Citerne et. al. Fundamental and higher order modes in microslot lines. *Proc. 5th European Microwave Conf. (Hamburg)*, pages 273–277, September 1975.

[20] R. Janaswami and D. H. Schaubert. Characteristic impedance of a wide slotline on low-permittivity substrates. *IEEE Trans. Microwave Theory Tech.*, 34:900–902, August 1986.

[21] J. B. Knorr and K. D. Kuchler. Analysis of coupled slots and coplanar strips on dielectric substrate. *IEEE Trans. Microwave Theory Tech.*, 23:541–548, 1975.

[22] M. R. Knox and P. P. Toulios. Integrated circuits for the millimetre through optical frequency range. *Proceedings of the symposium on submillimetre waves. Polytechnic, N.Y.*, pages 497–516, 1970.

[23] R. M. Knox. Dielectric waveguide integrated circuits - an overview. *IEEE Trans. Microwave Theory Tech.*, 24:806–814, November 1976.

[24] M. Koshiba and M. Suzuki. Vectorial wave analysis of dielectric waveguides for optical-integrated circuits using equivalent network approach. *Journal of Lightwave Technol.*, 4:656–664, July 1986.

[25] J. S. Kot and T. Rozzi. Rigorous modelling of single end coupled image and insular waveguide by transverse resonance diffraction. *Proc. 14th European Microwave Conf. (Liege)*, pages 424–429, September 1984.

[26] W. V. McLevige, T. Itoh, and R. Mittra. New waveguide structure for millimetre wave and optical integrated circuits. *IEEE Trans. Microwave Theory Tech.*, 23(10):788–794, October 1975.

[27] R. Mittra, Y. L. Hou, and V. Jamnejad. Analysis of open dielectric waveguides using mode matching technique and variational methods. *IEEE Trans. Microwave Theory Tech..*, 28:36–43, January 1980.

[28] Proceedings of the IEEE, editor. *Special Issue on Terahertz Technology*, New York, November 1992. IEEE.

[29] A. A. Oliner, S. T. Peng, T. I. Hsu, and A. Sanchez. Guidance and leakage properties of a class of open dielectric waveguide: Part 2 - new physical effects. *IEEE Trans. Microwave Theory Tech.*, 29:856–869, September 1981.

[30] S. T. Peng and A. A. Oliner. Guidance and leakage properties of a class of open dielectric waveguide: Part 1 - mathematical formulation. *IEEE Trans. Microwave Theory Tech.*, 29:843–855, September 1981.

[31] E. C. M. Pennings. *Bends in Optical Ridge Waveguides.* Delft University, Delft, 1990.

[32] B. M. A. Rahman and J. B. Davies. Vectorial-H finite element solution of GaAs/GaAlAs rib waveguide. *Proc. Inst. Elec. Eng., Pt. J*, 132:349–353, April 1985.

[33] M. J. Robertson, S. Ritchie, and P. Dayan. Semiconductor waveguides: analysis of optical propagation in single rib structures and directional couplers. *IEE Proc., Pt. J*, 132(6), December 1985.

[34] G. H. Robinson and J. L. Allen. Slotline application to miniature finite devices. *IEEE Trans. Microwave Theory Tech.*, 17:1097–1101, 1969.

[35] P. N. Robson and P. C. Kendal. *Rib Waveguide Theory by the Spectral Index Method.* John Wiley & sons Inc., Somerset, 1990.

[36] T. Rozzi, G. Cerri, M. N. Husain, and L. Zappelli. Variational analysis of the dielectric rib waveguide using the concept of transition function and including edge singularities. *IEEE Trans. Microwave Theory Tech.*, 39(2):247–257, February 1991.

[37] T. Rozzi and S. Hedges. Rigorous analysis and network modelling of the inset dielectric guide. *IEEE Trans. Microwave Theory Tech.*, 35:823–833, September 1987.

[38] T. Rozzi, R. De Leo, and A. Morini. Analysis of the microstrip loaded inset dielectric waveguide. *IEEE MTT's Int. Microwave Symp.*, pages 923–926, 1989.

[39] T. Rozzi, A. Morini, and G. Gerini. Analysis and applications of microstrip-loaded inset dielectric waveguide. *IEEE Trans. Microwave Theory Tech.*, 40:272–278, February 1992.

[40] T. Rozzi and G. H. In't Veld. Field and network analysis of interacting step discontinuities in planar dielectric waveguides. *IEEE Trans. Microwave Theory Tech.*, 27(4):303–309, April 1979.

[41] R. Sorrentino. *Numerical Methods for Passive Microwave and Millimeter-Wave Structures.* IEEE Press, New York, 1989.

[42] M. S. Stern. Semivectorial polarized finite difference method for optical waveguides with arbitrary index profile. *IEE Proc., Pt. J*, 135(1):56–63, February 1988.

[43] T. Trinh and R. Mittra. Coupling characteristics of planar dielectric waveguides of rectangular cross section. *IEEE Trans. Microwave Theory Tech.*, 29:875–880, September 1981.

[44] Y. Twu, A. Dienes, S. Wang, and J. R. Whinnery. High power coupled ridge waveguide semiconductor laser arrays. *Appl. Phys. Lett.*, 45:709–711, October 1984.

[45] A. Yariv. Coupled mode theory for guided wave optics. *IEEE J. Quantum Electron.*, 9:919, 1973.

[46] W. H. Zhou and T. Itoh. Analysis of trapped image guides using effective dielectric constant and surface impedances. *IEEE Trans. Microwave Theory Tech.*, 30:2163–2166, December 1982.

Chapter 7

Complete spectra of open waveguides

The continuum of an open waveguide describes radiation as simply, in principle, as the discrete spectrum of a classical guide describes any physical field in it. In fact, the knowledge of the modal spectra for open waveguides allows us to apply to discontinuity problems in open environments the same techniques originally developed for closed waveguides. However, this part of the spectrum, the continuum, has received little attention for most open guides. Therefore, this Chapter is devoted to an introduction of the continuum. The Chapter can be considered to be divided into two parts. In the first part we introduce the main characteristics of this part of the spectrum. In the second part we exemplify the procedure by deriving the continuum for some practical transmission lines such as coplanar, rib and inset waveguides.

In particular, in Section 7.1 we explore the main ideas which lead to finding the continuum and we also discuss its main features. In order to illustrate such a derivation we establish the continuum of a flange–backed rectangular waveguide slotted on its narrow wall as, in this example, the lack of a dielectric layer significantly simplifies the problem. In fact, the flange–backed rectangular waveguide does not possess any bound mode; moreover the fields can be considered as purely TE, without the need to introduce hybrid fields. Note that, in this case, the continuum is derived by considering the phase shift imposed on a plane wave incident from infinity by the presence of the slot and of the waveguide behind it.

In Section 7.2 the modes of open waveguides with non separable cross–sections are derived by means of an extension of the resonance equation for the electromagnetic field. The resonance equation is generalised according to considerations of functional analysis. By this approach, it is possible to derive the modal spectrum from the simultaneous diagonalization of the real and imaginary parts of the admittance of the structure. A variational interpretation of the solution of the generalised resonance equation gives additional insight into the modes of open waveguides. The generalised resonance equation, when applied to three–dimensional objects, provides the well known characteristic modes of these structures. The relationship between continuous spectrum, characteristic modes and leaky waves is also discussed.

In Section 7.3 the concepts previously introduced for scalar problems are generalised in order to consider fully hybrid fields.

The remaining Sections are devoted to illustrating, with some analytical details, the derivation of the continuum part of the spectrum for some practical waveguides, namely the CPW, the rib and the inset waveguides. The knowledge of this part of the spectrum allows us to consider, in the next Chapter, some discontinuity problems for three dimensional open waveguides.

7.1 Continuous spectrum of a slotted rectangular waveguide

The classical method for determining the electromagnetic field excited by a source in any closed waveguide is to evaluate the modal amplitudes excited by the source by means of the Lorentz theorem [1, pp.358-362] or its equivalents. The same process can be followed in open guides of one–dimensional cross–section, say, a dielectric slab [3, pp.303-306], [1, pp.485-495, pp.538-546], [23], or a two–dimensional separable one, such as an optical fibre. The power distributes itself among bound modes, if any, and a continuous orthonormal spectrum of real waves, bound at infinity. The proper, modal, spectrum of the guide is not to be confused with the "leaky modes" that are nonmodal, nonorthogonal, complex solutions of the wave equation (of the transverse resonance condition), growing at infinity and, as such, unsuitable for representing the field except in the immediate vicinity of the source.

For most guides of two–dimensional and essentially nonseparable cross–section, including the classical rectangular slotted waveguide, the evaluation of the continuous spectrum is a relatively new acquisition [19].

In the early stages, in order to calculate the pure LSE/LSM continua of the inset [18] and image [17] guide, a partial wave (spectral) decomposition of the field in a transverse direction of the guide cross–section was used. Each partial wave underwent a different phase–shift according to its transverse wavenumber due to transverse diffraction and the total field was then recomposed by superposition. In this scheme, each partial wave did not individually satisfy boundary and edge conditions in the transverse cross–section, resulting in various drawbacks.

In [20] the principle of a new approach was developed; it introduced the concept of a continuum of mutually orthonormal "packets" of waves, each packet travelling with the same propagation constant along the guide, each individually satisfying boundary and edge conditions on the cross–section. The latter formulation, being essentially self–consistent with transverse diffraction, is superior to the earlier one.

In the present paragraph, we consider the classical problem of a rectangular waveguide slotted in its narrow wall, that was studied many years ago by various authors as a "leaky" guide antenna and is described, for instance, in [2, ch. 14, 19, and 20] and [24]. The simplicity of the geometry allows insight into the operation of the method with modest analytical detail.

For the above structure, just a continuous spectrum exists; once this is derived we demonstrate, in the next Chapter, its application to a practical case by applying it to the problem of determining transmission and radiation properties of the junction between a flange–backed ordinary guide and one slotted on its narrow wall.

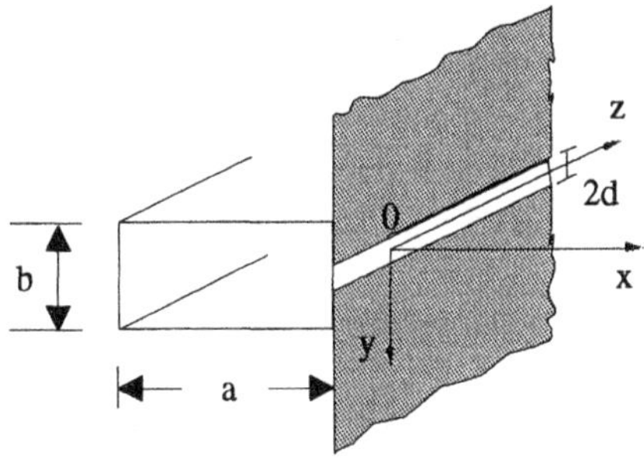

Figure 7.1 Geometry of a flange–backed rectangular waveguide with a symmetrical slot on its narrow wall.

7.1.1 Analysis

The geometry under study is shown in Figure 7.1. It consists of a classical rectangular waveguide with a slot symmetrically placed on its narrow wall (of zero thickness) backed by a perfectly conducting flange. If the excitation is pure TE, i.e. by the fundamental waveguide mode, this problem is describable in terms of a single TE potential $\mathbf{\Pi}_h = \mathbf{z}_0\Psi_h$. Clearly, no bound solutions are possible for this problem due to the presence of the slot.

We are looking for real solutions of the scalar wave equation for E_y, say,

$$\nabla_t^2 e_\nu + k_t^2 e_\nu = 0 \tag{7.1}$$

where $k_t^2 = k_x^2 + k_y^2 = k_0^2 - \beta^2$ $(0 \le k_t < \infty)$ is the transverse wavenumber and ν labels different modes corresponding to the same k_t; e_ν satisfies boundary and edge conditions pertaining to E_y on the cross–section S, which comprises that of the guide and the half–space $x \geq 0$ and is finite at infinity. Moreover, the following orthonormalization is imposed:

$$\int_S e_\nu(\mathbf{r}; k_t) e_\mu(\mathbf{r}; k'_t) dS = \delta_{\nu\mu}\delta(k_t - k'_t) \tag{7.2}$$

with $\mathbf{r} = (x, y)$.

The field we seek, e_ν, has different forms in the waveguide and in the air region and each is made up of a superposition of plane waves such that $k_x^2 + k_y^2 = k_t^2$. The form that holds in the interior of the waveguide is of the type

$$e_\nu(x, y; k_t) = N_\nu(k_t) \sum_{n=0,2}^{\infty} \tilde{E}_{\nu n}(k_t) c_n(y) \frac{\sin q_n(x+a)}{\sin q_n a} \quad : x < 0 \tag{7.3}$$

where $N_\nu(k_t)$ is a normalization constant, $\tilde{E}_{\nu n}$ is the discrete Fourier transform of the field in the slot pertaining to the mode n for that value of k_t; of course this field has to be identically zero on the flange plane. Moreover

$$c_n(y) = \begin{Bmatrix} \sqrt{\frac{2}{b}} \cos \frac{n\pi y}{b} & n = 2, 4, \ldots \\ \sqrt{\frac{1}{b}} & n = 0 \end{Bmatrix} \tag{7.4}$$

in view of the symmetrical location of the aperture on the sidewall and

$$q_n^2 = k_t^2 - (\frac{n\pi}{b})^2 \tag{7.5}$$

Form (7.3) gives the field excited within the guide by a given aperture distribution $E_n(y; k_t)$. It is noted that the field in the air half–space could expediently be written in terms of exactly the same solutions of the slot problem, seen in the context of the solution of the wave equation in elliptical coordinates. Although highly effective for this particular problem involving a slot in an infinite thin screen, the former approach would nonetheless be of much less general applicability than the one we are going to adopt in the following treatment. Namely, we will now write the corresponding form of the field excited in the air half–space, $x > 0$, as a superposition of plane waves both radiated from the slot and decaying away from it:

$$\begin{aligned} e_\nu(x,y;k_t) = N_\nu(k_t) \int_0^{k_t} \tilde{E}_\nu(k_y;k_t)c(y;k_y)\frac{\sin(k_x x+\alpha_\nu(k_t))}{\sin\alpha_\nu(k_t)}dk_y + \\ +N_\nu(k_t)\int_{k_t}^{\infty} \tilde{E}_\nu(k_y;k_t)c(y;k_y)e^{-|k_x|x}dk_y \end{aligned} \tag{7.6}$$

where

$$c(y;k_y) = \sqrt{\frac{2}{\pi}}\cos k_y y \tag{7.7}$$

is the continuous analogue of $c_n(y)$; $k_x^2 + k_y^2 = k_t^2$ and

$$\tilde{E}(k_y;k_t) = \int_{-d}^{d} E_\nu(y;k_t)c(y;k_y)dy = < c, E_\nu > \tag{7.8}$$

The above Fourier transform is taken just on the slot as the field vanishes outside the slot on the flange plane. In (7.6) we distinguish two types of behaviour in x of the plane wave components:

- $k_y < k_t$: standing waves in x in order to describe the radiation incident from infinity on the slotted flange and reflected back from it. In this context, the real quantity $\alpha_n(k_t)$ assumes the physical meaning of the phase shift imposed on the incident $e^{jk_x x}$ component by the presence of the slot and of the waveguide behind it. The factor $\sin\alpha_n(k_t) = 0$ corresponds to the flange being completely closed. Note that the process of placing the generator at infinity and considering standing waves is identical to that followed in deriving the continuum of a slab waveguide [3].

- $k_y > k_t$: in this case only attenuated waves away from the slot can exist in the air region.

Upon inspection of (7.3), (7.6) we note that $e_\nu(x,y;k_t)$ is completely defined once we have determined the field on the aperture $E_\nu(y;k_t)$, the phase shift $\alpha_\nu(k_t)$ and the normalization constant $N_\nu(k_t)$.

We shall now proceed to establish an eigenvalue equation where $E_\nu(y;k_t)$ is the eigenfunction and α_ν the eigenvalue. This is obtained by requiring the continuity of the H_z component, $h_\nu(x,y;k_t)$ corresponding to $e_\nu(x,y;k_t)$. Modal TE fields can be derived from TE potentials

$$\begin{aligned} \mathbf{\Pi}_\nu &= \mathbf{z}_0 \Psi_\nu e^{-j\beta z} \\ \mathbf{h}_{t\nu} &= -j\beta \nabla_t \Psi_\nu \\ \mathbf{e}_{t\nu} &= -\frac{\omega\mu_0}{\beta} \mathbf{z}_0 \times \mathbf{h}_{t\nu} \\ e_\nu &= \mathbf{y}_0 \cdot \mathbf{e}_{t\nu} \\ h_\nu &= k_t^2 \Psi_\nu \end{aligned} \tag{7.9}$$

Hence the relationship valid everywhere in the cross–section is

$$h_\nu = \frac{k_t^2}{j\omega\mu_0} \int e_\nu dx \tag{7.10}$$

It is noted that a vanishing integration constant is required in order to fulfil the boundary condition on the slot edge and at infinity.

By substituting (7.3), (7.6) into (7.10) and setting $x = 0$ we obtain the following eigenvalue equation for $\alpha_\nu(k_t)$

$$\begin{aligned} &\cot \alpha_\nu(k_t) \int_0^{k_t} \tilde{E}_\nu(k_y, k_t) \frac{c(y;k_y)}{k_x} dk_y = \\ &= \sum_{n=0}^{\infty} \tilde{E}_{\nu n}(k_t) c_n(y) \frac{\cot q_n a}{q_n} - \int_{k_t}^{\infty} \tilde{E}_\nu(k_y, k_t) \frac{c(y;k_y)}{|k_x|} dk_y \end{aligned} \tag{7.11}$$

Its eigenfunction is the aperture distribution $E_\nu(y; k_t)$ whose Fourier transform appears in the above equation. If we introduce the total transverse admittance operator for this cross-section, comprising the waveguide and the air region,

$$\hat{Y}(k_t) = \hat{Y}_{air}(k_t) + \hat{Y}_{wg}(k_t) \tag{7.12}$$

with

$$\begin{aligned} \hat{Y}_{wg}(k_t) E_\nu &\equiv \sum_{n=0}^{\infty} -j < E_\nu, c_n > c_n(y) \frac{\cot q_n a}{q_n} \\ \hat{Y}_{air}(k_t) E_\nu &\equiv \int_0^{k_t} < E_\nu, c > \frac{c(y; k_y)}{k_x} dk_y \\ &\quad + j \int_{k_t}^{\infty} < E_\nu, c > \frac{c(y; k_y)}{|k_x|} dk_y \end{aligned} \tag{7.13}$$

the eigenvalue equation (7.11) can be rewritten as

$$\cot \alpha_\nu(k_t) \mathrm{Re}\hat{Y}(k_t) E_\nu = \mathrm{Im}\hat{Y}(k_t) E_\nu \tag{7.14}$$

with

$$\hat{Y} = \mathrm{Re}\hat{Y} - j\mathrm{Im}\hat{Y} \tag{7.15}$$

or, equivalently,

$$\hat{Y}(k_t) E_\nu = \lambda_\nu \mathrm{Re}\hat{Y}(k_t) E_\nu \tag{7.16}$$

with

$$\lambda_\nu = 1 - j \cot \alpha_\nu \tag{7.17}$$

If $\mathrm{Re}\hat{Y}(k_t)$ vanishes, i.e. in the absence of radiation, then the standard transverse resonance condition is recovered, yielding the bound modes of the system. No such solution is possible in this context; only solutions of the type (7.3), (7.6) with normalization (7.2) are proper, modal, solutions. From (7.16) it is easy to see that two solutions E_ν, E_μ corresponding to the same value of k_t are orthogonal over the aperture as

$$< E_\mu, \mathrm{Re}\hat{Y} E_\nu > \;=\; \delta_{\mu\nu} \tag{7.18}$$

implying these two solutions do not exchange power due to the aperture. By multiplying $e_\nu(k_t)$ by $e_\mu(k_t')$ and integrating over the cross–section, under the condition that (7.14) holds we recover the normalization (7.2) with

$$N_\nu(k_t) \;=\; \sqrt{\frac{2k_t}{\pi}} \sin \alpha_\nu(k_t) \tag{7.19}$$

7.1.2 Discretization of the eigenvalue equation

An effective discretization of (7.14) into an ordinary matrix eigenvalue problem is now accomplished. We build in the edge condition by setting, for a symmetrically placed aperture,

$$E_\nu(y; k_t) \;=\; w(y) \sum_{m=0,2}^{N} \sqrt{\frac{\varepsilon_m}{\pi d}} \chi_m T_m(\frac{y}{d}) \tag{7.20}$$

$$\varepsilon_0 = 1; \quad \varepsilon_m = 2 \quad : m > 0 \tag{7.21}$$

where χ_m are the expansion coefficients and

$$w(y) \;=\; \frac{1}{\sqrt{1 - (\frac{y}{d})^2}} \tag{7.22}$$

The T_m's are the Chebyshev polynomials normalised so that [8, p.833]

$$\frac{\sqrt{\varepsilon_m \varepsilon_n}}{\pi d} < T_m, wT_n > \;=\; \frac{\sqrt{\varepsilon_m \varepsilon_n}}{\pi d} \int_{-d}^{d} T_m w T_n dy = \delta_{mn} \tag{7.23}$$

With this position, one substitutes to $\hat{Y}$ the discrete matrices

$$\begin{aligned} \mathrm{Im}Y_{mk} &= \frac{k_t^2}{\omega\mu_0}\left[\sum_{n=0,2,}^{\infty} \frac{\cot q_n a}{\sqrt{k_t^2 - (\frac{n\pi}{b})^2}} P_{mn} P_{kn} - \int_{k_t}^{\infty} \frac{P_m(k_y) P_k(k_y)}{\sqrt{k_y^2 - k_t^2}} dk_y\right] \\ \mathrm{Re}Y_{mk} &= \frac{k_t^2}{\omega\mu_0} \int_0^{k_t} \frac{P_m(k_y) P_k(k_y)}{\sqrt{k_t^2 - k_y^2}} dk_y \end{aligned} \tag{7.24}$$

where [8, p.836]

$$\begin{aligned} P_m(k_y) &= < \sqrt{\frac{\varepsilon_m}{\pi d}} T_m w, \sqrt{\frac{2}{\pi}} \cos k_y y >= \sqrt{2\varepsilon_m d} (-1)^{(m/2)} J_m(k_y d) \\ P_{mn} &= < \sqrt{\frac{\varepsilon_m}{\pi d}} T_m w, c_n >= \sqrt{\frac{\varepsilon_m \varepsilon_n \pi d}{b}} (-1)^{(m/2)} J_m(\frac{n\pi d}{b}) \end{aligned} \tag{7.25}$$

and (7.14) becomes a matrix eigenvalue equation of order $(N+1)/2$, capable of producing up to $(N+1)/2$ eigenvectors.

An interesting particular case is the "small aperture" approximation, where

$$E_\nu(y;k_t) \approx w(y)\sqrt{\frac{1}{\pi d}}(T_0=1) \tag{7.26}$$

independently of k_t and n, (7.14) becomes a scalar equation and we have

$$\cot\alpha(k_t) = \frac{\mathrm{Im}Y_{00}}{\mathrm{Re}Y_{00}} = \frac{\sum_{n=0,2,}^{\infty} \frac{\cot q_n a}{\sqrt{k_t^2-(\frac{n\pi}{b})^2}} P_{0n}^2 - \int_{k_t}^{\infty} \frac{P_0^2(k_y)}{\sqrt{k_y^2-k_t^2}}dk_y}{\int_0^{k_t} \frac{P_0^2(k_y)}{\sqrt{k_t^2-k_y^2}}dk_y} \tag{7.27}$$

which holds for

$$k_t \neq \frac{n\pi}{b} \tag{7.28}$$

The corresponding field outside and inside the guide is given by:

$$\begin{aligned} e(\mathbf{r};k_t) = \sqrt{\tfrac{2k_t d}{\pi}}\left[\int_0^{k_t} dk_y \sqrt{\tfrac{2}{\pi}} \cos k_y y J_0(k_y d) \sin(k_x x + \alpha)\right] + \\ + \sqrt{\tfrac{2k_t d}{\pi}}\left[\sin\alpha \int_{k_t}^{\infty} dk_y \sqrt{\tfrac{2}{\pi}} \cos k_y y J_0(k_y d) e^{(-|k_x|x)}\right] \qquad : x \geq 0 \end{aligned} \tag{7.29}$$

$$e(\mathbf{r};k_t) = \sqrt{\frac{2k_t d}{b}} \sin\alpha \sum_{n=0,2,}^{\infty} J_0(\tfrac{n\pi d}{b}) \frac{\sin q_n(x+a)}{\sin q_n a} c_n(y) \quad : x \leq 0 \tag{7.30}$$

It is noted that the modes given by the above expressions are independent of frequency and only dependent on the parameter k_t.

Once the continuous spectrum has been computed, it can be used for the solution of a variety of problems. In the next Chapter we show the application to the case of a slotted guide excited by a closed rectangular waveguide.

7.2 The generalised transverse equation

We have derived in the previous Section the continuum of an open waveguide. In the derivation some analytical details were necessary, with the possible effect of defocusing the understanding of the main features of this part of the spectrum. In this Section, we introduce the continuum from a different perspective, which is also more general, and with almost no analytical details. In order to do so, let us refer for a moment to the familiar case of closed waveguides. The modes of the latter waveguides of separable cross–section (e.g. rectangular, cylindrical, etc.) are well known from basic microwave courses [1]. These modes are real solutions of the two–dimensional Helmholtz equation considered in the plane transverse to the propagation direction; when the waveguide is closed, the eigenvalues of the above equation form a discrete set. A similar situation occurs also for three–dimensional closed resonators; when the boundary surfaces of the resonator are coordinate surfaces, the three–dimensional Helmholtz equation can be solved in closed form and, again, a discrete spectrum of eigenvalues is present.

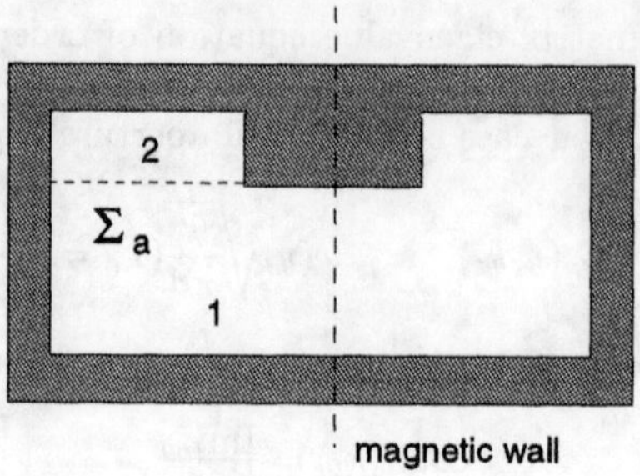

Figure 7.2 Example of a closed waveguide of non–separable cross–section.

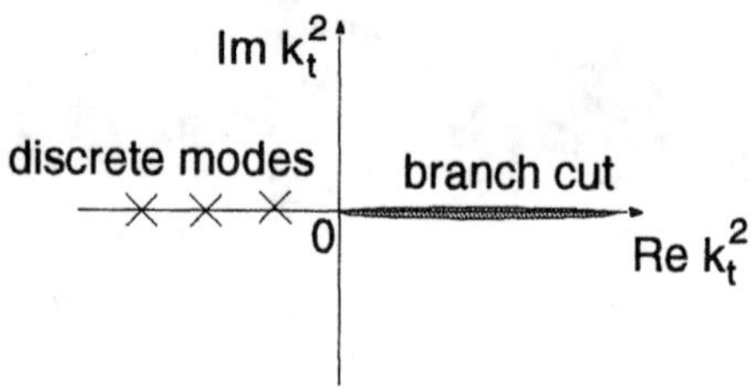

Figure 7.3 Complex k_t^2 plane. The branch cut on the real axis of k_t^2 is shown together with few discrete modes.

For closed waveguides, or resonators, of non–separable geometry, the solution procedure of the Helmholtz equation is still well known, even if it is more involved. In these cases, in fact, modes may be obtained by a transverse resonance procedure. As an example, let us examine the ridged waveguide of Figure 7.2. In this case, by using symmetry, we may consider just one half of the structure. By choosing as unknowns the tangential components of the electric field **E** on the separation surface Σ_a and by considering suitable admittance operators, the magnetic field is expressed in terms of the tangential components of **E**. In each of the two regions, that is the regions below and above Σ_a, we have $\mathbf{H}_1 = Y_1(\mathbf{E})$, $\mathbf{H}_2 = Y_2(\mathbf{E})$, respectively. After equating on Σ_a the tangential components of the magnetic fields, and by discretization, we obtain a matrix which depends on the transverse wavenumber k_t. The zeros of the determinant of this matrix, obtained for certain values of k_t, correspond to the resonance of the structure. Equivalently, we may find the eigenvalues of the above matrix; for certain values of k_t an eigenvalue is zero, then the determinant is zero, and the corresponding eigenvector provides the modal field distribution, i.e. the field distribution that can be sustained in the absence of sources. Following this transverse resonance procedure, we can rigorously obtain the whole modal spectrum of closed, non–separable, structures.

For open waveguides, apart from a few discrete modes at most, the spectrum becomes continuous (Figure 7.3). For open structures, when the coordinate surfaces coincide with the boundaries, the modal spectrum is well known [3], and has been described, for instance, in Chapter 3 for the case of the slab waveguide. Let us consider again the simple one–dimensional example of a dielectric slab waveguide. Its modal spectrum consists of a finite number of discrete (surface wave) modes and a continuous spectrum. The former are standing wave (in the direction normal to the surface) solutions with discrete eigenvalues. The continuous spectrum, on the other hand, has no discrete eigenvalue, but once a particular wavenumber is chosen the field is a standing wave solution consisting of the incoming wave and the outgoing wave in the direction normal to the dielectric surface. This standing wave remains finite at infinity and does not individually satisfy the radiation condition. A combination of these, however, represent radiation by any physical source, thus satisfying the radiation condition. Hence, such standing waves are "modes" of the structure.

Nevertheless, when dealing with open structures of non–separable cross–section, for a long time it was not quite clear how to proceed in order to obtain the spectrum as for the case of separable ones [21]. Most open waveguides currently used, both in microwaves and in optics, fall into the latter category (e.g. microstrips, slot lines, coplanar waveguides, inset waveguides, dielectric rib waveguides, etc.). In some cases it has been natural to extend the transverse resonance technique for bound modes into the complex plane, obtaining in this way the "leaky waves". This procedure, which is simple and elegant when applicable, has led to the understanding of some difficult problems in limited spatial domains with modest computer resources [22]. In an open environment, however, the Helmholtz equation together with its boundary condition is not self adjoint. As a consequence "leaky waves", which grow at infinity, are not part of the modal spectrum and are not suitable for a global representation of the field over the whole guide cross–section.

By properly extending the transverse resonance technique, the proper spectrum of real waves of open waveguides may be obtained as in the previous Section. The importance of such a real spectrum is evident; in fact, discontinuities present along the waveguide excite the whole spectrum, therefore generating radiation over wide angular distributions, surface waves, mode conversion, etc. Moreover, coupling and interference of waves incident from the exterior can easily be accounted for if the spectrum is known.

In this book we deal with open waveguides, but non–separable three–dimensional objects too possess modal fields and their knowledge greatly enhances the solution of antenna and scattering problems. The study of the above modes (also called *characteristic modes*), for three–dimensional objects, has been initiated with the fundamental work of Garbacz [5, 7]. In [10, 9] the method of moments has been introduced in order to obtain a quantitative formulation of the problem as well as an efficient numerical algorithm; in [11] the technique has been suitably extended also to apertures in metallic bodies. Successively, characteristic modes have been successfully used for the synthesis and optimization of antennas [12, 6, 16, 14].

In the following, we describe the derivation of the continuous spectrum of non–separable open waveguides by functional analysis considerations. This method, when applied to three–dimensional objects, gives the modal characterization of the structure (characteristic modes); when applied to waveguides (open or closed), it yields the spectrum (both continuous or discrete). Therefore, it represents the generalization of the transverse resonance approach to open, non–separable problems.

7.2.1 The generalised transverse equation

As already mentioned in the previous paragraph, for closed structures it is possible to obtain the discrete set of modes by searching for the resonance of the structure.

This approach can be rigorously extended to open structures, not only with a view to recovering the bound modes (as was done in the previous Chapter), but also to finding the modes of the continuum. To be specific let us consider again the case of the slotted waveguide as in Figure 7.1. In this case the problem is naturally formulated in terms of the equivalence theorem and of admittance operators. With reference to Figure 7.1 we define the following inner product on the aperture surface, Σ_a, as

$$< B, C > = \int_{\Sigma_a} B^* C \, d\ell \tag{7.31}$$

where the $*$ denotes complex conjugate; moreover we define an analogous inner product $< B, C >_\infty$ on the surface at infinity. Since we are considering a uniform guiding structure, the z–dependence has been separated out, and the integrals along Σ_a , and the surface at infinity Σ_∞ are simple line integrals.

Since the structure is open, radiation phenomena of the electromagnetic energy may be present. Due to radiation, the admittance operator involved is complex. In the following, in order to avoid unnecessary analytical burden, we refer to the scalar TE or TM cases, even though the hybrid case is fairly readily recovered by using dyadics. If we try to account for radiation by using a complex k_z we also get complex values of k_t in the 2–D Helmholtz equation

$$\nabla_t^2 \tau + k_t^2 \tau = 0 \tag{7.32}$$

where $k_0^2 = \omega^2 \mu \varepsilon$; $k_t^2 = k_x^2 + k_y^2 = k_0^2 - k_z^2$ and τ represents the generic field (or potential). On the other hand, it is possible to observe that the continuous spectrum corresponds in the complex plane (Figure 7.3) to a branch cut from the origin to infinity. If we require real fields finite at infinity, this cut can only be made along the real axis. But a branch cut along the real axis for $k_t \le k_0$ corresponds to a real value of k_z, i.e. absence of attenuation in the propagation direction. This absence of attenuation can only be realised if the outgoing waves (radiation from the structure to infinity) are compensated by incoming waves (energy coming from infinity to the structure). This arrangement corresponds to standing waves for the radiative part ($k_t \le k_0$) of the field, a condition which was actually implemented in the previous Section, when determining the continuous spectrum of the slotted rectangular waveguide. In that case, the characteristics of the standing waves were described in terms of the phase shift a plane wave undergoes when it is reflected from, say, the slotted plane. Then, by equating the tangential components of the magnetic fields on Σ_a, a generalised eigenvalue equation is obtained. The solution of the generalised eigenvalue equation produces the continuous spectrum.

It is possible to arrive at the same conclusions, in a very direct way, by considering equation (7.32) from the viewpoint of functional analysis, while considering k_t real ($0 < k_t < \infty$). Referring to the wave equation (7.32) the boundary condition corresponding to the Sommerfeld radiation condition (waves propagating toward infinity) is given by

$$\frac{\partial v}{\partial r} - j k_t v = 0 \tag{7.33}$$

where v denotes the outgoing part of τ. While (7.32) is self–adjoint in the Hermitian sense, (7.33) is not; in fact, the adjoint boundary condition is

$$\frac{\partial u}{\partial r} + jk_t u = 0 \tag{7.34}$$

where u denotes the part of τ which corresponds to a wave coming in from infinity toward the guide. Therefore, if we want to use the eigenfunction expansion method for our diffraction problem, we have to complement it with the adjoint problem, requiring waves coming *from* infinity as well as waves going *to* infinity [13, pp.299-300].

Let a denote the waves coming from infinity and b the waves reflected by our scatterer. In terms of some scattering operator S we have

$$b = Sa \tag{7.35}$$

where, obviously, the operator S is unitary ($SS^\dagger = I$), and $S^\dagger$ is the adjoint of S, but not self adjoint ($S \neq S^\dagger$). We are therefore faced with the "shifted" eigenvalue problem [13]

$$\begin{aligned} Su_n &= \sigma_n v_n \\ S^\dagger v_n &= \sigma_n u_n \end{aligned} \tag{7.36}$$

where u_n, v_n are the eigenfunctions, while σ_n are their common singular values [13]. Note that:

- the u_n form a basis appropriate to representing the incoming waves (a);
- the v_n form a basis appropriate to representing the outgoing waves (b).

Analogous considerations hold for a three–dimensional scatterer if we expand the fields in terms of multipoles corresponding to incident (e^{jkr}) and reflected (e^{-jkr}) spherical waves.

As already stated, when apertures on a metallic screen are present, the natural representation of diffraction problems is in terms of admittance operators. For example, with reference to Figure 7.1 we can enforce the continuity of the tangential components of the electric field by considering the appropriate magnetic currents. Accordingly, our problem has been divided into two parts (a region corresponding to the waveguide, and a region corresponding to the air half–space). We can express the magnetic field inside each region in terms of the magnetic currents, and then equate the tangential components of the magnetic field on the aperture Σ_a. The equation thus obtained contains an admittance operator Y, the sum of the two operators corresponding to the two regions, which is not self–adjoint. Let us separate its real and imaginary parts

$$Y = G + jB \tag{7.37}$$

where G , B are symmetrical, real and self–adjoint; they can be obtained by Y and by its complex conjugate Y^*, as

$$\begin{aligned} G &= (Y + Y^*)/2 \\ B &= (Y - Y^*)/2j \end{aligned} \tag{7.38}$$

When sources are present, the general radiation/scattering problem can be described in terms of the operator Y as,

$$Y(\phi) = \psi \tag{7.39}$$

where ϕ represents the field on Σ_a and ψ the excitation. Note that Y depends on the geometry of the problem, dielectric constants and on k_t. Note also that, since G, B are self-adjoint, they admit a spectral decomposition of the following type

$$\begin{aligned} G\xi_k &= \mu_k\xi_k \\ B\zeta_k &= \nu_k\zeta_k \end{aligned} \tag{7.40}$$

where μ_k, ν_k are the eigenvalues (real and positive) and ξ_k, ζ_k are the eigenfunctions. Naturally, the inner product $< \xi_k, G\xi_k >$ taken over the volume of the body corresponds to the radiated energy, while $< \zeta_k, B\zeta_k >$ corresponds to the reactive energy. The two operators, being real symmetric with G positive definite, can be diagonalised simultaneously by a common basis of eigenfunctions, say ϕ_n. Hence, we can write

$$\begin{aligned} G\phi_n &= \mu_n\phi_n \\ B\phi_n &= \nu_n\phi_n \end{aligned} \tag{7.41}$$

By substituting the first into the second of the above eigenvalue equations we obtain

$$\chi_n G\phi_n = B\phi_n \tag{7.42}$$

where $\chi_n = \nu_n/\mu_n$. Equation (7.42) is the *generalised transverse resonance condition for open structures* (non self–adjoint problems). From (7.42), for each k_t value, it is possible to obtain a discrete number of eigenfunctions ϕ_n $(n = 1, 2, \ldots)$ which represent the modes of our structure. Observe that only the ratio between ν_n and μ_n appears in (7.42). The possibility of differently normalizing the quantity $< \phi_n, G\phi_n >$ corresponds, for the three–dimensional case, to the different choices adopted in [12, 6, 16, 14]. In the following the choice $< \phi_n, G\phi_n >= 1$ is adopted. Since G, B are real, symmetrical, operators all the eigenvalues χ_n are real, and all the eigenfunctions ϕ_n can be chosen to be real. The eigenfunctions, when suitably normalised, satisfy the following orthogonality relationships

$$\begin{aligned} < \phi_m, G(\phi_n) > &= \delta_{mn} \\ < \phi_m, B(\phi_n) > &= \chi_n\delta_{mn} \\ < \phi_m, Y(\phi_n) > &= \lambda_n\delta_{mn} \end{aligned} \tag{7.43}$$

where δ_{mn} is Kronecker delta ($\delta_{mn} = 0, m \neq n; \delta_{mn} = 1, m = n$) and $\lambda_n = 1 + \jmath\chi_n$. By using the generalised Green's theorem [15, pp.870-874], it is also possible to show that for the inner product taken over the sphere at infinity $(<,>_\infty)$, the following orthonormality relationship holds:

$$< \phi_m, G(\phi_n) >_\infty = \delta_{mn} \tag{7.44}$$

Therefore, the eigenfunctions ϕ_n simultaneously diagonalise the operators G, B, Y. It is also noted that the solutions of (7.42) also minimise the following Rayleigh quotient

$$\chi = \frac{< \phi, B(\phi) >}{< \phi, G(\phi) >} \tag{7.45}$$

This functional corresponds to the ratio between the stored reactive energy and the radiated energy. Accordingly, the eigenfunctions correspond to the field distributions which minimise the ratio between the reactive and radiative energy. In terms of the quality factor Q, this corresponds to saying that Q is minimum; or, in terms of Lagrangians, minimization of (7.45) yields a minimum of the Lagrangian of the system. Therefore for open waveguides, to each k_t value there corresponds a multiplicity of eigenvalues and eigenfunctions; each eigenfunction describes a different field distribution over the discontinuous interface and generates in the whole space a field which is a solution of the wave equation (7.32); these are the modal fields. The modal solutions, denoted as $\tau_n(\mathbf{r}; k_t)$ in the following, are easily calculated from the eigenfunctions. At a given frequency part of the spectrum is radiating (propagating) ($k_t < k$), while the remaining part contributes only to the reactive (localised) energy ($k_t > k$). Moreover, it is possible to show that modes corresponding to different k_t values are orthogonal; that is over the unbounded section S, we have

$$\int_S \tau_n(\mathbf{r}; k_t)\tau_m(\mathbf{r}; k'_t)dS = \delta_{nm}\delta(k_t - k'_t) \tag{7.46}$$

In addition, the modes are also complete since

$$\int_0^\infty \sum_n \tau_n(\mathbf{r}; k_t)\tau_n(\mathbf{r}'; k_t)dk_t = \delta(\mathbf{r} - \mathbf{r}') \tag{7.47}$$

The completeness follows from the fact that the operators G, B are self–adjoint and G is positive definite [15, pp.774-778]. Actually G is only positive semidefinite, but when it vanishes no radiation is present and the classical formulation used for bound modes is recovered.

Relationships (7.44), (7.46) and (7.47) allow modal expansion of a given field over a section $z = const$ in a manner similar to that employed in closed waveguides. They represent the starting point for handling discontinuity problems in open structures by modal analysis.

Characteristic modes for three–dimensional objects

The above modal theory is also applicable to three–dimensional objects. Actually, it was introduced in [5] for the purpose of studying scattering phenomena. A clear exposition of the theory in the framework of functional analysis, as well as its discretization by the method of moments, was developed by Harrington and Mautz in [10, 9, 11]. What was overlooked in these papers is the possibility of obtaining the continuous spectrum of open waveguides. In other words, if the theory of characteristic modes is applied to a two–dimensional problem for all real values of k_t ($0 \le k_t < \infty$), the complete modal spectrum is obtained. To conclude, the spectrum of two–dimensional or three–dimensional structures can be obtained from a resonance equation. In a closed environment the zeros of the eigenvalues correspond to the natural frequencies of the structure. In an open environment (7.42) becomes the resonance equation and for each value of k_t (for 2–D problems) or k (for 3–D problems) we have a set of eigenvalues (characteristic values), eigenvectors (characteristic vectors) providing the modal characterization of the structure.

Leaky waves

In the case of "leaky waves" the procedure consists in determining a *complex* k_t value such that

$$\begin{aligned} Y(k_t)\phi_n &= \lambda_n(k_t)\phi_n \\ \lambda_n &= 1 + \jmath\chi_n = 0 \end{aligned} \tag{7.48}$$

This cannot happen, however, for real values of k_t since for real k_t χ_n is also real. Therefore one is forced to consider complex values of k_t corresponding to waves growing toward infinity (nonmodal solutions). It should also be observed that, while the transverse resonance for closed waveguides is a rigorous procedure, when it is applied to the leaky wave formalism it becomes an approximation, albeit an effective one in its domain of validity. As stated in [22] this approximation holds if

- The geometry of the guide/antenna is capable of supporting waves of complex type. This is equivalent to saying that at least a solution of (7.48) exists, e.g. a dielectric slab guide.
- The field contribution due to a complex eigenvalue is predominant in the near field. This condition corresponds to saying that the near field is representable principally in terms of the leaky wave.

It should be noted that, rigorously speaking, the field has to be expanded in terms of the continuous spectrum as

$$\int_0^\infty \sum_n A_n(k_t)\tau_n(\mathbf{r}; k_t)dk_t \tag{7.49}$$

where $A_n(k_t)$ is the modal amplitude. However this integral can be deformed in the complex k_t plane. When the contribution to the above integral is principally due to the poles corresponding to leaky waves then the second condition holds and the leaky wave formalism becomes a useful approximation.

Some remarks

Having recognised the relationship between the continuous spectrum and characteristic modes, all the results developed for the latter can be almost immediately employed to determine the spectrum of open waveguides and vice versa.

The relationship between $\alpha_n(k_t)$, as obtained in the previous Section, and χ_n is simply given by

$$-\cot\alpha_n(k_t) = \chi_n \tag{7.50}$$

Thus, by means of simple considerations of functional analysis, we may directly derive a generalization of the resonance equation. This equation allows computation of the continuous spectrum of open waveguides of non–separable cross–section.

The generalised equation has been obtained from the simultaneous diagonalization of the reactance and of the radiation conductance by means of a set of common fields. These fields minimise the Lagrangian (immittance) of the system and the ratio between the reactive and radiative energies.

7.3 Fully hybrid fields

We have seen in the two previous Sections how to compute the continuous spectrum for open waveguides of nonseparable cross–section. However, in order to keep things simple we have dealt with scalar cases, even if most of the waveguides used in common practice (e.g. microstrip, slotline inset guide, coplanar waveguide, etc.) require a fully hybrid field. This case is considered in detail in this Section.

7.3.1 Characteristics of the spectrum

With reference to Figure 7.4 or to Figure 7.6 we are seeking solutions of the transverse wave equation for the electric field transverse to y

$$\begin{array}{ll} \nabla_t^2 \mathbf{e}_\nu + (k_t^2 + v^2)\mathbf{e}_\nu = 0 & \mathbf{e}_\nu = (e_{\nu x}, e_{\nu z}); \\ v^2 = (\varepsilon_r - 1)k_0^2 & \mathbf{e}_\nu = \mathbf{e}_\nu(\mathbf{r}; k_t); \\ \mathbf{r} = (x, y); & k_t^2 = k_0^2 - \beta^2 \end{array} \tag{7.51}$$

that satisfy the boundary condition on the metal surfaces, the radiation condition at infinity and the appropriate edge conditions at the sharp corners as well as being continuous together with the magnetic fields at the air–dielectric interface.

As illustrated in the previous Sections it is well known that the spectrum on the k_t^2-plane is as shown in Figure 7.3, i.e., it comprises discrete values for $\mathrm{Re}k_t^2 <$ 0, guided modes, as well as a continuum of values for $k_t^2 > 0$, corresponding to propagating (radiative) modes for $k_t < k_0$ and nonpropagating modes for $k_t > k_0$.

As we have seen in the first example of this Chapter, the slotted rectangular waveguide, discrete modes may not exist in some cross–sections lacking confinement; they may also be degenerate, as different transverse field distributions share the same k_t^2–value. Continuous modes will essentially belong to the latter kind, in view of the 2D–nature of the cross–section.

To further illustrate the above point, consider the simple scalar example of separable cross–section, namely the half–space over a ground plane. In cylindrical coordinates, scalar eigenfunctions of the wave equation satisfying boundary and radiation conditions are [15] the Bessel functions of argument $k_t r$,

$$\phi_\nu(r, \theta; k_t) = \sqrt{\frac{2}{\pi}}\sqrt{k_t}\sin \nu\theta J_\nu(k_t r) \tag{7.52}$$

with $\nu = 1, 2, 3, \ldots$. They also satisfy the orthonormality conditions

$$\begin{array}{c} \int\int \phi_\nu(k_t)\phi_\mu(k_t)d\mathbf{r} = \\ \frac{2}{\pi}\sqrt{k_t k_t'}\int_0^\pi \sin\nu\theta \sin\mu\theta d\theta \int_0^\infty J_\nu(k_t r)J_\mu(k_t' r) r dr = \delta_{\mu\nu}\delta(k_t - k_t') \end{array} \tag{7.53}$$

Hence, given any $k_t \geq 0$, a whole angular spectrum exists that is characterised by the discrete eigenvalue ν.

Let us now return to consideration of waveguides of nonseparable cross sections such as the microstrip line, the slotline and the inset dielectric guide, i.e. to waveguides which are essentially constituted by a ground plane with perturbations. The eigenfunction e_ν in the open region, $(y > 0)$, may be described as a superposition of functions such as (7.52) with sine and cosine angular dependence, i.e.

$$e_\nu(\mathbf{r}; k_t) = \sum_{m=0,1,} \sqrt{\frac{2}{\pi}}\sqrt{k_t}\left(s_m^{(\nu)} \sin m\theta + c_m^{(\nu)} \cos m\theta\right) J_m(k_t r) \tag{7.54}$$

with real s_m, c_m.

It is seen, therefore, that the simultaneous excitation of a whole "wave packet" globally satisfying boundary, edge and radiation conditions is essentially required.

Finally, we impose on these packets a normalization over the whole guide cross–section S of the type

$$\int\int_S e_\mu(\mathbf{r}; k_t) e_\nu(\mathbf{r}; k_t') ds = \delta_{\mu\nu}\delta(k_t - k_t') \tag{7.55}$$

for scalar fields or

$$\int\int_S \mathbf{e}_\nu(\mathbf{r}; k_t) \times \mathbf{h}_\mu(\mathbf{r}; k_t') \cdot \mathbf{z}_0 ds = \delta_{\mu\nu}\delta(k_t - k_t') \tag{7.56}$$

for vector fields. By using the asymptotic limit of the Bessel's functions and the wave equation, it can be shown that two scalar fields characterised by (μ, k_t) and (ν, k_t') are indeed orthonormal according to (7.55). This result is generalised to the vector case (7.56) by using the vector development as in [1]

7.3.2 Completeness and eigenvalue equation

By completeness of the spectrum of the guide the following dyadic identity is understood

$$\sum_{\nu=1}^{s} \mathbf{e}_\nu(\mathbf{r})\mathbf{e}_\nu(\mathbf{r}') + \int_0^\infty \sum_\nu \mathbf{e}_\nu(\mathbf{r}; k_t)\mathbf{e}_\nu(\mathbf{r}'; k_t) dk_t = \delta(\mathbf{r} - \mathbf{r}')\underline{\underline{\mathbf{I}}} \tag{7.57}$$

with $\underline{\underline{\mathbf{I}}}$ representing the unit dyad, $\underline{\underline{\mathbf{I}}} = \mathbf{x}_0\mathbf{x}_0' + \mathbf{z}_0\mathbf{z}_0'$. Equation (7.57) is also identical to the contour integral of the dyadic Green's function $\underline{\underline{\mathbf{g}}}$ of the transverse wave equation (7.51) over the path c in the complex k_t^2–plane of Figure 7.3, enclosing all its singularities [4]

$$\oint_c \underline{\underline{\mathbf{g}}}(\mathbf{r}, \mathbf{r}'; k_t^2) dk_t^2 = \delta(\mathbf{r} - \mathbf{r}')\underline{\underline{\mathbf{I}}} \tag{7.58}$$

Singularities are either poles, corresponding to discrete modes, or branch cuts, corresponding to the continuum. As shown next, for each value of k_t^2 the Green's function can be written as the sum of products of a travelling–wave solution in the upper half–space $(y > 0)$, $\mathbf{V}^+$, and a standing wave solution $\mathbf{V}^-$ in the lower half–space $(y < 0)$, i.e.,

$$\underline{\underline{\mathbf{g}}}(\mathbf{r}, \mathbf{r}'; k_t^2) = \sum_\nu \frac{\mathbf{V}_\nu^+(\mathbf{r}; k_t^2)\mathbf{V}_\nu^-(\mathbf{r}'; k_t^2)}{\lambda_\nu(k_t^2)} \tag{7.59}$$

with

$$\mathbf{V}_\nu^+(x, 0) = \mathbf{V}_\nu^-(x, 0) = \mathbf{E}_\nu(x) \tag{7.60}$$

$\lambda_\nu(k_t^2)$ is, within a constant, the transverse admittance of the νth solution for the value of k_t^2.

Vanishing of the transverse admittance gives rise to the discrete modes; the branch cut singularity, where λ_ν does not vanish, gives rise to the continuum. After isolating the residues of the pole singularities by an appropriate deformation of the path c around the positive k_t^2–axis, using (7.57)-(7.59), one obtains the completeness relationship in the form required, namely

$$\begin{aligned}\frac{2}{\pi}\sum_{\nu=1}^{s}\frac{\mathbf{V}_\nu^+(\mathbf{r})\mathbf{V}_\nu^-(\mathbf{r}')}{\lambda_\nu'(k_t^2)}+\frac{2}{\pi}\int_0^\infty k_t\mathrm{Re}\sum_\nu\frac{\mathbf{V}_\nu^+(\mathbf{r};k_t^2)\mathbf{V}_\nu^-(\mathbf{r}';k_t^2)}{\lambda_\nu(k_t^2)}dk_t &=\\ \sum_{\nu=1}^{s}\mathbf{e}_\nu(\mathbf{r})\mathbf{e}_\nu(\mathbf{r}')+\int_0^\infty\sum_\nu\mathbf{e}_\nu(\mathbf{r};k_t)\mathbf{e}_\nu(\mathbf{r}';k_t)dk_t\end{aligned} \tag{7.61}$$

where $\lambda_\nu'(k_t^2)$ denotes the derivative of λ_ν evaluated at the pole location. The finite summation is the contribution of the discrete modes, the integral that of the continuum. The amplitudes $V_\nu^\pm$ are complex quantities, always representable as Fourier integrals, if the problem is intrinsically nonseparable.

In order to render homogeneous the edge conditions satisfied by the unknown fields, we assume definite even parity with respect to x of, say, E_x and $j\partial E_z/\partial x$ which are both real for a lossless guide and use these quantities as the unknowns, retaining $\tilde{\mathbf{E}}_\nu$ to denote their cosine Fourier transform (even to x problem). Hence

$$\mathbf{V}_\nu^+(\mathbf{r};k_t^2) = \int_0^\infty \tilde{\mathbf{E}}_\nu(k_x;k_t)\sqrt{\frac{2}{\pi}}\cos k_x x e^{-jk_y y}dk_x \tag{7.62}$$

with

$$\begin{aligned} k_y &= \sqrt{k_t^2-k_x^2}, & & k_x\le k_t\\ k_y &= -j\gamma, & \gamma=\sqrt{k_x^2-k_t^2} \quad & k_x\ge k_t\end{aligned} \tag{7.63}$$

$\tilde{\mathbf{E}}_\nu$ is a real quantity in the absence of losses and, if a ground plane is located at $y=-h$, the standing wave $\mathbf{V}_\nu^-$ in the lower half–space is also real and given by

$$\mathbf{V}_\nu^-(\mathbf{r};k_t) = \sum_n \mathbf{E}_{\nu n}\phi_n(x)\frac{\sin q_n(y+h)}{\sin q_n h} \tag{7.64}$$

The sum over n denotes an actual summation in the case of a finite region, as for example for the inset guide; it denotes an integral over k_x in the cases of a microstrip line or of a slotline; for a rib waveguide it denotes a sum over discrete components and an integral over the continuum. $\phi_n(x)$ is the appropriate normalised transverse distribution.

$\mathbf{E}_\nu$ is closely related to the total complex dyadic admittance operator $\hat{\mathbf{Y}}(k_t)$ at the interface $y=0$:

$$\hat{\mathbf{Y}}(x,y=0;x',y'=0;k_t) = \hat{\mathbf{Y}}^+ + \hat{\mathbf{Y}}^- \tag{7.65}$$

$\hat{\mathbf{Y}}^+,\hat{\mathbf{Y}}^-$ are the contributions of the upper and lower half–spaces, respectively, which link the electric and magnetic fields transverse to y at the interface, namely

$$\pm\begin{pmatrix} H_z^\pm(x,0)\\ -j\int H_x^\pm(x,0)dx\end{pmatrix} = \begin{pmatrix}\hat{Y}_{11}^\pm & \hat{Y}_{12}^\pm\\ \hat{Y}_{12}^\pm & -\hat{Y}_{22}^\pm\end{pmatrix}\cdot\begin{pmatrix}E_x\\ j\frac{\partial E_z}{\partial x}\end{pmatrix} = \hat{\mathbf{Y}}\cdot\begin{pmatrix}E_x\\ \partial_x E_z\end{pmatrix} \tag{7.66}$$

where the upper signs hold for $y=0^+$, the lower ones for $y=0^-$. In the above we have considered the propagating continuum, i.e. $k_0>k_t$. For $k_t>k_0$, we redefine the above equation in the following way

$$\pm\begin{pmatrix} H_z^\pm(x,0)\\ \int H_x^\pm(x,0)dx\end{pmatrix} = \begin{pmatrix}\hat{Y}_{11}^\pm & j\hat{Y}_{12}^\pm\\ j\hat{Y}_{12}^\pm & -\hat{Y}_{22}^\pm\end{pmatrix}\cdot\begin{pmatrix}E_x\\ \partial_x E_z\end{pmatrix} \tag{7.67}$$

Identifying now $\mathbf{E}_\nu$ with $(E_x, \partial_x E_z)$ the argument proceeds in the same way.

In the upper region

$$\begin{aligned}
\hat{Y}_{11}^{+} &= \int_0^\infty y_e^+ \cos^2\theta + y_h^+ \sin^2\theta \; k_x >< k_x dk_x \\
\hat{Y}_{12}^{+} &= \int_0^\infty \frac{y_e^+ - y_h^+}{k_x} \sin\theta\cos\theta \; k_x >< k_x dk_x \\
\hat{Y}_{22}^{+} &= \int_0^\infty \frac{y_e^+ \sin^2\theta + y_h^+ \cos^2\theta}{k_x^2} \; k_x >< k_x dk_x
\end{aligned} \tag{7.68}$$

with

$$\begin{array}{lll}
\cos\theta = \frac{k_x}{\sqrt{k_x^2+\beta^2}} & \sin\theta = \frac{\beta}{\sqrt{k_x^2+\beta^2}} & \\
y_e^+ = \frac{\omega\varepsilon_0}{k_y} & y_h^+ = \frac{k_y}{\omega\mu_0} & k_x \le k_t \\
y_e^+ = \frac{j\omega\varepsilon_0}{\gamma} & y_h^+ = \frac{\gamma}{j\omega\mu_0} & k_x \ge k_t
\end{array} \tag{7.69}$$

For the convenience of writing operators, we have also adopted the bracket notation, where

$$k_x > \;=\; \sqrt{\frac{2}{\pi}} \cos k_x x \tag{7.70}$$

and $< k_x$ denotes the operation of multiplying by the above function of x' and integrating in x' over the aperture at the interface, so that

$$\tilde{\mathbf{E}}_\nu \;=\; < k_x, \mathbf{E}_\nu > \tag{7.71}$$

It is noted that (7.69) is real or pure imaginary according as $k_x < k_t$ or $k_x > k_t$. An analogous definition holds in the lower half–space, where, because of the ground plane at $y = -h$, we have

$$\begin{aligned}
y_{en}^- &= -j\frac{\omega\varepsilon_0}{q_n} \cot q_n h \\
y_{hn}^- &= -j\frac{q_n}{\omega\mu_0} \cot q_n h \\
q_n^2 &= k_t^2 + v^2 - k_n^2
\end{aligned} \tag{7.72}$$

k_n is the x–directed wavenumber in the lower half–space

$$\begin{aligned}
k_n > &= \phi_n(x) \\
\tilde{\mathbf{E}}_{\nu n} &= < k_n, \mathbf{E} > \\
\hat{Y}_{11}^{-} &= \sum_n y_{en}^- \cos^2\theta_n + y_{hn}^- \sin^2\theta_n \; k_n >< k_n
\end{aligned} \tag{7.73}$$

etc. As the terms in (7.73) are purely imaginary so is (7.73).

The continuity at $y = 0$ of the magnetic fields transverse to y, which are linked to $\hat{\mathbf{Y}}$ by (7.66), constrains $\mathbf{E}_\nu$ to be a real eigenvector of the complex operator $\hat{\mathbf{Y}}(k_t)$. In order for this to be possible, $\mathbf{E}_\nu$ must be an eigenvector common to its real and imaginary parts, with eigenvalue $\cot\alpha_\nu(k_t)$, namely

$$\mathrm{Im}\hat{\mathbf{Y}}(k_t) \cdot \mathbf{E}_\nu \;=\; \cot\alpha_\nu(k_t) \mathrm{Re}\hat{\mathbf{Y}}(k_t) \cdot \mathbf{E}_\nu \tag{7.74}$$

with

$$\hat{\mathbf{Y}} = \text{Re}\hat{\mathbf{Y}} - j\text{Im}\hat{\mathbf{Y}} \quad (7.75)$$

where

$$\text{Re}\hat{Y}_{11} = \int_0^{k_t} y_e^+ \cos^2\theta + y_h^+ \sin^2\theta k_x >< k_x dk_x \quad (7.76)$$

and

$$\text{Im}\hat{Y}_{11} = \int_{k_t}^{\infty} y_e^+ \cos^2\theta + y_h^+ \sin^2\theta k_x >< k_x dk_x - \sum_n y_{en}^- \cos^2\theta_n + y_{hn}^- \sin^2\theta_n k_n >< k_n \quad (7.77)$$

etc. Equivalently, we rewrite (7.74) as

$$\hat{\mathbf{Y}} \cdot \mathbf{E}_\nu = \lambda_\nu \text{Re}\hat{\mathbf{Y}} \cdot \mathbf{E}_\nu \quad (7.78)$$

with

$$\lambda_\nu(k_t) = 1 - j\cot\alpha_\nu(k_t) \quad (7.79)$$

The fundamental equation (7.78) reduces to the usual resonance condition for discrete modes when $\lambda_\nu = 0$. Two eigenvectors $\mathbf{E}_\nu(k_t), \mathbf{E}_\mu(k_t)$ of (7.78) are orthonormal with weight Re $\{\hat{\mathbf{Y}}\}$ in the interval $0 \le x < \infty$, i.e.

$$< \mathbf{E}_\mu \cdot \text{Re}\hat{\mathbf{Y}} \cdot \mathbf{E}_\nu > = \int_0^{\infty}\int_0^{\infty} \mathbf{E}_\mu \cdot \text{Re}\hat{\mathbf{Y}} \cdot \mathbf{E}_\nu dx dx' = \delta_{\mu\nu} \quad (7.80)$$

The one–dimensional orthonormality (7.80) over the interface $y = 0$ should not be confused with the orthonormality (7.56) of the continuous modes over the cross–section S; moreover, the dot product (7.80) involves the scalar product of a dyadic with two vectors in the x-z plane and results in a scalar, whereas the operation defined in (7.73) is the Fourier transform of the x-z vector and results in a vector. The eigenvectors of (7.78) above are closely linked to the continuous modes of the open guides $\mathbf{e}_\nu(x, y; k_t)$. It turns out that, within a proportionality constant, we have

$$\sin\alpha_\nu \tilde{\mathbf{E}}_\nu(k_x, k_t) \propto \tilde{\mathbf{e}}_\nu(k_x, 0; k_t) \quad (7.81)$$

The modes are now simply obtained from $\tilde{\mathbf{E}}_\nu$ through (7.71), upon application of (7.62), (7.64), and (7.79) in the completeness relationship (7.61). In considering the latter, it is recalled that $\mathbf{V}_\nu^+$ is complex because of the exponential y–dependence, $\mathbf{V}_\nu^-$ is real and moreover, one has

$$\text{Re}\frac{e^{-jk_y y}}{1 - j\cot\alpha_\nu} = \begin{cases} \sin\alpha_\nu \sin(k_y y + \alpha_\nu) & k_x \le k_t \\ \sin^2\alpha_\nu e^{-\gamma y} & k_x > k_t \end{cases} \quad (7.82)$$

By introducing the above equation into (7.61), we finally recover the normalised modes as

$$\mathbf{e}_\nu(\mathbf{r}; k_t) = \begin{cases} \sqrt{\frac{2k_t}{\pi}} \sin\alpha_\nu \sum_n \mathbf{E}_{\nu n}\phi_n(x) \frac{\sin q_n(y+h)}{\sin q_n h} & y \le 0 \\ \sqrt{\frac{2k_t}{\pi}} \left[\begin{array}{c} \int_0^{k_t} \tilde{\mathbf{E}}_\nu k_x > \sin(k_y y + \alpha_{nu}) dk_x + \\ \sin\alpha_\nu \int_{k_t}^{\infty} \tilde{\mathbf{E}}_\nu k_x > e^{-\gamma y} dk_x \end{array} \right] & y > 0 \end{cases} \quad (7.83)$$

The field $\mathbf{e}_\nu$ is obviously continuous on the interface $y = 0$, whereas satisfaction of edge and boundary conditions is guaranteed by an appropriate choice of $\mathbf{E}_\nu$. From (7.66), continuity of the transverse to y magnetic fields at $y = 0$ requires just the satisfaction of the eigenvalue equation (7.78). Moreover, the orthogonality condition (7.61) can be checked out explicitly [20].

7.3.3 Solution of the eigenvalue equation

Efficiently solving the eigenvalue equation (7.78) and finding the ν-spectrum for each value of k_t $(0 \leq k_t < \infty)$ is the key to practical application of the continuum. Although various techniques, recursive and variational, are available, we shall mention here just the Ritz–Galerkin variational approach in the space domain, which is applicable also when upper and lower half spaces are of different cross–sections and, as such, not amenable to the same Fourier transform. The latter requires an orthonormal scalar basis $\{f_p\}$ over the aperture at the interface with respect to the weight function $W(x)$, i.e.,

$$< p, W, q > \;\; \equiv \;\; < f_p, W, f_q >= \delta_{pq} \tag{7.84}$$

The weight function $W(x)$ incorporates all edge conditions appropriate to the configuration under study. We then assume a finite expression of the unknown $\mathbf{E}_\nu = (E_{\nu x}, E_{\nu z})$ as

$$\mathbf{E}_\nu(x) \;\; = \;\; W(x) \sum_{p=0}^{N-1} \mathbf{E}_{\nu p} f_p(x) \tag{7.85}$$

and form the $2N \times 2N$ matrices $(p, q = 0, \ldots, N-1)$ of elements

$$\begin{aligned} G_{pq} &= < f_p, \mathrm{Re}\hat{\underline{\underline{\mathbf{Y}}}} \cdot f_q > \\ B_{pq} &= < f_p, \mathrm{Im}\hat{\underline{\underline{\mathbf{Y}}}} \cdot f_q > \end{aligned} \tag{7.86}$$

The eigenvalue equation (7.78) then reduces to the matrix equation

$$\hat{\mathbf{B}} \cdot \mathbf{E}_\nu \;\; = \;\; \cot \alpha_\nu \hat{\mathbf{G}} \cdot \mathbf{E}_\nu \tag{7.87}$$

The choice of the order of the expansion N is, of course, as low as possible, compatible with the near field detail of the scattering problem where the continuum is being used. As the eigenvalue equation contains the continuous parameter k_t, its practical solution requires interpolation.

7.3.4 Summary of the procedure for determining the continuum

In summary the following sequence of steps is to be accomplished for finding the continuum, orthonormalised according to (7.56), of a nonseparable cross–section:

1. Establish an appropriate basis for expressing the unknown field transverse to y;
2. For a fixed k_t form and discretise the dyadic admittance matrix (7.65) by means of the above functions and solve the eigenvalue equation (7.87);

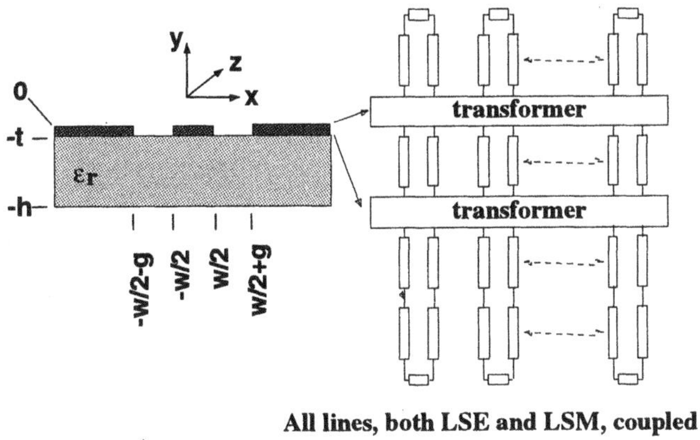

Figure 7.4 Transverse equivalent network of a coplanar waveguide.

3. Where required, obtain all necessary field components over the cross–section through the use of suitable potentials;

4. Repeat step 2 at a number of values of k_t and interpolate numerically between them.

7.4 Continuous spectrum of coplanar waveguide

As has been pointed out several times, the mode spectra of open structures consist of a discrete number of bound or surface modes and a continuum of radiation modes. In practical applications, the guide parameters are chosen such that only one discrete mode exists in the frequency band of operation, although it is also possible that the structure does not possess even a single discrete mode, for example the case of a metallic rod in a homogeneous medium. However, in both cases a continuum of radiation modes exist that radiate transversally, and either propagate, or are evanescent along the guide axis (z). The existence of these modes ensures the necessary completeness of a modal representation of the fields in an analogous manner to that of the evanescent modes of a closed waveguide and therefore, if we are to contemplate analysing discontinuities in open CPW structures using modal techniques, it is essential to identify these modes.

We have seen in the previous Sections how the continuum may be recovered from various conceptual viewpoints, which may further be regarded as a generalisation of the transverse resonance method. This is the approach that will be briefly discussed here.

In a conventional transverse resonance analysis of an open nonseparable structure, a transmission line model, similar to that of Figure 7.4, is envisaged. Throughout the transverse cross–section of the guide, a common $e^{-j\beta z}$ dependence is imposed on the fields, where z lies along the guide axis and β is the longitudinal propagation constant. In each of the separable subregions of the guide, the electric fields transverse to y are expressed as a superposition of known separable solutions to the wave equation with expansion coefficients determined from the transverse electric fields at the various interfaces between the regions. Obviously, continuity of

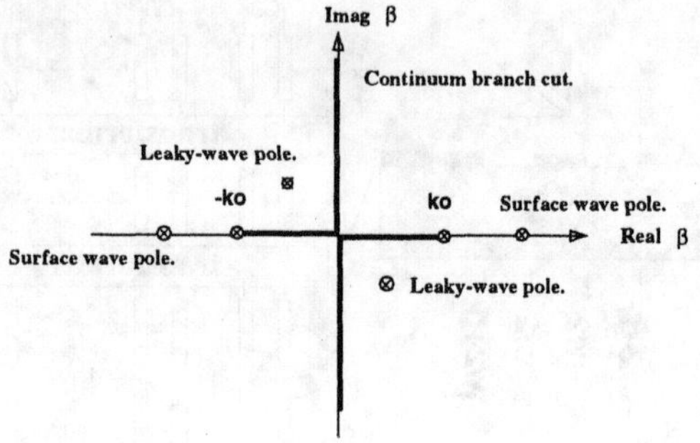

Figure 7.5 Poles and branch cuts in the complex β plane.

the transverse electric fields has already been built into the model. Subsequently, an expression for the total transverse admittance seen at an arbitrary reference plane is derived from the electric fields and the admittances of the individual transmission lines used in the model. The transverse admittance is a function of β and, on the premise that the discrete modes exist in the absence of a source in the transverse plane, a null is sought in the transverse admittance. Hence the propagation constants of the discrete modes can be identified and the fields subsequently found by solving the homogeneous equation using these values of β. It is immediately apparent, that requiring a null in the transverse admittance is equivalent to requiring the correct continuity of the transverse magnetic fields throughout the transverse plane.

For a bound mode with fields that decay exponentially away from the guide, the admittance seen on either side of the reference plane is reactive in the absence of losses and hence the possibility of a resonance, i.e. a true null in the transverse admittance exists. This can only occur if all the transverse transmission lines in the open regions are below cutoff. If some of these lines are permitted to support propagating waves, then it may still be possible to find a null in the transverse admittance even though the admittance looking into the open region is now complex. However, the discrete values of β for which this can occur are also complex, and correspond to fields that are exponentially increasing in the open region and are consequently non-physical solutions. These are the leaky wave modes, which although they may be useful in certain situations, form no part of the proper modal spectrum.

The completeness of modal field representations due to the presence of arbitrary sources requires, in addition to any bound modes, the inclusion of the proper radiation modes which exist for a continuous range of β and possess fields that radiate transversally but are bounded at infinity. Figure 7.5 shows the contour in the complex β plane along which these values lie and which may be succinctly expressed as $0 < k_t < \infty$, where k_t is the transverse wavenumber defined by $k_t^2 = k_0^2 - \beta^2$ and k_0 is the free space wavenumber.

Therefore, if β is constrained to lie on this contour, which means that the fields in the open region consist of a superposition of standing waves as well as decaying waves, we may again seek nulls in the transverse admittance. Standing waves in open regions however are not uniquely defined by their amplitudes on a boundary and the requirement for boundedness at infinity for an arbitrary phase shift in each standing wave is still present. Fortunately, it has been shown by considering modal orthogonality, that for a given mode, the same value of phase shift is introduced in every standing wave and thus this phase shift is characteristic of the particular mode in question.

Now equating the transverse admittance to zero, results in an eigenvalue equation rather than a homogeneous equation as is the case for the bound modes, where the eigenvalue corresponds to the cotangent of this characteristic phase shift. Therefore, for every continuous value of β on the above contour, an infinite set of modes may be found satisfying an eigenvalue equation and characterised by its standing wave phase shift and the electric fields at the interfaces between the subregions of the transverse cross–section.

Distinguishing the different modes for a given value of k_t, by a subscript, ν, the modal orthonormality is correctly expressed by equation (7.56)

7.4.1 The radiation spectrum of coplanar waveguide with finite thickness metallisation

Referring to Figure 7.4, the electric fields for $y > 0$ may be expressed as a Fourier transform

$$\begin{bmatrix} E_{\nu x}(x,y;k_t) \\ \partial_x E_{\nu z}(x,y;k_t) \end{bmatrix} = \int_0^{k_t} \phi(k_x,x)\sin\left(k_y y + \alpha_\nu(k_t)\right)\tilde{\mathbf{e}}_\nu(k_t,k_x)dk_x + \int_{k_t}^{\infty} \phi(k_x,x)e^{-|k_y|y}\sin\left(\alpha_\nu(k_t)\right)\tilde{\mathbf{e}}_\nu(k_t,k_x)dk_x \tag{7.88}$$

The factors $e^{-j\beta z}$ and $e^{j\omega t}$ have been suppressed for brevity. The coefficients $\tilde{\mathbf{e}}_\nu(k_t,k_x)$ are derived from the electric field distribution in the aperture, $(w/2 < x < w/2 + g;\ y = 0)$, as

$$\begin{aligned} \tilde{\mathbf{e}}_\nu(k_t,k_x)\sin\alpha_\nu &= \int_{-\infty}^{\infty} \begin{bmatrix} E_{\nu x}(x,0;k_t) \\ \partial_x E_{\nu z}(x,0;k_t) \end{bmatrix} \phi(k_x,x)dx \\ \phi(k_x,x) &= \sqrt{\frac{2}{\pi}}\sin(k_x x) \end{aligned} \tag{7.89}$$

Similar superpositions are used for the slots, dielectric slab and air region beneath it. From these expressions, it has been shown, in the previous Sections, that the transverse eigenproblem may be written in the following form:

$$\left(\cot\alpha_\nu(k_t)\begin{bmatrix} \mathbf{B}_{11} & \mathbf{B}_{12} \\ \mathbf{B}_{21} & \mathbf{B}_{22} \end{bmatrix} - \begin{bmatrix} \mathbf{Y}_{11} & \mathbf{Y}_{12} \\ \mathbf{Y}_{21} & \mathbf{Y}_{22} \end{bmatrix}\right) \times \begin{pmatrix} \begin{bmatrix} E_{\nu x}(x,0;k_t) \\ \partial_x E_{\nu z}(x,0;k_t) \end{bmatrix} \\ \begin{bmatrix} E_{\nu x}(x,-t;k_t) \\ \partial_x E_{\nu z}(x,-t;k_t) \end{bmatrix} \end{pmatrix} = \begin{pmatrix} 0 \\ 0 \\ 0 \\ 0 \end{pmatrix} \tag{7.90}$$

where:

$$\begin{aligned} \mathbf{Y}_{11} &= \sum_n \phi_n(x)\mathbf{Y}_n^b\phi_n(x')\cot(q_n t) - \int_{k_t}^{\infty} \phi(x;k_x)\tilde{\mathbf{Y}}^a(k_x)\phi(x';k_x)dk_x \\ \mathbf{Y}_{12} = \mathbf{Y}_{21} &= -\sum_n \phi_n(x)\mathbf{Y}_n^b\phi_n(x')\mathrm{cosec}(q_n t) \end{aligned} \tag{7.91}$$

$$\mathbf{Y}_{22} = \sum_n \phi_n(x)\mathbf{Y}_n^b\phi_n(x')\cot(q_n t) - \mathrm{Im}\left[\int_{k_t}^{\infty}\phi(x;k_x)\tilde{\mathbf{Y}}^c(k_x)\phi(x';k_x)dk_x\right]$$

and

$$\begin{aligned}
\mathbf{B}_{11} &= \int_0^{k_t}\phi(x;k_x)\tilde{\mathbf{Y}}^a(k_x)\phi(x';k_x)dk_x \\
\mathbf{B}_{22} &= \mathrm{Re}\left[\int_0^{k_t}\phi(x;k_x)\tilde{\mathbf{Y}}^c(k_x)\phi(x';k_x)dk_x\right] \\
\mathbf{B}_{12} = \mathbf{B}_{21} &= 0
\end{aligned} \tag{7.92}$$

with

$$\begin{aligned}
\tilde{\mathbf{Y}}^a(k_x) &= \frac{-j\omega\mu_0}{|k_y|}\begin{pmatrix} k_t^2 & \beta \\ \beta & \frac{k_0^2-k_x^2}{k_x^2}\end{pmatrix} \\
\tilde{\mathbf{Y}}^c(k_x) &= \frac{-j\omega\mu_0}{|k'_y|}g(k'_y,h)\begin{pmatrix} {k'}_t^2 & \beta \\ \beta & \frac{\varepsilon_r k_0^2-k_x^2}{k_x^2}\end{pmatrix} \\
\tilde{\mathbf{Y}}_n^b(k_x) &= \frac{-j\omega\mu_0}{|q_n|}\begin{pmatrix} k_t^2 & \beta \\ \beta & \frac{k_0^2-(\frac{n\pi}{g})^2}{(\frac{n\pi}{g})^2}\end{pmatrix}
\end{aligned} \tag{7.93}$$

and

$$g(k'_y,h) = \frac{k_y + jk'_y\tan(k'_y h)}{k'_y + jk_y\tan(k'_y h)} + \frac{\sec^2(k_y h)(\varepsilon_r - 1)\varepsilon_r k_0^2}{[k'_y + jk_y\tan(k'_y h)][\varepsilon_r k_y + jk'_y\tan(k'_y h)]} \tag{7.94}$$

while

$$\begin{aligned}
k_y^2 &= k_t^2 - k_x^2 \\
q_n^2 &= k_t^2 - \left(\frac{n\pi}{g}\right)^2 \\
{k'}_t^2 &= k_t^2 + (\varepsilon_r - 1)k_0^2 \\
{k'}_y^2 &= {k'}_t^2 - k_x^2 \\
\phi_n(x) &= \frac{2}{\sqrt{\pi}}\sin\left(\frac{n\pi x}{g}\right)
\end{aligned} \tag{7.95}$$

Equation (7.90) is a generalised eigenvalue equation of the form $Ax = \lambda Bx$, which has been solved using Galerkin's method in the space domain.

To ensure rapid convergence, weighted Gegenbauer polynomials of order 1/6 can be used as the basis functions as these intrinsically satisfy the known singular behaviour of the fields at the 90 degree metallic corners,

$$f_n(x) = \left[1 - \left(\frac{2x}{g}\right)^2\right]^{-\frac{1}{3}} C_n^{\frac{1}{6}}\left(\frac{2x}{g}\right) \tag{7.96}$$

However, in contrast to modelling the fundamental discrete mode, recovering the N most significant radiation modes for a given value of k_t will require using a minimum

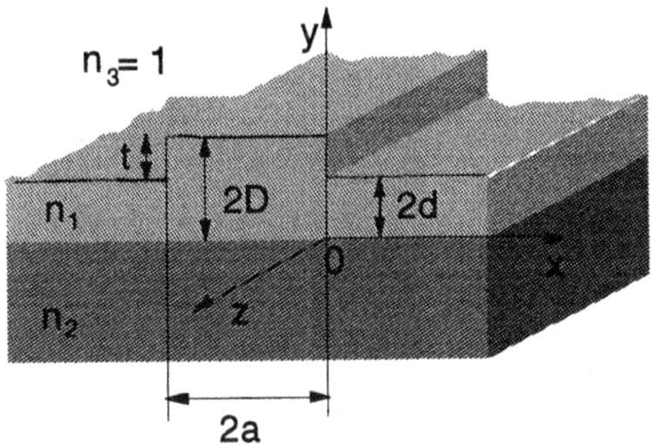

Figure 7.6 Cross–section of the infinite rib waveguide.

of N, and preferably at least $2N$, basis terms. Consequently, even though the modes need only be determined once, it is desirable to seek efficient methods of evaluating the elements of the matrix eigenvalue equation. One alternative is to use a simple trigonometric basis set, resulting in a more rapid evaluation of the modes, albeit at the expense of convergence.

7.5 Radiation modes of dielectric rib waveguide

The state of the art in millimetric and integrated optical technology is such that fairly sophisticated circuits containing a number of components designed to perform complex functions are now realizable. Rib waveguide is, possibly, the most widely used transmission medium in this connection, which motivated a considerable modelling effort in order to characterise its guided modes, as already noted in the previous Chapters.

Realistic components, however, imply the presence of discontinuities, whose effect is not only to alter the propagation characteristics of the fundamental mode, but also to excite radiative and reactive unbound fields in the substrate and air regions. Radiative fields can travel a long way in both regions and cause interaction with other components sharing the same substrate or cladding, particularly in multilevel, multiple guide, configurations. It is therefore important to ascertain what these fields are in rib guide. The complete spectrum of multilayer slab waveguides is well known and has been studied in Chapter 3. It consists of a few bound modes, if any, and a continuum of radiative and reactive air and substrate modes. The latter are real fields, finite at infinity that, under appropriate orthonormalization, constitute together with the bound modes a complete description of any physical field around the guide. For instance, in the presence of discontinuities, they describe the field in a manner conceptually analogous to a modal expansion in a closed guide.

The object of this Section is to introduce the radiative and reactive spectra for the rib guide. Generally, works on the step discontinuity in rib guide either consider just the fundamental mode, which is fair enough in the absence of serious radiative effects, or approach the problem by means of numerical methods. Finite elements, the beam propagation method or the method of lines are used to investigate other kinds of discontinuities such as bends or Y–junctions. In this book, in Chapter

5, the curvature was examined by means of the EDC transformation of the rib waveguide into a slab waveguide, and then applying the concept of the "local modes" to the study of the bend, taking into account the complete spectrum of the slab waveguide. In general, while useful results are produced by numerical methods for isolated discontinuities, the presence of multiple discontinuities causes occupation of memory and computer time to grow with the cube of the dimensions of the structure. Moreover, with a view to interpreting results and deriving simplified models, it is important to develop analytical tools that can be applied to this class of problem.

The analytical development is complicated by the non–separable nature of the rib guide cross–section, involving diffraction transverse to x (see Figure 7.6). In the following, the continuous spectrum of rib waveguide is found, by using the concept of "wave packets" that are individual solutions of the transverse diffraction problem satisfying boundary, edge and radiation conditions. Each packet is, in general, labelled by two indices, namely, the continuous transverse wavenumber k_t, and a discrete one, ν, identifying possible degenerate solutions pertaining to the same value of k_t. Each packet is fully characterised by a single "phase shift" $\alpha_\nu(k_t)$, which is found by solving an eigenvalue equation arising from field continuity at the discontinuous interface of the transverse step. "Wave packets" (continuous modes) are amenable to orthonormalization and, together with the bound modes, constitute a complete spectrum of modes for the rib guide.

It is remarked that the full hybrid nature of the field in rib guide has not yet been dealt with in this Section. The hybrid nature plays a significant role close to modal cutoff of the fundamental mode and in those situations where the discontinuity may cause considerable cross–polarization. Where the rib aspect ratio $(t/2a)$ is sufficiently low, typically ≤ 0.4, the minor hybrid content of the field does not seem to warrant the additional complication involved. In the following, LSE polarization is assumed, but this assumption is removable if required.

When determining the spectrum, use is made of the concept of "transition function", previously introduced in Chapter 5, in order to generate accurate first order variational solutions for the bound modes that considerably reduce the computational load.

7.5.1 Continuous spectrum of the infinite rib guide

As two indices are needed in order to represent the complete spectrum of a closed two–dimensional waveguide, two indices are also needed in this open problem. Moreover, the complete spectrum must reproduce the whole range of β–values on the complex plane: the real values correspond to radiating waves, the imaginary ones to reactive effects. We use a continuous variable that labels the continuous spectrum, namely $k_t = (k_0^2 - \beta^2)^{1/2}$, with $0 \leq k_t < \infty$, (air continuous modes) or $\sigma_t = (k_0^2\varepsilon_2 - \beta^2)^{1/2}$, with $0 \leq \sigma_t \leq v = k_0\sqrt{\varepsilon_2 - 1}$ (substrate continuous modes). The whole continuous spectrum is therefore constituted by a propagating part ($k_t \leq k_0$ corresponding to $0 \leq \beta(k_t) \leq k_0$ and $0 \leq \sigma_t \leq v$ corresponding to $k_0 < \beta(k_t) \leq k_0\sqrt{\varepsilon_2}$), and a non-propagating one ($k_t > k_0$ corresponding to $\mathrm{Im}\beta(k_t) > 0$).

To a given k_t–value (or σ_t–value) may further correspond more than one field distribution, each satisfying the boundary conditions, that will be labelled by the discrete index ν. Hence, the unbound part of a typical field component, say H_y, can be written in terms of this spectrum as

$$
\begin{aligned}
H_y(x,y) &= \int_0^{\infty} \sum_{\nu=1,2,\cdots} A_\nu(k_t) H_{\nu y}(x,y;k_t)dk_t + \\
&\quad \int_0^{v} \sum_{\nu=1,2,\cdots} B_\nu(\sigma_t) H_{\nu y}(x,y;\sigma_t)d\sigma_t
\end{aligned} \tag{7.97}
$$

Each component of the continuum ("wave packet") is therefore completely labelled by two indices, the continuous index k_t (or σ_t) and the discrete index ν.

Finding the continuous spectrum of the rib waveguide, shown in Figure 7.6, starts from the knowledge of the complete spectra of the two slabs constituting the structure. They include the surface waves, $\phi_s(y)$ for $x < 0$, $\psi_s(y)$ for $x > 0$, and the continua of the air regions $\phi_e(\rho, y)$, $\phi_o(\rho, y)$, $\psi_e(\rho, y)$, $\psi_o(\rho, y)$, e=even, o=odd and of the substrate regions $\phi_{sb}(\sigma, y)$, $\psi_{sb}(\sigma, y)$.

Considering transverse propagation in the x–direction, the presence of the discontinuity at $x = 0$ produces scattering among the slab modes. As in a discontinuity problem, a combination of these modes is required in order to satisfy all boundary and edge conditions imposed by the rib corner. Starting from these physical considerations, we suppose that there is only one surface wave in both slabs and electric wall symmetry at the rib plane of symmetry $x = -a$; we fix a value of k_t and of the discrete index ν. We then choose a y–directed potential that is constituted by the superposition of slab modes of the following form (suppressing the factor $e^{-j\beta z}$):

$$
\begin{aligned}
\Pi_\nu^h(x,y;k_t) &= \frac{P_\nu^-}{k_0^2\varepsilon_{e^-}}\phi_s(y)\frac{\cos[q_s(k_t)(x+a)]}{\cos[q_s(k_t)a]} + \\
&\quad \sum_{m=e,o}\int_0^{\infty}\frac{Q_{\nu m}^-(\rho)}{k_0^2-\rho^2}\phi_m(\rho,y)\frac{\cos[q(\rho,k_t)(x+a)]}{\cos[q(\rho,k_t)a]}d\rho + \\
&\quad \int_0^{v}\frac{S_\nu^-(\sigma)}{k_0^2\varepsilon_2-\sigma^2}\phi_{sb}(\sigma,y)\frac{\cos[p(\sigma,k_t)(x+a)]}{\cos[p(\sigma,k_t)a]}d\sigma \\
&\quad -a \le x \le 0
\end{aligned} \tag{7.98}
$$

while for $x > 0$ we have:

$$
\begin{aligned}
\Pi_\nu^h(x,y;k_t) &= \frac{P_\nu^+}{k_0^2\varepsilon_{e^+}}\psi_s(y)\frac{\cos[k_x(k_t)x+\alpha_\nu(k_t)]}{\cos[\alpha_\nu(k_t)]} + \\
&\quad \sum_{m=e,o}\int_0^{k_t}\frac{Q_{\nu m}^+(\rho)}{k_0^2-\rho^2}\psi_m(\rho,y)\frac{\cos[q(\rho,k_t)x+\alpha_\nu(k_t)]}{\cos[\alpha_\nu(k_t)]}d\rho + \\
&\quad \sum_{m=e,o}\int_{k_t}^{\infty}\frac{Q_{\nu m}^+(\rho)}{k_0^2-\rho^2}\psi_m(\rho,y)e^{-\gamma(\rho,k_t)x}d\rho + \\
&\quad \int_0^{v}\frac{S_\nu^+(\sigma)}{k_0^2\varepsilon_2-\sigma^2}\psi_{sb}(\sigma,y)\frac{\cos[p(\sigma,k_t)x+\alpha_\nu(k_t)]}{\cos[\alpha_\nu(k_t)]}d\sigma
\end{aligned} \tag{7.99}
$$

where e, o denotes parity of the continua of the slabs and, moreover:

$$
\begin{aligned}
k_x(k_t) &= \sqrt{k_t^2 + k_0^2(\varepsilon_{e^+} - 1)} \\
q_s(k_t) &= \sqrt{k_t^2 + k_0^2(\varepsilon_{e^-} - 1)}
\end{aligned}
$$

$$
\begin{aligned}
p(\sigma, k_t) &= \sqrt{k_t^2 - \sigma^2 + v^2} \\
q(\rho, k_t) &= \sqrt{k_t^2 - \rho^2} \\
\gamma(\rho, k_t) &= \sqrt{\rho^2 - k_t^2} \\
v &= k_0\sqrt{\varepsilon_2 - 1}
\end{aligned} \tag{7.100}
$$

$P_\nu^-, P_\nu^+, Q_{\nu m}^-, Q_{\nu m}^+, S_\nu^-, S_\nu^+$ are as yet unknown amplitudes and $\varepsilon_{e-}, \varepsilon_{e+}$ the effective dielectric constants of the slabs for $x < 0$, $x > 0$ respectively.

The form of (7.99), (7.100) requires some comment. The x–dependence in (7.99) arises from the even plane of symmetry at $x = -a$, that in (7.100) implies a standing wave in the external region. The latter can be seen as the combination of a wave $e^{jk_x x}$ generated by a source at $x = \infty$ and travelling towards the discontinuity and a reflected wave $e^{-j(k_x x + 2\alpha_\nu k_t)}$. The phase shift $\alpha_\nu(k_t)$ is of special significance.

It is noted that it is taken as a function of just $k_t = \sqrt{k_x^2 + k_y^2}$ and not of k_x, k_y separately. This is an essential requirement in order that fields transverse to x derived from the same potential and travelling with the same phase velocity in the z–direction be continuous at the transverse step discontinuity, that is, they form a "wave packet" individually satisfying the boundary conditions. In fact, taking into account that the fields in the LSE case are given by:

$$
\begin{aligned}
E_{\nu x}(x, y; k_t) &= \omega\mu\beta(k_t)\Pi_\nu^h(x, y; k_t) \\
E_{\nu y}(x, y; k_t) &= 0 \\
E_{\nu z}(x, y; k_t) &= -j\omega\mu\partial_x\Pi_\nu^h(x, y; k_t) \\
H_{\nu x}(x, y; k_t) &= \partial_{xy}^2\Pi_\nu^h(x, y; k_t) \\
H_{\nu y}(x, y; k_t) &= \left(k_0^2\varepsilon_r + \partial_y^2\right)\Pi_\nu^h(x, y; k_t) \\
H_{\nu z}(x, y; k_t) &= -j\beta(k_t)\partial_y\Pi_\nu^h(x, y; k_t)
\end{aligned} \tag{7.101}
$$

we apply the continuity of the $H_{y\nu}(x, y, k_t)$ component at the interface $x = 0$. Denoting the interface field $H_{y\nu}(x, y, k_t) \equiv h_{y\nu}(y, k_t)$, we recover, from orthogonality of the slab mode functions, the following expressions for the amplitude coefficients P, Q, S at either side of $x = 0$:

$$
\begin{aligned}
P_\nu^- &= \int_{-\infty}^{\infty} h_{\nu y}(y; k_t)\phi_s(y)dy \\
&= \langle h_{\nu y}(y; k_t), \phi_s(y) \rangle \\
P_\nu^+ &= \langle h_{\nu y}(y; k_t), \psi_s(y) \rangle \\
Q_{\nu m}^- &= \langle h_{\nu y}(y; k_t), \phi_m(y) \rangle \\
Q_{\nu m}^+ &= \langle h_{\nu y}(y; k_t), \psi_m(y) \rangle \\
S_\nu^- &= \langle h_{\nu y}(y; k_t), \phi_{sb}(y) \rangle \\
S_\nu^+ &= \langle h_{\nu y}(y; k_t), \psi_{sb}(y) \rangle
\end{aligned}
$$

The analytical expressions of these coefficients have been reported in Chapter 6. As these coefficients are now expressed in terms of $h_{y\nu}$, we apply the continuity of $E_{z\nu}(x, y, k_t)$ at $x = 0$. From this equation, we obtain an eigenvalue equation for the as yet undetermined phase shift $\alpha_\nu(k_t)$, that is, the last unknown quantity in the expressions (7.99), (7.100) for the potential. This is:

$$\tan\alpha_\nu(k_t)\left\{\begin{array}{c}\frac{P_\nu^+}{k_0^2\varepsilon_{e+}}\psi_s(y)k_x+\\ \sum_{m=e,o}\int_0^{k_t}\frac{Q_{\nu m}^+(\rho)}{k_0^2-\rho^2}\psi_m(\rho,y)(\rho,k_t)d\rho+\\ \int_0^v\frac{S_\nu^+(\sigma)}{k_0^2\varepsilon_2-\sigma^2}\psi_{sb}(\sigma,y)p(\sigma,k_t)d\sigma\end{array}\right\}=$$

$$\begin{array}{c}-\sum_{m=e,o}\int_{k_t}^{\infty}\frac{Q_{\nu m}^+(\rho)}{k_0^2-\rho^2}\psi_m(\rho,y)\gamma(\rho,k_t)d\rho+\\ \frac{P_\nu^-}{k_0^2\varepsilon_{e-}}\phi_s(y)q_s(k_t)\tan\left[q_s(k_t)a\right]+\\ \sum_{m=e,o}\int_0^\infty\frac{Q_{\nu m}^-(\rho)}{k_0^2-\rho^2}\phi_m(\rho,y)q(\rho,k_t)\tan\left[q(k_t)a\right]d\rho+\\ \int_0^v\frac{S_\nu^-(\sigma)}{k_0^2\varepsilon_2-\sigma^2}\phi_{sb}(\sigma,y)p(\sigma,k_t)\tan\left[p(\sigma,k_t)a\right]d\sigma\end{array} \tag{7.102}$$

On inspection and bearing in mind the definitions of P, Q and S, (7.102) can be written more compactly as the following operator eigenvalue equation

$$\tan\alpha_\nu(k_t)\mathrm{Re}\left\{\tilde{Z}(k_t)\right\}\cdot h_{y\nu}(y;k_t) \;=\; \mathrm{Im}\left\{\tilde{Z}(k_t)\right\}\cdot h_{y\nu}(y;k_t) \tag{7.103}$$

where $\tilde{Z}(k_t) = \mathrm{Re}\left\{\tilde{Z}(k_t)\right\} - j\mathrm{Im}\left\{\tilde{Z}(k_t)\right\}$ is the transverse impedance integral operator acting on the field $h_{y\nu}(y,k_t)$ whose kernel $Z = Z(y,y';k_t)$ is defined below:

$$\mathrm{Re}\{Z\} = \omega\mu\left\{\begin{array}{c}\frac{k_t}{k_0^2\varepsilon_{e+}}\psi_s(y)\psi_s(y')+\\ \sum_{m=e,o}\int_0^{k_t}\frac{q(\rho,k_t)}{k_0^2-\rho^2}\psi_m(\rho,y)\psi_m(\rho,y')d\rho+\\ \int_0^v\frac{p(\sigma,k_t)}{k_0^2\varepsilon_2-\sigma^2}\psi_{sb}(\sigma,y)\psi_{sb}(\sigma,y')d\sigma\end{array}\right\} \tag{7.104}$$

$$\mathrm{Im}\{Z\} = \omega\mu\left\{\begin{array}{c}-\sum_{m=e,o}\int_{k_t}^\infty\frac{\gamma(\rho,k_t)}{k_0^2-\rho^2}\psi_m(\rho,y)\psi_m(\rho,y')d\rho+\\ \frac{q_s(k_t)}{k_0^2\varepsilon_{e-}}\phi_s(y)\phi_s(y')\tan\left[q_s(k_t)a\right]+\\ \sum_{m=e,o}\int_0^\infty\frac{q(\rho,k_t)}{k_0^2-\rho^2}\phi_m(\rho,y)\phi_m(\rho,y')\tan\left[q(k_t)a\right]d\rho+\\ \int_0^v\frac{p(\sigma,k_t)}{k_0^2\varepsilon_2-\sigma^2}\phi_{sb}(\sigma,y)\phi_{sb}(\sigma,y')\tan\left[p(\sigma,k_t)a\right]d\sigma\end{array}\right\}$$

We may cast (7.103) in a variational form, by multiplying by $h_{y\nu}(y;k_t)$, integrating in y and obtaining:

$$\tan\alpha_\nu(k_t) = \frac{<h_{y\nu}(y;k_t),\mathrm{Im}\left\{\tilde{Z}(k_t)\right\},h_{y\nu}(y;k_t)>}{<h_{y\nu}(y;k_t),\mathrm{Re}\left\{\tilde{Z}(k_t)\right\},h_{y\nu}(y;k_t)>} \tag{7.105}$$

The above problem is amenable to a standard Ritz–Galerkin approach. Thus, we choose an appropriate orthonormal basis $f_k(y)$ in order to expand $h_{y\nu}(y;k_t)$:

$$h_{y\nu}(y;k_t) = \sum_{k=1}^{N}H_{k\nu}f_k(y) \tag{7.106}$$

Substituting the above in (7.103) turns the latter integral equation into the following matrix eigenvalue problem:

$$\tan\alpha_\nu(k_t)\mathrm{Re}\left\{\mathbf{Z}(k_t)\right\}_{N\times N}[\mathbf{H}_\nu]_{N\times 1} = \mathrm{Im}\left\{\mathbf{Z}(k_t)\right\}_{N\times N}[\mathbf{H}_\nu]_{N\times 1} \tag{7.107}$$

where $\mathbf{Z}(k_t) = \mathrm{Re}\{\mathbf{Z}(k_t)\} - j\mathrm{Im}\{\mathbf{Z}(k_t)\}$ is the transverse impedance matrix of the rib guide, defined elementwise by:

$$\mathrm{Re}\{\mathbf{Z}(k_t)\}_{ik} = \omega\mu \left\{ \begin{array}{c} \frac{k_t}{k_0^2\varepsilon_{e+}} P_i^+ P_k^+ + \\ \sum_{m=e,o} \int_0^{k_t} \frac{q(\rho,k_t)}{k_0^2-\rho^2} Q_{im}^+(\rho) Q_{km}^+(\rho) d\rho + \\ \int_0^v \frac{p(\sigma,k_t)}{k_0^2\varepsilon_2-\sigma^2} S_i^+(\sigma) S_k^+(\sigma) d\sigma \end{array} \right\} \tag{7.108}$$

$$\mathrm{Im}\{\mathbf{Z}(k_t)\}_{ik} = \omega\mu \left\{ \begin{array}{c} -\sum_{m=e,o} \int_{k_t}^{\infty} \frac{\gamma(\rho,k_t)}{k_0^2-\rho^2} Q_{im}^+(\rho) Q_{km}^+(\rho) d\rho + \\ \frac{q_s(k_t)}{k_0^2\varepsilon_{e-}} P_i^- P_k^- \tan[q_s(k_t)a] + \\ \sum_{m=e,o} \int_0^{\infty} \frac{q(\rho,k_t)}{k_0^2-\rho^2} Q_{im}^-(\rho) Q_{km}^-(\rho) \tan[q(k_t)a]\, d\rho + \\ \int_0^v \frac{p(\sigma,k_t)}{k_0^2\varepsilon_2-\sigma^2} S_i(\sigma) S_k(\sigma) \tan[p(\sigma,k_t)a]\, d\sigma \end{array} \right\}$$

$\tan\alpha_\nu(k_t)$ is the eigenvalue of (7.107), $\mathbf{H}_\nu$ is the eigenvector and $\mathrm{Re}\{\mathbf{Z}(k_t)\}$ is a weight matrix.The eigenvector is just the n-th component of the interface field $h_{y\nu}(y;k_t)$ in the basis (7.106). For this kind of problem, the following orthogonality relationship at the interface $x = 0$ holds (T denotes transposition):

$$\mathbf{H}_\nu^T \cdot \mathrm{Re}\{\mathbf{Z}(k_t)\} \cdot \mathbf{H}_\nu^T = \delta_{\nu\mu} \tag{7.109}$$

The physical interpretation of the above equation is that two field distributions pertaining to the same k_t, but corresponding to different indices μ, ν, are also orthogonal in y over the interface at $x = 0$, i.e. they do not exchange power at this discontinuous interface. In order to actually set up and solve (7.107) while keeping the matrix size as low as possible, we must effect a prudent choice of the expansion basis for the field at the interface, that includes all known physical features of the solution. In fact, we use a single function, the "transition function" introduced in Chapter 5, a choice that makes this problem scalar. The transition function consists of the distribution of the component h_{yt} of the guided mode of a slab of intermediate height, that maximises the coupling integrals between the slabs to the left and to the right of the transverse discontinuity and the intermediate slab itself. With this choice $N = 1$ in (7.106), and (7.107) becomes a scalar equation.

7.6 Continuum of the inset guide

The complete modal spectrum of the inset dielectric guide (IDG) may be identified by integrating its transverse characteristic Green's function around a closed contour in the complex k_t^2-plane. This integral reduces to a sum of pole residues and a branch cut integral from $0 < k_t < \infty$, corresponding to the discrete and continuous modes respectively.

As noticed previously, the characteristic Green's function may be represented in the form:

$$G_E(\mathbf{r}_t, \mathbf{r}_t'; k_t) = \frac{j}{\omega\mu} \sum_\nu \frac{\mathbf{E}_\nu^\pm(\mathbf{r}_t; k_t)[\mathbf{E}_\nu^\pm(\mathbf{r}_t; k_t)]^\dagger}{\int_{-\infty}^{\infty} \left[\mathbf{E}_\nu^\dagger(\mathbf{r}_t; k_t) \times \mathbf{H}_\nu^\pm(\mathbf{r}_t; k_t)\right]_{y=0-}^{y=0+} dx} \tag{7.110}$$

where $\mathbf{r}_t = (x, y)$, and the $\pm$ superscripts apply for $y > 0$ and $y < 0$ respectively. $\mathbf{E}_\nu^+(\mathbf{r}; k_t)$ and $\mathbf{E}_\nu^-(\mathbf{r}; k_t)$ are vectors satisfying the homogeneous Helmholtz equation

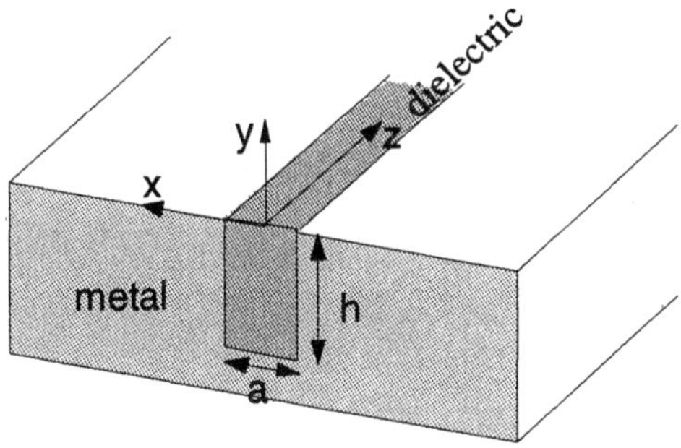

Figure 7.7 Cross–section of the infinite inset guide.

and the boundary conditions for $y > 0$ and $y < 0$ (see the IDG geometry shown in the relevant Section in Chapter 6). In the lossless case, the adjoint field, indicated by the superscript †, is a reverse propagating wave analogue of the original field. Continuity requires that

$$\mathbf{E}_\nu^+(x,0;k_t) = \mathbf{E}_\nu^-(x,0;k_t) \tag{7.111}$$

The fields in the open region may be derived from the following pair of y–directed Hertzian potentials:

$$\begin{aligned}
\mathbf{\Pi}_e^+(\mathbf{r}_t;k_t) &= \mathbf{y}_0 \int_0^\infty \frac{\phi_e(x;k_x)e^{-jk_y y}}{j\omega\varepsilon_0\sqrt{k_x^2+\beta^2}}\tilde{I}(k_t,k_x)dk_x \\
\mathbf{\Pi}_h^+(\mathbf{r}_t;k_t) &= \mathbf{y}_0 \int_0^\infty \frac{\phi_h(x;k_x)e^{-jk_y y}}{j\omega\mu_0\sqrt{k_x^2+\beta^2}}\tilde{V}(k_t,k_x)dk_x \\
\phi_e(x;k_x) &= \sqrt{\frac{2}{\pi}}\cos(k_x x) \\
\phi_h(x;k_x) &= \sqrt{\frac{2}{\pi}}\sin(k_x x)
\end{aligned} \tag{7.112}$$

where $k_x^2 + k_y^2 = k_t^2$ and the functions $\phi_{e,o}(x;k_x)$ have been chosen to give an E_x component that is an odd function of x.

Moreover, the magnetic fields may be expressed by means of an admittance operator as:

$$\begin{pmatrix} H_z(\mathbf{r}_t;k_t) \\ \int H_z(\mathbf{r}_t;k_t)dx \end{pmatrix} = -\int_0^\infty \int_0^\infty \underline{\underline{\mathbf{Y}}}^+(x,x';k_t) \begin{pmatrix} E_x(\mathbf{r}_t;k_t) \\ \partial_x E_z(\mathbf{r}_t;k_t) \end{pmatrix} dk_x dx' \tag{7.113}$$

with

$$\hat{Y}^+(x,x';k_t) = \frac{\phi_h(x;k_x)}{\omega\mu k_y} \begin{pmatrix} k_t^2 & j\beta \\ -j\beta & \frac{k_0^2-k_x^2}{k_x^2} \end{pmatrix} \phi_h(x';k_x) \tag{7.114}$$

As always it is convenient to deal with quantities that are all of the same parity with respect to x.

In a similar manner, the fields in the slot may be derived from the following y directed Hertzian potentials:

$$\begin{aligned}
\mathbf{\Pi}_e^-(\mathbf{r}_t;k_t) &= \mathbf{y}_0 \sum_{n=1,3,\cdots}^{\infty} \frac{\phi_{en}(x)\cos\left[q_n(y+h)\right]}{j\omega\varepsilon\sqrt{\left(\frac{n\pi}{a}\right)^2+\beta^2}\cos(q_n h)} I_n(k_t) \\
\mathbf{\Pi}_h^-(\mathbf{r}_t;k_t) &= \mathbf{y}_0 \sum_{n=1,3,\cdots}^{\infty} \frac{\phi_{hn}(x)\sin\left[q_n(y+h)\right]}{j\omega\mu\sqrt{\left(\frac{n\pi}{a}\right)^2+\beta^2}\sin(q_n h)} V_n(k_t) \\
\phi_{en}(x) &= \frac{2}{\sqrt{a}}\cos\left(\frac{n\pi}{a}x\right) \qquad (7.115)\\
\phi_{hn}(x) &= \frac{2}{\sqrt{a}}\sin\left(\frac{n\pi}{a}x\right) \qquad (7.116)
\end{aligned}$$

where $q_n^2 = (\varepsilon_r - 1)k_0^2 + k_t^2 - (n\pi/a)^2$. The transverse to y fields in the slot may be related by

$$\begin{pmatrix} H_z(\mathbf{r}_t;k_t) \\ \int H_x(\mathbf{r}_t;k_t)dx \end{pmatrix} = -\int_0^{\infty} \sum_{n=1,3,\cdots}^{\infty} \underline{\underline{\mathbf{Y}}}^-(x,x';k_t) \begin{pmatrix} E_x(\mathbf{r}_t;k_t) \\ \partial_x E_z(\mathbf{r}_t;k_t) \end{pmatrix} dx' \qquad (7.117)$$

where

$$\underline{\underline{\mathbf{Y}}}^-(x,x';k_t) = \frac{\phi_{hn}(x)}{\omega\mu q_n} \begin{pmatrix} k_t^2 + (\varepsilon_r - 1)k_0^2 & j\beta \\ -j\beta & \frac{\varepsilon_r k_0^2 - \left(\frac{n\pi}{a}\right)^2}{\left(\frac{n\pi}{a}\right)^2} \end{pmatrix} \cot(q_n h)\phi_{hn}(x') \qquad (7.118)$$

By using (7.113) and (7.114) the denominator of the transverse characteristic Green's function, the transverse admittance, may be expressed as

$$\begin{aligned}
&-j\omega\mu_0 \int_{-\infty}^{\infty} \left[\mathbf{E}_\nu^\dagger(\mathbf{r}_t;k_t) \times \mathbf{H}_\nu^\pm(\mathbf{r}_t;k_t)\cdot\mathbf{y}_0\right]_{y=0-}^{y=0+} = \\
&-j\omega\mu_0 \int_{-\infty}^{\infty}\int_{-\infty}^{\infty} \left(E_{\nu x}(x,0;k_t) \quad \partial_x E_{\nu z}(x,0;k_t) \right)^\dagger \\
&\left[\underline{\underline{\mathbf{Y}}}^+(x,x';k_t) - \underline{\underline{\mathbf{Y}}}^-(x,x';k_t)\right] \begin{pmatrix} E_{\nu x}(0,x';k_t) \\ \partial_x E_{\nu z}(0,x';k_t) \end{pmatrix} dx dx'
\end{aligned} \qquad (7.119)$$

which may be written, without loss of generality as

$$-jP_\nu(k_t)(1 - j\cot\alpha_\nu(k_t)) \qquad (7.120)$$

This constrains the vector $\begin{pmatrix} E_{\nu x}(0,x';k_t) \\ \partial_x E_{\nu z}(0,x';k_t) \end{pmatrix}$ and $\cot\alpha_\nu(k_t)$ to be respectively eigenvector and eigenvalue of the following equation

$$\begin{aligned}
&\cot\alpha_\nu(k_t)\omega\mu_0 \int_{-\infty}^{\infty}\int_{-\infty}^{\infty} \left(E_{\nu x}(x,0;k_t) \quad \partial_x E_{\nu z}(x,0;k_t) \right)^\dagger \\
&\mathrm{Re}\left[\underline{\underline{\mathbf{Y}}}(x,x';k_t)\right] \begin{pmatrix} E_{\nu x}(0,x';k_t) \\ \partial_x E_{\nu z}(0,x';k_t) \end{pmatrix} dx dx' = \\
&\omega\mu_0 \int_{-\infty}^{\infty}\int_{-\infty}^{\infty} \left(E_{\nu x}(x,0;k_t) \quad \partial_x E_{\nu z}(x,0;k_t) \right)^\dagger \\
&\mathrm{Im}\left[\underline{\underline{\mathbf{Y}}}(x,x';k_t)\right] \begin{pmatrix} E_{\nu x}(0,x';k_t) \\ \partial_x E_{\nu z}(0,x';k_t) \end{pmatrix} dx dx'
\end{aligned} \qquad (7.121)$$

where

$$\begin{aligned}
\underline{\underline{\mathbf{Y}}}(x,x';k_t) &= \left[\underline{\underline{\mathbf{Y}}}^{+}(x,x';k_t) - \underline{\underline{\mathbf{Y}}}^{-}(x,x';k_t)\right] \\
P_\nu(k_t) &= \delta_{\nu\mu}\omega\mu_0 \int_{-\infty}^{\infty}\int_{-\infty}^{\infty} \left(E_{\mu x}(x,0;k_t) \quad \partial_x E_{\mu z}(x,0;k_t) \right)^\dagger \\
&\quad \mathrm{Re}\left[\underline{\underline{\mathbf{Y}}}(x,x';k_t)\right] \begin{pmatrix} E_{\nu x}(0,x';k_t) \\ \partial_x E_{\nu z}(0,x';k_t) \end{pmatrix} dx dx'
\end{aligned} \tag{7.122}$$

For convenience $P_\nu(k_t)$ may be normalised to $\delta_{\mu\nu}$.

In addition to the poles that give rise to the discrete modes of the guide, the multivalued nature of the transverse admittance gives rise to branch point singularities in the transverse characteristic Green's function. The integral in the k_t^2 plane reduces then to a sum of residues and a branch cut integral from $0 \leq k_t^2 < \infty$. The integral around the branch cut identifies the continuum modes of the guide [4, 3]. The transverse electric fields of the continuum modes are subsequently given by:

$$\begin{aligned}
\begin{pmatrix} E_x(\mathbf{r}_t;k_t) \\ \partial_x E_z(\mathbf{r}_t;k_t) \end{pmatrix} &= \sqrt{\tfrac{2k_t}{\pi}} \int_0^{k_t} \phi_h(x;k_x)\sin(k_y y + \alpha_\nu(k_t))\tilde{\mathbf{e}}(k_x)dk_x + \\
&\quad \int_0^{k_t} \phi_h(x;k_x)e^{-|k_y|y}\sin(\alpha_\nu(k_t))\tilde{\mathbf{e}}(k_x)dk_x \\
&\qquad : y > 0
\end{aligned} \tag{7.123}$$

and

$$\begin{pmatrix} E_x(\mathbf{r}_t;k_t) \\ \partial_x E_z(\mathbf{r}_t;k_t) \end{pmatrix} = \sum_{n=1,3,\cdots}^{\infty} \phi_{hn}(x;k_x)\frac{\sin(q_n(y+h))}{\sin(q_n h)}\sin(\alpha_\nu(k_t))\tilde{\mathbf{e}}_{\nu n} \quad : y < 0 \tag{7.124}$$

where the coefficient vectors are the continuous and discrete Fourier transforms respectively of the interface fields defined by the eigenvalue equation. The latter may be solved in the usual way by Galerkin's method.

Bibliography

[1] R. E. Collin. *Field Theory of Guided Waves.* IEEE Press, New York, 1990.

[2] R. E. Collin and F. J. Zucker (Editors). *Antenna Theory.* McGraw-Hill, New York, 1969.

[3] L. Felsen and N. Marcuvitz. *Radiation and Scattering of Waves.* Prentice Hall, Englewood Cliffs, NJ, 1973.

[4] B. Friedman. *Principles and Techniques of Applied Mathematics.* John Wiley & Sons Inc., New York, 1956.

[5] R. J. Garbacz. Modal expansions for resonance scattering phenomena. *Proc. IEEE*, 53:856–864, August 1965.

[6] R. J. Garbacz and D. M. Pozar. Antenna shape synthesis using characteristic modes. *IEEE Trans. Antennas Propagat.*, 30:340–350, May 1982.

[7] R. J. Garbacz and R. H. Turpin. A generalized expansion for radiated and scattered fields. *IEEE Trans. Antennas Propagat.*, 19:348–358, May 1971.

[8] I. S. Gradshteyn and I. M. Ryzhik. *Table of integrals, series and products.* Academic Press, New York, 1980.

[9] R. F. Harrington and J. R. Mautz. Computation of characteristic modes for conducting bodies. *IEEE Trans. Antennas Propagat.*, 19:629–639, September 1971.

[10] R. F. Harrington and J. R. Mautz. Theory of characteristic modes for conducting bodies. *IEEE Trans. Antennas Propagat.*, 19:622–628, September 1971.

[11] R. F. Harrington and J. R. Mautz. Characteristic modes for aperture problems. *IEEE Trans. Microwave Theory Tech.*, 33:500–505, June 1985.

[12] N. Inagaki and R. J. Garbacz. Eigenfunctions of composite Hermitian operators with application to discrete and continuous radiating systems. *IEEE Trans. Antennas Propagat.*, 30:571–575, July 1982.

[13] C. Lanczos. *Linear Differential Operators.* Van Nostrand, New York, 1961.

[14] D. Liu, R. J. Garbacz, and D. M. Pozar. Antenna synthesis and optimization using generalized characteristic modes. *IEEE Trans. Antennas Propagat.*, 38:862–868, June 1990.

[15] P. M. Morse and H. Feshback. *Methods of Theoretical Physics.* McGraw-Hill, New York, 1950.

[16] D. M. Pozar. Antenna synthesis and optimization using weighted Inagaki modes. *IEEE Trans. Antennas Propagat.*, 32:159–165, February 1984.

[17] T. Rozzi and J. Kot. The complete spectrum of image line. *IEEE Trans. Microwave Theory Tech.*, 37:868–874, May 1989.

[18] T. Rozzi and L. Ma. Mode completeness, normalization and Green's function of the inset dielectric guide. *IEEE Trans. Microwave Theory Tech.*, 36:542–551, March 1988.

[19] T. Rozzi and M. Mongiardo. Continuous spectrum of a flange-backed slotted waveguide with application. *IEEE Trans. Microwave Theory Tech.*, 40:2042–2049, November 1992.

[20] T. Rozzi and P. Sewell. The continuous spectrum of open waveguides of nonseparable cross section. *IEEE Trans. Antennas Propagat.*, 40:1283–1291, November 1992.

[21] V. V. Shevchenko. *Continuous Transition in Open Waveguides.* Golem Press, Boulder, Colorado, 1971.

[22] T. Tamir. Leaky-wave antennas. In R. E. Collin and F. J. Zucker, editors, *Antenna theory.* McGraw-Hill, New York, 1969.

[23] T. Tamir and A. A. Oliner. Guided complex waves, I field at an interface. *Proc. IEE*, B-110:325, February 1963.

[24] C. H. Walter. *Travelling Wave Antennas.* McGraw-Hill, New York, 1965.

Chapter 8

Some examples and applications

In the previous Chapters we have introduced the complete spectrum of open waveguides, considering how to compute both guided modes and continuum. Once the spectrum is derived we are in the position to solve a wide variety of problems: in abstract terms we can say that, once the complete spectrum is known, we have knowledge of the Green's function, and thus a complete characterization of that particular structure.

From a practical viewpoint, the knowledge of the complete spectrum allows us to derive simple equivalent circuits of waveguide discontinuities in much the same manner as in closed waveguides. In this way we can, for example, characterise an open microstrip discontinuity; or we can provide an equivalent circuit for a microstrip step discontinuity, etc. This possibility of producing rigorous and effective equivalent circuits of open waveguide discontinuities is, of course, of paramount importance for CAD of microwave circuits.

Moreover, the same methodology can also be applied to antenna problems. As an example a microstrip patch antenna can be considered as a step discontinuity interacting with an open end microstrip. If both discontinuities are characterised by their multi–mode equivalent circuits we can rigorously describe the radiation process taking place in this structure.

As a matter of fact, it is quite impossible to try to describe all the most significant applications of the complete spectra of open waveguides in just one Chapter. As a consequence we have chosen to describe just a few of these applications with the purpose of providing at least some examples.

Like the previous one, this Chapter too may be considered to be divided into two parts: the first part deals with somewhat simpler structures and is used to introduce the methodology. In the second part we provide some examples of applications to practical problems such as the characterization of air–bridges in coplanar waveguides, of via–holes in microstrip lines and of step discontinuities in dielectric rib waveguides.

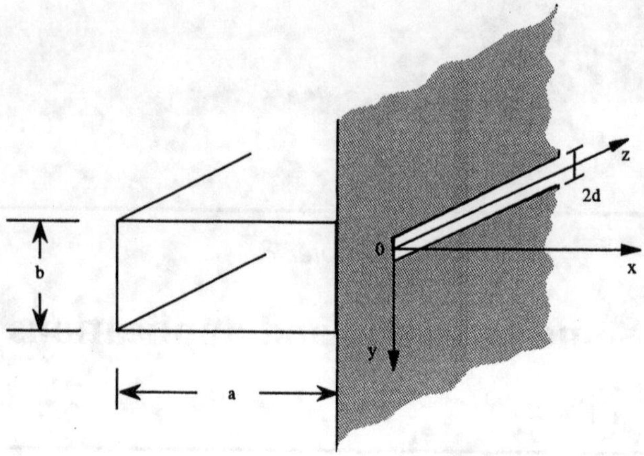

Figure 8.1 Geometry of the transition between a closed and a slotted waveguide.

8.1 Slotted rectangular waveguide

With reference to Figure 8.1, let us consider the problem of the junction between a flange–backed ordinary waveguide and a slotted one, whose spectrum has been derived in the previous Chapter. We neglect higher order mode excitation in the closed guide as well as excitation of the open region for $z < 0$. At $z = 0$, the E_y component of the incident TE_{10} waveguide mode can be decomposed in terms of the continuous spectrum of the slotted guide

$$e_{y_{10}} = \sqrt{\frac{2}{ab}} \sin \frac{\pi x}{a} = \int_0^{\infty} A(k_t) e(x, y; k_t) dk_t \tag{8.1}$$

with (see equations (7.29 and 7.30) on page 309)

$$\begin{gathered} A(k_t) = \int_{-a}^{0} dx \int_{-\frac{b}{2}}^{\frac{b}{2}} dy \sqrt{\tfrac{2}{ab}} \sin \tfrac{\pi x}{a} e(x, y; k_t) = \\ \sqrt{\tfrac{2k_t}{\pi}} \sin \alpha_\nu \sum_{n=0,2,..}^{\infty} \tilde{E}_n(k_t) \int_{-a}^{0} \sqrt{\tfrac{2}{a}} \sin \tfrac{\pi x}{a} \tfrac{\sin q_n (x+a)}{\sin q_n a} dx \int_{-\frac{b}{2}}^{\frac{b}{2}} \sqrt{\tfrac{1}{b}} c_n(y) dy \end{gathered} \tag{8.2}$$

and, by using in the above the appropriate expression for the transform of the field (7.26) over the aperture

$$\tilde{E}_n = \sqrt{\frac{2}{\pi d}} \int_{-d}^{d} \sqrt{\frac{2\varepsilon_n}{b}} \frac{\cos \frac{n\pi y}{b}}{\sqrt{1 - (\frac{y}{d})^2}} dy = \sqrt{\frac{\varepsilon_n \pi d}{b}} J_0(\frac{n\pi d}{b}) \tag{8.3}$$

we get

$$A(k_t) = 2\pi \sqrt{k_t a \frac{d}{b}} \frac{\sin \alpha_\nu}{(k_t a)^2 - \pi^2} \tag{8.4}$$

The E_y field at a distance z down the guide is then given by

$$\begin{gathered} E_y(r, k_0) = \int_0^{k_0} dk_t A(k_t) e(r, k_t) e^{-jz\sqrt{k_0^2 - k_t^2}} + \\ + \int_{k_0}^{\infty} dk_t A(k_t) e(r, k_t) e^{-z\sqrt{k_t^2 - k_0^2}} \end{gathered} \tag{8.5}$$

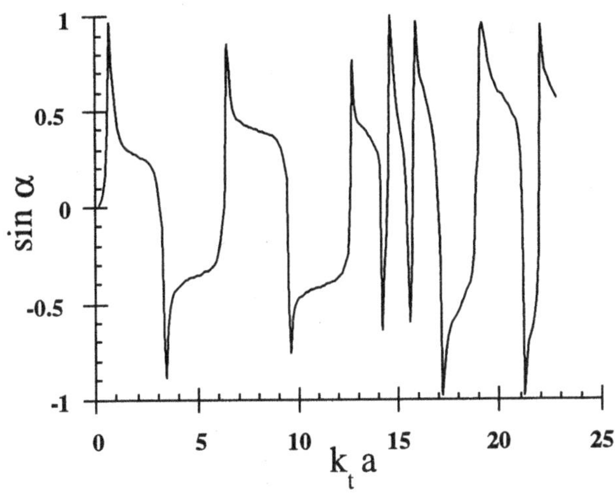

Figure 8.2 Behaviour of the phase shift α as a function of k_ta for a slotted waveguide. The zeros of $\sin\alpha$ *correspond to the resonances of the guide.*

Table 8.1 Waveguide resonances corresponding to the zeros of $\sin\alpha$ *(a/b = 2.25).*

m	n=0	n=2
1	3.14	14.48
2	6.28	15.47
3	9.42	16.99
4	12.57	18.91
5	15.71	21.13
6	18.85	23.56
7	21.99	26.14

8.1.1 Numerical results

The behaviour of $\sin\alpha$ as obtained from (7.27) is reported in Figure 8.2. It should be noted that α depends only on the guide aspect ratio a/b, normalised slot aperture d/a, and on $k_t a$. In this case $d/a = 1/22.86$. As previously stated, the real quantity α represents the phase shift imposed on the incident $e^{jk_x x}$ component by the presence of the slot and of the waveguide behind it. When k_t is zero, that is the wave is directed along z, $\sin\alpha$ is also zero. Moreover, when waveguide resonances occur, that is when k_t is equal to the cutoff frequencies of the unperturbed guide, $\sin\alpha$ is again zero. These locations are given in Table 8.1.

Once α has been determined it is possible to calculate the modal fields inside and outside the guide according to (7.29).

When the slotted waveguide is excited from a closed one, the incident field is decomposed in terms of the continuous spectrum of the slotted waveguide. Each component of the continuous spectrum is excited with a different amplitude, as given in (8.4). Figure 8.3 shows the amplitude of these components, $A(k_t)$, when the fundamental mode is incident on a slotted guide with $d/a = 1/22.86$.

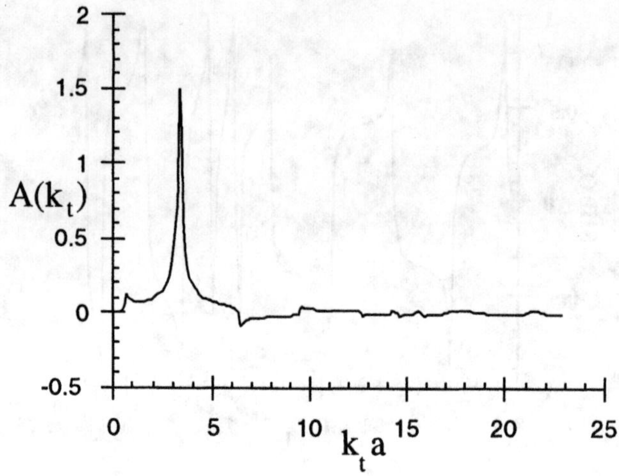

Figure 8.3 Amplitude of the modal fields (continuous spectrum) excited by an incident TE_{10}. The slot width is $d/a = 1/22.86$

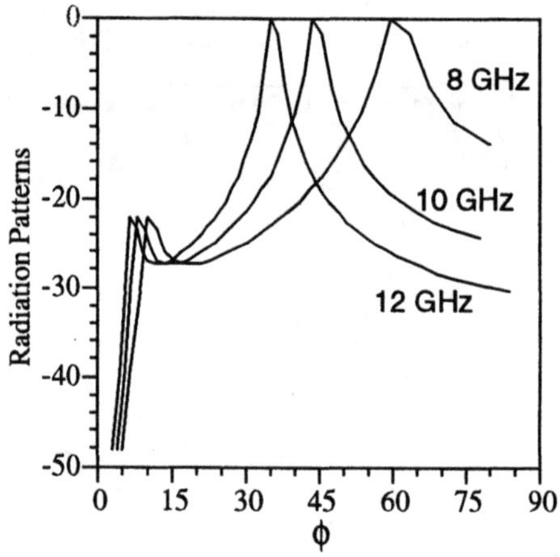

Figure 8.4 Radiation patterns on the plane xz ($\phi = \mathrm{arctg}(b/k_t)$) for different frequencies. The slot width is $d/a = 1/22.86$.

At any given frequency each mode, characterised by a k_t value, has a propagation constant β either real (propagating mode) or imaginary (evanescent mode). For a fixed k_t the angle with the z–axis formed by the direction of propagation of the mode in the air region is given by $\phi = \mathrm{arctg}(k_t/b)$, while $A(k_t)$ determines its amplitude. A plot relating the latter quantity to the angle ϕ is shown in Figure 8.4 for different frequencies.

As expected, as the frequency increases the maximum modal amplitude moves toward endfire. This can be explained by recalling the interpretation of the field inside the waveguide as a superposition of two plane waves. Each plane wave impinges on the narrow wall at an angle with respect to the z direction that increases as the frequency decreases.

From the preceding figures the appearance of "leaky wave" is obvious corresponding to a peak of $A(k_t)$ in Figure 8.3 and to a preferred angle of radiation as shown in Figure 8.4.

The far field of the radiation mode of the slotted guide can be obtained by a steepest descent calculation. The radiative part of a continuous mode is given by the first of the two integrals appearing in (7.6). By substituting in this formula the following standard transformation

$$\begin{array}{ll} x = r\cos\theta & y = r\sin\theta \\ k_x = k_t\cos\theta & k_y = k_t\sin\theta \end{array} \tag{8.6}$$

we obtain

$$\begin{aligned} e_\nu &= \frac{N_\nu}{\sqrt{2\pi}}\left[\int_{-k_t}^{k_t} \tilde{E}_\nu \cos k_y y \cos k_x x dk_y + \cot\alpha_\nu \int_{-k_t}^{k_t} \tilde{E}_\nu \cos k_y y \sin k_x x dk_y\right] \\ &= \frac{N_\nu k_t}{\sqrt{2\pi}}\left[\mathrm{Re}\int_c \tilde{E}_\nu e^{-k_t r}\cos\theta d\theta + \cot\alpha_\nu Im\int_c \tilde{E}_\nu e^{-k_t r}\cos\theta d\theta\right] \end{aligned} \tag{8.7}$$

where C denotes the appropriate path in the complex θ plane defined by (8.6) and

$$\begin{aligned} \mathbf{k}_t &= (k_x, k_y) \\ \mathbf{r} &= (x, y) \\ \mathbf{k}_t \cdot \mathbf{r} &= k_x r\cos(\Theta - \theta) \end{aligned} \tag{8.8}$$

Hence by saddle point integration at $\Theta = \theta$, we obtain the sought separation of radial and angular dependences, namely:

$$e_\nu \approx N_\nu \sqrt{\frac{k_t}{r}}\tilde{E}_\nu(\theta)\cos\theta \frac{\sin(k_t r - \pi/4 + \alpha_\nu)}{\sin\alpha_\nu} \tag{8.9}$$

This far field is plotted in Figure 8.5 as a function of the angle θ.

8.2 Self and mutual impedance of dipoles in presence of an open waveguide

The knowledge of self and mutual impedance of a system of sources and receivers in the presence of an open transmission line is often required in a variety of applications [4, 3, 24]. Since open waveguides, such as microstrip line, slotted line, etc., can significantly couple electromagnetic energy to/from open space, it is of considerable interest to evaluate the interaction of an elementary current source, say an electric dipole, placed at a certain distance from the transmission line.

However, the rigorous analysis of the above problem by the standard integral equation technique (moment method solution), often becomes a formidable task. As an example, let us consider the case of a dipole over a slot line. Even assuming assigned currents on the dipole, we have still to set up an integral equation having the electric field along the slot as unknown. Since the slot is in principle infinitely long, a large number of terms are expected to be necessary to express the electric field. As a consequence, the above approach is neither numerically efficient, nor theoretically elegant.

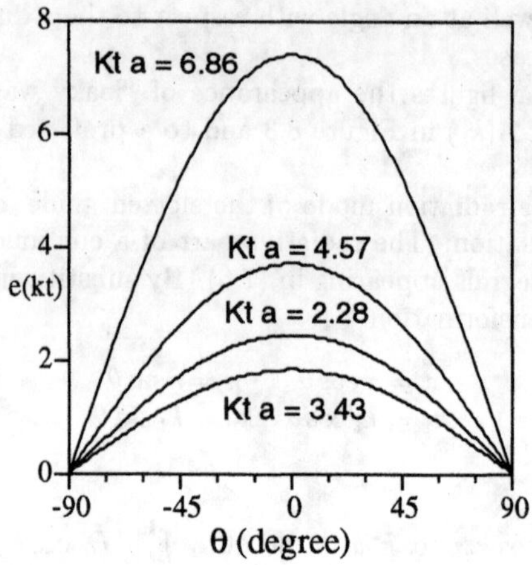

Figure 8.5 Far field of the slotted waveguide on the plane xy obtained by steepest descent calculation. The electric field amplitude is reported on the y–axis.

The problem of the coupling of elementary sources to apertures on shielding planes also arises in EMC. Similarly, a rigorous treatment can be effected, for instance, by following a moment method approach, where the electric field over the whole aperture is taken as unknown. Unfortunately, as the electrical dimensions of the aperture increase, so does the number of unknowns to be used in order to represent the electric field. Again, when the aperture shape becomes similar to a slot (i.e. when we have essentially a guiding structure), the above approach becomes ineffective.

On the other hand, approaches based on direct discretization of Maxwell's equation, such as FEM, FD, FDTD, TLM, etc., are even less efficient, for we have to tackle difficulties related to the large dimensions of the coupling regions (resulting in a very large number of mesh points), as well as to the terminating (absorbing) boundary conditions.

It is interesting to note that determining the coupling of a dipole source to the field of a closed waveguide is, instead, a straightforward exercise [5]. Actually, given the currents on the dipole, the field inside the guide can be found directly from modal superposition, each discrete mode being excited with its proper amplitude according to Lorentz' reciprocity theorem. An analogous approach, i.e. based on the reciprocity theorem and the complete modal spectrum, can also be adopted for open waveguides. In fact, also for open waveguides with non–separable cross–section we have just derived a continuum of real, mutually orthogonal modes, satisfying all boundary and edge conditions. In terms of these modes, it is possible to express the field anywhere in the open waveguide and even to write the Green's function for the structure, with a view to studying excitation and perturbation problems like the one described.

The modes of open waveguides also take into account the two different mechanisms of coupling between dipoles in the presence of the guide, both contributing to their mutual impedance. On the one hand, there is coupling through the fundamental mode (the bound mode) of the open waveguide; on the other there is

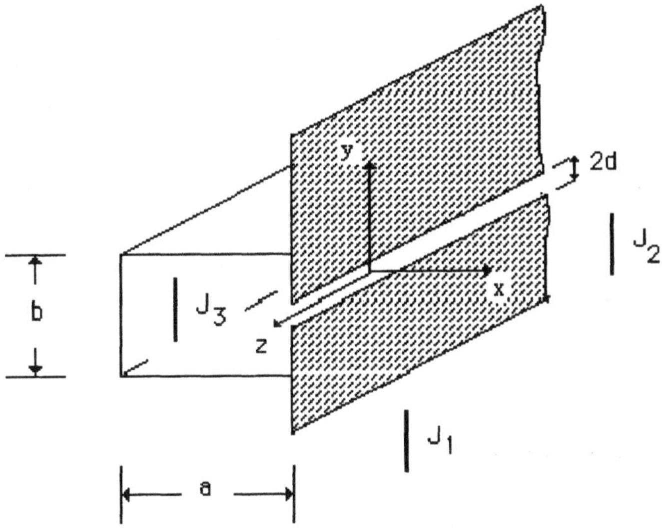

Figure 8.6 Geometry of a slotted waveguide with dipoles.

interference through the fields radiated by the two probes, in the presence of the waveguide boundary conditions. By this approach, the contributions are separated out in the model, allowing accurate treatment and physical insight, as well as the development of simple equivalent circuits.

8.2.1 Analysis

Let us consider emitting and receiving dipoles located near an open waveguide, e.g. a slotted ground plane, backed by a rectangular groove, as illustrated in Figure 8.6. This figure shows two electric dipoles J_1, J_2, respectively, placed at (x_1, y_1, z_1), (x_2, y_2, z_2) and taken as parallel to the slot plane for simplicity. Moreover, one or more dipoles could also be located inside the enclosure, e.g. J_3.

It is required to evaluate the impedance of the system of dipoles. When the two exterior dipoles are electrically far from the slot, its presence is not felt and the treatment is straightforward. The influence of the slot, however, compounds matters as some of the transmission takes place through the intervening length of open waveguide and radiation from the whole slot takes place. In order to describe this situation we can make use of the block "equivalent network" for the two dipoles, say J_1 and either J_2 or J_3, as illustrated in Figure 8.7.

The "network elements" here are admittance–type integral operators of the kind used when treating interacting cascaded discontinuities for the planar dielectric waveguides. In particular, $\hat{Y}_1$ represents the load of the semi–infinite region to the left $(z > z_1)$ of the first dipole and to the right $(z < z_2)$ of the second one. When

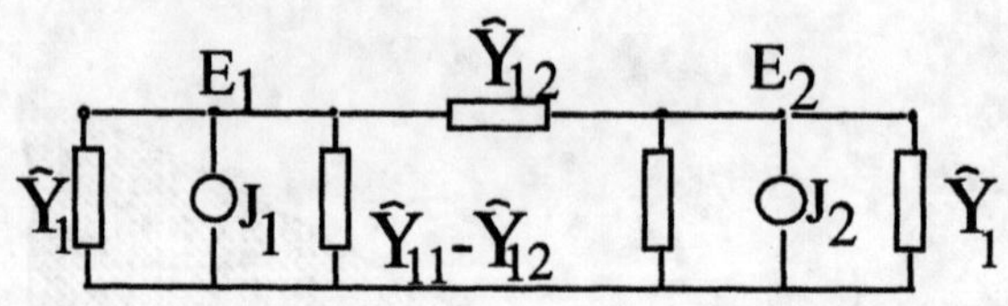

Figure 8.7 Equivalent network used for the evaluation of self and mutual impedances. The network elements represent admittance–type integral operators.

considering only TE modes the kernel $Y_1(x, y; x', y')$ is of the form

$$Y_1 = \int_0^\infty dk_t \frac{\beta}{\omega\mu_0} \sum_\nu e_\nu(x, y; k_t) e_\nu(x', y'; k_t) \tag{8.10}$$

while Y_{11}, Y_{12} have the following expressions

$$Y_{11} = \int_0^\infty dk_t \frac{\beta}{\omega\mu_0} \text{ctg}(\beta\ell) \sum_\nu e_\nu(x, y; k_t) e_\nu(x', y'; k_t)$$

$$Y_{12} = \int_0^\infty dk_t \frac{\beta}{\omega\mu_0} \frac{1}{\sin(\beta\ell)} \sum_\nu e_\nu(x, y; k_t) e_\nu(x', y'; k_t) \tag{8.11}$$

with $\beta^2 = k_0^2 - k_t^2$. The above "two–port" admittance operators contain the mode functions $e_\nu(x', y'; k_t)$. J_1, J_2 denote the y–directed current sources impressed on the dipoles: their distributions are assumed here of the type

$$J_i = J_{0i} f(x, y, z) \tag{8.12}$$

From the "equivalent network" of Figure 8.7, one derives the relationship between the fields E and the sources J on the transverse planes where the dipoles are located, namely

$$\begin{bmatrix} J_1 \\ J_2 \end{bmatrix} = \begin{bmatrix} \hat{Y}_{11} + \hat{Y}_1 & \hat{Y}_{12} \\ \hat{Y}_{12} & \hat{Y}_{11} + \hat{Y}_1 \end{bmatrix} \cdot \begin{bmatrix} E_1 \\ E_2 \end{bmatrix} = \hat{Y}_T \cdot \begin{bmatrix} E_1 \\ E_2 \end{bmatrix} \tag{8.13}$$

or

$$\begin{bmatrix} E_1 \\ E_2 \end{bmatrix} = \hat{Z}_T \cdot \begin{bmatrix} J_1 \\ J_2 \end{bmatrix} \tag{8.14}$$

where $\hat{Z}_T$ denotes the inverse of $\hat{Y}_T$ in (8.13).

The elements of these operators are straightforward to evaluate in terms of the modal components of the slotted guide. In the general case, components relative to both TE and TM polarizations ought to be considered. The "dot–product" denotes here the action of the integral operator on a function of space. Self and mutual impedances z_{11}, z_{12} of the dipoles are then evaluated from (8.13) as follows: we form the "dot product"

$$[J_1 \;\; J_2] \cdot \begin{bmatrix} E_1 \\ E_2 \end{bmatrix} = [J_1 \;\; J_2] \cdot \begin{bmatrix} \hat{Z}_{11} & \hat{Z}_{12} \\ \hat{Z}_{12} & \hat{Z}_{22} \end{bmatrix} \cdot \begin{bmatrix} J_1 \\ J_2 \end{bmatrix} \tag{8.15}$$

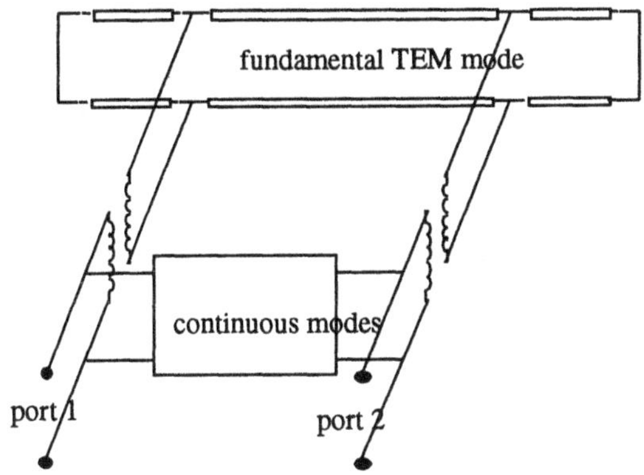

Figure 8.8 Equivalent circuit of dipole mutual coupling near an open waveguide supporting a fundamental mode and a continuous spectrum.

This is evaluated by integration over the length of the dipoles, and it reduces to the ordinary matrix multiplication

$$[J_{01} \quad J_{02}] \cdot \begin{bmatrix} z_{11} & z_{12} \\ z_{12} & z_{11} \end{bmatrix} \cdot \begin{bmatrix} J_{01} \\ J_{02} \end{bmatrix} \tag{8.16}$$

It is emphasised that the "dot products" in (8.15) are taken in the functional sense, i.e. they involve integration over the length of the dipole (in general), whereas (8.16) is an ordinary matrix multiplication and it defines the sought impedances.

8.2.2 Equivalent circuit

In many cases the open waveguide can support a fundamental mode (e.g. microstrip, slotline, etc.). Once this fundamental propagating mode and the continuous spectrum are known, we are able to compute the coupling coefficients between the two dipoles and the open waveguide. Then, following the steps outlined in the previous Section, we recover the equivalent circuit shown in Figure 8.8.

8.2.3 Theoretical and experimental results

The above model has been compared with experimental results realised by using a slotted ground plane. We have considered a slot (300 mm long, 1 mm wide) on a ground plane (made by cutting a copper sheet of size 600 × 600 mm^2 and thickness 0.2 mm). Two probes were realised from small 50 Ω coaxial cables. The probes, 10 mm long and located at 50 mm from the slotline shorts, are electrically connected to the ground plane as shown in Figure 8.9.

The measurements, in the frequency range 100 MHz-2 GHz, were carried out by means of a *hp*8510 network analyzer. Figure 8.10 show a comparison between theoretical and experimental values of $|s_{11}|$ and $|s_{12}|$, respectively; the dashed line represents the theoretical simulations as obtained by considering only the fundamental slot line mode, while the continuous line takes into account the continuous spectrum; dots are the experimental results.

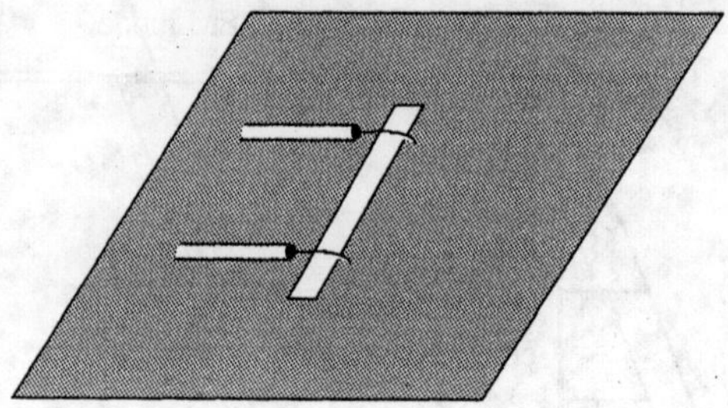

Figure 8.9 Experimental set–up. Two coaxial probes coupled through a slotted line.

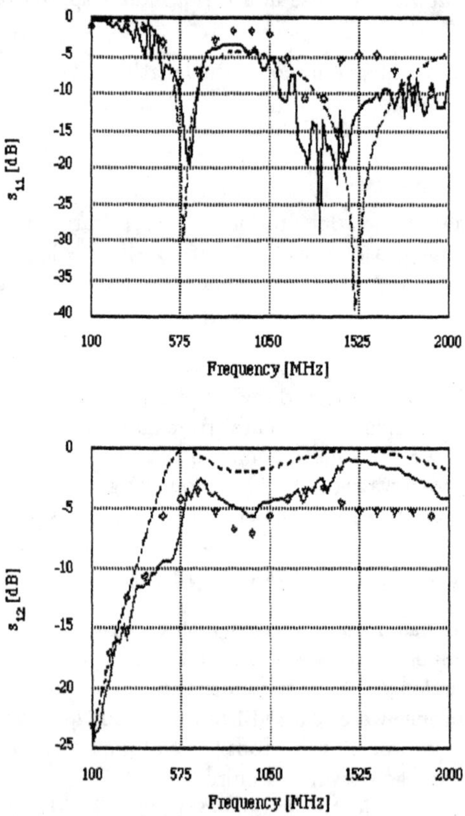

Figure 8.10 Theoretical and experimental results for the $|s_{11}|, |s_{12}|$ of a dipole over a slotted ground plane.

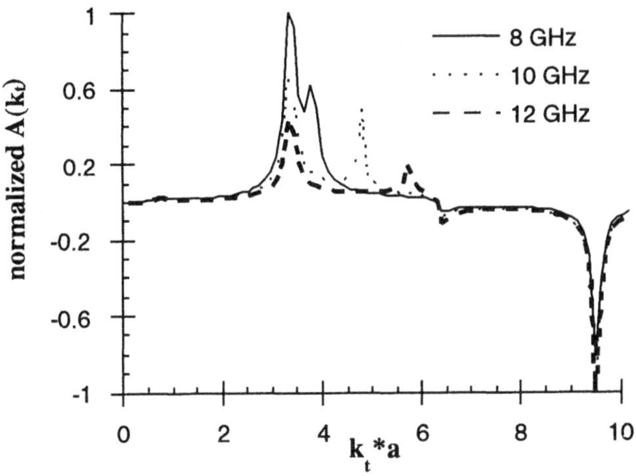

Figure 8.11 Amplitude of the modal components excited by the dipole in the slotted waveguide (WR90) for three different frequencies. At the frequency of 8 GHz k_0a = 3.83, while at 10 GHz k_0a = 4.79 and at 12 GHz k_0a = 5.75.

As a further example, the case of a dipole inside a slotted waveguide (as in Figure 8.6) has been theoretically investigated. The dipole, with current distribution given by

$$J_i = J_{0i}\delta(x - x_i)\delta(z - z_i)$$

is supposed to extend over the whole waveguide height, therefore generating only TE modes. Due to the presence of the slot, the dipole excites a continuous spectrum. The amplitude $A(k_t)$ of the various components of the spectrum is shown in Figure 8.11 for three different frequencies.

The perturbation of the discrete spectrum introduced by the slot is apparent in Figure 8.11. In fact, in the case of a closed waveguide, only discrete modes are present. These modes give contributions only for

$$k_t a = \sqrt{(m\pi)^2 + \left(\frac{n\pi a}{b}\right)^2}$$

while, in the present case, a continuous spectrum is present, so that we can also observe other values of $k_t a$ contributing. The real part of $A(k_t)$ is obtained for $k_t < k_0$ and corresponds to radiated power, while the imaginary part of $A(k_t)$ corresponds to the energy stored near the dipole.

The frequency dependence of the real part of the impedance of the dipole radiating inside the slotted waveguide (a WR90) has been reported in Figure 8.12. It can be observed that before cut–off (occurring at 6.56 GHz) the real part of the impedance is almost zero since no propagation can take place in the waveguide. On the other hand, near and above cut–off propagation is present giving a real part of the impedance different from zero.

In Figure 8.13 the mutual admittance between two dipoles placed at the centre (x = -11.43 mm) of a slotted WR90 waveguide is reported. Also in this case the

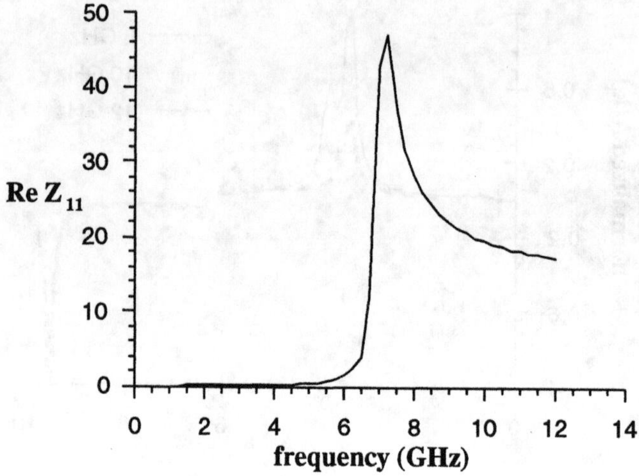

Figure 8.12 Real part of the impedance of a dipole placed inside a slotted WR90 waveguide.

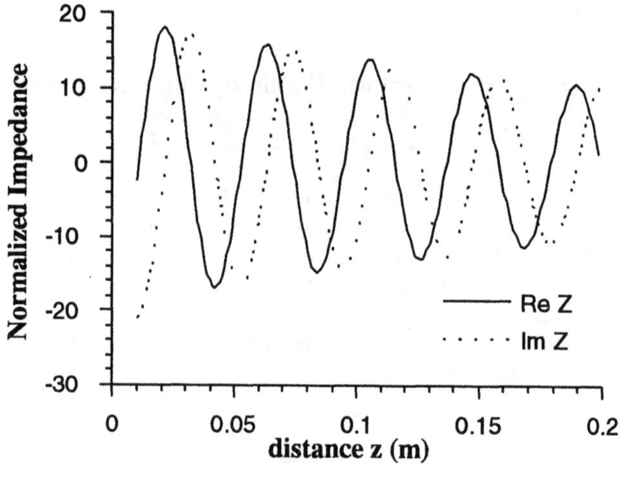

Figure 8.13 Real and imaginary parts of the mutual impedance of two dipoles placed inside a slotted WR90 waveguide.

dipoles extend for the whole waveguide height, while the slot width is d = 0.5 mm. In Figure 8.13 the mutual admittance is evaluated at a frequency of 10 GHz varying the position of the dipoles in the propagation direction z. As expected, the oscillating behaviour of the real and the imaginary parts of the mutual admittance between the two dipoles is attenuated as the distance between the dipoles is increased. This effect is due to the presence of the slot that allows a "leakage" of energy into the free space region.

The effect of the x–position on a single dipole inside the slotted guide has also

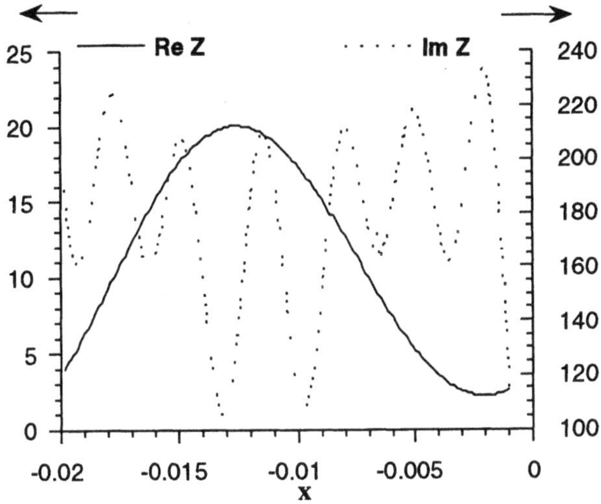

Figure 8.14 Real and imaginary parts of the impedance of a dipole placed inside a slotted WR90 waveguide.

been investigated and some results are reported in Figure 8.14. Again the dipole has the same height as the guide and the computation has been performed assuming a frequency of 10 GHz for a standard WR90 waveguide and a slot width of d = 0.5 mm.

In conclusion, in this Section we have evaluated self and mutual impedances of a system of dipoles in the proximity of an open waveguide by using the modes of the open waveguide. In this way, the mutual coupling due to propagation through the guide as well as to radiation in open space is rigorously taken into account. The theoretical model has been compared with experimental results. This model avoids the use of a large set of basis functions to express the field in the guide, as in standard moment method approaches. As a consequence, the evaluation of coupling with external sources/receivers in MIC/MMIC, in printed antenna feed systems, and in compatibility problems, becomes numerically more efficient.

8.3 Spectrum of a slotted plane

In the previous Chapter we have recognised the relationship between the continuous spectrum and the characteristic modes; most of the results developed for the latter can be almost immediately employed to determine the spectrum of open waveguides and vice versa. As an example, in [17, 18] the case of a slotted screen (Figure 8.15) has been studied in terms of the characteristic modes. On the other hand, in Section 7.1, the case of the slotted guide of Figure 7.1 has been studied. With minor modification it is possible to apply the theory developed in Section 7.1 in order to study the problem of Figure 8.15. We will now compare the results. The knowledge of the continuous spectrum corresponds to the knowledge, for each k_t, of the eigenvalues, eigenvectors of (7.42). In this case (Figure 8.15), these eigenvalues depend on the geometry but not on frequency. A rigorous Galerkin procedure has

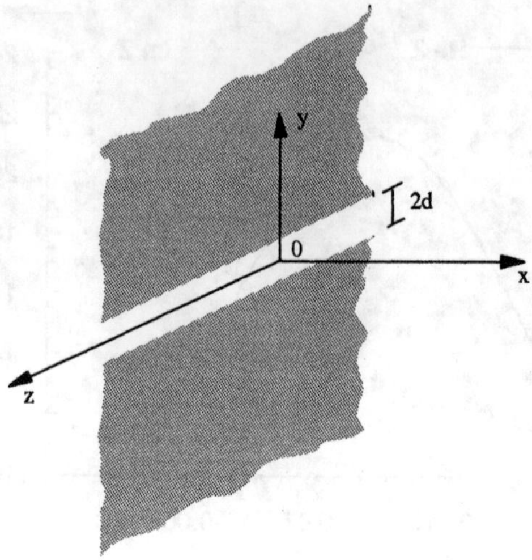

Figure 8.15 Geometry of a conducting plane with an infinitely long slot of width 2d.

been applied to solve (7.42) for both the TE and TM case. An example of the convergence of the eigenvalues with respect to the number of basis functions is given in Table 8.2 for the TE case and Table 8.3 for the TM case. Both cases have been computed for $k_t 2d = \pi$ and may be compared with the results of [17, 18]. The comparison shows that, by employing Chebyshev polynomials, very few terms are necessary to achieve a good convergence. Moreover, by increasing the number of terms, the results practically do not change. In Table 8.2, a slight disagreement is noted with the first column of [17, Table I]; this is probably due to the fact that in [17] convergence has not yet been achieved. In [17, 18] useful approximations for the χ_n have been found. These approximations have been checked against the rigorous results and are reported in Figure 8.16 for the TE case and Figure 8.17 for the TM case. Often, equation (7.42) is solved in terms of the phase shift $\alpha_\nu(k_t)$ which a plane wave undergoes when it is reflected from the slotted screen. The relationship between $\alpha_\nu(k_t)$ and χ_n is simply given by $-\cot\alpha_\nu(k_t) = \chi_n$. It is interesting to note that the oscillations in Figure 8.16 are similar to a Fresnel phenomenon (field perpendicular to the edge). It should also be noted that better results are obtained for the approximation relative to the TE case.

Figures 8.16 and 8.17 provide all the information necessary to evaluate the spectrum of the slotted conducting plane. They also describe the electrical behaviour of the slotted conducting plane. The phase shift $\alpha_\nu(k_t) = 90°$ corresponds to $\sin\alpha_\nu(k_t) = 1$ and is obtained for wide slots in the TE case. Here the χ_n's are positive since the energy stored near the slot is essentially capacitive. When increasing $k_t 2d$, $\sin\alpha_\nu(k_t)$ tends to unity, and therefore χ_n tends to zero. The phase shift $\alpha_\nu(k_t) = 270°$ corresponds to $\sin\alpha_\nu(k_t) = -1$ and is obtained for wide slots in the

Table 8.2 Convergence of the eigenvalues for the TE case. The electric field on the aperture has been expanded in terms of Chebyshev polynomials and N refers to the number of basis functions used.

N	χ_1	χ_2	χ_3	χ_4	χ_5	χ_6
1	.4453					
2	.4413	.4453				
3	.0252	.4113	1.481			
4	.0252	.3033	1.481	9.58		
5	.0210	.3033	1.457	9.58	162.57	
6	.0211	.3023	1.457	9.51	162.57	5137.82
7	.0211	.3023	1.457	9.51	161.74	5137.82
8	.0211	.3023	1.457	9.51	161.74	5120.99
9	.0209	.3023	1.457	9.51	161.62	5120.99
10	.0209	.3022	1.457	9.51	161.62	5115.81

Table 8.3 Convergence of the eigenvalues for the TM case. The derivative of the electric field on the aperture, $\partial E_z/\partial y$, has been expanded in terms of Chebyshev polynomials. N refers to the number of basis functions used.

N	χ_1	χ_2	χ_3	χ_4	χ_5	χ_6
1	-.0597					
2	-.0597	-.6092				
3	-.0229	-.6092	-8.1384			
4	-.0229	-.5096	-8.1384	-159.8632		
5	-.0220	-.5096	-7.8127	-159.8632	-5063.6068	
6	-.0220	-.5070	-7.8127	-157.4882	-5063.6068	-243298.4952
7	-.0219	-.5070	-7.7943	-157.4882	-5030.5213	-243298.4952
8	-.0219	-.5065	-7.7943	-157.1237	-5030.5213	-242626.5292
9	-.0219	-.5065	-7.7839	-157.1237	-5016.7864	-242626.5292

TM case. Here, χ_n's are negative since the energy stored near the slot is essentially inductive.

Observe that the first eigenvalue for the TE case is different from zero when k_t is zero. It is therefore possible to find a mode having a zero value of k_t and, as such, propagating just in the z direction. This is the TEM mode of the structure.

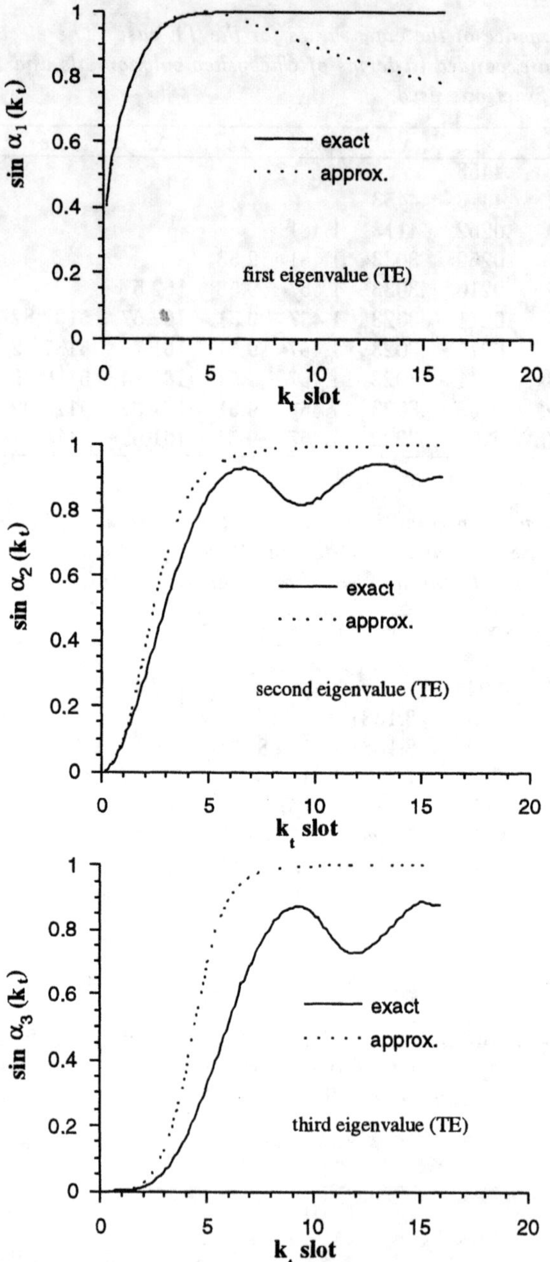

Figure 8.16 Behaviour of the first three eigenvalues of a slotted screen for the TE case (the slot width is 2d). The continuous curves refer to the rigorous results obtained by considering six Chebyshev basis functions, while the approximate results (dashed lines) are taken from Kabalan.

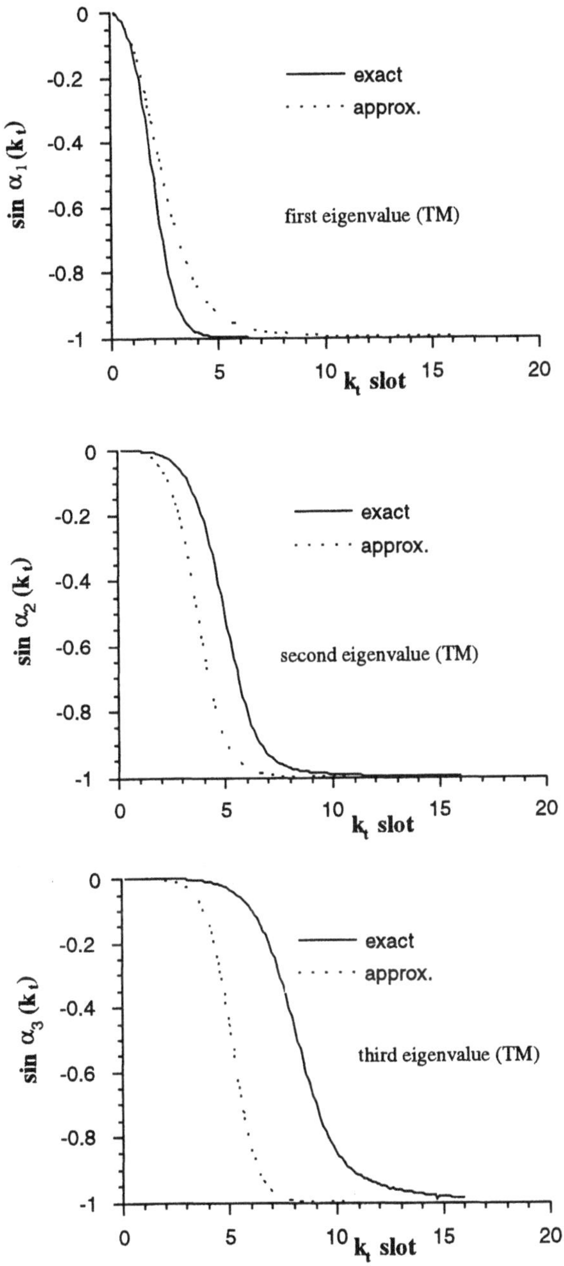

Figure 8.17 Behaviour of the first three eigenvalues of a slotted screen for the TM case (the slot width is 2d). The continuous curves refer to the rigorous results obtained by considering six Chebyshev basis functions, while the approximate results (dashed lines) are taken from Kabalan.

8.4 Characterization of air–bridges in coplanar waveguide

Coplanar waveguide (CPW) is rapidly becoming a preferred medium for MMIC applications due to the ease with which both shunt and series mounted components can be incorporated to form complete circuits without the need for via holes. In addition, the field confinement and losses of CPW are preferable to those of microstrip and its characteristic impedance is lower than that of slotline, in particular allowing the use of 50 Ω lines. Being a 3–conductor structure it supports two types of fundamental mode, the "coplanar" and the "slotline" modes. For practical operation, the latter must be suppressed, which is achieved by using ohmic bridges between the two exterior conductors.

In the following, we present an analysis of air–bridges in open CPW, including the effects of finite thickness metallization, using the complete modal spectrum of the CPW. This technique allows us to extract and preprocess the common features of CPW circuitry without consideration of the particular problem in hand, that is, it is a suitable basis upon which to build a CAD package. Using this approach, scattering parameters, radiation and substrate leakage patterns of isolated bridges can be derived for both types of fundamental mode incidence and compared, where possible, to those obtained from other methods. Furthermore, pairs of bridges, coupled by both the near and far fields, can also be analyzed.

Modern circuit design increasingly relies upon CAD facilities, utilizing a wide range of computing systems, from the humble PC to multi–processor, vectorial machines. Naturally, the choice of technique upon which a CAD package is based, is not only dependent upon the host machine, but also the type of problem and the required accuracy.

In general, "the one off" analysis of a complex structure may be undertaken relatively quickly and with good accuracy using a powerful computer and a numerical technique such as FDTD. However, the repeated analysis of geometrically simple structures, as obviously happens in a typical CAD situation, can be rendered far more efficient by adopting more analytical methods, such as modal techniques. In the latter approach, the extraction and preprocessing, once and for all, of features common to a wide range of problems minimises the effort subsequently needed for each particular example. The cost of this approach is that it may require more time to develop and may be less flexible.

Clearly, these two methodologies are complementary, each better suited to a different class of problem, albeit with some overlap. In fact, in the overlap region, a compromise, i.e. a semi–analytical method, such as the space domain integral equation, (SDIE), or moment method may prove very effective.

When dealing with truly open structures, the situation is further complicated, particularly for purely numerical methods. Techniques which are well proven for closed structures encounter difficulties when dealing with open ones. Obviously, an open structure may always be deformed into a closed one by introducing artificial walls at an "appropriate" distance from the region propagating the majority of the power. Whilst this device may be considered in certain circumstances, particularly when the circuit exists in a packaged environment, it is clearly inappropriate in many others, that is when either the deformation of the situation introduces unacceptable errors or else consideration of the radiated fields is essential, from the point of view of compatibility or the design of antenna elements or unshielded feeding networks.

Various methods have been extended to the open case as will now be briefly discussed. Quasi–static methods have been applied to many planar circuit geometries and provide a first order approximation to the solution, [12]. However, with the increasing demands being placed upon circuit performance and on operating frequency, it is important to seek more accurate solutions.

The SDIE approach does retain the open nature of the structure and has been applied to many practical components, [12, 13]. Its effectiveness for many problems, however, is constrained by the necessity to model the fields over large interaction regions. This is a significant limitation, so that the fields at some distance before and after a discontinuity are often considered to consist solely of incident, reflected and transmitted fundamental modes. This assumption is only valid for fairly well spaced planar circuits, for substrate modes and radiation fields only decay as the square root of the distance and, if significantly excited by the discontinuity, will invalidate the assumption above.

Finite difference, finite element and transmission line methods have also been applied to open discontinuities, [13], and while they succeed in generating useful results, it must be recalled that they suffer from certain disadvantages. Again, the open nature of the structure necessitates introducing an artificial boundary, either physical or upon which to apply an absorbing boundary condition, or the use of a graded mesh, both of which obviously reduce the emphasis that is placed upon the radiation field and make it difficult to provide far field radiation and substrate leakage patterns. A classical limitation of these methods, now being negotiated, arises in dealing with field singularities. Either the number of nodes becomes prohibitive or else errors are introduced. Finally, it must be said that purely numerical techniques give little help towards the understanding of the physical mechanisms involved and optimization, let alone synthesis, of a whole structure is a lot harder.

In the following it will be demonstrated how the modal method, a powerful approach for the analysis of closed waveguide problems, may now be extended to the open situation, with particular reference to air–bridges in open coplanar waveguide.

8.4.1 Discontinuities in the open case

The class of problem that we are considering in this Section is those discontinuities that consist of a number of finite regions supporting magnetic and electric currents. Examples such as air–bridges, radiating strips, via–holes and stubs are shown in Figure 8.18. It is well known, [5, 8, 10], that if the complete mode spectrum of a structure is available then modal Green's functions may be constructed, allowing the scattered fields due to magnetic and electric discontinuity currents to be determined. The modal representation of these fields ensures that the boundary conditions of the basic guide are intrinsically satisfied and consequently, in contrast to the other methods, it is only necessary to directly model the fields and currents actually on the discontinuity. In fact, the SDIE method is analogous, except that the basic guiding structure is regarded as a dielectric slab, (or a grounded slab), and thus the metallization, (or apertures in the ground plane), of the actual guide itself must be treated as a source and its boundary conditions imposed explicitly for each discontinuity problem. There is of course, no theoretical reason why one could not start with the Green's function of free space and additionally impose the boundary

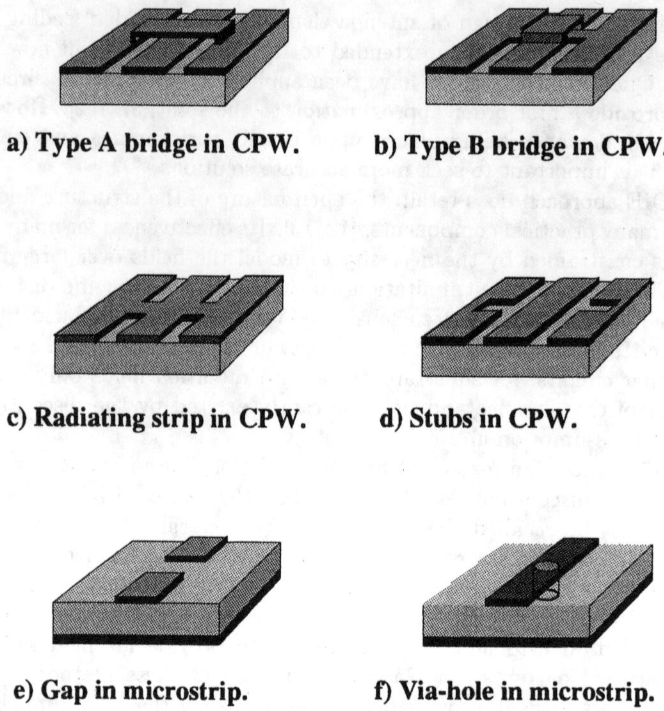

Figure 8.18 Typical discontinuities in open planar circuitry.

conditions of the slab. In practice, a compromise is made between the simplicity of the modes used to model the fields and the number of boundary conditions and sources it is necessary to include in the formulation of each discontinuity problem.

By using the modes of the actual guide, that is CPW in this case, rather than just those of the dielectric slab, it is proposed that we extract and preprocess that part of the problem common to all discontinuities in CPW; subsequent analysis of a particular discontinuity need then only be concerned with the conditions specific to that problem. Naturally, the cost is that the modes of the CPW are more complex than those of the slab, but, they need only be determined once for each operating frequency, stored, and retrieved whenever required.

Therefore, once the complete mode spectrum has been determined, waveguide discontinuities may be analyzed using modal techniques. In principle, there is little difference between the general method currently used for closed waveguide problems and that applied here to the open case. Consequently, it is unnecessary to go beyond a brief summary.

We are considering the class of circuit elements typified by those shown in Figure 8.18, including bridges in CPW, radiating strips and stubs in CPW and slotline and gaps in microstrip and CPW. Each may be characterised by a system of magnetic and electric discontinuity currents over a finite region denoted by $\mathbf{M}(\mathbf{r})$ and $\mathbf{J}(\mathbf{r})$ respectively ($\mathbf{r}$ represents an appropriate spatial coordinate vector). On the imposition of excitation fields, $\mathbf{E}_i(\mathbf{r})$ and $\mathbf{H}_i(\mathbf{r})$, the total fields, i.e. the sum of the

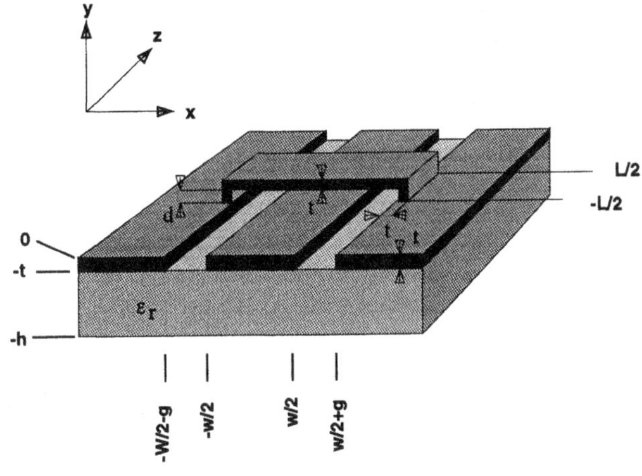

Figure 8.19 The air–bridge under consideration, (g = 10 μm, w = 15 μm, h = 100 μm, t = 3 μm, ε_r=12.9).

incident and scattered fields, must satisfy the boundary conditions of the discontinuity. Again, it is stressed that only the boundary conditions and currents actually on the discontinuity are involved in the analysis, those of the basic guide having already been incorporated into the modal representation. Consequently, the current system is the solution of an integral equation of the type

$$\begin{pmatrix} \mathbf{E}_i(\mathbf{r}) \\ \mathbf{H}_i(\mathbf{r}) \end{pmatrix} = \begin{pmatrix} \mathbf{G}_{11}(\mathbf{r},\mathbf{r}') & \mathbf{G}_{12}(\mathbf{r},\mathbf{r}') \\ \mathbf{G}_{21}(\mathbf{r},\mathbf{r}') & \mathbf{G}_{22}(\mathbf{r},\mathbf{r}') \end{pmatrix} \cdot \begin{pmatrix} \mathbf{J}(\mathbf{r}) \\ \mathbf{M}(\mathbf{r}) \end{pmatrix} = \begin{pmatrix} 0 \\ 0 \end{pmatrix} \tag{8.17}$$

where $\mathbf{G}_{ij}(\mathbf{r},\mathbf{r}')$ are dyadic Green's functions constructed from the modes in the classical manner.

Equation (8.17) may again be solved using Galerkin's method in the space domain. Basis functions that incorporate any appropriate field singularities ensure rapid convergence, usually with only a handful of terms, and in fact often a single term, even without the singularities, gives excellent results.

No limitations have been placed upon the geometry of the discontinuity and in particular, there is no difficulty in coping with nonplanar problems such as CPW bridges. Further, for a given guide and operating frequency, it usually only requires minor computational effort to alter the dimensions of the discontinuity.

8.4.2 Numerical analysis of air–bridges in CPW

As a significant example, we analyse a class of CPW air–bridges using the modal technique. The bridge under consideration is detailed in Figure 8.19 and is used to suppress the unwanted "slotline" type mode of the guide excited by asymmetrical discontinuities such as bends and stubs. Previous allowances for bridges have been made by supplementing the equivalent network of the circuit obtained using a full wave SDIE technique without bridges, with their quasi–static lumped element models, [6], and in addition, FDM and TLM analyses of single bridges have been reported, [16, 15], although as yet, due to the small effect of each one, the only experimental data available has been for a cascade of bridges [20].

In the present analysis, we consider both the effect of single bridges on the "coplanar" type mode and also the rejection of the "slotline" type mode for various frequencies and bridge dimensions. Further, the effect of two coupled bridges on each mode is determined.

For coplanar mode (CM) incidence, the currents on the bridge are modelled as follows:

$$
\begin{aligned}
J_x(x,d,z) &= x\left[\sum_{n=0,2,\cdots} A_n(d)\cos\left(\frac{n\pi z}{L}\right) + \sum_{n=1,3,\cdots} B_n(d)\sin\left(\frac{n\pi z}{L}\right)\right] \\
J_y\left(\frac{w}{2}+g,y,z\right) &= \left(\frac{w}{2}+g\right) \\
&\quad \left[\sum_{n=0,2,\cdots} A_n(d)\cos\left(\frac{n\pi z}{L}\right) + \sum_{n=1,3,\cdots} B_n(d)\sin\left(\frac{n\pi z}{L}\right)\right] \\
J_z(x,d,z) &= \left[\sum_{n=2,4,\cdots} C_n(d)\sin\left(\frac{n\pi z}{L}\right) + \sum_{n=1,3,\cdots} D_n(d)\cos\left(\frac{n\pi z}{L}\right)\right] \\
J_z\left(\frac{w}{2}+g,y,z\right) &= \left[\sum_{n=2,4,\cdots} C_n(d)\sin\left(\frac{n\pi z}{L}\right) + \sum_{n=1,3,\cdots} D_n(d)\cos\left(\frac{n\pi z}{L}\right)\right]
\end{aligned}
\tag{8.18}
$$

and similarly for the outer surface of the bridge.

Note that the appropriate continuity has been ensured at the edges and that the bridge has been considered to consist of a perfect conductor of finite thickness. It is found that A_n, B_n, C_n and D_n are all significantly smaller than the constant A_0 term so that in order to improve the efficiency, just a single term may be taken. In fact, this is done for slotline mode (SM) incidence, using currents of the form:

$$
\begin{aligned}
J_x(x,d,z) &= A_0(d)e^{-\gamma z} \\
J_y\left(\frac{w}{2}+g,d,z\right) &= A_0(d)e^{-\gamma z} \\
J_x(x,d+t,z) &= A_0(d+t)e^{-\gamma z} \\
J_y\left(\frac{w}{2}+g,d+t,z\right) &= A_0(d+t)e^{-\gamma z}
\end{aligned}
\tag{8.19}
$$

where the exponential is included to model the known evanescent behaviour of the fields under the bridge.

Single bridges

The dimensions of the guide, shown in Figure 8.19, correspond to a characteristic impedance of 50 Ω and have been chosen so as to allow comparison, where possible, with previous analyses using the FDM and TLM methods, [16, 15] As a first step, it is necessary to consider the modal convergence with both of the parameters k_t and ν. Figure 8.20 shows the convergence of the Green's function kernels with k_t after spatially integrating with the currents. Clearly, rapid convergence is seen,

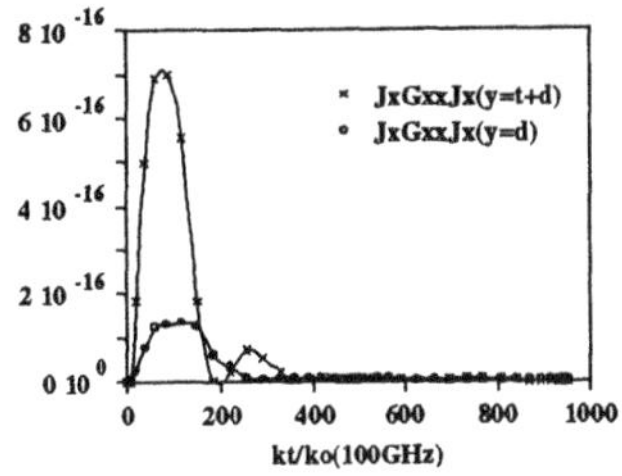

Figure 8.20 Green's function convergence with k_t for coplanar mode incidence at the frequency of 40 GHz; L = 30 μm, d = 3 μm.

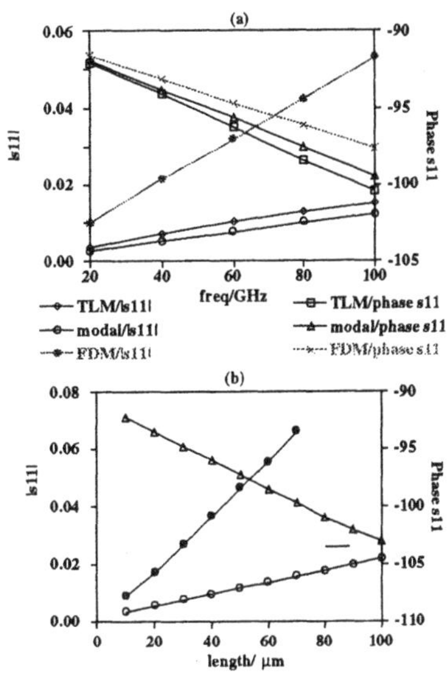

Figure 8.21 Reflection coefficient for coplanar mode incidence; d = 3 μm; a) vs. frequency, L = 30 μm; b) vs. bridge length, at the frequency of 50 GHz.

good accuracy requiring only a moderate number of points to be evaluated. Convergence with ν has been considered in terms of the modal eigenvalues $|\cot \alpha_\nu|$, that is the summation over ν includes all modes with $|\cot \alpha_\nu|$ less than some prescribed limit. This is permissible, as modes with large $\cot \alpha_\nu$ have fields that are small in the vicinity of the bridge, (analogous to considering a series of Bessel functions of increasing order evaluated near to the coordinate origin). After a series of trials, it was found that $|\cot \alpha_\nu|_{max}$ could be set at 100 with good accuracy.

Figure 8.21 shows the reflection from the bridge under CM incidence as a function of bridge length and frequency. It is observed that the results from the modal, TLM and FDM analyses all exhibit the same general characteristics, i.e. that $|s_{11}|$ is an almost linear function of both length and frequency. This is physically consistent with the fact that the bridge appears to the CM as a shunt connected capacitance, modelled to a first approximation by a parallel plate capacitor formed by the centre

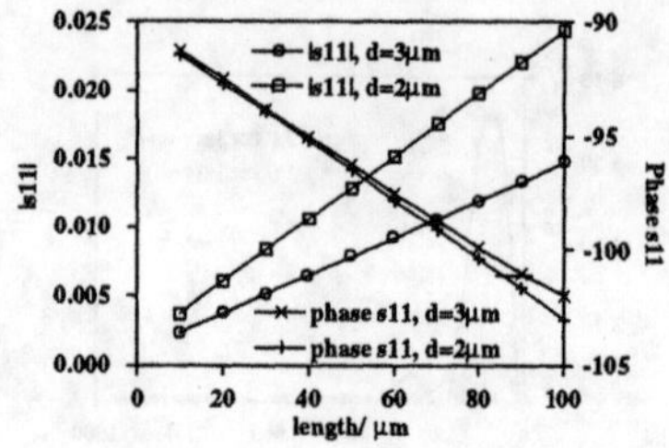

Figure 8.22 The effect of the bridge height; coplanar mode incidence, frequency of 40 GHz, d = 2 μm, 3 μm.

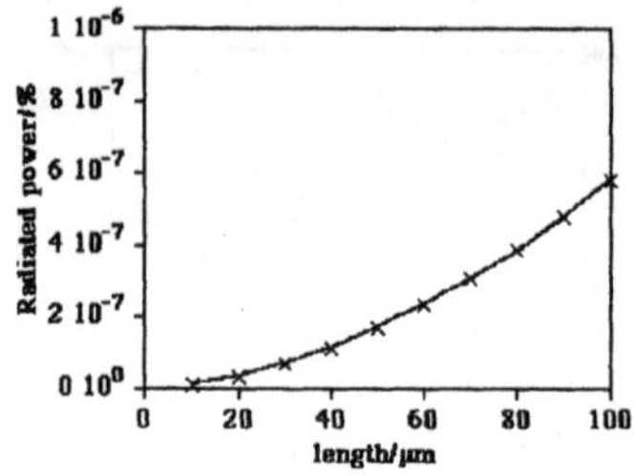

Figure 8.23 Radiated power vs. bridge length for coplanar mode incidence, at 40 GHz, with d = 3 μm.

strip of the CPW and the underside of the bridge, with an additional term due to fringing fields at the ends (hence the offset for zero length).

However, it is apparent that while the modal and TLM results are in reasonable accordance, those from the FDM are 9-16 dB higher (the effect of the different substrate heights has been checked and found unimportant). Before proceeding with further results, let us consider the discrepancies between the various methods. Firstly, it must be recognised that the air–bridge problem is by no means a trivial one for any method being a structure that is not only open in 3–dimensions, but is also dimensionally of the order of 10^{-3} of a wavelength. Therefore, it would be optimistic to expect perfect agreement. Secondly, all techniques have "method related" parameters, i.e. mesh size for the numerical ones and the limits on k_t and ν for the modal method. Reference [15] reveals that the results presented have not completely converged with mesh size, with a tendency for $|s_{11}|$ to decrease. In fact, as $|s_{11}|$ is of the order -40 to -60 dB, the ability of certain numerical techniques to resolve the difference in fundamental mode amplitude in loaded and unloaded guides is becoming progressively more difficult. Therefore in the light of these observations, we conclude that the TLM and modal analyses certainly agree within realistic expectations, whereas the FDM results differ somewhat from the other two. Reducing the bridge height, Figure 8.22, increases the magnitude of the reflection as an inverse law, again consistent with the parallel plate behaviour and finally, it is observed in Figure 8.23 that the radiated power is of insignificant magnitude for CM excitation.

We shall now discuss the effectiveness of the bridge at suppressing the slotline

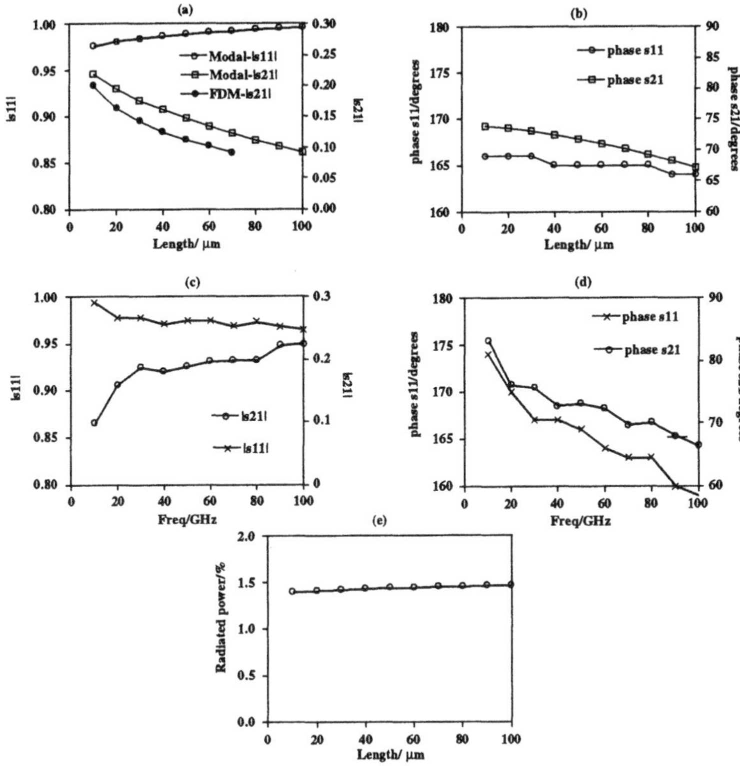

Figure 8.24 S–parameters under slotline mode incidence; d = 3 μm; a) $|s_{11}|$, $|s_{21}|$ vs. bridge length, frequency 50 GHz. b) Phase of s_{11} and s_{21} vs. bridge length, frequency 50 GHz. c) $|s_{11}|$ and $|s_{21}|$ vs. frequency, L = 30 μm. d) Phase of s_{11} and s_{21} vs. frequency, L = 30 μm. e) Radiated power vs. bridge length, frequency 50 GHz.

mode. Figure 8.24 shows the transmission properties of a bridge under SM incidence, demonstrating that the suppression is significant but not as low as would be desired. The behaviour with length is approximately exponential as the fundamental mode under the bridge is below cutoff over the frequency range of interest. Increasing the bridge length is apparently not a useful means of increasing the suppression of the SM, as clearly the "leading edge" of the bridge causes the largest reflection. Similarly, as seen in Figure 8.25, no advantage is obtained by reducing the bridge height, having only a minor effect upon the results. Finally as expected, the radiated field is now more important, falling between 1 and 2 % of the incident energy, as shown in Figure 8.24e.

For compatibility purposes and consideration of interelement interaction, it is desirable to establish radiation patterns and in particular, the substrate leakage behaviour. Modal and SDIE methods can easily provide this information; other methods with some difficulty.

Sample far field patterns, derived from an asymptotic technique, (steepest de-

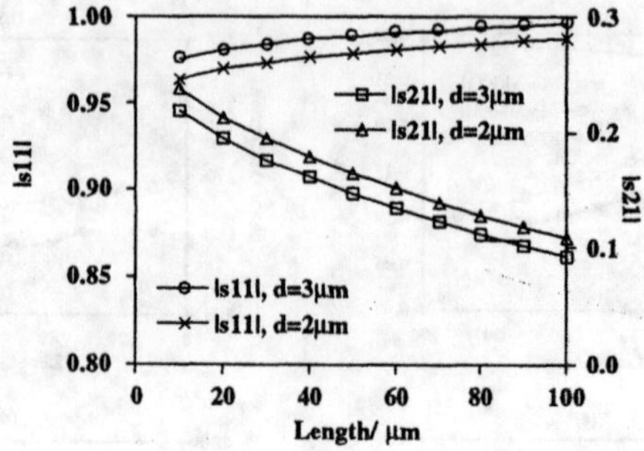

Figure 8.25 The effect of the bridge height for slotline mode incidence, frequency = 50 GHz, L = 30 μm.

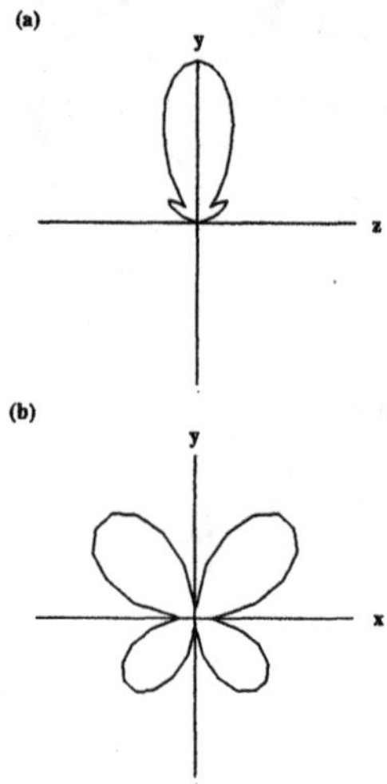

Figure 8.26 Far fields scattered from a bridge under coplanar mode incidence, L = 30 μm, d = 3 μm, (E_x values normalised to maximum).

scent path), are shown in Figures 8.26-8.28 and it was also found that the radiation and substrate leakage increase with frequency for both types of excitation. The latter results for SM incidence clearly show that substrate leakage could cause coupling to other circuit elements further along the guide.

Coupled bridges

Given the limited effectiveness of a single bridge at suppressing the slotline mode, consideration was given to pairs of bridges at spacings that prohibit neglect of near and radiation field coupling, the extension to the analysis being straightforward. Figure 8.29 compares $|s_{11}|$ for CM incidence and $|s_{21}|$ for SM incidence of a single bridge of length 60 μm, two 30 μm bridges assuming monomode coupling and two 30 μm bridges with full multimode coupling. For CM incidence, it is seen that the results of the monomodal and multimodal models differ only slightly. This is

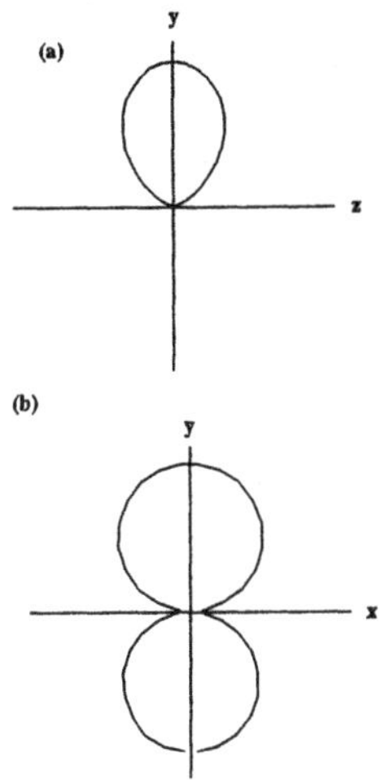

Figure 8.27 Far fields scattered from a bridge under slotline mode incidence, L = 30 μm, d = 3 μm, (E_x values normalised to maximum).

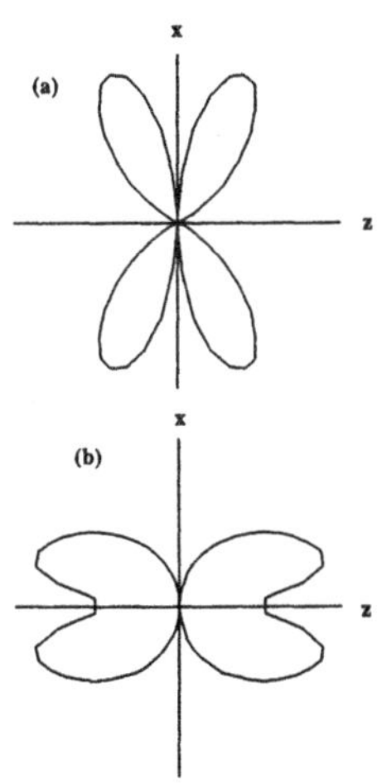

Figure 8.28 Substrate leakage patterns, L = 30 μm, d = 3 μm, a) Coplanar mode incidence, frequency 40 GHz; b) Slotline mode incidence, frequency 50 GHz, (H_z values normalised to maximum).

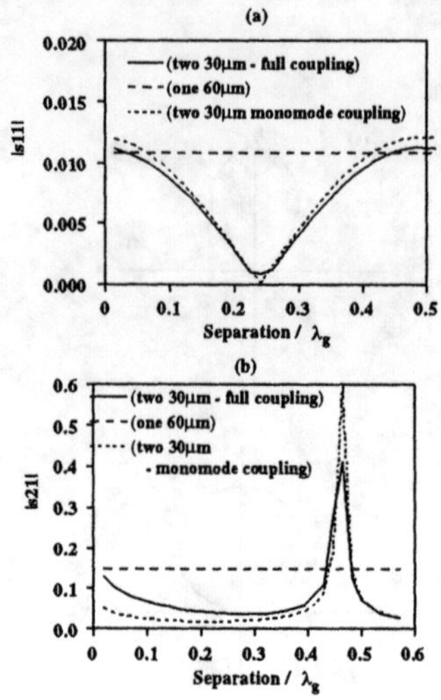

Figure 8.29 S–parameters for coupled bridges; a) Coplanar mode incidence, frequency 40 GHz, d = 3 μm. b) Slotline mode incidence, frequency 50 GHz, d = 3 μm.

expected, as it has already been shown that in this case the higher order modes are only weakly excited by each bridge and therefore are of minor importance for their coupling. However, this is not so for the SM incidence and for smaller separations the significant difference between the results from the two models is clearly attributable to higher order mode coupling.

In assessing the effectiveness of the coupled bridges, it is noted that unless practically unrealistic separations are used to create frequency dependent fundamental mode resonances, the suppression of the SM by a pair of 30 μm bridges is worse than that of a single bridge of their combined length. The reflection of the CM can indeed be reduced, but again for practical values of the separation, ($< 0.1\lambda_g$), the advantage is minor. Therefore on balance, it would appear that coupled bridges have little to offer over the, albeit far from ideal, behaviour of single bridges.

This final example does however show that the modal technique can easily incorporate near and radiation field interactions as required.

Computational expenditure

The introduction discussed the use of CAD packages on smaller, rather than larger, computers and it is, therefore, appropriate to provide an idea of the computational effort required to obtain the preceding results. This effort consists of two distinct parts:

1. derivation of the modes;

2. analysis of the bridges.

The former is by far the more demanding and, for each frequency, it requires time of the order of a minute on a CRAY Y-MP 8/432 (although, as yet, numerical efficiency has not been sought and run times could surely be significantly reduced in the future). But again, it is stressed that this is a "one off" procedure, yielding modes which may be used for the analysis of a wide range of problems, for example, all the different sets of bridge dimensions considered above.

The analysis of the bridges themselves used an HP 9000 series 375 mini–computer and required about 10 minutes to analyze the first example for a given bridge height and then typically 30–40 seconds to recover both S–parameters and radiation patterns for each different bridge length.

Further comments

In conclusion modal techniques can be advantageously used for the analysis of open transmission media. With respect to existing methods, they offer the considerable advantage that the field boundary conditions are built into the solution of different discontinuity problems. As a significant example, we have analyzed isolated and coupled pairs of a type of air–bridge in CPW. For the former, good general agreement is found with existing methods; for the latter, no comparable results seem to be available.

Radiation effects of the bridges in air and into the substrate are also highlighted, which would be difficult to obtain otherwise and further, the importance of including higher order mode coupling for strongly radiating discontinuities has been demonstrated.

8.5 Via–hole grounds in microstrip lines

The current trend toward improved performance is pushing the use of higher frequencies in MMICs. It is now common to find via–holes operating at frequencies of 40 GHz, while studies of circuits with vias operating even above 100 GHz are under way. Whereas at low frequencies the radiation effect of the via discontinuity can be neglected, as the frequency increases, radiation and surface wave effects of the via, and the possible mutual coupling to other discontinuities, become more and more important.

As an application of the complete spectrum of microstrip line, it is shown how the equivalent circuit of a via–hole may be derived. In this equivalent circuit, both reactive and resistive effects due to radiation and surface waves are taken into account; while the element values of the equivalent circuit are obtained in simple form and are easy to evaluate by a desktop computer. This model is useful in order to take into account the high frequency behaviour of the via hole discontinuity which is important in high frequency MMICs applications, since it is found that radiation, while negligible at low frequencies, increases significantly at higher frequencies.

The characterization of the via–hole discontinuity in microstrip line has recently received considerable attention. In its simplest form, this characterization is effected by means of quasi–static approximations leading to models useful for low frequency applications [28, 22, 11, 19]. As such, these models are useless when investigating the radiation behaviour of the via discontinuity. Another common approach, useful to model with higher accuracy its frequency behaviour , is to consider the via enclosed in a waveguide; i.e. to use the equivalent waveguide model of the microstrip line [9, 14]. Also in this case, radiation and surface wave effects are neglected.

Rigorous full–wave characterizations have also been performed; these are generally based on purely numerical methods, such as FDTD or TLM, [21, 1, 27, 7]. In these cases, apart for possible problems with absorbing boundaries conditions, it is difficult to provide simple equivalent circuits to model radiation phenomena. Full–wave characterizations based on the moment method [26], or on mode matching techniques by enclosing the via in a cavity [25] have also appeared. By enclosing the via, radiation is neglected; anyway the computation volume involved in a numerical/moment method full–wave analysis prevents its application to practical CAD.

As a consequence, at present, the only prediction of via hole radiation is provided by the early work of [2]. Unfortunately, this model considers the radiation from a current distribution in a dielectric slab, not taking into account the presence of the microstrip line and the via dimensions. Consequently, the above model gives only a rough estimate of the via radiation and it is inadequate for design purposes in the high frequency range.

The equivalent circuit presented here accounts for radiative effects since it is based on the full rigorous spectrum of open microstrip. Moreover, since radiation corresponds to that spectral region where the transverse wavenumber is less than the free–space wavenumber, the computation time required is very limited, leading to a simple calculation of the radiation resistance.

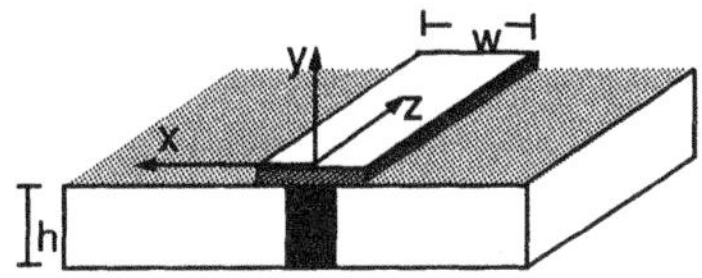

Figure 8.30 Geometry of the microstrip via–hole ground with the via modelled as a flat metallic strip.

8.5.1 Analysis

The configuration under study consists of a symmetrically placed conducting post connecting ground plane and strip conductor; the cone shaped via is assumed to be represented by an equivalent flat post having the same perimeter at its mid height section, and the current **J** on the surface of the post to be uniform and y–directed.

Application of Lorentz' reciprocity principle yields the amplitudes of the forward, C_0^+, $C^+(k_t)$, and backward, C_0^-, $C^-(k_t)$, waves of the fundamental mode and of each wave packet of the continuous spectrum as

$$\begin{aligned} C_0^+ &= -\frac{<\mathbf{J},\underline{\mathbf{e}}_0>}{P_0} = C_0^- \\ C^+(k_t) &= -\frac{<\mathbf{J},\underline{\mathbf{e}}(k_t)>}{P(k_t)} = C^-(k_t) \end{aligned} \tag{8.20}$$

where:

- k_t is the transverse wavenumber;
- β_0 is the propagation constant of the guided mode;
- P_0 is the normalization coefficient of the guided mode;
- $P(k_t)$ is the normalization coefficient of the continuous spectrum;
- **J** is the short circuit current;
- $\mathbf{e}_0$ is the field distribution of the guided mode;
- $\mathbf{e}(k_t)$ is the field distribution of the continuous spectrum.

Retracing the conceptual step of the closed waveguide case [5, section 8-5], the total field in the guide is the sum of an incident and scattered field and must vanish on the perfectly conducting strip.

The resulting integral equation for the current is

$$\mathbf{e}_0 e^{-j\beta_0 z} + C_0^- \mathbf{e}_0 e^{j\beta_0 z} + \int_0^\infty C^-(k_t)\mathbf{e}(k_t)e^{j\beta(k_t)z}dk_t = 0 \tag{8.21}$$

on the strip. In the above appears:

$$\beta^2(k_t) = k_0^2 - k_t^2 \tag{8.22}$$

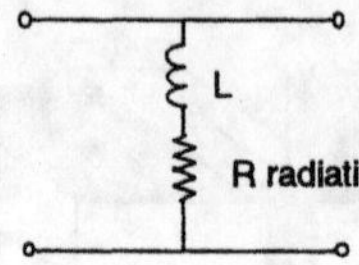

Figure 8.31 Rigorous equivalent circuit of the via–hole ground with the radiative losses explicitly accounted.

In (8.21), C_0^-, given by (8.20), for unit incidence assumes the meaning of reflection coefficient R of the fundamental mode, so, at $z = 0$, (8.21) can be written as

$$(1+R)\mathbf{e}_0 = \int_0^\infty \frac{<\mathbf{J}, \underline{\mathbf{e}}(k_t)>}{P(k_t)} \mathbf{e}(k_t) dk_t \qquad (8.23)$$

The shunt impedance modelling the post in the line is related to the reflection coefficient R by the expression

$$\frac{Z}{Z_0} = -\frac{1+R}{2R} \qquad (8.24)$$

After multiplying (8.24) by $\mathbf{J}$, integrating over the flat post and dividing by P_0^2 one obtains

$$\frac{Z}{Z_0} = \frac{1}{2\pi} \int_0^\infty \frac{\beta(k_t)}{\beta_0} \frac{P_0}{P(k_t)} \left[\frac{<\mathbf{J}, \mathbf{e}(k_t)>}{<\mathbf{J}, \mathbf{e}_0>} \right]^2 dk_t \qquad (8.25)$$

This integral contains two contributions: one is due to the radiative part of the spectrum, expressed by:

$$\frac{R}{Z_0} = \frac{1}{2\pi} \int_0^{k_t} \frac{\beta(k_t)}{\beta_0} \frac{P_0}{P(k_t)} \left[\frac{<\mathbf{J}, \mathbf{e}(k_t)>}{<\mathbf{J}, \mathbf{e}_0>} \right]^2 dk_t \qquad (8.26)$$

The other, due to the reactive effects, is expressed by:

$$\frac{j\omega L}{Z_0} = \frac{1}{2\pi} \int_{k_t}^\infty \frac{\beta(k_t)}{\beta_0} \frac{P_0}{P(k_t)} \left[\frac{<\mathbf{J}, \mathbf{e}(k_t)>}{<\mathbf{J}, \mathbf{e}_0>} \right]^2 dk_t \qquad (8.27)$$

where the energy storage around the small post is assumed as essentially inductive. In the above formulae the field distribution of the guided mode as well as that of the continuous spectrum is evaluated in the manner described in the previous Chapters. Rigorously, the latter quantities need to be computed for each different spot frequency; however, for practical purposes, it is sufficient to evaluate the impedance in (8.25) at midband, and the values of the inductance and resistances appearing in Figure 8.31. These values can then be used with a modest error also at different frequencies.

8.5.2 Some results

In order to test the accuracy of the new equivalent circuit several comparisons with other full wave analyses and with measured data have been performed. As an example, the equivalent circuit was used to simulate a practical via–hole on GaAs. This

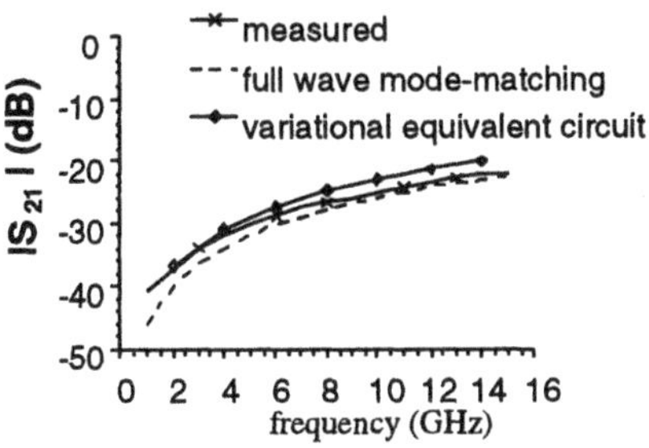

Figure 8.32 Comparison of measured data with the proposed equivalent circuit for a via–hole on GaAs. Also shown are the results relative to the mode–matching full–wave approach. The substrate thickness is $d = 200\ \mu m$, $\varepsilon_r = 12$, the via diameter is $80\ \mu m$ and the microstrip line is $570\ \mu m$ wide.

via has also been simulated in [25] by a 3–D mode matching technique which, compared with the proposed one, is quite time–consuming. The very good agreement with both the measured data and the theoretical simulation of [25] is apparent from Figure 8.32. It is also noted that, while the mode–matching simulation of [25] was applied to a boxed structure, the present one was developed for an open microstrip. As a consequence, the rigorous equivalent circuit introduced here directly accounts for the radiative losses.

Hence, it is possible to investigate the behaviour of the radiation resistance with frequency, as shown in Figure 8.33. From the latter figure it is confirmed that the radiation resistance increases with frequency following the quadratic law predicted by [2]. However, in our case we have taken into account both the presence of the microstrip line and the finite via radius, while in [2] the radiation resistance was calculated for a simple, infinitesimal, dipole without any microstrip line. As a consequence the values of radiation resistance we find are one or two orders of magnitude lower than those reported in [2]. The effect of the via radius on the radiation resistance has also been computed according to our model and the results are shown in Figure 8.34. Finally, Figure 8.35 shows the variation of the radiation resistance with the thickness of the dielectric substrate. This variation is seen to follow a cubic law in agreement with [2], although the absolute values differ as noted before.

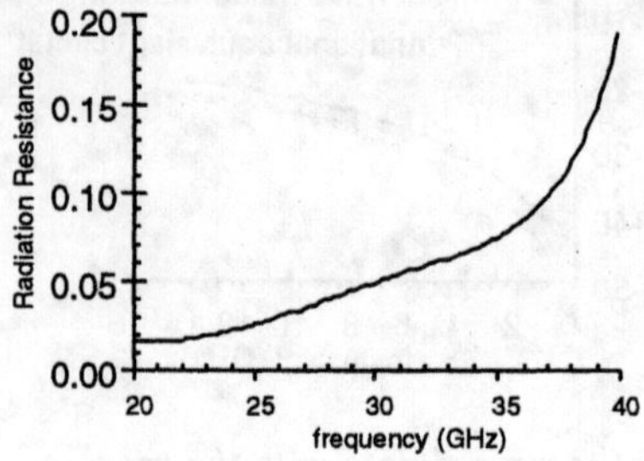

Figure 8.33 Radiation resistance (in mΩ) of a via hole as a function of frequency. Via and substrate dimensions are those of previous figure.

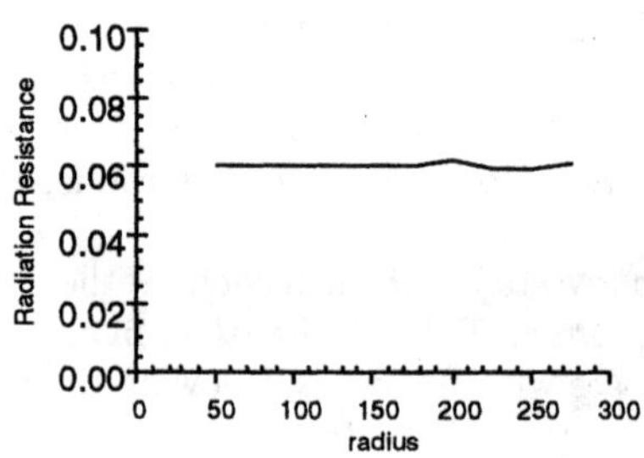

Figure 8.34 Radiation resistance (in mΩ) of a via hole as a function of the via radius (in μm).

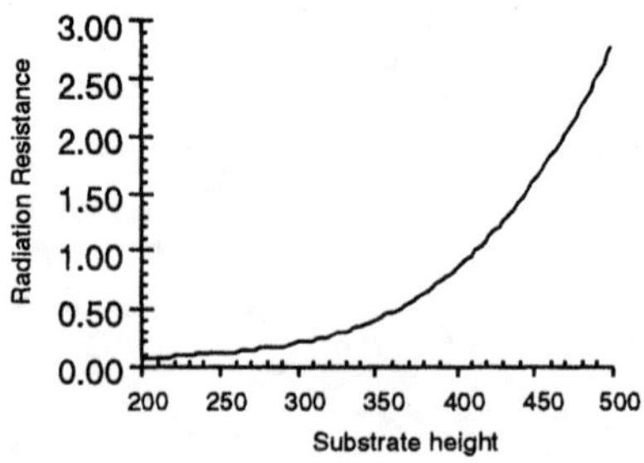

Figure 8.35 Radiation resistance (in mΩ) of a via hole as a function of the substrate height (in μm).

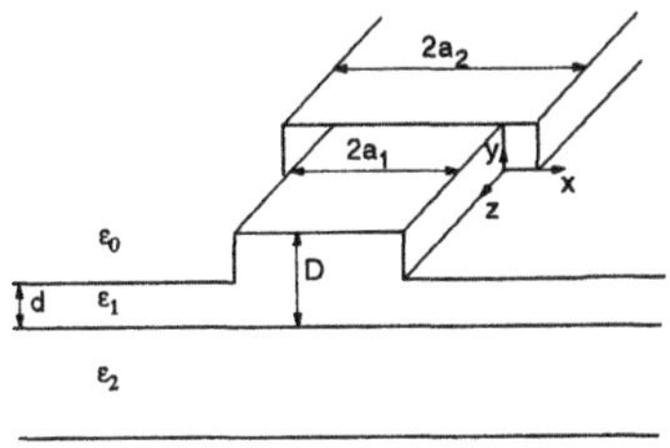

Figure 8.36 Change in rib width between two rib waveguides.

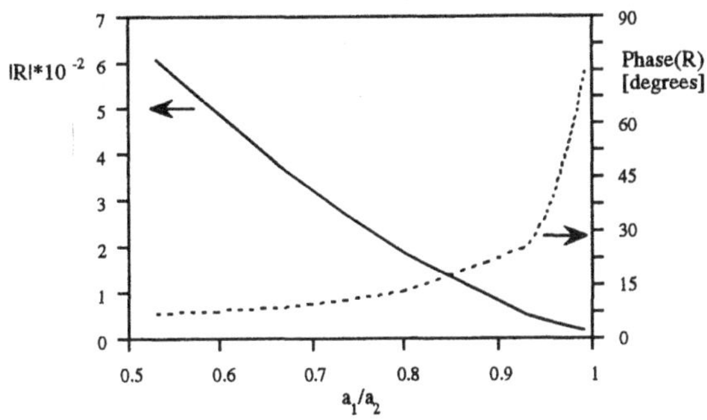

Figure 8.37 Change in rib width between two rib waveguides. Modulus and phase of the reflection coefficient vs. a_1/a_2.

8.6 Step discontinuities in rib guide

Knowledge of the continuous spectrum of rib waveguide is now applied to the practically important examples of step discontinuities. Step discontinuities were examined in the literature [23] but the presence of a rib guide continuum is generally not accounted for. The continuum, nevertheless, plays the same role as that of higher order modes in closed guide, when they are excited by a step discontinuity. In the following, results are presented for the reflection coefficient of a step in rib width and for three different abrupt terminations, that is, the abrupt termination of the rib, of the rib and guiding layer and of the whole guide. In the last case, the forward and backscattered radiated field is also evaluated.

8.6.1 E–plane step

The first case under consideration is the discontinuity in the E–plane of an LSE–polarised rib guide due to a change of rib width occurring at $z = 0$, as shown in Figure 8.36. By standard techniques, we recover the integral equation for the discontinuity in terms of the transverse magnetic field distribution H_y at the step:

$$\begin{gathered}\tfrac{1+R}{1-R}A^- e_x^-(x,y) - \\ \int_0^\infty A^-(k_t)e_x^-(x,y;k_t)dk_t + \int_0^V A^-(\sigma_t)e_x^-(x,y;\sigma_t)d\sigma_t = \\ A^+e_x^+(x,y) + \int_0^\infty A^+(k_t)e_x^+(x,y;k_t)dk_t + \int_0^V A^+(\sigma_t)e_x^+(x,y;\sigma_t)d\sigma_t\end{gathered} \tag{8.28}$$

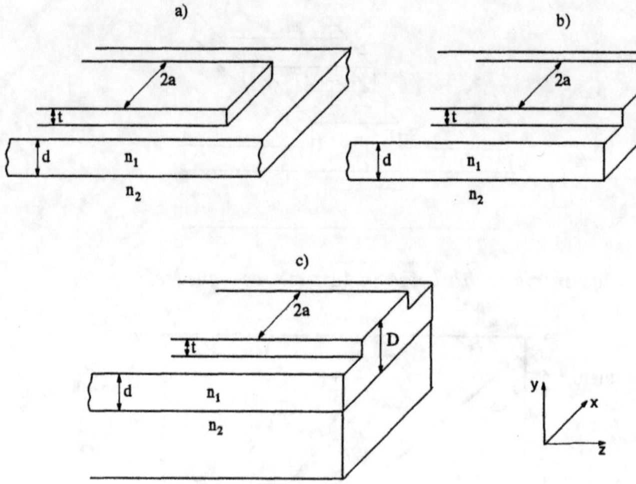

Figure 8.38 Abruptly terminated rib waveguide: a) just the rib is terminated; b) the rib and supporting slab are terminated; c) the whole guide is terminated.

with

$$\begin{array}{ll} A^- =< H_y, e_x^-(x,y) > & A^+ =< H_y, e_x^+(x,y) > \\ A^-(k_t) =< H_y, e_x^-(x,y;k_t) > & A^+(k_t) =< H_y, e_x^+(x,y;k_t) > \\ A^-(\sigma_t) =< H_y, e_x^-(x,y;\sigma_t) > & A^+(\sigma_t) =< H_y, e_x^+(x,y;\sigma_t) > \end{array} \quad (8.29)$$

where $<,>$ stands for the integration over the cross–section, $e_x^-(x,y)$, $e_x^-(x,y;k_t)$, $e_x^-(x,y;\sigma_t)$, $e_x^+(x,y)$, $e_x^+(x,y;k_t)$, $e_x^+(x,y;\sigma_t)$ denote the x–components of the electric field of the bound modes and of the continua, to the left and to the right of the discontinuity, respectively, and the field is properly normalised. Using the "transition function" as a single term expansion reduces the above equation (8.28) to a scalar one: we take as H_y the transverse distribution of the bound mode of an ideal rib guide of width intermediate between a_1 and a_2, so as to ensure maximum conservation of power at the step discontinuity. By multiplying (8.28) by H_y and integrating in the transverse section, a relatively simple expression of the reflection coefficient is obtained:

$$R = \frac{A^{+2} - A^{-2} + \int_0^\infty \left[A^{-2}(k_t) + A^{+2}(k_t)\right] dk_t + \int_0^V \left[A^{-2}(\sigma_t) + A^{+2}(\sigma_t)\right] d\sigma_t}{A^{+2} + A^{-2} + \int_0^\infty \left[A^{-2}(k_t) + A^{+2}(k_t)\right] dk_t + \int_0^V \left[A^{-2}(\sigma_t) + A^{+2}(\sigma_t)\right] d\sigma_t} \quad (8.30)$$

Now, the above overlapping integrals can be evaluated analytically. The results obtained for modulus and phase of the reflection coefficient at the step discontinuity of Figure 8.36 are shown in Figure 8.37, assuming the following parameter values : $\varepsilon_1 = 3.44$, $\varepsilon_2 = 3.40$, $l = 1.15\,\mu$m, $2a_2 = 3\,\mu$m, $d = 0.6\,\mu$m and $D = 1.0\,\mu$m.

8.6.2 Abruptly terminated rib

In the same manner, we have examined the abruptly terminated rib waveguide shown in Figure 8.38. In particular, in Figure 8.38a just the rib is terminated; in

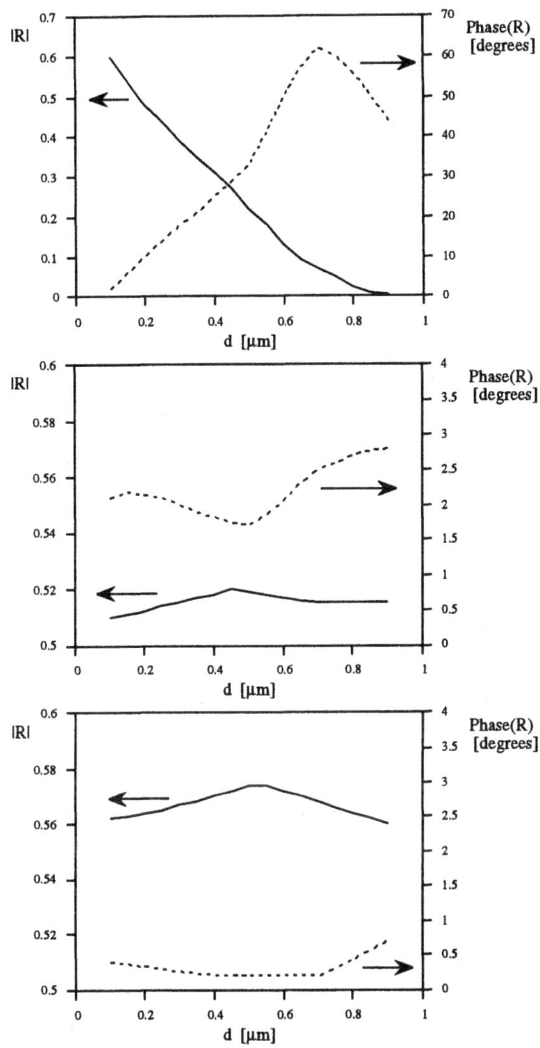

Figure 8.39 Abruptly terminated rib waveguide: modulus and phase of the reflection coefficients vs. outer slab thickness d for the corresponding cases of previous figure.

Figure 8.38b the guide and the surrounding slab are terminated; finally, in Figure 8.38c the whole guide is terminated. Correspondingly, the region $z > 0$ is represented by the spectrum of an asymmetrical slab waveguide of thickness d for the case of Figure 8.38a, by the spectrum of an air–substrate space for the case of Figure 8.38b and by the spectrum of the air space in the core of Figure 8.38c. For the case of Figure 8.38a we use as "transition function" the bound mode of a rib waveguide whose height $\overline{d}$ is intermediate between those of the guide D and the slab d to the left and to the right of the step respectively. Again $\overline{d}$ is chosen so as to ensure maximum conservation of power. In the case were no guidance is present in region $z \geq 0$, i.e. for cases of Figure 8.38b and 8.38c , the above criterion breaks down. For these cases, we choose as "transition function" the bound mode of the rib waveguide whose height is the same as the height of the "transition slab". Although this choice is somewhat arbitrary, its use is justified by the insensitivity of the results to slight variations of the variational trial function.

The reflection coefficients for the abruptly terminated rib waveguide are shown in Figure 8.39. In the case of Figure 8.39a where only the rib is terminated at $z = 0$, the reflection coefficient decreases as d increases. This is to be expected, since, as d increases, the rib waveguide approaches the asymmetrical slab guide at $z \geq 0$ and most of the incident wave from the rib guide will transmit through to the asymmetrical slab guide for $z \geq 0$. As the confinement increases, i.e. d decreases, the reflection coefficient of the configuration of Figure 8.38a approaches that of Figure 8.38b: this is so because, as d decreases, the slab at $z \geq 0$ will not support surface waves and there is no guiding. The reflection coefficient of the configuration of Figure 8.38c where the guide is terminated by the air half space is slightly higher than those of Figure 8.38b or Figure 8.38a with a slab below cutoff. This suggests that some of the incident wave leaks through the substrate. However, the difference in the reflection coefficients of the two different terminations is not large.

Finally, the forward and backscattered radiation patterns of the far field for the configuration of Figure 8.38c, calculated by means of saddle point integration, are shown in Figure 8.40, where the curves with negligible value have not been reported. Figures 8.40a-b show the forward radiation pattern in the azimuthal and transverse directions (see Figure 8.38c). We notice a narrow lobe in Φ as well as in Ψ, at about $\Phi \approx 0$, $\Psi \approx 0$. Figure 8.40c-d show the backward radiation pattern. In the Φ–plane we notice that the maximum radiation occurs at about $\Phi \approx 160$, $\Psi \approx 20$. Obviously, this radiation is symmetrical with respect to the plane $\Phi = \pi$. In the Ψ–plane, the maximum is at about $\Psi \approx 25$, $\Phi \approx 160$. In the latter radiation pattern, the air region is represented by $0 \leq \Psi \leq 90$, the substrate region by $270 \leq \Psi \leq 360$. Note, however, that the radiation patterns are normalised to the maximum value and the backscattered radiation in this problem is about three orders of magnitude less than the forward radiation. We note the presence of backscattered radiation over a wide range of angles, not unlike the case of a step discontinuity in slab guide; this last result would be particularly difficult to obtain without a knowledge of the rib guide radiation modes.

In conclusion, by considering the complete spectrum of the rib waveguide and by using a relatively simple approximation of this continuum, it is easy to examine, for most cases of common interest, several step discontinuities also including radiative effects.

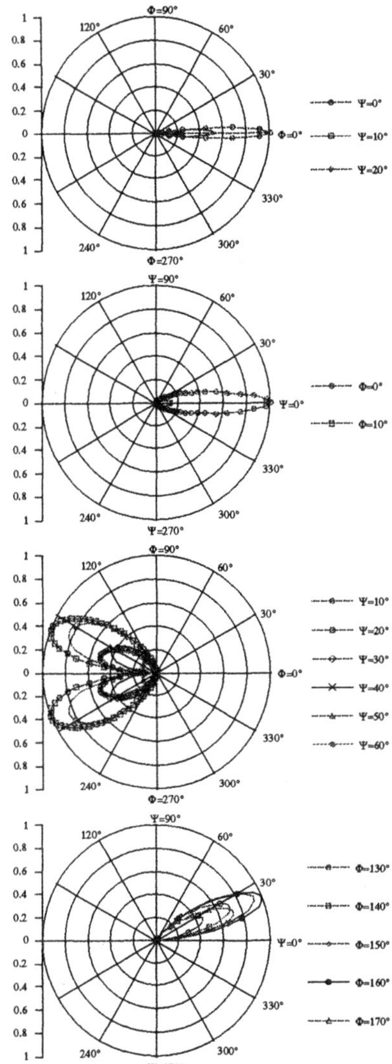

Figure 8.40 Forward and backscattered radiation patterns of the far field for the third configuration of abruptly terminated rib guide: a) Φ–plane and b) Ψ–plane for forward radiation; c) Φ–plane and d) Ψ–plane for backward radiation.

Bibliography

[1] W. D. Becker, P. Harms, and R. Mittra. Time domain electromagnetic analysis of a via in a multilayer computer chip package. *IEEE MTT-S Digest*, pages 1129–1232, 1992.

[2] C. R. Brewitt-Taylor. Radiation from via-holes in MMICs. *Europ. Microwave Conf.*, pages 433–436, 1987.

[3] C. H. Butler, Y. Rahmat-Samii, and R. Mittra. Electromagnetic penetration through apertures in conducting surfaces. *IEEE Trans. Antennas Propagat.*, 26:82–93, January 1978.

[4] C. H. Butler and K. R. Umashankar. Electromagnetic excitation of a wire through an aperture -perforated conducting screen. *IEEE Trans. Antennas Propagat.*, 24:456–462, July 1976.

[5] R. E. Collin. *Field Theory of Guided Waves.* IEEE Press, New York, 1990.

[6] N. I. Dib, P. B. Katehi, and G. Ponchak. Analysis of shielded CPW discontinuities with air-bridges. *Proc. IEEE MTT-S, Boston*, pages 469–472, 1991.

[7] C. Eswarappa and W. J. R. Hoefer. Time domain analysis of shorting pins in microstrip using 3-D SCN TLM,. *IEEE MTT-S Digest*, pages 917–920, 1993.

[8] L. Felsen and N. Marcuvitz. *Radiation and Scattering of Waves.* Prentice Hall, Englewood Cliffs, NJ, 1973.

[9] K. L. Finch and N. G. Alexopoulos. Shunt posts in microstrip transmission lines. *IEEE Trans. Microwave Theory Tech.*, 38:1585–1594, July 1990.

[10] B. Friedman. *Principles and Techniques of Applied Mathematics.* John Wiley & Sons Inc., New York, 1956.

[11] M. E. Goldfarb and R. A. Pucel. Modeling via hole grounds in microstrip. *IEEE Microwave and Guided Wave Letters*, 1:135–137, June 1991.

[12] K. C. Gupta, R. Garg, and I. J. Bahl. *Microstrip lines and Slotlines.* Artech House, Norwood, MA, 1979.

[13] T. Itoh. *Numerical Techniques for microwave and millimeter-wave passive structures.* Wiley, New York, 1989.

[14] R. H. Jansen. A full-wave electromagnetic model of cylindrical and conical via hole grounds for use in interactive MIC/MMIC design,. *IEEE MTT-S Digest*, pages 1233–1236, 1992.

[15] Hang Jin and R. Vahldieck. Calculation of frequency-dependent S-parameters of CPW air bridges considering finite metallisation thickness and conductivity. *Proc. IEEE MTT-S, Alburquerque*, pages 207–210, 1992.

[16] W. Heinrich K. Beilenhoff and H. L. Hartnagel. The scattering behaviour of air bridges in coplanar MMICs. *Proc. European Microwave Conf., Stuttgart, Germany*, pages 1131–1135, 1991.

[17] K. Y. Kabalan, R. F. Harrington, H. A. Auda, and J. R. Mautz. Characteristic modes for slots in a conducting plane, TE case. *IEEE Trans. Antennas Propagat.*, 35:162–168, February 1987.

[18] K. Y. Kabalan, R. F. Harrington, J. R. Mautz, and H. A. Auda. Characteristic modes for a slot in a conducting plane, TM case. *IEEE Trans. Antennas Propagat*, 35:331–335, March 1987.

[19] P. Kok and D. D. Zutter. Capacitance of a circular symmetric model of a via hole including finite ground plane thickness. *IEEE Trans. Microwave Theory Tech.*, 39:1229–1234, July 1991.

[20] N. H. L. Koster, S. Koblowski, R. Bertenburg, S. Heinen, and I. Wolff. Investigations on air-bridges used for MMICs in CPW technique. *Proc. European Microwave Conf., Wembley, UK*, pages 666–671, 1989.

[21] S. Maeda, T. Kashiwa, and I. Fukai. Full wave analysis of propagation characteristics of a through hole using the finite difference time-domain method. *IEEE Trans. Microwave Theory Tech.*, 38:2154–2159, December 1991.

[22] J. P. Quine, H. F. Webster, H. H. Glascock, and R. O. Carlson. Characterization of via connections in silicon circuit boards. *IEEE Trans. Microwave Theory Tech.*, 36:21–27, January 1988.

[23] B. M. A. Rahman and J. B. Davies. Analysis of optical waveguides and some discontinuity problems. *Proc. Inst. Elec. Eng., Pt. J*, 135:349–353, October 1988.

[24] Y. Rahmat-Samii. Electromagnetic pulse coupling through an aperture into a two-parallel plate region. *IEEE Trans. Electromagn. Compat.*, 20:436–442, August 1978.

[25] R. Sorrentino, F. Alessandri, M. Mongiardo, G. Avitabile, and L. Roselli. Full-wave modeling of via-hole grounds in microstrip by three-dimensional mode matching technique. *IEEE Trans. Microwave Theory Tech.*, 40:2228–2234, December 1992.

[26] W. J. Tsai and J. T. Aberle. Analysis of a microstrip line terminated with a shorting pin. *IEEE Trans. Microwave Theory Tech.*, 40:645–651, April 1992.

[27] S. Visan, O. Picon, and V. Foud Hanna. 3D characterization of air bridges and via holes in conductor backed coplanar waveguides for MMIC applications. *IEEE MTT-S Digest*, pages 709–712, 1993.

[28] T. Wang, R. F. Harrington, and J. Mautz. Quasi-static analysis of a microstrip via through a hole in a ground plane. *IEEE Trans. Microwave Theory Tech.*, 36:1008–1013, June 1988.

Index

www.ingramcontent.com/pod-product-compliance
Lightning Source LLC
LaVergne TN
LVHW011905200926
848484LV00002B/539